AF358638

Fundamentals and Applications of Fluid Mechanics and Acoustics in Biomedical Engineering

Fundamentals and Applications of Fluid Mechanics and Acoustics in Biomedical Engineering

Guest Editors

Iris Little
Ephraim Gutmark

Basel • Beijing • Wuhan • Barcelona • Belgrade • Novi Sad • Cluj • Manchester

Guest Editors

Iris Little
Department of Pediatrics
Cincinnati Children's
Hospital Medical Center
Cincinnati
United States

Ephraim Gutmark
Department of Aerospace
Engineering and Engineering
Mechanics
University of Cincinnati
Cincinnati
United States

Editorial Office
MDPI AG
Grosspeteranlage 5
4052 Basel, Switzerland

This is a reprint of the Special Issue, published open access by the journal *Bioengineering* (ISSN 2306-5354), freely accessible at: www.mdpi.com/journal/bioengineering/special_issues/ 9H4HU51Z72.

For citation purposes, cite each article independently as indicated on the article page online and using the guide below:

Lastname, A.A.; Lastname, B.B. Article Title. *Journal Name* **Year**, *Volume Number*, Page Range.

ISBN 978-3-7258-2748-0 (Hbk)
ISBN 978-3-7258-2747-3 (PDF)
https://doi.org/10.3390/books978-3-7258-2747-3

Contents

About the Editors

Iris Little

Iris Little has been an Associate Professor of Pediatrics, Division of Endocrinology, at the Cincinnati Children's Hospital Medical Center since 2010. She received her MD degree from Johns Hopkins University School of Medicine, Baltimore, MD. She is the director of the Turner Syndrome (TS) and Growth Hormone Centers. She is a full member of the Pediatric Endocrine Society and the Endocrine Society. Her research efforts encompass broad areas across pediatric endocrinology, with a focus on TS. Her main research effort involves the application of computational fluid dynamics (CFD) to assess anomalous aortic flow in TS, with the potential to noninvasively predict cardiovascular morbidity and mortality and guide interventions. Stemming from her expertise in the application of CFD techniques to biological systems, she leads the development of a novel airway clearance device that has resulted in several patents. She has a sustained publication record in high-impact-factor journals, very recently including prominent authorship in the International Guidelines for the Care of TS. She has an extensive track record of mentorship of fellows, PhD candidates, residents, and medical students. For every year in the last 9 years, she was selected as one of the top doctors by the Cincinnati Magazine and by Castle-Connolly. She received the Turner Syndrome Strategies to Improve Care Outcomes Distinguished Research Award and the Karolinska Award by the Lawson Wilkins Pediatric Endocrine Society.

Ephraim Gutmark

Ephraim (Effie) Gutmark has been a distinguished Professor of Aerospace Engineering and Engineering Mechanics and an Ohio Regents Eminent Scholar at the University of Cincinnati (UC) since 2000. He is an Affiliate Professor of Otolaryngology at the UC Medical Center. He is an Affiliate Professor of Mechanics at the Royal Institute of Technology in Sweden. He was a chaired Professor of Mechanical Engineering and the Chairman of the Mechanical Engineering Department at Louisiana State University between 1995 and 2000. Before, he worked as a senior research scientist at the Naval Air Warfare Center in California. The research performed in the Gas Dynamics and Propulsion Laboratory under his direction includes aeroacoustics, biomedical fluid dynamics, advanced propulsion systems, combustion control, scramjet propulsion, turbomachinery, flight control, and flow structure interactions. Since 2000, 15 post-doc, 55 PhD, and 70 MS students were or currently are part of his research team. He is an editor of the journal *Flow, Turbulence, and Combustion.*

He is a Fellow of the American Institute of Aeronautics and Astronautics (AIAA), a Fellow of the American Physical Society (APS), and a Fellow of the American Society of Mechanical Engineers (ASME), and he was an Einstein Fellow, Germany, in 2016–2019. He is the recipient of the 2021 AIAA Aeroacoustics Award, the 2013 ASME Fluids Engineering Award, and the Hanin International Lifetime Achievement Aerospace Award 2018; he is a Fulbright Specialist, the recipient of the 2022 Rieveschl Award for Distinguished Scientific Research, and the 2018 university best PhD mentoring award. His work on voice research was featured on the Discovery series of BBC International. He has published 317 papers in refereed archival journals, 636 reviewed conference papers, and is a co-inventor of 80 US and international patents and patent applications.

bioengineering

Editorial

Fundamentals and Applications of Fluid Mechanics and Acoustics in Biomedical Engineering

Iris Little [1] and Ephraim Gutmark [2,*]

[1] Cincinnati Children's Hospital Medical Center, Cincinnati, OH 45229, USA; iris.little@cchmc.org
[2] Department of Aerospace Engineering and Engineering Mechanics, University of Cincinnati, Cincinnati, OH 45221, USA
* Correspondence: ephraim.gutmark@uc.edu

The field of biomedical engineering has experienced important recent advances in experimental, computational, and analytical research in fluid mechanics and acoustics. New imaging modalities, advanced instrumentation, and efficient computational methodologies have enabled these advances. Computational tools are leveraging advances in modern engineering techniques and mathematical tools. The findings contribute to a better understanding of physiological processes in the circulatory and respiratory systems, as well as phonation, and have already led to new therapies that aim to promote better human health. These transformative achievements include the early diagnosis of diseases, monitoring of their progression, individualized interventions, surgical planning, effective drug delivery, and new medical devices.

Synergy between computational tools and experimental diagnostics is important for the development of both experimental and computational approaches. Accurate experimental data are crucial for validation of Computational Fluid Dynamics (CFD) and for tuning each step of the CFD process to obtain reliable results. CFD can help in designing the experimental setup and provide a deeper understanding of experimental results by adding information that cannot be measured.

This Special Issue focuses on original research papers and on comprehensive reviews on the "Fundamentals and Novel Applications of Fluid Mechanics and Acoustics in Biomedical Engineering". It highlights recent developments in these areas and identifies directions and topics that require further scientific research.

Manuscripts for this Special Issue were solicited from a large group of the top researchers in this field. Submissions to this thematic issue were subjected to a rigorous peer-review process following the standards and policies of the *Bioengineering* journal. Each paper was reviewed by several internationally recognized experts in the field. Thirteen papers were eventually accepted following revisions, resulting in the high standards of this issue.

The 13 papers included in this issue are grouped into two categories. The first group includes seven contributions covering upper airway flow, phonation, sleep apnea, and nasal flow; the remainder are concerned with cardiovascular flow and blood rheology.

Airway aerodynamics and acoustics: OSA, phonation, and rhinology.

Seven papers in this issue deal with the impact of flow and acoustics of laryngeal dynamics and phonation on Obstructive Sleep Apnea (OSA), and on rhinology.

Kraxberger et al. [1] presented an experimental and computational study of the impact of supraglottal vocal tract length on the vibrations of the silicone vocal folds model. They showed that when the vocal tract length is increased, it results in lower acoustic eigenfrequencies. When these frequencies are near the mechanical eigenmodes of the folds, the vibration frequency of the folds will align with the acoustic frequency. Conversely, when the vocal tract length is decreased, the fold vibration frequency is uncoupled with

check for
updates

Citation: Little, I.; Gutmark, E. Fundamentals and Applications of Fluid Mechanics and Acoustics in Biomedical Engineering. *Bioengineering* **2024**, *11*, 1125. https://doi.org/10.3390/bioengineering11111125

Received: 31 October 2024
Revised: 5 November 2024
Accepted: 6 November 2024
Published: 8 November 2024

the acoustic modes. This indicates that for certain conditions, the length of the folds can influence the fold vibration frequency and therefore impact phonation.

Further support for the Kraxberger et al. findings was provided by Näger et al. [2], who performed flow field measurements using Particle Imaging Velocimetry (PIV) in the same experimental model. The measured velocity fields were used to calculate the acoustic source terms. They observed a strong interaction between the fundamental acoustic resonance frequency and the vocal fold vibration frequency when the acoustic resonance frequency was not too high. At low acoustic frequencies, the vocal fold vibrations matched the fundamental acoustic modes. The supraglottal aerodynamic pulsating frequency matched that of the folds' vibrations. When such coupling occurs, the vocal efficiency, signal-to-noise ratio, harmonics-to-noise ratio, and cepstral peak prominence increase. This led to the conclusion that under such conditions it is possible to phonate longer and with higher quality. The range of the vocal lengths tested in this study was limited to frequencies associated mostly with children and female singers.

While the synthetic vocal folds tested by Kraxberger et al. [1] and Näger et al. [2] were cast from a single layer of silicone, Tur et al. [3] aimed to more closely represent the function of physiological larynges by adding artificial ligament fibers to a multi-layer silicone larynx model. The impact of elongation, abduction, and adduction on the laryngeal dynamics was assessed and compared to results of ex vivo and in vivo tests. The models replicated the vibration fundamental frequency, subglottal pressure, symmetry, and glottal gap. They also demonstrated the impact of various ligament parameters. This demonstrated the ability to replicate the phonation behavior of professional female singers.

A different aspect of the aeroelastic influence of the glottal flow on the folds' dynamics is elucidated by Jiang et al. [4] They used flow-structure-interaction (FSI) simulations to investigate the impact of the laryngeal flow separation vortices (FSVs) on the folds' vibrations. Their simulations showed that during the closing phase of the cycle, the larynx acquires a divergent shape that causes the flow to separate from the folds, generating recirculating flow between the jet and the larynx walls (called FSVs). These vortices produce negative pressure of nearly 30% of the subglottal pressure. This increases the aeroelastic energy transfer from the airflow to the vocal folds by 32%. They quantified the contribution of the FSV by comparing the impact of their presence to the case when they are artificially removed from the flow field. They showed that the vibration amplitude and flow rate were increased by 20% and the closing speed, skewness quotient, and Maximum Flow Declination Rate (MFDR) by up to 40% when FSVs were present. Their results demonstrate the importance of FSVs on vocal fold dynamics. The authors emphasize that since their model is two-dimensional, the impact of FSVs on a realistic three-dimensional model will be lower.

An application of FSI simulations for studying Obstructive Sleep Apnea (OSA) and snoring is described by Li et al. [5] They studied the dynamic response of the uvulopalatal flexible structures to respiratory flow using a simplified 3D model of the soft palate. The study included three levels of flow rate and changes in the ratio between the oral and nasal flow rates. High flow rates resulted in vortex shedding behind the soft palate that led to high-amplitude tip displacement vibrations, especially when the oral and nasal flow rates were equal or when only nasal flow was present. The large deformation was attributed to pressure differences between the oral and nasal sides of the palate. Low and medium flow rates had small wake fluctuations and low tip displacement.

In a related study, Palomares et al. [6] investigated the impact of snoring associated vibro-acoustic loading from vibrating soft tissue, catecholamine exposure, and hypoxia associated with OSA on platelet activation. They hypothesized that these factors could result in an increased risk of thrombotic stroke associated with OSA and snoring. They exposed platelets to increased sound intensity and duration and showed that a low frequency of 200 Hz had higher impact on platelet activation than higher frequency of 900 Hz. They also showed increased platelet activation by epinephrine (increased catecholamines) and hypoxia. Aspirin, which inhibits platelet activation, had no added effect on these observations.

OSA, and treatment of OSA, such as continuous positive airway pressure (CPAP), are affected by the nasal flow and nasal resistance. A comprehensive summary of the status of the computational versus experimental approaches to upper airway flow investigation is provided in Johnsen's [7] review of nasal airflow. The paper describes the field of computational rhinology, reviews the published literature on in vitro and in silico nasal air flow, and presents results on Large Eddy Simulations (LES) computational rhinometry research. The paper also analyzes the significant disagreement between computations and in vivo rhinomanometry (RMM) data. Three possible CFD modeling deficiencies are rejected, namely, the wrong choice of the turbulent model, poor special or temporal resolution, and ignoring transient effects. Other potential reasons that could include airway tissue compliance or nasal hair effects should be considered.

Cardiovascular flows and blood rheology.

The other six papers of this issue discuss the impact of the aortic valve morphology on its function, the rheological properties of blood, and the possible effect of hemodynamic forces on cardiovascular disease.

An FSI computational study by Sundström and Tretter [8] investigated the impact of bicuspid aortic valve (BAV) commissural angle on valve function and proximal aortic hemodynamics. Their simulations showed that with an asymmetric commissural angle of $120°$, the aortic opening area is reduced, and the ejected flow is swirling and recirculating, resulting in high wall shear stresses on the proximal ascending aorta. In a more symmetric commissural angle of $180°$, these patterns are less pronounced. The asymmetry may thus lead to increased aortic dilatation and valvular deterioration, highlighting the clinical importance of considering the commissural angle.

The clinical importance of FSI assessment of aortic flow in patients with bicuspid aortic valve (BAV) led Sundström and Laudato [9] to test a Machine-Learning (ML) algorithm that will accelerate the process of segmenting a patient-specific four-dimensional phase-contrast magnetic resonance imaging (4D-PCMRI) of the thoracic aorta. They used the imaging of six subjects, three with non-stenotic tricuspid aortic valves (TAV) and three with non-stenotic functionally bicuspid aortic valves (BAV). Using TotalSegmentator-based segmentation they compared the flow field features of the two groups. They showed strong swirling motion in the proximal ascending aorta of the TAV cases compared to BAV, resulting in higher tangential shear stresses.

Sundström et al. [10] described the use of blood speckle imaging (BSI) based on echocardiographic data to compare pre- and post-operative flow patterns following subaortic membrane resection and aortic valve repair. The data indicated that the flow had less regurgitation following surgery, resulting in changes in the wall shear stresses. While the time-averaged value of wall shear stress (TAWSS) remained unchanged, the oscillatory shear index (OSI) was reduced. This indicated lower risk for aortic wall and leaflet damage.

The role of shear stresses on arterial diseases, such as atherosclerosis, was also investigated in the carotid artery by Wild et al. [11]. They developed carotid artery bifurcation models with "healthy" and "predisposed" geometries and used CFD to compare between the resulting two flow fields. They showed that in the "healthy" geometry, a hairpin vortical structure develops in the internal carotid artery (ICA) sinus and persists during a significant part of the cardiac cycle. This structure appears earlier in the cycle in the "predisposed" geometry and persists for a much shorter duration, followed by less organized structures. This change in flow behavior results in lower wall shear stresses (WSS) and a weaker favorable streamwise pressure gradient in the "predisposed" geometry, making it more prone to atherosclerotic plaque formation.

Computations of hemodynamics in the large blood vessels, such as the aorta, assume that the blood behaves as a Newtonian fluid. Others use different rheological models to represent non-Newtonian behavior. Fuchs et al. [12] compared three non-Newtonian models: Casson, Quemada, and Walburn–Schneck to Newtonian viscosity flow in the human thoracic aorta. They compared several features of the flow: (i) magnitude of the viscosity relative to the Newtonian case; (ii) wall shear stress (WSS); (iii) WSS-related quantities,

Oscillatory Shear Index (OSI), Time-Averaged WSS (TAWSS) and Relative Residence Time (RRT); (iv) size of retrograde flow, when blood flows backwards; and (v) transport of small particles in the thoracic aorta. The main differences were in instantaneous WSS at low shear rates (near walls or stagnation zones), where space-averaged WSS differed by ~10% and the temporal derivative of WSS by up to 20%. Transport of particles was also impacted.

The non-Newtonian rheological properties of blood are the result of the elastic behavior of the red blood cells (RBC). Jafarinia et al. [13] proposed that since the electrical conductivity of blood is related to hemodynamics, it can become a safe, low-cost diagnostic method. They developed two anisotropic electrical conductivity models to describe two three-dimensional flows: a straight, rigid pipe and an idealized aorta geometry. In the rigid pipe, the two models matched experimental results. In the simplified Aorta model, the two models gave different results. Due to the lack of experimental data, it is not possible for now to ascertain the accuracy of the two models.

This Special Issue highlights the growing interest in flow and acoustic phenomena related to human biology and health and will encourage further scientific contributions and discussions on the fundamental physics and applications of this new and exciting topic.

Conflicts of Interest: The authors declare no conflicts of interest.

References

1. Kraxberger, F.; Näger, C.; Laudato, M.; Sundström, E.; Becker, S.; Mihaescu, M.; Kniesburges, S.; Schoder, S. On the Alignment of Acoustic and Coupled Mechanic-Acoustic Eigenmodes in Phonation by Supraglottal Duct Variations. *Bioengineering* **2023**, *10*, 1369. [CrossRef] [PubMed]
2. Näger, C.; Kniesburges, S.; Tur, B.; Schoder, S.; Becker, S. An Investigation of Acoustic Back-Coupling in Human Phonation on a Synthetic Larynx Model. *Bioengineering* **2023**, *10*, 1343. [CrossRef] [PubMed]
3. Tur, B.; Gühring, L.; Wendler, O.; Schlicht, S.; Drummer, D.; Kniesburges, S. Effect of Ligament Fibers on Dynamics of Synthetic, Self-Oscillating Vocal Folds in a Biomimetic Larynx Model. *Bioengineering* **2023**, *10*, 1130. [CrossRef] [PubMed]
4. Jiang, W.; Zheng, X.; Farbos de Luzan, C.; Oren, L.; Gutmark, E.; Xue, Q. The Effects of Negative Pressure Induced by Flow Separation Vortices on Vocal Fold Dynamics during Voice Production. *Bioengineering* **2023**, *10*, 1215. [CrossRef] [PubMed]
5. Li, P.; Laudato, M.; Mihaescu, M. Time-Dependent Fluid-Structure Interaction Simulations of a Simplified Human Soft Palate. *Bioengineering* **2023**, *10*, 1313. [CrossRef] [PubMed]
6. Palomares, D.; Tran, P.; Jerman, C.; Momayez, M.; Deymier, P.; Sheriff, J.; Bluestein, D.; Parthasarathy, S.; Slepian, M. Vibro-Acoustic Platelet Activation: An Additive Mechanism of Prothrombosis with Applicability to Snoring and Obstructive Sleep Apnea. *Bioengineering* **2023**, *10*, 1414. [CrossRef] [PubMed]
7. Johnsen, S. Computational Rhinology: Unraveling Discrepancies between In Silico and In Vivo Nasal Airflow Assessments for Enhanced Clinical Decision Support. *Bioengineering* **2024**, *11*, 239. [CrossRef] [PubMed]
8. Sundström, E.; Tretter, J. Impact of Variation in Commissural Angle between Fused Leaflets in the Functionally Bicuspid Aortic Valve on Hemodynamics and Tissue Biomechanics. *Bioengineering* **2023**, *10*, 1219. [CrossRef] [PubMed]
9. Sundström, E.; Laudato, M. Machine Learning-Based Segmentation of the Thoracic Aorta with Congenital Valve Disease Using MRI. *Bioengineering* **2023**, *10*, 1216. [CrossRef] [PubMed]
10. Sundström, E.; Jiang, M.; Najm, H.; Tretter, J. Blood Speckle Imaging: An Emerging Method for Perioperative Evaluation of Subaortic and Aortic Valvar Repair. *Bioengineering* **2023**, *10*, 1183. [CrossRef] [PubMed]
11. Wild, N.; Bulusu, K.; Plesniak, M. Vortical Structures Promote Atheroprotective Wall Shear Stress Distributions in a Carotid Artery Bifurcation Model. *Bioengineering* **2023**, *10*, 1036. [CrossRef] [PubMed]
12. Fuchs, A.; Berg, N.; Fuchs, L.; Prahl Wittberg, L. Assessment of Rheological Models Applied to Blood Flow in Human Thoracic Aorta. *Bioengineering* **2023**, *10*, 1240. [CrossRef] [PubMed]
13. Jafarinia, A.; Badeli, V.; Krispel, T.; Melito, G.; Brenn, G.; Reinbacher-Köstinger, A.; Kaltenbacher, M.; Hochrainer, T. Modeling Anisotropic Electrical Conductivity of Blood: Translating Microscale Effects of Red Blood Cell Motion into a Macroscale Property of Blood. *Bioengineering* **2024**, *11*, 147. [CrossRef] [PubMed]

 *bioengineering*

Article

On the Alignment of Acoustic and Coupled Mechanic-Acoustic Eigenmodes in Phonation by Supraglottal Duct Variations

Florian Kraxberger [1,*,†], Christoph Näger [2,†], Marco Laudato [3], Elias Sundström [3], Stefan Becker [2], Mihai Mihaescu [3], Stefan Kniesburges [4] and Stefan Schoder [1,†]

1 Institute of Fundamentals and Theory in Electrical Engineering (IGTE), Graz University of Technology, Inffeldgasse 18/I, 8010 Graz, Austria; stefan.schoder@tugraz.at

2 Institute of Fluid Mechanics (LSTM), Friedrich-Alexander-Universität Erlangen-Nürnberg, Cauerstraße 4, 91058 Erlangen, Germany; christoph.naeger@fau.de (C.N.); stefan.becker@fau.de (S.B.)

3 Department of Engineering Mechanics, FLOW Research Center, KTH Royal Institute of Technology, Osquars Backe 18, 10044 Stockholm, Sweden; laudato@kth.se (M.L.); elias@kth.se (E.S.); mihai@mech.kth.se (M.M.)

4 Division of Phoniatrics and Pediatric Audiology, Department of Otorhinolaryngology, Head & Neck Surgery, Friedrich-Alexander-Universität Erlangen-Nürnberg, Waldstraße 1, 91054 Erlangen, Germany; stefan.kniesburges@uk-erlangen.de

* Correspondence: kraxberger@tugraz.at

† These authors contributed equally to this work.

Abstract: Sound generation in human phonation and the underlying fluid–structure–acoustic interaction that describes the sound production mechanism are not fully understood. A previous experimental study, with a silicone made vocal fold model connected to a straight vocal tract pipe of fixed length, showed that vibroacoustic coupling can cause a deviation in the vocal fold vibration frequency. This occurred when the fundamental frequency of the vocal fold motion was close to the lowest acoustic resonance frequency of the pipe. What is not fully understood is how the vibroacoustic coupling is influenced by a varying vocal tract length. Presuming that this effect is a pure coupling of the acoustical effects, a numerical simulation model is established based on the computation of the mechanical-acoustic eigenvalue. With varying pipe lengths, the lowest acoustic resonance frequency was adjusted in the experiments and so in the simulation setup. In doing so, the evolution of the vocal folds' coupled eigenvalues and eigenmodes is investigated, which confirms the experimental findings. Finally, it was shown that for normal phonation conditions, the mechanical mode is the most efficient vibration pattern whenever the acoustic resonance of the pipe (lowest formant) is far away from the vocal folds' vibration frequency. Whenever the lowest formant is slightly lower than the mechanical vocal fold eigenfrequency, the coupled vocal fold motion pattern at the formant frequency dominates.

Keywords: voice production; fluid-structure-acoustic interaction; mechanical-acoustical eigenvalue simulation; vocal fold motion; finite element model

Citation: Kraxberger, F.; Näger, C.; Laudato, M.; Sundström, E.; Becker, S.; Mihaescu, M.; Kniesburges, S.; Schoder, S. On the Alignment of Acoustic and Coupled Mechanic-Acoustic Eigenmodes in Phonation by Supraglottal Duct Variations. *Bioengineering* **2023**, *10*, 1369. https://doi.org/10.3390/bioengineering10121369

Academic Editor: Chiara Giulia Fontanella

Received: 6 October 2023
Revised: 15 November 2023
Accepted: 23 November 2023
Published: 28 November 2023

1. Introduction

The human voice is physically created in a complex process characterized by fluid-structure-acoustic interaction [1]. In this process, the vocal folds are excited to vibrate by the airflow of the lungs $\dot{V}$. During vocal fold vibration, the superficial tissue of the vocal fold moves in a wave-like manner, exhibiting a vertical phase difference. This motion generates what is known as a "mucosal wave", with a frequency that plays a role in determining the pitch of the voice. This vibration leads to a modulation of the airflow, forming a pulsating free jet in the vocal tract. The sound that constitutes the voice thereby arises aero-acoustically from the turbulent free jet region [2–4], as well as vibro-acoustically by sound radiation from the vibrating vocal fold surface [5]. This sound is filtered by the vocal tract and radiated through the mouth and nares, resulting in the voice pattern. A linear behavior between the sound source and filter was assumed for a long time, i.e., changes in the source due to the filter were neglected [6]. However, this simplified representation is not always valid, especially when a resonance frequency of the vocal tract or the trachea region f_R is close to the vibration frequency of the vocal folds f_o [7]. This behavior was also observed in a patient study recently [8]. It has been studied using a lumped-mass description of tissue mechanics, quasi-steady flow, and one-dimensional acoustics in [9]. In doing so, frequency deviations and maxima jump of the threshold pressure occur when the mechanical oscillation frequency is slightly above a vocal tract resonance. Both the trachea and the vocal tract may produce those same effects [10,11]. In [12], the assumption that vocal tract formants interact with the voice source was analyzed in vivo by investigating the data collected by a study consisting of twelve classical singers. The analysis used transnasal high-speed videoendoscopy, electroglottography, and audio recordings. However, the presented data partially corroborates that vowel transitions may result in level-two interactions (using the nomenclature of Titze [7]). The authors of [13] reported that under certain conditions, e.g., singing voice, the fundamental frequency of the vocal folds can go up and interfere with the formant frequencies. So, acoustic feedback from the vocal tract filter to the vocal fold motion becomes strong and non-negligible. Again, a multi-mass model was used to confirm the findings [9]. Due to the complexity of the problem, often only simple metrics such as the variation of f_0 or the change in the threshold of transglottal pressure for oscillation have been studied. Therefore, in [14,15], first experimental measurements are conducted to gain further insight into the interaction between vocal tract acoustics, structural dynamics, and aerodynamics for different vocal tract lengths. The variation of the vocal tract length thereby changes the acoustic properties of the vocal tract, allowing a systematic investigation of the relationship between flow and acoustics.

Within the present contribution, the experimental results reported in [14,15] are extended substantially, and accompanying simulations are conducted to report the details on the non-negligible mechanical-acoustic back-coupling (feedback of the vocal tract resonances on the vibration) which is leading to the so-called "non-linear" filtering property at certain conditions (e.g., singing). Thereby, nonlinear effects can be, e.g., a nonlinear behavior of the vocal fold material, large mechanical deformations, or a contact between the vocal folds during phonation. For this purpose, experimental investigations will be performed on a simplified, synthetic larynx model using laser scanning vibrometry and high-speed camera (HSC) records. From a computational perspective, the simplified lumped-mass representation and the one-dimensional wave equation presented in [9] is significantly extended to a three-dimensional finite element model, being able to represent any given upper airway geometry [16]. This numerical simulation model is able to describe the most efficient vocal fold motion mode and the phenomena of the linear filter range, as well as the non-linear interaction and coupling of modes potentially leading to deviations in the oscillation frequency and the maxima jumps. Nevertheless, the governing equations of the acoustic and structural dynamics fields are linear, meaning that no hard contact between the vocal folds or nonlinear mechanical material laws is incorporated. Furthermore, the similarity of these numerically computed modes with experimental data is used to explain the details of the mechanical-acoustic feedback. The findings exhibit that the presented simulation approach yields similar results to the measurements which enables to use the simplified linear numerical model to gain insights into the linear coupling effects between mechanic and acoustic fields. In the final discussion, the limitations are explained and the connection of the numerical and experimental findings to other voice parameters like the sound pressure level and vocal efficiency are drawn.

The paper is organized as follows. Section 2 describes the experimental setup. In Section 3, the numerical model of the quadratic mechanical-acoustic eigenvalue system is presented. Section 4 reports the experimental results of the vocal fold motion and the numerical results of the eigenmode analysis. The application results are discussed in Section 5 providing also the model limitations. Conclusions are drawn in Section 6.

2. Experimental Study

The synthetic vocal folds are based on the vocal fold M5 geometry model, proposed by Scherer et al. [17], with the detailed geometry presented in [14] and were cast from a single layer of silicone. The specimens were made of a three-component addition-cure silicone (Smooth-on, Inc., Macungie, PA, USA). The compound consists of a two-part Ecoflex 0030$^©$ (A + B) silicone rubber and three-parts silicone thinner (T) assembling the mixture 113 as described in [18]. The experimental test rig is shown schematically in Figure 1. The flow through the setup with a volume discharge of $\dot{V}$ is generated by a mass flow generator [19] and takes place from left to right as drawn by the arrow at the inflow. First, the flow is acoustically preconditioned by an acoustic silencer [20,21] before entering the subglottal tube and then the vocal folds. The vocal folds are marked in red and are located between the subglottal channel and the first section of the simplified vocal tract, both having a rectangular cross-section of $\Delta y \times \Delta x = 18\,\text{mm} \times 15\,\text{mm}$ that anatomically corresponds to the lateral–longitudinal orientation of a human larynx. The vocal tract consists of two sections: the first is a rectangular channel with the same dimensions as the vocal folds and the subglottal channel. Connected to this, there is a second section that has a circular cross-section. The second section consists of two telescopic tubes that enable continuously varying the length of the vocal tract, allowing its acoustic resonance frequencies to be adjusted. The measurements are performed for different vocal tract lengths L in the range $L \in [170, 930]$ mm.

Additionally, the vocal fold motion was investigated by laser scanning vibrometry (LSV) in [14]. Before the vocal folds section, a curved subglottal channel is placed with a small optical window to record the vocal folds' surface motion (see Figure 1).

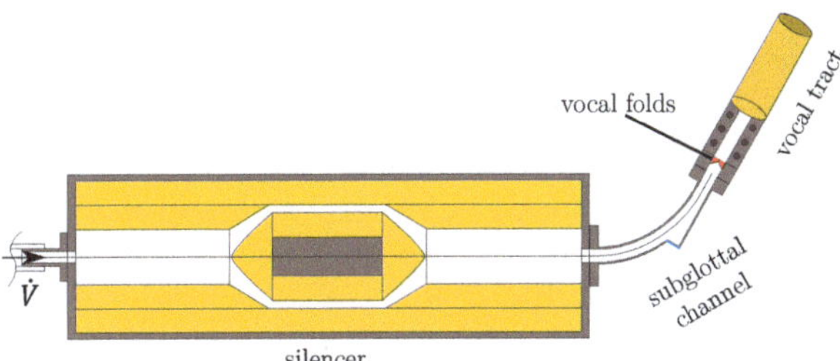

Figure 1. The experimental setup. The vocal folds are marked in red and are located between the vocal tract and the curved subglottal channel. A silencer is placed in front of the vocal folds to dampen the sound in the inflow. The flow direction is from left to right.

This approach can examine the surface motion on the subglottal and supraglottal sides using two LSVs. As depicted in Figure 2, the LSV measurement positioning is illustrated. A total of 748 measurement points are measured, 374 on each vocal fold surface. For the high-speed camera recordings, the camera is set up at position 2 to record the surface motion by video. A wall pressure sensor is placed in the subglottal channel below the vocal folds to synchronize the measurements at a specific start time of each measurement. Its signal generates a trigger signal that determines the start time.

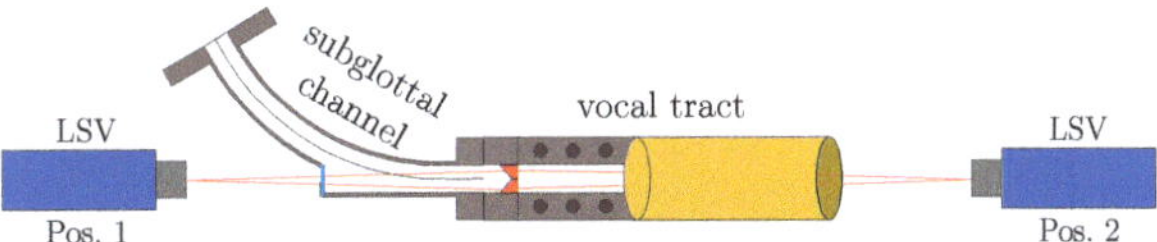

Figure 2. Schematic of the experimental setup for the LSV measurements.

3. Numerical Model

This section describes the numerical model using the Finite Element (FE) method. All numerical simulations have been performed using *Ansys Mechanical 2022 R2* [22,23].

3.1. Governing Equations

The governing equations of the acoustic and solid mechanic field, as well as their physical coupling conditions, are discussed in the following.

3.1.1. Acoustic Field

Using the linear acoustic wave equation [24] (Equation (5.28)) to describe the acoustic field in the context of human phonation is state of the art and this approach is commonly used in the literature, e.g., [1,3,16,25]. Therefore, the acoustic field in a 3D acoustic domain Ω_a is governed by

$$\frac{1}{c_0^2}\frac{\partial^2 p}{\partial t^2} - \Delta p = 0, \tag{1}$$

where p is the acoustic pressure, $c_0 = \sqrt{K/\rho_0}$ is the isentropic speed of sound computed from the bulk modulus K, and the ambient fluid mass density ρ_0, t is time, and $\Delta = \nabla \cdot \nabla$ is the Laplacian. For the acoustic domain, all boundaries except the inlet and outlet surfaces Γ_{in} and Γ_{out}, and the interface boundary to the mechanic domain Γ_{IF}, as depicted in Figure 3 are considered sound hard, i.e., homogeneous Neumann boundary conditions are imposed on the pressure. This is justified by the fact that the specific acoustic impedance of air (a fluid) is several orders of magnitude smaller than the specific acoustic impedance of the duct wall of the experimental setup made of acrylic glass, and the large impedance jump of three orders of magnitude, is approximated by a sound-hard boundary condition. At the surface Γ_{in}, a homogeneous Dirichlet boundary condition is applied on the pressure, which models a sound soft boundary [24] (Chapter 5.4). At Γ_{out}, the radiation impedance is adjusted to the test rig [26]. For air at 22 °C (experimental condition), which is the medium in the acoustic domain Ω_a, typical values are $c_0 = 346.25\,\text{m/s}$ and $\rho_0 = 1.225\,\text{kg/m}^3$.

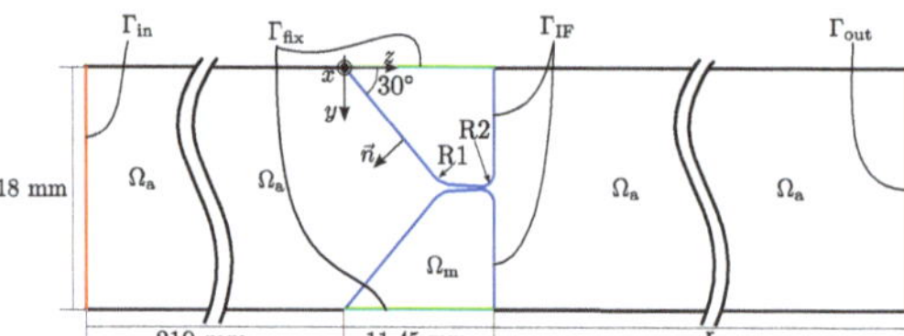

Figure 3. Schematic sketch of the longitudinal cut of the investigated geometry. Ω_a and Ω_m are the acoustic and mechanic domains, respectively, Γ_{IF} is the interface surface (blue), Γ_{fix} is the fixed surface of the mechanic domain (green), and Γ_{in} and Γ_{out} are inlet and outlet surfaces (red), respectively. The length L is varied.

3.1.2. Mechanic Field

Newton's second law (conservation of momentum) in differential form states that

$$\vec{f} = \rho\frac{\partial^2 \vec{d}}{\partial t^2}, \tag{2}$$

where $\vec{f} = \nabla \cdot \sigma + \vec{g}$ is the force density computed from the stress tensor σ and external body forces $\vec{g}$, ρ is the material's mass density, and $\vec{d} = (d_1, d_2, d_3)$ is the displacement. Thus, Equation (2) can be reformulated to

$$\rho\frac{\partial^2 \vec{d}}{\partial t^2} - \nabla \cdot \sigma = \vec{g}. \tag{3}$$

The external body forces are assumed to be zero. Furthermore, a linear elastic stress strain s relationship is assumed $\sigma = \mathbf{C} : \mathbf{s}$ in concludence with [27], where the stiffness tensor $\mathbf{C}$ depends on the Young's modulus E and the Poisson ratio ν, as defined in [23] (Equations (2)–(4)). The silicone rubber material used in the experiments was characterized in [18], called "mixture 113". From [18] (Table 3), we know that for the used silicone mixture $\nu = 0.499$ and $\rho = 976 \, \text{kg/m}^3$. The Young's modulus E will be described in Section 3.3.

3.1.3. Finite Element Formulation of Mechanic-Acoustic Coupling

Using the FE method, the following matrix-vector equation is set up for each finite element, of the coupled mechanic-acoustic problem [23] (Equations (8)–(32))

$$\begin{bmatrix} \mathbf{M}_e & 0 \\ \mathbf{M}^{\text{fs}} & \mathbf{M}_e^{\text{p}} \end{bmatrix} \frac{\partial^2}{\partial t^2} \begin{pmatrix} \mathbf{d}_e \\ \mathbf{p}_e \end{pmatrix} + \begin{bmatrix} \mathbf{C}_e & 0 \\ 0 & \mathbf{C}_e^{\text{p}} \end{bmatrix} \frac{\partial}{\partial t} \begin{pmatrix} \mathbf{d}_e \\ \mathbf{p}_e \end{pmatrix} + \begin{bmatrix} \mathbf{K}_e & \mathbf{K}^{\text{fs}} \\ 0 & \mathbf{K}_e^{\text{p}} \end{bmatrix} \begin{pmatrix} \mathbf{d}_e \\ \mathbf{p}_e \end{pmatrix} = \begin{bmatrix} \mathbf{F}_e \\ 0 \end{bmatrix}, \tag{4}$$

with $\mathbf{d}_e$ and $\mathbf{p}_e$, being mathematical vectors collecting the unknowns at the degrees of freedom from one element, for (mechanic) displacement and (acoustic) pressure, respectively. The size of $\mathbf{d}_e$ is $3N_{\text{dof},e}$ and of $\mathbf{p}_e$ it is $N_{\text{dof},e}$, where $N_{\text{dof},e}$ is the number of degrees of freedom. Considering one node of the mesh, either 3 degrees of freedom for the displacement or one degree of freedom for the pressure are sought, depending on the location of the respective node in the computational domain. Thereby, $\mathbf{M}_e$, $\mathbf{C}_e$ and $\mathbf{K}_e$ are standard element mass, damping and stiffness matrices of the mechanics domain, respectively [23] (Section 2.2). Furthermore, $\mathbf{M}_e^{\text{p}}$, $\mathbf{C}_e^{\text{p}}$, and $\mathbf{K}_e^{\text{p}}$, are the standard element mass, damping and stiffness matrices of the acoustic domain [23] (Section 8.2). $\mathbf{C}_e^{\text{p}}$ accounts for the damping of the open duct radiation [26]. The coupling terms are $\mathbf{M}^{\text{fs}}$ for the kinematic coupling condition and $\mathbf{K}^{\text{fs}}$ for the dynamic coupling condition, which are described in [23] (Section 8.4). At the interface between air and the vocal folds, the continuity requires that the normal component of the mechanical surface velocity is equal to the normal component of the acoustic particle velocity. This is the so-called kinematic coupling condition and can be rewritten in an acoustic pressure formulation as [24] (Section 8.1)

$$\vec{n} \cdot \left(\frac{\partial^2 \vec{d}}{\partial t^2} + \frac{1}{\rho_0} \nabla p \right) = 0, \tag{5}$$

where $\vec{n}$ is the outward pointing normal vector at the interface Γ_{IF}, which is depicted in Figure 3. Furthermore, the continuity of the normal component of the forces must be satisfied [24] (Section 8.1), i.e.,

$$\vec{n} \cdot \sigma + \vec{n} p = 0. \tag{6}$$

Finally, the forcing vector $\mathbf{F}_e$ is zero due to no externally applied element forces.

3.2. Eigenvalue Problem

In order to recast the system of Equation (4) into an eigenvalue problem, a harmonic ansatz for pressure and displacement is introduced as

$$p = \Re\left\{ \tilde{p} e^{i\omega t} \right\}, \qquad \vec{d} = \Re\left\{ \tilde{\vec{d}} e^{i\omega t} \right\}, \tag{7}$$

where $\omega = 2\pi f$ is the angular frequency and "i" is the imaginary unit. Therewith, using zero forcing on the right-hand side of the system of equations, Equation (4) is reformulated to

$$-\omega^2 \begin{bmatrix} \mathbf{M}_e & 0 \\ \mathbf{M}^{\text{fs}} & \mathbf{M}_e^{\text{p}} \end{bmatrix} \begin{pmatrix} \tilde{\mathbf{d}}_e \\ \tilde{\mathbf{p}}_e \end{pmatrix} + i\omega \begin{bmatrix} \mathbf{C}_e & 0 \\ 0 & \mathbf{C}_e^{\text{p}} \end{bmatrix} \begin{pmatrix} \tilde{\mathbf{d}} \\ \tilde{\mathbf{p}}_e \end{pmatrix} + \begin{bmatrix} \mathbf{K}_e & \mathbf{K}^{\text{fs}} \\ 0 & \mathbf{K}_e^{\text{p}} \end{bmatrix} \begin{pmatrix} \tilde{\mathbf{d}}_e \\ \tilde{\mathbf{p}}_e \end{pmatrix} = \begin{bmatrix} 0 \\ 0 \end{bmatrix}, \tag{8}$$

which is a quadratic eigenvalue problem in ω. Using the substitution method described in [23] (Section 15.16.6), the quadratic eigenvalue problem is transformed into an equivalent system of generalized eigenvalue problems. This system is solved with *Ansys Mechanical* [22].

3.3. Material Damping Model

Rupitsch et al. [18] characterized the damping behavior of the same material used for the present study. Therein, they introduced a model for a complex-valued frequency-dependent equivalent Young's modulus $E(f)$ as

$$E(f) = E_R(f) + iE_I(f), \qquad \tan\delta(f) = 2\xi(f) = \frac{E_I(f)}{E_R(f)},$$

$$E_R(f) = A_R + B_R f + C_R \log_{10}\left(\frac{f + 1\,\mathrm{Hz}}{1\,\mathrm{Hz}}\right), \tag{9}$$

$$E_I(f) = A_I + B_I f + C_I \log_{10}\left(\frac{f + 1\,\mathrm{Hz}}{1\,\mathrm{Hz}}\right),$$

where $\{A_R, B_R, C_R\}$ are coefficients of the real part $E_R(f)$, and $\{A_I, B_I, C_I\}$ are coefficients of the imaginary part $E_I(f)$ of $E(f)$. Thereby, A_R is equivalent to the static Young's modulus, which can be obtained by a tensile test. From real and imaginary parts $E_R(f)$ and $E_I(f)$, the damping $\xi(f)$ and a loss factor $\tan\delta(f)$ can be obtained. For the silicone mixture material of interest, Rupitsch et al. [18] found parameters for $E(f)$, as listed in Table 1, by fitting parameters of a FE simulation to measurements results from a vibration transmission analyzer.

Table 1. Model parameters for frequency-dependent Young's Modulus $E(f)$ determined by Rupitsch et al. ("mixture 113") [18].

A_R	B_R	C_R	A_I	B_I	C_I	ν
7.02 kPa	1.09×10^1	8.02×10^2	4.05×10^3	1.07×10^1	-1.21×10^3	0.499

The model of Rupitsch et al. [18] can be interpreted as an extension to the standard Rayleigh damping model [24] (Section 3.7.2). However, Rupitsch's model is not available in Ansys Mechanical; therefore, the material model introduced in [18] is approximated in an $f_1 = 140\,\mathrm{Hz}$ with a Rayleigh damping model. Note, that this Rayleigh damping model is only an approximation. Using the model of Rupitsch et al. [18] would lead to a non-linear eigenvalue simulation [28] instead of the linear one proposed, limiting the applicability of the model to frequencies around $f_1 = 140\,\mathrm{Hz}$. However, the iterative algorithm increases the computational cost significantly, and therefore a simple Rayleigh approximation of Rupitsch's model has been found as a suitable tradeoff balancing computational cost and accuracy for the present investigation. Thus, the damping matrix $\mathbf{C}_e$ in the mechanic domain in Equation (8) is composed of a linear combination of mass and stiffness matrices

$$\mathbf{C}_e = \alpha \mathbf{M}_e + \beta \mathbf{K}_e. \tag{10}$$

The Rayleigh model coefficients α and β are obtained from $\xi(f)$ and $E_R(f)$ at a frequency of interest f_1 using a small frequency deviation Δf [24] (p. 113), such that

$$\beta(f_1) = \frac{4\Delta f \xi(f_1)}{(f_1 + \Delta f)^2 - (f_1 - \Delta f)^2} \tag{11}$$

$$\alpha(f_1) = 2(f_1 + \Delta f)\xi(f_1) - \beta(f_1 + \Delta f)^2$$

In Table 2, the obtained values for $\alpha(f_1)$ and $\beta(f_1)$ and Young's modulus $E_R(f_1)$ are listed for the operating frequency $f_1 = 140\,\mathrm{Hz}$.

Table 2. Rayleigh parameters α and β for the operating point $f_1 = 140\,\text{Hz}$ as well as Young's modulus evaluated by approximating the dynamic model of Rupitsch [18].

f_1	$\alpha(f_1)$	$\beta(f_1)$	$E_R(f_1)$
140 Hz	126.2313	1.631×10^{-4}	10.26 kPa

The Rayleigh approximation of the Rupitsch model will use a frequency-independent Youngs modulus. In Figure 4, the model of Rupitsch et al. is compared to the standard Rayleigh damping model in the operating point $f_1 = 140\,\text{Hz}$ exemplarily. For further calculations, an operating point in the experimentally estimated fundamental frequency f_0 range is selected [14].

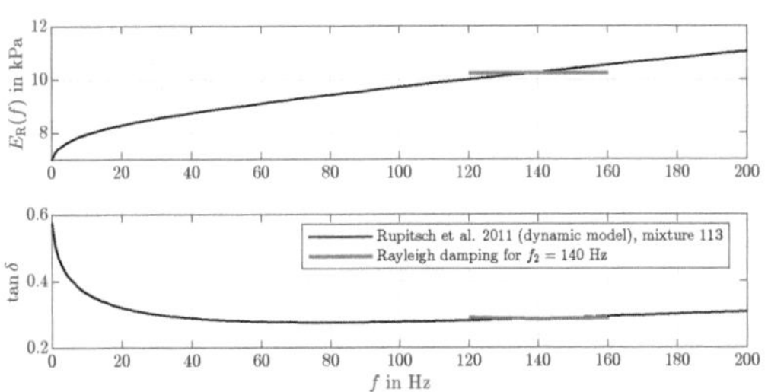

Figure 4. Comparison of the dynamic model of [18] with Rayleigh damping in frequency bands of $\pm 20\,\text{Hz}$ around the operating point $f_1 = 140\,\text{Hz}$.

3.4. Simulation Setup

In Figure 3, a sketch of the longitudinal cut of the investigated 3D-geometry is depicted. It shows the acoustic and mechanic domains Ω_a and Ω_m, respectively. For the numerical simulations, the length L is varied from 200 mm to 900 mm in steps of 20 mm, see Figure 3.

3.5. Mesh Convergence Study

The convergence of the FE model with decreasing element size (using second-order finite elements) was assessed. In a similar manner as in [29], a relative frequency error $Err^{L_2}_{\text{rel},f}$,

$$Err^{L_2}_{\text{rel},f} = \sqrt{\frac{\sum_{i=1}^{N_{\text{modes}}} \left(f^{(i)}_{\text{mode,ref}} - f^{(i)}_{\text{mode,num}} \right)^2}{\sum_{i=1}^{N_{\text{modes}}} \left(f^{(i)}_{\text{mode,ref}} \right)^2}} \tag{12}$$

was defined, where $f^{(i)}_{\text{mode,ref}}$ are the modal frequencies of a reference (benchmark) simulation using a very fine mesh, and $f^{(i)}_{\text{mode,num}}$ are the modal frequencies of the mesh that is tested against the reference. The number of investigated modes is $N_{\text{modes}} = 20$ accounting for the range of interest. The parameters of the investigated meshes and the result of the mesh convergence study are listed in Table 3. The approximate element size has been chosen based on an upper frequency limit of 8.5 kHz, resulting in a wavelength λ that is discretized with 10, 20, or 40 elements in the subglottal and supraglottal regions (also called "duct" in Table 3). The region around the vocal folds is discretized much finer and unstructured due to the small glottal gap that must be discretized sufficiently fine. For the later numerical study, mesh 2 of Table 3 was used, as an error $Err^{L_2}_{\text{rel},f} < 0.5\%$ was seen as an acceptable compromise between computational speed and accuracy. The discretization error reduces monotonically. Approximately, it reduces by two orders of magnitude for each grid refinement order. This correlates with the use of a second-order spatial discretization schemes.

Table 3. Mesh convergence analysis of the numerical model for the coupled mechanic-acoustic system. The supraglottal length L was 1050 mm for all meshes (see Figure 3).

Mesh	Approx. Elem. Size (in mm)		Wavelength at f = 8.5 kHz		$Err^{L_2}_{\text{rel},f}$
	Duct	VT	Duct	VT	
mesh 1	4.0	1.4	$\lambda/10$	$\lambda/29$	0.0086
mesh 2	2.0	0.7	$\lambda/20$	$\lambda/58$	0.0020
mesh 3 (reference)	1.0	0.35	$\lambda/40$	$\lambda/115$	—

4. Results

4.1. Experimental Results

From the LSV results of the fluid flow perturbation, the oscillation frequency of the vocal folds can be estimated using the discrete Fourier transform. The measurements were conducted in the range $L \in [170, 930]$ mm with an increment of 10 mm between the measurements. The results of this analysis, i.e., the primary oscillation frequency of the vocal folds f_0 as a function of the duct length, are reported in Figure 5. The background of this figure is colored by the acoustic input impedance of the vocal tract computed via a transmission line model [30]. For this purpose, the VT is divided into sections of equal cross-section, whereby the frequency-dependent vocal tract input impedance Z_{in} can be calculated by a concatenation of matrix multiplications. The maxima of the reactance of Z_{in} correspond to the VT resonance frequencies [26], which coincides for the current duct configuration with the maximum of the amplitude of Z_{in}. Additionally, the fundamental frequency f_0 based on particle image velocimetry data from [31] is displayed in Figure 5. In doing so, the time series signals are Fourier transformed and the most dominant peak in the amplitude spectrum is picked defining the respective f_0 at a corresponding duct length L. Figure 5 shows clearly that the trends of the experiments are reproducible. For duct length below $L = 360$ mm, a regime is detected where the acoustic resonance of the duct does not influence the fundamental frequency f_0. At around $L = 400$ mm, the oscillation shape of the vocal folds is suddenly completely different explained by mode switching of the silicon vocal fold oscillation. In a study by Sundberg et al. [8], pitch jumps and voice breaks occured for singers through long tube with different resistances [8]. For larger lengths and below the actual alignment with the acoustic resonance frequency of the duct, the fundamental frequency is gradually shifted towards the acoustic resonance frequency with increasing duct length until a length of about $L = 600$ mm. Above this duct length, alignment takes place with the acoustic resonance frequency of the duct.

Figure 6a–h shows eight snapshots from the synchronized LSV and HSC measurements for a duct length of $L = 200$ mm. The LSV measurement points were phase-synchronized based on a subglottal pressure sensor signal. The sensor recorded the f_0-periodic pressure time series from which the LSV records were triggered. These are equidistantly distributed over one vibration period. The surface velocities are in the range of about 2 m/s. It can be seen well that during the opening process (Figure 6a), the velocities are maximum in the positive direction. This can be explained by the fact that before this, in the closed state, high subglottal pressure quickly builds up upstream of the vocal folds, accelerating the vocal folds upward while simultaneously opening them. The HSC images also show a clear convergent-divergent deformation of the glottis. While in images (a) and (b), there is still a convergent glottal duct shape, in image (c), the tips of the vocal folds begin to oscillate toward the upper and lower edges, respectively, so that in images (d) and (e), a divergent glottal duct shape can be seen. From image (f), a reverse process occurs again to the initial position (a). This convergent-divergent glottal duct deformation is essential to phonation to efficiently drive self-sustaining vocal fold oscillations [32].

Bandpass filtering can extract individual oscillation modes from the LSV data. The result of such filtering with cut-on and cut-off frequencies of 130 Hz and 180 Hz, respectively, can be seen in Figure 7.

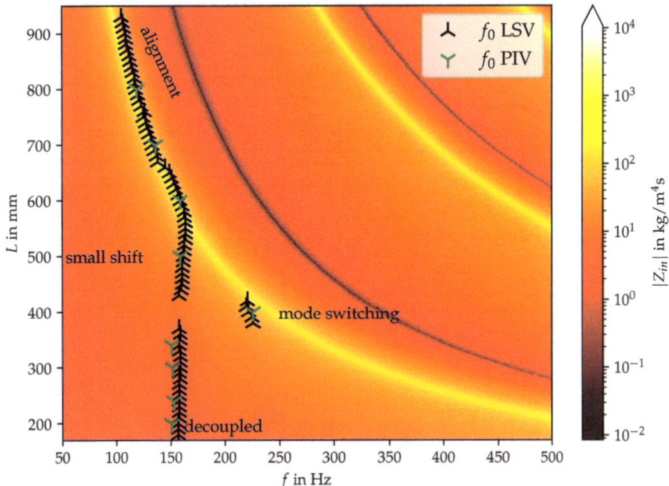

Figure 5. Vocal tract input impedance as a function of frequency f and length L. Superimposed are the primary oscillation frequencies of the vocal folds f_0 for the individual measurements.

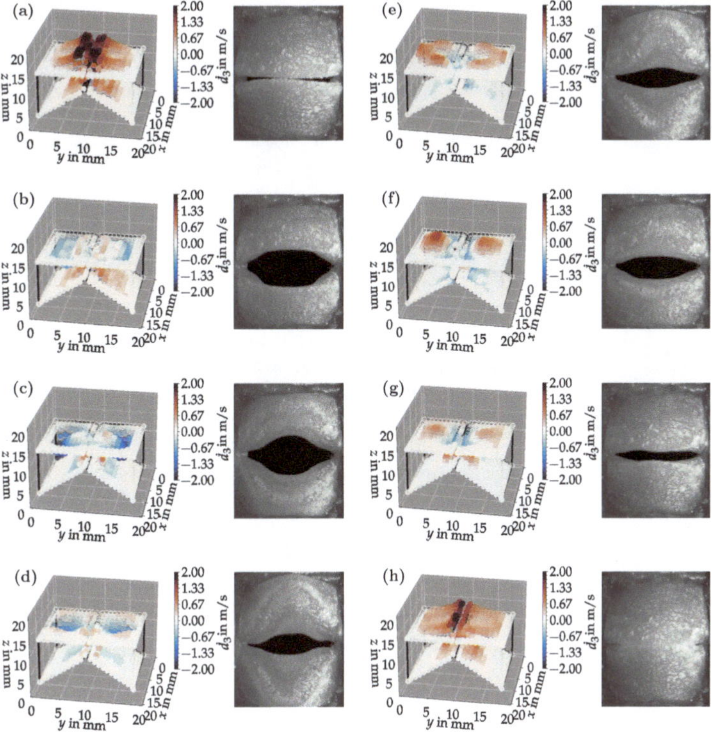

Figure 6. Time-synchronized LSV surface velocity and HSC measurements for a duct length of $L = 200$ mm. The deflection of the points as well as the color in the LSV plots are proportional to the measured surface velocity. The sequence (**a**–**h**) shows the oscillation of one complete period.

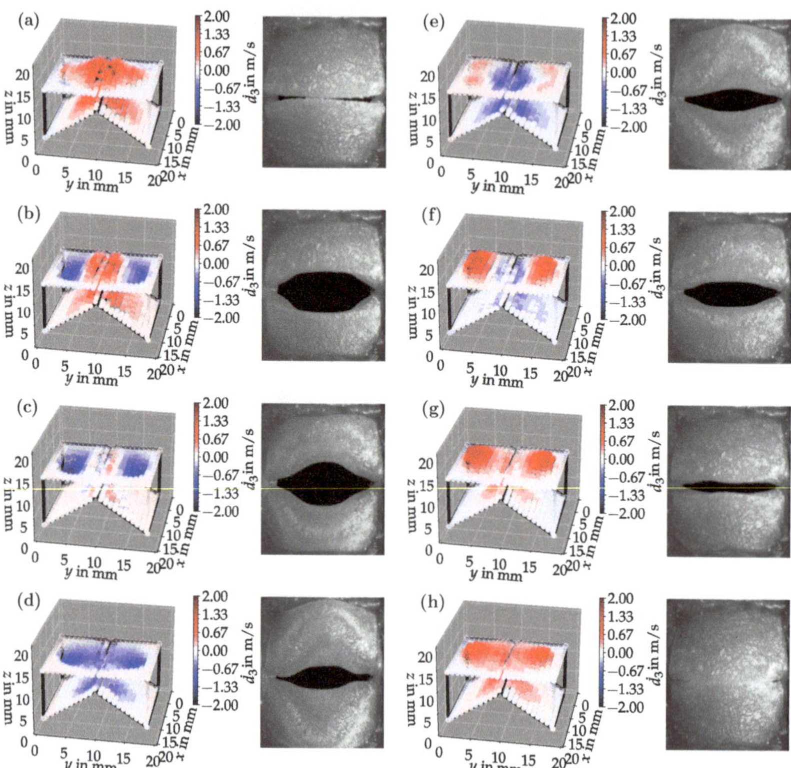

Figure 7. Bandpass filtered LSV data synchronized with the HSC recordings for a duct length of $L = 200$ mm. The oscillation at the fundamental frequency (1. mode) can be seen. The sequence (**a–h**) shows the oscillation of a complete period.

The filter has isolated the lowest oscillation frequency f_0 in the surface velocity spectrum, illustrating the oscillation at this first peak. Figure 8 shows a schematic representation of the motion in this lowest oscillation mode. It can be seen that immediately after the opening of the glottis, the surface velocities on the vocal fold top change from positive to negative direction (Figure 7a–c). Figure 8 summarizes, in (a)–(c), the opening of the vocal folds and illustrates the surface velocities accordingly. This leads to a rotation of the upper surface, changing the convergent glottis shape to a divergent shape. As soon as the maximum angle of divergence is reached, the opposite effect takes place, causing counter-rotation and a convergent glottis shape again (Figure 7d–f) and in Figure 8 illustrations (c)–(e). It can be concluded that the first mode is crucial for the divergent-convergent nature of vocal fold oscillation. Thus, it is an important component of self-sustained oscillation.

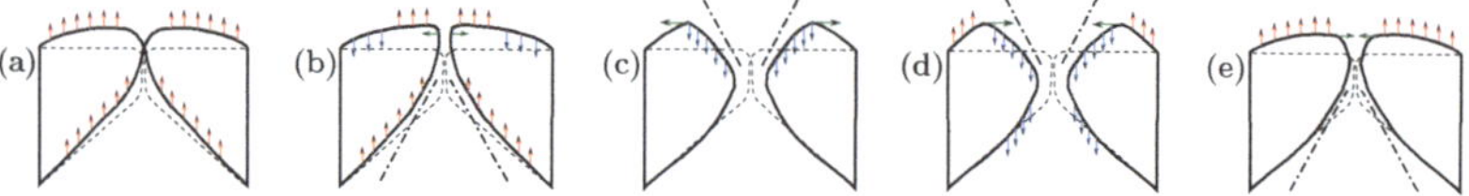

Figure 8. Schematic representation of vocal fold movement. The schemas (**a–e**) show different timesteps of one oscillation cycle. The red and blue arrows indicate the direction of surface velocity, and the green arrows indicate the direction of motion as revealed by the HSC images. The changing surface velocities on the vocal fold top cause rotation, responsible for the convergent-divergent glottal motion.

4.2. Numerical Results

Using the FE-based eigenvalue solver provided in Ansys Mechanical [22], the eigenvalues ω have been evaluated for (i) the acoustic system and (ii) the coupled mechanic-acoustic system. For the coupled mechanic-acoustic systems, the viscoelastic material at an operating condition of $f_1 = 140\,\mathrm{Hz}$ was used, as described in Section 3.3. The radiation impedance of the open end was modeled such that the fundamental frequencies of the acoustic modes match the test rig ones.

In Table 4, the first ten mode shapes from the numerical simulations are visualized for four duct lengths $L = \{200, 400, 700, 800\}$ mm. A view analogous to LSV Pos. 2 and HSC (see Figure 2) has been used, which shows the superior surface view of the deformed VFs superimposed by a color proportional to the displacement in z-direction d_3.

At $L = 200\,\mathrm{mm}$, mode index 1 represents a mechanical mode opening and closing the vocal folds with in-phase displacement in z-direction at a frequency of $f = 81.29\,\mathrm{Hz}$. This mode possibly provides both aerodynamic constriction of the airflow as well as an effective coupling to the acoustic plane wave with its in-phase displacement. The same is valid for $L \in \{400\,\mathrm{mm}, 700\,\mathrm{mm}, 800\,\mathrm{mm}\}$. Mode index 2, which is at $f = 81.98\,\mathrm{Hz}$ for $L = 200\,\mathrm{mm}$ does not contribute to acoustic radiation, because the z-component of the displacement is not in phase for both vocal folds hence no acoustic wave is scattered to the far field. The same is valid for $L \in \{400\,\mathrm{mm}, 700\,\mathrm{mm}, 800\,\mathrm{mm}\}$. At $L = 200\,\mathrm{mm}$, mode index 3 represents an acoustically ineffective mechanical mode due to the displacement being out of phase within the vocal folds for a frequency of $f = 125.19\,\mathrm{Hz}$. The same holds for $L \in \{400\,\mathrm{mm}, 700\,\mathrm{mm}\}$. However, at $L = 800\,\mathrm{mm}$, an additional coupled mechanical-acoustic motion pattern arises, which is similar to mode index 1. Based on the mode shape morphology with a symmetric phase of the displacement z-component (where the symmetry axis is the glottis center line), this mode is acoustically effective. For mode index 4 at $L = 200\,\mathrm{mm}$ being at $f = 125.24\,\mathrm{Hz}$ the asymmetric displacement z-component suggests an acoustically non-efficient mechanical mode. The same is valid for $L \in \{400\,\mathrm{mm}, 700\,\mathrm{mm}\}$. Due to the additional coupled mechanical-acoustic motion pattern at $L = 800\,\mathrm{mm}$ however, mode index 4 of this length is the same as mode index 3 for $L = 700\,\mathrm{mm}$, which is again acoustically ineffective. Mode index 5 at $L = 200\,\mathrm{mm}$ presents an acoustically effective vibration pattern at a frequency of $f = 137.88\,\mathrm{Hz}$ due to the symmetry of the displacement z-component. Identical vibration patterns are present at $L = 400\,\mathrm{mm}$ and $L = 700\,\mathrm{mm}$, where in the latter case only a sign change is present overall. In contrast to that, mode index 5 at $L = 200\,\mathrm{mm}$ at $f = 124.60\,\mathrm{Hz}$ is acoustically ineffective due to the asymmetric displacement z-component. For mode index 6 at $L = 200\,\mathrm{mm}$ being at $f = 138.06\,\mathrm{Hz}$, as well as mode index 6 at $L = 400\,\mathrm{mm}$, acoustically ineffective vibration patterns are present. In contrast to that, at $L = 700\,\mathrm{mm}$ an acoustically effective additional coupled mechanical-acoustic motion pattern arises, which is visually very similar to mode index 3 at $L = 800\,\mathrm{mm}$. Furthermore, at $L = 800\,\mathrm{mm}$, an acoustically effective mode is present. Mode indices 7 and 8 present acoustically ineffective vocal fold vibration patterns for all four VT length configurations. For mode index 9 at $L = 200\,\mathrm{mm}$, which is at a frequency of $f = 154.04\,\mathrm{Hz}$, an acoustically effective motion pattern is present. However, for $L \in \{400\,\mathrm{mm}, 700\,\mathrm{mm}, 800\,\mathrm{mm}\}$, the modes are acoustically ineffective due to the asymmetries in the displacement z-component. Finally, for mode index 10 at $L = 200\,\mathrm{mm}$ being at $f = 154.23\,\mathrm{Hz}$, an acoustically inefficient mode is present. The same holds for $L = 400\,\mathrm{mm}$. However, at $L \in \{700\,\mathrm{mm}, 800\,\mathrm{mm}\}$, the z-component of the displacement is symmetric at the glottis center line, indicating acoustically effective vocal fold vibration modes at these length configurations.

Summarizing the above, from Table 4 it is evident that the first ten mode shapes do not change significantly comparing the cases $L = 200\,\mathrm{mm}$ and $L = 400\,\mathrm{mm}$. However, at $L = 700\,\mathrm{mm}$, an additional mode shape arises at mode 6, which shifts the subsequent mode indices. The same additional mode shape also arises in the case of $L = 800\,\mathrm{mm}$, which is mode 3, resulting once more in a shift of the mode indices. Regarding the effective contributions of individual modes to phonation, it is important that the top surfaces of both VFs have the same phase throughout. Therefore, for the cases $L = 200\,\text{mm}$ and $L = 400\,\text{mm}$, the following modes can be considered effective for phonation: Modes 1, 5, and 9, with other modes having only minor contributions to phonation. The other modes 2, 3, 4, 6, 7, 8, 10 are counterbalancing displacements and velocity patterns. This counterbalancing pattern is not able to couple efficiently to a plane acoustic wave inside the acoustic duct, which is in the case of the lowest formant a requirement for an effective coupling of the mechanical and acoustic field. For instance, mode 2 is a parallel motion of the left and right vocal fold which does not constrict the gap between the vocal folds and thus being a very inefficient mode from the whole fluid-structure-acoustic interaction perspective. This behavior in a higher order shape occurs for mode 4, 6, and 8. For the case $L = 700\,\text{mm}$, the following modes can be considered effective for phonation: Modes 1, 5, 6, 10. Finally, for case $L = 800\,\text{mm}$, the following modes can be considered effective for phonation: Modes 1, 3, 6, 10. It can be clearly seen, that in the case of $L = 700\,\text{mm}$ and $L = 800\,\text{mm}$ an additional mode ("phonation-effective mode") enters being the mode shape with mode number 6 and 3 respectively. This phonation-effective mode is the coupled mechanical-acoustic motion pattern when the acoustic back-coupling is active. When this mode occurs, the mode numbers above this additional mode are shifted by one $k \leftarrow k + 1$. This behavior can be seen that the mode 5 in the case of $L = 700\,\text{mm}$ is mode 6 for the case of $L = 800\,\text{mm}$.

Table 4. Numerically obtained mode shapes for different VT lengths L. The color indicates the displacement component d_3 in z-direction, while the x and y-components of the displacement are visualized by a scaled geometry deformation. Negative Displacement $d_3 < 0$ is colored blue, positive displacement $d_3 > 0$ is colored red, and $d_3 = 0$ is colored green. The displacement component d_3 is normalized to the maximum absolute value.

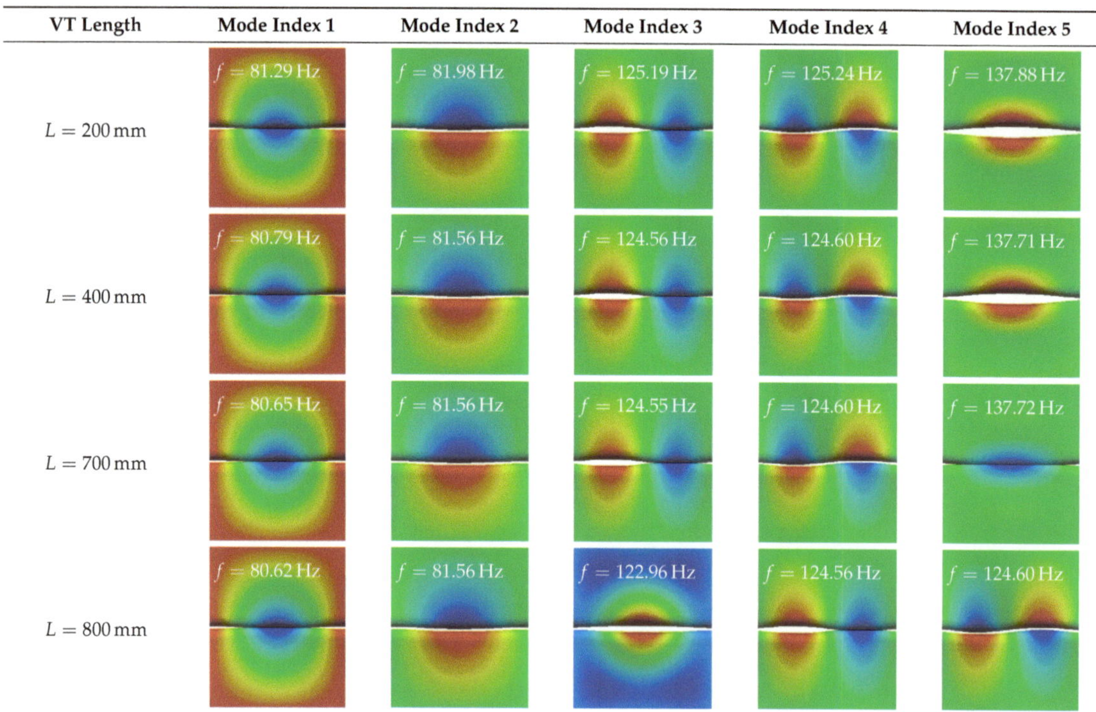

VT Length	Mode Index 1	Mode Index 2	Mode Index 3	Mode Index 4	Mode Index 5
$L = 200\,\text{mm}$	$f = 81.29\,\text{Hz}$	$f = 81.98\,\text{Hz}$	$f = 125.19\,\text{Hz}$	$f = 125.24\,\text{Hz}$	$f = 137.88\,\text{Hz}$
$L = 400\,\text{mm}$	$f = 80.79\,\text{Hz}$	$f = 81.56\,\text{Hz}$	$f = 124.56\,\text{Hz}$	$f = 124.60\,\text{Hz}$	$f = 137.71\,\text{Hz}$
$L = 700\,\text{mm}$	$f = 80.65\,\text{Hz}$	$f = 81.56\,\text{Hz}$	$f = 124.55\,\text{Hz}$	$f = 124.60\,\text{Hz}$	$f = 137.72\,\text{Hz}$
$L = 800\,\text{mm}$	$f = 80.62\,\text{Hz}$	$f = 81.56\,\text{Hz}$	$f = 122.96\,\text{Hz}$	$f = 124.56\,\text{Hz}$	$f = 124.60\,\text{Hz}$

Table 4. *Cont.*

VT Length	Mode Index 6	Mode Index 7	Mode Index 8	Mode Index 9	Mode Index 10
$L = 200\,\text{mm}$	$f = 138.06\,\text{Hz}$	$f = 143.44\,\text{Hz}$	$f = 143.52\,\text{Hz}$	$f = 154.04\,\text{Hz}$	$f = 154.23\,\text{Hz}$
$L = 400\,\text{mm}$	$f = 137.91\,\text{Hz}$	$f = 142.89\,\text{Hz}$	$f = 142.98\,\text{Hz}$	$f = 153.89\,\text{Hz}$	$f = 154.12\,\text{Hz}$
$L = 700\,\text{mm}$	$f = 139.18\,\text{Hz}$	$f = 137.90\,\text{Hz}$	$f = 142.87\,\text{Hz}$	$f = 142.95\,\text{Hz}$	$f = 153.98\,\text{Hz}$
$L = 800\,\text{mm}$	$f = 137.75\,\text{Hz}$	$f = 137.91\,\text{Hz}$	$f = 142.89\,\text{Hz}$	$f = 142.98\,\text{Hz}$	$f = 153.96\,\text{Hz}$

The additional modes are further investigated by connecting them with the acoustic impedance of the vocal tract in the following. Based on visualizations of the mechanic displacement, modes have been identified that are similar to those of the experimental investigations. Figure 9 supplements Figure 5 by adding the simulation results of the linear eigenfrequency (mode index 5, 6 and 9) of the coupled mechanic-acoustic eigenmode simulation for varying duct length. For a duct length of less than 550 mm, the acoustic and the mechanical eigenmodes are decoupled from each other, suggesting that acoustic effects do not significantly influence the silicon vocal fold vibration pattern, as it is also visible in Table 4. Between 550 mm and 700 mm, the coupled mechanical-acoustic mode is aligned with the acoustic mode, which is the effect of the additional phonation-effective mode depicted in Table 4. In this regime, the acoustic back-coupling strongly influences the vibration frequency of the vocal fold motion. This indicates that acoustic effects of the fluid are essential for the realistic determination of the oscillation frequency. For longer ducts, the physical shape of the mode with index 9 is now the mode with index 10. Virtually shifting oscillation frequency of a fixed mode index 9 to a lower frequency (where mode 5 is aligned with the acoustic mode) compared to the uncoupled situation for a duct length below 550 mm due to the additional phonation-effective mode. In contrast to modes index 5 and index 9, mode index 6 does not couple at all to the acoustic mode because the oscillation pattern of the vocal folds is not acoustically effective (as indicated in the discussion of Table 4 by the counteracting the motion in z-direction). Regarding a more detailed analysis and the evolution of the modes with varying lengths, the modal assurance criterion (MAC) is evaluated for the simulated and measured modes.

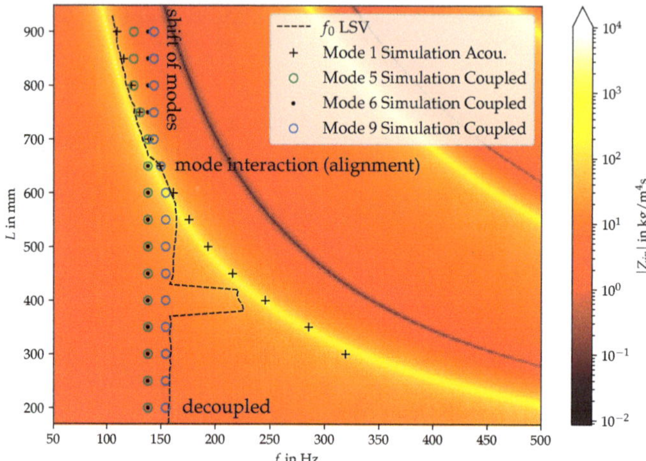

Figure 9. Numerically evaluated eigenfrequencies of the acoustic and coupled mechanic-acoustic system. The mechanical damping has been modeled by Rayleigh damping, which approximates the Rupitsch model at the operating frequency of $f = 140\,\text{Hz}$.

The modal assurance criterion (MAC) is a measure of the similarity between different mechanical modes introduced in [33]. Given complex mode shapes in the form of two mode shape vectors $\boldsymbol{\Phi}_k$ and $\boldsymbol{\Psi}_l$, it computes as follows [33]

$$MAC(\boldsymbol{\Phi}_k, \boldsymbol{\Psi}_l) = \frac{\left|\boldsymbol{\Phi}_k^\mathrm{T}\boldsymbol{\Psi}_l^*\right|^2}{\left(\boldsymbol{\Phi}_k^\mathrm{T}\boldsymbol{\Phi}_k^*\right)\left(\boldsymbol{\Psi}_l^\mathrm{T}\boldsymbol{\Psi}_l^*\right)} \cdot 100\,\%. \tag{13}$$

Thereby, k and l are the indices of the respective modes, i.e., the k-th mode shape of $\boldsymbol{\Phi}$ is compared with the l-th mode shape of $\boldsymbol{\Psi}$. Hence, the MAC enables (i) comparisons of simulated mode shapes with measured mode shapes, (ii) comparisons of mode shapes originating from two different simulations, or (iii) self-comparisons of mode shapes from one simulation or one measurement (Auto-MAC) showing the self-similarity of different mode shapes of one configuration. However, the MAC requires that the mode shape pairs $\boldsymbol{\Phi}_k$ and $\boldsymbol{\Psi}_l$ are evaluated at identical coordinates. Therefore, the simulation result (i.e., the displacement field) is interpolated using the FE basis functions to the measurement points.

Firstly, the self-comparisons of mode shapes from one simulation as Auto-MAC is studied and illustrated in Figure 10. Therein, the MAC has been evaluated for the two LSV positions separately (see Figure 2), and the mean MAC values of the two measurement plane values are assessed. The three illustrations show the Auto-MAC of the simulated modes for the duct length 200 mm, 400 mm and 700 mm, in the three subfigures (a), (b), and (c), respectively. These lengths correspond to the three regimes (decoupling, alignment, shift) of mode 9 in Figure 9. As intended by the MAC, it shows a strong correlation of the modes with itself by the black squares in the diagonals of the matrix plots in Figure 10. For all individual modes, a low similarity is present concerning the other modes, which is expressed by the relatively low Auto-MAC values in the off-diagonal elements. This behavior is present in all three regimes investigated. Qualitatively, the three sub-figures look very similar, which expresses that the modes as a solution of the numerical system have persistent shapes with a length variation. Especially for length 200 mm and 400 mm and according to Table 4, the mode shapes are very similar (no coupling of the lower modes to the mechanic field) and therefore this is expected for the Auto-MAC values. This may allow us to use these shapes in a potential model order reduction over a wide range of operating points that are, respectively, length variations. From Figure 10, the mode indices and motions contributing to phonation show also correlations to each other, being the mode index numbers 1, 5, and 9 for length 200 mm and 400 mm. For length 700 mm, the phonation-effective mode (index 6) is present, showing that a high MAC value is present between mode index 1 and index 6.

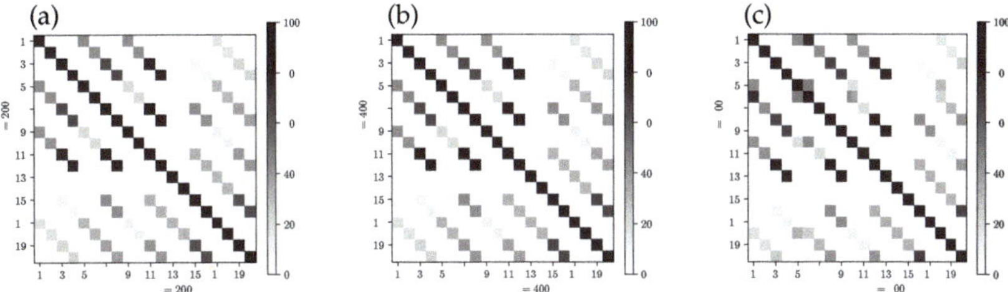

Figure 10. Auto-MAC of the simulation results. (**a**) Auto-MAC for $L = 200$ mm, (**b**) Auto-MAC for $L = 400$ mm, and (**c**) Auto-MAC for $L = 700$ mm.

In Figure 11, the MAC is depicted for the modes $k \in \{1, 5, 10\}$, along VT length variations using the simulation result data. With the k-th mode shape obtained from the VT length L, the MAC is computed for two reference mode shapes: (i) the k-th mode shape of $L = 200$ mm denoted as $\mathbf{\Psi}_k^{200\,\text{mm}}$, and (ii) the $(k-1)$-th mode shape for of $L = 200$ mm denoted as $\mathbf{\Psi}_{k-1}^{200\,\text{mm}}$. As depicted in Figure 11, a shift in mode indices is evident, i.e., for each mode index k, there is a critical VT length L_{crit} at which the $MAC(\mathbf{\Phi}_k^L, \mathbf{\Psi}_k^{200\text{mm}})$ switches from close to 100 % to close to 0 %, and the $MAC(\mathbf{\Phi}_k^L, \mathbf{\Psi}_{k-1}^{200\text{mm}})$ simultaneously raises from close to 0 % to close to 100 %. Therefrom it is evident that an index shift occurs, i.e., the k-th mode for $L < L_{\text{crit}}$ is the $(k-1)$-th mode for $L > L_{\text{crit}}$. Together with Figure 9 and Table 4, it can be concluded, that L_{crit} depends on the mode index k, and it is located at the intersection between the acoustic modes and the coupled acoustic-mechanic modes.

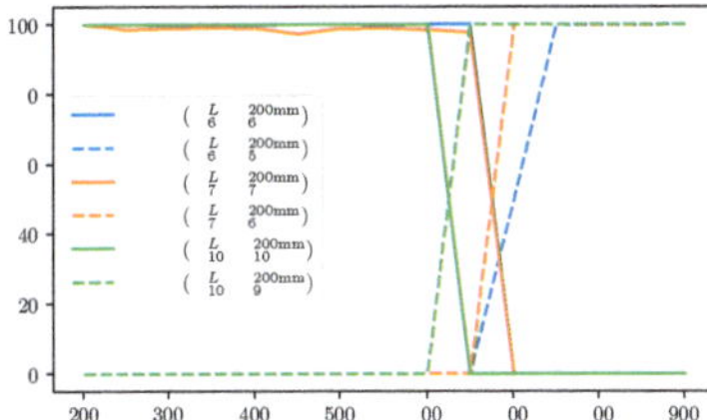

Figure 11. Comparison of MAC values for mode indices $k \in \{6, 7, 10\}$ across VT length variations. The solid lines indicate the MAC based on the mode k of length 200 mm as reference, whereas the dashed lines indicate that the MAC is computed based on the mode $k-1$ as reference to illustrate a switch in the mode number.

4.3. Comparison of Experimental and Numerical Results by Modal Assurance Criterion

To quantify the agreement between the measured surface velocities from the LSV measurements with the numerical simulations, the MAC is evaluated for the measured VT lengths. The result quantity of the simulation is the displacement $\vec{d}$, the measurement quantity of the LSV is the velocity $\dot{\vec{d}} = (\dot{d}_1, \dot{d}_2, \dot{d}_3) = \partial \vec{d}/\partial t$. Furthermore, the simulation results are available in the frequency domain a priori, while the measurements are delivered in the time domain. Hence, the measured velocity field is Fourier transformed, such that a frequency resolution of $\Delta f = 10$ Hz is achieved. Then, the measured velocity field is compared to the simulation result (displacement) at the measurement points. The simulated displacement results are therefore multiplied by $i\omega$ being the Fourier transform of $\partial/\partial t$. Thereby, $\omega = 2\pi f_m$ with f_m being the mode frequency, respectively. The MAC matrices are depicted for both LSV positions depicted in Figure 2 separately.

In Figures 12–14, the MAC for $L = \{200, 400, 700\}$ mm is depicted, respectively. The modes identified by FE simulations are frequency-wise closely together; therefore, individual modes are hard to identify in the measurements. However, a qualitative assessment of mode shape morphology similarities is possible nevertheless. In Figure 12, it is visible that simulated modes 1 and 9 are mainly representing the experimental components at the measured fundamental frequency, providing evidence on the phonation effective modes in the decoupled regime. In Figure 13, the simulated mode 13 at 220 Hz explains the main characteristics of the experimental results, indicating a dominating mode shift. Furthermore, more modes add minor contributions to the overall vibration behavior compared to the decoupled regime. From Figure 14 one can conclude that the simulated modes 1, 5, and 6 (phonation-effective mode) are shifted to lower frequencies around 140 Hz while also being similar to higher-frequency components at integer multiples (i.e., 280 Hz and 420 Hz). These comparisons to experimental modes provide a clear picture of which modes are essential for the phonation in the decoupled and coupled mechanical-acoustic regime.

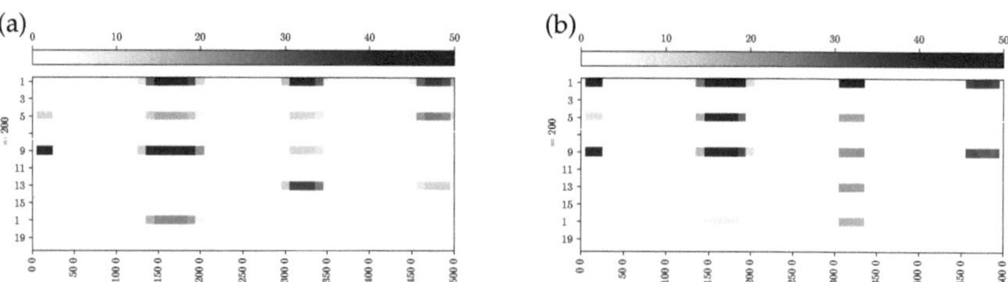

Figure 12. MAC for measured (LSV) and simulated modes with $L = 200$ mm for (**a**) subglottal LSV Pos. 1 and (**b**) supraglottal LSV Pos. 2.

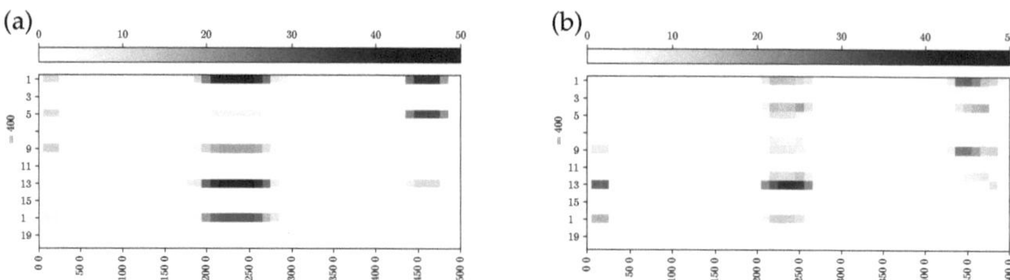

Figure 13. MAC for measured (LSV) and simulated modes with $L = 400$ mm for (**a**) subglottal LSV Pos. 1 and (**b**) supraglottal LSV Pos. 2.

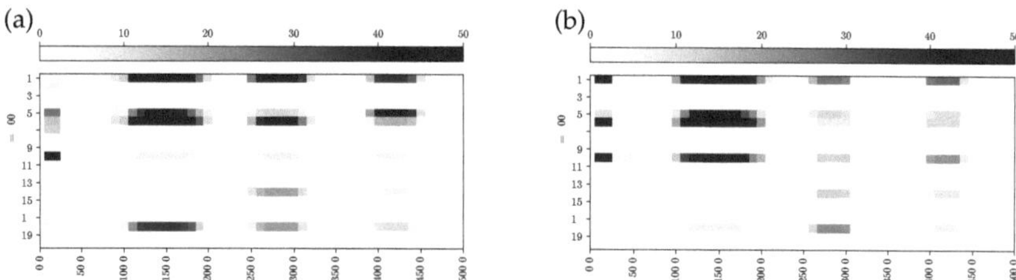

Figure 14. MAC for measured (LSV) and simulated modes with $L = 700$ mm for (**a**) subglottal LSV Pos. 1 and (**b**) supraglottal LSV Pos. 2.

5. Discussion

5.1. Model Limitations

Comparing Figure 5 and Figure 9 it can be clearly seen that the established model provides a clear condition when the acoustic-structure back-coupling is present. The frequency alignment and the decoupled regime can be predicted for varying duct length. The limitations of the linear eigenfrequency analysis are that the non-linear effects of the vocal fold contact and the multi-harmonic or chaotic oscillation behavior cannot be modeled accordingly. Therefore, the regimes (as indicated in Figure 5) "small shift" and "mode switching" cannot be explained by the coupled mechanical-acoustical model. Nevertheless, the model is useful to learn about the interaction of the mechanical and acoustical fields.

5.2. Oscillation Pattern and Relation to Other Voice Parameters

Initially, a harmonic oscillation pattern was assumed by the eigenvalue computation (see Equation (7)). In this paragraph, the limitations of this assumption are discussed using aspects of nonlinear dynamics, as introduced by Herzel et al. [34]. Figure 15 shows the phase space diagram of the oscillation in the z-direction (inferior-superior direction) of the left vocal fold's superior edge (according to the coordinate system defined in Figure 7). Additionally, the time series of the velocity in z-direction $\dot{d}_3$, the time series of the flow derivative is displayed over one period. The velocity in z-direction $\dot{d}_3$ is measured at the point 7.5 mm and 7.1 mm (superior edge). Given the measured velocity $\dot{d}_3$, the displacement in z-direction d_3 is estimated by integration of the high-pass filtered (110 Hz) velocity signal $\dot{d}_3$. The initial conditions $d_3(t = 0) = 0$. In Figure 15, the displacement is normalized by the initial vocal fold gap $a = 0.2$ mm and the velocity by a typical mucosal wave speed of about $c_m = 1$ m/s [35]. For normalized velocities $\dot{d}_3/c_m$ being plotted over the normalized displacement d_3/a, the phase space diagram orbits are turning in the clockwise direction for the left vocal fold (as indicated by the star as starting point and the circle as ending point marker). Additional markers for the second orbit and the notch in the magenta and black curve are marked by a square sign. The notch in the black and magenta curve is an inflection of the velocity, where the superior edge undergoes a wiggle-like motion going from deceleration to acceleration during the opening phase of the vocal fold. The orbit of $(d_3/a, \dot{d}_3/c_m)$ for the duct length 200 mm has a somewhat elliptical shape, with slight deviations from orbit to orbit. This can also be seen from the evolution of the $\dot{d}_3$ over a period of the oscillation. Despite some minor high-frequency patterns, the velocities time signal has a dominant harmonic content at the oscillation frequency f_0. According to Figure 6, the highest surface velocity corresponds to the phase of closed vocal folds and is indicated in Figure 15. Additionally, to the high-speed camera visualization, the glottal flow [31] was correlated to the time series of the velocity. The flow derivative is used to interpret the vocal fold motion, the opening and closing phase, and regarding [36], it shows a connection of intra-glottal vortices causing a more rapid closing resulting in an increased sound pressure level [37]. The flow rate derivative was estimated based on two-dimensional particle image velocimetry data [31] averaging ten lines perpendicular to the glottal jet at ten different streamwise positions starting at the vocal folds and assuming full coherence in the third dimension. In doing so, the time series of varying glottal flow rate $\dot{V}(t)$ was estimated and its time derivative $dV(t)/dt$ computed. The estimated flow rate derivative was normalized by the mean glottal flow $\overline{V}$ and the fundamental frequency f_0. For the length 200 case, the flow derivative increases monotonically over half a period and declines approximately the other half of the period. This configuration is considered to be the base configuration and results in a sound pressure level (SPL) in front of the duct of 82.5 dB [31].

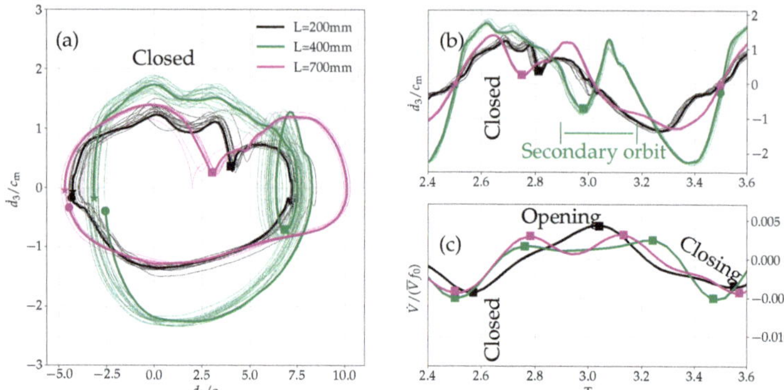

Figure 15. (**a**) Phase space diagram showing velocity $\dot{d}_3$ and d_3 displacement trajectories of the superior edge of the left vocal fold in the inferior-superior direction for all considered vocal tract lengths. The smaller orbit within the outer orbit (in the clockwise direction) for the $L = 400\,\text{mm}$ case indicates periodic doubling and onset of chaotic behavior. The $L = 200\,\text{mm}$ and $L = 700\,\text{mm}$ cases show more regular orbits and with a pronounced notch for $L = 700\,\text{mm}$. The star marker (*) is the starting point and the circle marker (o) is the end point of one orbit. (**b**) Time history of the inferior-superior velocity $\dot{d}_3$ from PSV measurements. The square marker (■) indicates the periodic doubling in the $L = 400\,\text{mm}$ case and the kink in the $L = 200\,\text{mm}$ and $L = 700\,\text{mm}$ case. (**c**) Time history of the flow rate derivative $\dot{V}(t)$ extracted from the PIV measurements of [31].

At a length of 400 mm the orbit shows a secondary loop indicating a periodic doubling at this length. A regular large orbit moves into a secondary orbit during each period. This is also visible in the velocity evolution over a period. The secondary orbit is indicated by the green square marker. It is indicated as secondary orbit and has nearly half the velocity amplitude of the dominating orbit. Variations from orbit to orbit are visible by the green dotted curves. Compared to the $L = 200\,\text{mm}$ case, the orbit variations of the $L = 400\,\text{mm}$ case are larger. In general, this indication of periodic doubling strongly violates the assumption of a harmonic ansatz, supporting the arguments that such a behavior cannot be explained by the eigenvalue simulation. It displays a multi-harmonic nature of the vocal fold oscillation at these conditions. The flow derivative for the length 400 case has a comparable small and extended positive part of the time series, spanning about $0.7T$ and a rapid negative closing dip twice as high as the positive part. This behavior of a sharp negative dip was also observed in the investigations conducted in [36]. Additionally, the flow derivative has a secondary oscillation which occurs at the same time as the periodic doubling. For this case, the SPL in front of the duct is 95 dB [31].

The orbit of $(z, \dot{d}_3)$ for the length $L = 700\,\text{mm}$ has a elliptical shape with a kink in the positive part of the positive velocity fluctuations (indicated by a magenta square sign). Also for this case, slight deviations occur from orbit to orbit, which are comparable to the $L = 200\,\text{mm}$ case. In this case, the positive part of the positive velocity fluctuations is oscillating, since it does stay positive no secondary orbit is formed. This is already a strong deviation from a pure harmonic velocity evolution. In connection to the closed vocal fold, the kink in the phase space diagram happens shortly after the vocal folds have closed. In this case, the flow derivative is positive for $0.6T$ and has a relative sharp negative dip compared to the length $L = 200\,\text{mm}$ case. The evolution of the flow derivative looks very similar to the one from the length $L = 400\,\text{mm}$ case. Where the kink occurs in the phase space diagram, the flow derivative shows a secondary oscillation. In this case, the SPL is 84 dB [31]. Comparing the three cases, it appears that the negative dip in the flow derivative is positively correlated with the SPL, as well as the inertance effect, as described by Titze [35], which gives a low supraglottal pressure that produces a push on vocal the vocal folds during closing. Due to the correlation with SPL, the vocal efficiency at the length

$L = 400\,$mm case is about one order of magnitude higher than the length $L = 200\,$mm case [31]. Furthermore, the value of the vocal efficiency for the length $L = 700\,$mm case is increased [31], showing a positive correlation of the strength of the additional wiggle in the flow derivative and the vocal efficiency. In relation to Figure 8, the motion in the inferior-superior direction of the schematic and the phase space diagram are consistent.

6. Conclusions

The fluid-structure-acoustic interaction process, like human phonation process, is one of the most challenging physical phenomena. To enhance the understanding of this type of process, an experimental apparatus was designed to mechanically align the acoustic of the duct and the coupled mechanic-acoustic mode of the coupled vibroacoustic setup consisting of single mold silicone vocal folds and a straight duct. The experiments showed that when increasing the supraglottal duct of the apparatus, the acoustic eigenfrequency decreases monotonically. In the case when the acoustic eigenfrequency of the duct came into the range of the fundamental (mechanical) vibration frequency of the silicone vocal folds their vibration frequency deviated from it. This effect is dominant and strong deviations occur, when the acoustic eigenfrequency of the duct is lower than the corresponding mechanical eigenfrequency.

Regarding these experimental findings, the vocal folds motion of the uncoupled and the coupled mechanical-acoustic eigenvalue problem are investigated. The purpose of the simulation is to show that for a length smaller than the critical length (crossing of the acoustic and the mechanical eigenfrequency), the mechanical eigenmodes of the vocal folds in the neighborhood of the fundamental frequency are not influenced by the acoustic (compressible) subsystem. Whereas, for a length longer than the critical length, the combined system is of importance and the vibration frequency of the vocal folds is aligned with the acoustic mode frequency. The results demonstrate that changing the vocal tract length has an influence on the frequency of the mode arising by the coupled mechanic-acoustic field. Furthermore, the quantitative comparison between numerical and experimental results by means of the MAC exhibits a strong correlation of the coupled mechanic-acoustic mode, indicating a strong contribution of this mode on phonation. It was found that a changing vocal tract length allows for a changing frequency of the coupled mode that greatly contributes to phonation. As a consequence of this analysis, the findings report the importance of the interaction and the back-coupling of the acoustic onto the mechanical structure in certain regimes. Whereas under normal conditions, the back-coupling can be neglected as reported in numerous studies before. Finally, the use of the eigenmode analysis is an elegant way of investigating the dependence of the modes on each other. This may allow us to use these shapes in a potential model order reduction over a wide range of operating points that are, respectively, length variations. Relating to the acoustic-structure interaction, one recent publication by Manconi et al. [38] included the characterization of this interaction by a dispersion curve. This approach could be transferred to human phonation to display the dispersion relation of the mucosal wave, but, as a prerequisite, equally spaced surface displacement data is necessary. However, the measurement data capturing all nonlinear effects are not equally spaced. Hence, the findings presented in [38] provide an interesting approach for future investigations.

Finally, the MAC measure between the simulated and experimental modes showed which mode shapes effectively contribute to the phonation and which modes do not contribute. As numerical and experimental results are in good agreement, the model can be used to provide explanatory insight for acoustic contributions of individual modes. This analysis provided a clear picture of both the coupled (source-filter interaction) and the decoupled (normal phonation) mechanical-acoustic regime. Furthermore, the results showed strong correlations of the obtained vocal fold motion characteristics with previously found correlations to other voice parameters like vocal efficiency and the sound pressure level.

Author Contributions: Conceptualization, S.S., C.N. and F.K.; methodology, S.S., C.N. and F.K.; software, S.S., C.N., E.S. and F.K.; validation, S.S., C.N. and F.K.; formal analysis, C.N., S.S. and F.K.; investigation, S.S., C.N. and F.K.; resources, S.S., S.B. and S.K.; data curation, S.S., C.N. and F.K.; writing—original draft preparation, S.K., M.L., E.S., M.M., S.S., C.N. and F.K.; writing—review and editing, S.K., M.L., E.S., M.M., S.S., C.N., S.B. and F.K.; visualization, S.S., C.N. and F.K.; supervision, S.S., S.K., S.B. and M.M.; project administration, S.S.; funding acquisition, S.S. and S.B. All authors have read and agreed to the published version of the manuscript.

Funding: F.K. received funding from the Austrian Research Promotion Agency (FFG) under Bridge Project No. 39480417. C.N. received funding from the Deutsche Forschungsgemeinschaft (DFG, German Research Foundation) under Project No. 446965891. Supported by TU Graz Open Access Publishing Fund.

Institutional Review Board Statement: Not applicable.

Informed Consent Statement: Not applicable.

Data Availability Statement: The data presented in this study are available on request from the corresponding author.

Acknowledgments: Open Access Funding by the Graz University of Technology.

Conflicts of Interest: The authors declare no conflict of interest.

References

1. Döllinger, M.; Zhang, Z.; Schoder, S.; Šidlof, P.; Tur, B.; Kniesburges, S. Overview on state-of-the-art numerical modeling of the phonation process. *Acta Acust.* **2023**, *7*, 25. [CrossRef]
2. Schoder, S.; Maurerlehner, P.; Wurzinger, A.; Hauser, A.; Falk, S.; Kniesburges, S.; Döllinger, M.; Kaltenbacher, M. Aeroacoustic sound source characterization of the human voice production-perturbed convective wave equation. *Appl. Sci.* **2021**, *11*, 2614. [CrossRef]
3. Maurerlehner, P.; Schoder, S.; Freidhager, C.; Wurzinger, A.; Hauser, A.; Kraxberger, F.; Falk, S.; Kniesburges, S.; Echternach, M.; Döllinger, M.; et al. Efficient numerical simulation of the human voice: simVoice—A three-dimensional simulation model based on a hybrid aeroacoustic approach. *e i Elektrotech. Informationstechnik* **2021**, *138*, 219–228. [CrossRef]
4. Schoder, S.; Kraxberger, F.; Falk, S.; Wurzinger, A.; Roppert, K.; Kniesburges, S.; Döllinger, M.; Kaltenbacher, M. Error detection and filtering of incompressible flow simulations for aeroacoustic predictions of human voice. *J. Acoust. Soc. Am.* **2022**, *152*, 1425–1436. [CrossRef] [PubMed]
5. Zhang, Z. Cause-effect relationship between vocal fold physiology and voice production in a three-dimensional phonation model. *J. Acoust. Soc. Am.* **2016**, *139*, 1493–1507. [CrossRef] [PubMed]
6. Fant, G. *Acoustic Theory of Speech Production: With Calculations Based on X-ray Studies of Russian Articulations*; Number 2; Walter de Gruyter: Berlin, Germany, 1971. [CrossRef]
7. Titze, I.R. Nonlinear source–filter coupling in phonation: Theory. *J. Acoust. Soc. Am.* **2008**, *123*, 2733–2749. [CrossRef]
8. Sundberg, J.; Lã, F.; Granqvist, S. Fundamental frequency disturbances in female and male singers' pitch glides through long tube with varied resistances. *J. Acoust. Soc. Am.* **2023**, *154*, 801–807. [CrossRef]
9. Lucero, J.C.; Lourenço, K.G.; Hermant, N.; Van Hirtum, A.; Pelorson, X. Effect of source–tract acoustical coupling on the oscillation onset of the vocal folds. *J. Acoust. Soc. Am.* **2012**, *132*, 403–411. [CrossRef]
10. Zhang, Z.; Neubauer, J.; Berry, D.A. Aerodynamically and acoustically driven modes of vibration in a physical model of the vocal folds. *J. Acoust. Soc. Am.* **2006**, *120*, 2841–2849. [CrossRef]
11. Zhang, Z.; Neubauer, J.; Berry, D.A. Influence of vocal fold stiffness and acoustic loading on flow-induced vibration of a single-layer vocal fold model. *J. Sound Vib.* **2009**, *322*, 299–313. [CrossRef]
12. Echternach, M.; Herbst, C.T.; Köberlein, M.; Story, B.; Döllinger, M.; Gellrich, D. Are source-filter interactions detectable in classical singing during vowel glides? *J. Acoust. Soc. Am.* **2021**, *149*, 4565–4578. [CrossRef] [PubMed]
13. Migimatsu, K.; Tokuda, I.T. Experimental study on nonlinear source–filter interaction using synthetic vocal fold models. *J. Acoust. Soc. Am.* **2019**, *146*, 983–997. [CrossRef] [PubMed]
14. Näger, C.; Lodermeyer, A.; Becker, S. Charakterisierung der Stimmlippenvibration an einem synthetischen Larynx-Modell mittels Laser-Scanning-Vibrometrie. In Proceedings of the DAGA 2022, Stuttgart, Germany, 21–24 March 2022; pp. 927–930.
15. Näger, C.; Kniesburges, S.; Becker, S. Investigation of Acoustic Back-coupling in Human Phonation via Particle Image Velocimetry. In Proceedings of the Forum Acusticum 2023, Torino, Italy, 11–15 September 2023 .
16. Falk, S.; Kniesburges, S.; Schoder, S.; Jakubaß, B.; Maurerlehner, P.; Echternach, M.; Kaltenbacher, M.; Döllinger, M. 3D-FV-FE aeroacoustic larynx model for investigation of functional based voice disorders. *Front. Physiol.* **2021**, *12*, 616985. [CrossRef] [PubMed]
17. Scherer, R.C.; Shinwari, D.; De Witt, K.J.; Zhang, C.; Kucinschi, B.R.; Afjeh, A.A. Intraglottal pressure profiles for a symmetric and oblique glottis with a divergence angle of 10 degrees. *J. Acoust. Soc. Am.* **2001**, *109*, 1616–1630. [CrossRef] [PubMed]

18. Rupitsch, S.J.; Ilg, J.; Sutor, A.; Lerch, R.; Döllinger, M. Simulation based estimation of dynamic mechanical properties for viscoelastic materials used for vocal fold models. *J. Sound Vib.* **2011**, *330*, 4447–4459. [CrossRef]
19. Durst, F.; Heim, U.; Ünsal, B.; Kullik, G. Mass flow rate control system for time-dependent laminar and turbulent flow investigations. *Meas. Sci. Technol.* **2003**, *14*, 893. [CrossRef]
20. Lodermeyer, A.; Becker, S.; Döllinger, M.; Kniesburges, S. Phase-locked flow field analysis in a synthetic human larynx model. *Exp. Fluids* **2015**, *56*, 1–13. [CrossRef]
21. Lodermeyer, A.; Bagheri, E.; Kniesburges, S.; Näger, C.; Probst, J.; Döllinger, M.; Becker, S. The mechanisms of harmonic sound generation during phonation: A multi-modal measurement-based approach. *J. Acoust. Soc. Am.* **2021**, *150*, 3485–3499. [CrossRef]
22. ANSYS, Inc. *Ansys Mechanical, Release 2022 R2*; ANSYS, Inc.: Canonsburg, PA, USA, 2022.
23. ANSYS, Inc. *Theory Reference for the Mechanical APDL and Mechanical Applications*; Kohnke, P., Ed.; Technical Report; ANSYS, Inc.: Canonsburg, PA, USA, 2009.
24. Kaltenbacher, M. *Numerical Simulation of Mechatronic Sensors and Actuators*, 3rd ed.; Springer: Berlin/Heidelberg, Germany, 2015. [CrossRef]
25. Dabbaghchian, S.; Arnela, M.; Engwall, O.; Guasch, O. Simulation of vowel-vowel utterances using a 3D biomechanical-acoustic model. *Int. J. Numer. Methods Biomed. Eng.* **2021**, *37*, e3407. [CrossRef]
26. Story, B.H.; Laukkanen, A.M.; Titze, I.R. Acoustic impedance of an artificially lengthened and constricted vocal tract. *J. Voice* **2000**, *14*, 455–469. [CrossRef]
27. Murray, P.R.; Thomson, S.L. Vibratory responses of synthetic, self-oscillating vocal fold models. *J. Acoust. Soc. Am.* **2012**, *132*, 3428–3438. [CrossRef] [PubMed]
28. Kraxberger, F.; Museljic, E.; Kurz, E.; Toth, F.; Kaltenbacher, M.; Schoder, S. The Nonlinear Eigenfrequency Problem of Room Acoustics with Porous Edge Absorbers. In Proceedings of the Forum Acusticum 2023, European Acoustics Association, Torino, Italy, 11–15 September 2023.
29. Kraxberger, F.; Kurz, E.; Weselak, W.; Kubin, G.; Kaltenbacher, M.; Schoder, S. A Validated Finite Element Model for Room Acoustic Treatments with Edge Absorbers. *Acta Acust.* **2023**, *7*, 1–19. [CrossRef]
30. Sondhi, M.; Schroeter, J. A hybrid time-frequency domain articulatory speech synthesizer. *IEEE Trans. Acoust. Speech Signal Process.* **1987**, *35*, 955–967. [CrossRef]
31. Näger, C.; Kniesburges, S.; Tur, B.; Schoder, S.; Becker, S. Investigation of Acoustic Back-coupling in Human Phonation on a Synthetic Larynx Model. *Bioengineering* **2023**, *10*, 1343. [CrossRef]
32. Titze, I.R.; Martin, D.W. *Principles of Voice Production*; Acoustical Society of America: Melville, NY, USA, 1998.
33. Pastor, M.; Binda, M.; Harčarik, T. Modal assurance criterion. *Procedia Eng.* **2012**, *48*, 543–548. [CrossRef]
34. Herzel, H.; Berry, D.; Titze, I.R.; Saleh, M. Analysis of vocal disorders with methods from nonlinear dynamics. *J. Speech Lang. Hear. Res.* **1994**, *37*, 1008–1019. [CrossRef] [PubMed]
35. Titze, I.R. How can vocal folds oscillate with a limited mucosal wave? *JASA Express Lett.* **2022**, *2*, 105201. [CrossRef] [PubMed]
36. Sundström, E.; Oren, L.; de Luzan, C.F.; Gutmark, E.; Khosla, S. Fluid-Structure Interaction Analysis of Aerodynamic and Elasticity Forces During Vocal Fold Vibration. *J. Voice* **2022**. [CrossRef]
37. Titze, I.R. Theoretical analysis of maximum flow declination rate versus maximum area declination rate in phonation. *J. Speech Lang. Hear. Res.* **2006**, *49*, 439–447. [CrossRef]
38. Manconi, E.; Mace, B.R.; Garziera, R. Wave Propagation in Laminated Cylinders with Internal Fluid and Residual Stress. *Appl. Sci.* **2023**, *13*, 5227. [CrossRef]

Article

An Investigation of Acoustic Back-Coupling in Human Phonation on a Synthetic Larynx Model

Christoph Näger [1,*], Stefan Kniesburges [2], Bogac Tur [2], Stefan Schoder [3] and Stefan Becker [1]

[1] Institute of Fluid Mechanics, Friedrich-Alexander-Universität Erlangen-Nürnberg, Cauerstraße 4, 91058 Erlangen, Germany; stefan.becker@fau.de
[2] Division of Phoniatrics and Pediatric Audiology, Department of Otorhinolaryngology, Head & Neck Surgery, University Hospital Erlangen, Medical School, Friedrich-Alexander-Universität Erlangen-Nürnberg, Waldstrasse 1, 91054 Erlangen, Germany
[3] Aeroacoustics and Vibroacoustics Group, Institute of Fundamentals and Theory in Electrical Engineering, Graz University of Technology, Inffeldgasse 16, 8010 Graz, Austria; stefan.schoder@tugraz.at
* Correspondence: christoph.naeger@fau.de

Abstract: In the human phonation process, acoustic standing waves in the vocal tract can influence the fluid flow through the glottis as well as vocal fold oscillation. To investigate the amount of acoustic back-coupling, the supraglottal flow field has been recorded via high-speed particle image velocimetry (PIV) in a synthetic larynx model for several configurations with different vocal tract lengths. Based on the obtained velocity fields, acoustic source terms were computed. Additionally, the sound radiation into the far field was recorded via microphone measurements and the vocal fold oscillation via high-speed camera recordings. The PIV measurements revealed that near a vocal tract resonance frequency f_R, the vocal fold oscillation frequency f_o (and therefore also the flow field's fundamental frequency) jumps onto f_R. This is accompanied by a substantial relative increase in aeroacoustic sound generation efficiency. Furthermore, the measurements show that f_o-f_R-coupling increases vocal efficiency, signal-to-noise ratio, harmonics-to-noise ratio and cepstral peak prominence. At the same time, the glottal volume flow needed for stable vocal fold oscillation decreases strongly. All of this results in an improved voice quality and phonation efficiency so that a person phonating with f_o-f_R-coupling can phonate longer and with better voice quality.

Keywords: human phonation; source–filter interaction; particle image velocimetry; synthetic larynx model; transmission line model; aeroacoustic source computation

Citation: Näger, C.; Kniesburges, S.; Tur, B.; Schoder, S.; Becker, S. An Investigation of Acoustic Back-Coupling in Human Phonation on a Synthetic Larynx Model. *Bioengineering* **2023**, *10*, 1343. https://doi.org/10.3390/bioengineering10121343

Academic Editor: Andrea Cataldo

Received: 28 September 2023
Revised: 12 November 2023
Accepted: 19 November 2023
Published: 22 November 2023

1. Introduction

The human voice is generated by a complex physiological process that is described by the fluid–structure–acoustic interaction (FSAI) between the tracheal fluid flow, structural vibration of laryngeal tissue (i.e., the vocal folds), and the sound generation and modulation in the larynx and vocal tract [1–3]. In this process, the two vocal folds are aerodynamically stimulated to vibrate by the airflow that arises from the lungs. In turn, this vibration leads to a modulation of the airflow, generating a pulsating jet flow in the supraglottal region, which is above the vocal folds [1].

Within this dynamic process of tissue–flow interaction, the basic sound of the human voice is generated by the highly complex 3D field of aeroacoustic sound sources which are produced by the turbulent jet flow in larynx [4]. Moreover, vibroacoustical sound generation also occurs by sound radiation from the vocal fold surface [5]. The generated basic sound is further filtered by the vocal tract and radiated through the mouth, exhibiting the typical spectral characteristics of the human voice composed of tonal harmonic components of the fundamental frequency and additional tonal components, called formants, originating from resonance effects in the vocal tract [6].

In early times, a linear behavior between sound source (vocal folds) and filter (vocal tract) was assumed within the linear source–filter theory, which excluded the influence of the acoustic filter signal back on the source [7]. However, this simplified representation turned out to be invalid, especially when the fundamental oscillation frequency of the vocal folds f_0 is close to a resonance frequency f_R of the vocal tract [8]. Acoustic back-coupling has been studied using theoretical modeling (e.g., [9,10]), simulations (e.g., [11–13]), in vivo studies (e.g., [14,15]), and ex vivo and in vitro experiments (e.g., [16–18]). However, due to the complexity of the problem, most studies were restricted to metrics such as f_0-variation and the change in the subglottal oscillation threshold pressure. The direct acoustic–tissue or acoustic–flow interaction have not yet been investigated.

In this context, Particle Image Velocimetry (PIV) enables us to measure the unsteady flow field in the aeroacoustic source region to gain a deep insight into the FSAI process of phonation. This technique has already been applied successfully to study aerodynamic effects in synthetic as well as ex vivo larynx models that showed typical vocal fold vibrations similar to phonation. In this context, classical planar low-frequency PIV measurements allowed to study the basic features of the supra- and intraglottal aerodynamics [19–22]. Being still 2D, the emergence of high-speed PIV techniques made it then possible to directly analyze aeroacoustic source terms that were computed from time-resolved PIV data obtained directly in the source region above the vocal folds [5,23]. These data provided the distribution and dynamics of aeroacoustic sources based on state-of-the-art aeroacoustic analogies, such as the Lighthill analogy [24] or the Perturbed Convective Wave Equations [25]. With the rising availability of 3D tomographic PIV measurements, even first studies of volumetric parameters as the Maximum Flow Declination Rate were investigated using ex vivo canine models [26].

However, none of the studies described above have investigated the effects of supraglottal acoustics on the glottal aerodynamics and the aeroacoustic source field yet. Therefore, the present study provides highly resolved data of the entire process to analyze the complete FSAI between vocal tract acoustics and laryngeal aerodynamics. Based on high-speed PIV measurements in combination with aerodynamic and acoustic pressure data, as well as high-speed visualizations of the vocal fold dynamics in a synthetic larynx model [22], the influence of the resonance effects formed in the vocal tract on the supraglottal flow field and the vocal fold motion is studied. Different vocal tract models have been applied with an incremental increase in length. This length change produced acoustic properties of the vocal tract that shifted its resonance frequency down to the fundamental frequency of the vocal folds. This procedure enabled us to systematically study the relationship between laryngeal flow and supraglottal acoustics.

2. Materials and Methods

2.1. Basic Experimental Setup

Synthetic vocal folds were cast from a single layer of Ecoflex 00-30 silicone (Smooth-On, Macungie, PA, USA) with a static Young's modulus of 4.4 kPa. Their shape was based on the M5 model [27,28], and is displayed in Figure 1. The vocal folds were glued into their mounting, positioned between the subglottal and supraglottal channels. In the prephonatory posturing of the vocal folds, the glottis was completely closed. The subglottal channel had a length of 210 mm and a rectangular cross-section of 18 mm × 15 mm, which is within the dimensional range found in vivo [1]. Furthermore, this length was chosen small enough to prevent the interaction of the vocal folds' oscillation with the subglottal acoustic resonances (see the description in Lodermeyer et al. [21] based on the results by Zhang et al. [29]). The supraglottal channel had a rectangular section of 18 mm × 15 mm and a length of 80 mm in the region directly downstream of the vocal folds. Attached to it, a circular cross-section tube with a diameter of 32 mm followed. An additional tube with a diameter of 34 mm made it possible to adjust the vocal tract length and thereby resonance frequency continuously via telescoping. This basic setup is shown schematically in Figure 2, and is based on the setup previously described by, e.g., Kniesburges et al. [22]. A mass flow generator with a supercritical valve [30] produced a constant volume flow $\dot{V}$ through the setup. Between the mass flow generator and the subglottal channel, a silencer was placed for conditioning the flow and attenuating sound propagation from the supply hose to the vocal fold position. Several measurements were performed for different vocal tract lengths in the interval $L \in [200, 800]$ mm to study the influence of the vocal tract acoustic resonance frequencies on the supraglottal flow field and vocal fold oscillation. The mass flow rate for each length was set to the corresponding minimum required for vocal fold oscillation with complete glottis closure.

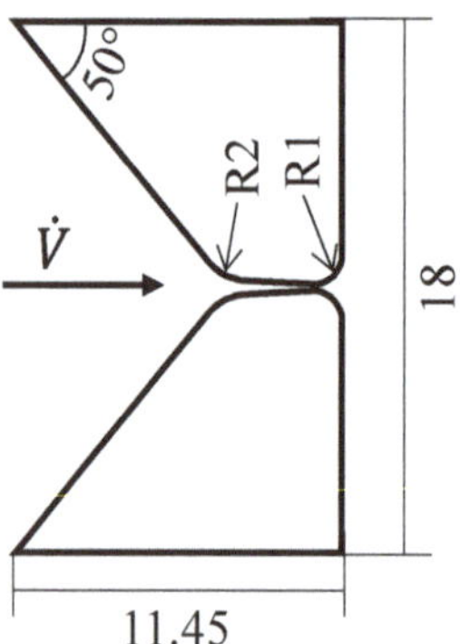

Figure 1. 2D representation of the vocal fold model used, with its dimensions given in mm. The flow direction in the experiment is indicated.

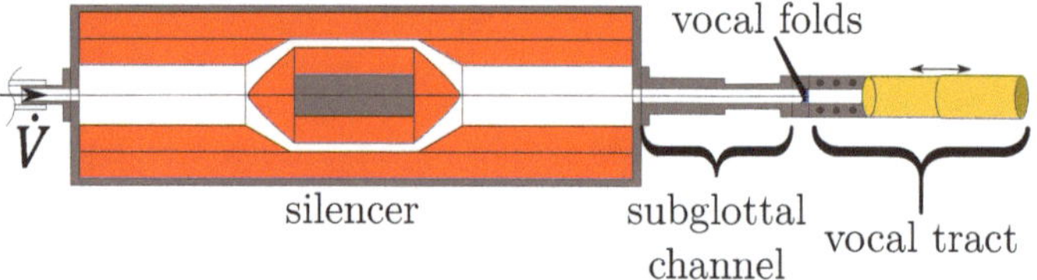

Figure 2. 2D cut through the experimental setup. The vocal fold position is indicated between the vocal tract and the subglottal channel. A silencer is placed upstream to attenuate emerging sound in the inflow hose. The flow direction is from left to right.

2.2. Measurement Setup

Multiple measurement tasks were performed. The transglottal pressure was recorded via two pressure sensors: in the subglottal channel, a Kulite XCQ-093 sealed gauge pressure sensor (Kulite Semiconductor, Leonia, NJ, USA) was flush-mounted into the channel wall 50 mm upstream of the glottis. In the supraglottal channel, a Kulite XCS-093 open gauge pressure sensor (Kulite Semiconductor, Leonia, NJ, USA) was mounted the same way at a distance of 50 mm downstream of the glottis. The sound radiation from the vocal tract end was recorded in our anechoic chamber by a Brüel and Kjaer 4189-L-001 1/2''-microphone (Brüel and Kjaer, Nærum, Denmark) at a distance of 1 m perpendicular to the channel outlet. Microphone and wall pressure signals were sampled by a National Instruments PXIe 6356 multifunctional card (National Instruments, Austin, TX, USA) with a resolution of 16 bit and a sample rate of 44.1 kHz. The vocal fold movement was recorded using a Photron FASTCAM SA-X2 high-speed camera (Photron, Tokyo, Japan) at a frame rate of 10 kHz. From the microphone recordings, additional related parameters like the signal-to-noise ratio (SNR), harmonics-to-noise-ratio (HNR), as well as the cepstral peak prominence (CPP), were extracted using the Glottis Analysis Tools [31,32] (GAT; University Hospital Erlangen, Erlangen, Germany).

The planar flow velocity in the supraglottal region in the coronal plane midway along the vocal fold length was measured with a 2D-2C planar PIV setup. This setup is shown in Figure 3. The measurement region of interest was a rectangle with dimensions of 45 mm × 18 mm and chosen similarly to the previous study by Lodermeyer et al. [5]. For seeding purposes, a PIVlight30 seeding generator (PIVtec GmbH, Göttingen, Germany) based on Laskin nozzles for atomization of the seeding fluid was applied. *PIVtec PIVfluid*, which is a propylene glycol mixture based on double de-ionized water and other components, was used as a seeding fluid, resulting in a mean particle diameter of 1.2 μm. The resulting Stokes number was $St = 0.033 < 0.1$, yielding an acceptable flow tracing accuracy [33,34]. The particles were illuminated by a laser light sheet with a thickness of approximately 0.5 mm. The laser used in this study was a double-pulse, frequency-doubled Nd:YLF *Continuum Terra PIV* high-speed laser (Continuum, San Jose, CA, USA) with a wavelength of 527 nm and a repetition rate of 2×5 kHz. The offset between the two pulses was set to 4 μs, realized by an ILA synchronizer (ILA, Jülich, Germany). A *Vision Research Phantom v2511* high-speed camera (Vision Research Inc., Wayne, NJ, USA) in combination with a *Canon Macro Lens EF 180 mm Ultrasonic* lens (Canon, Tokyo, Japan) was applied to record the distribution of the illuminated seeding particles.

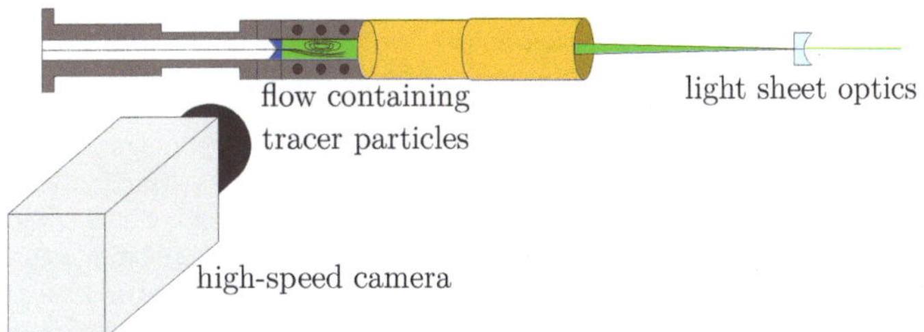

Figure 3. The setup for the PIV measurements. The flow was visualized using tracer particles and laser double pulses at a repetition rate of 2×5 kHz. The rectangular section of the vocal tract provided optical access through a glass window, allowing the flow to be recorded with a high-speed camera.

Two image pre-processing steps were applied to increase the signal-to-noise ratio in the recorded images. In the regions downstream of the vocal folds, a background removal via proper orthogonal decomposition (POD) proposed by Mendez et al. [35] was applied. This method is suited for removing background noise without moving walls, but showed poor background removal in the region close to the vocal folds. Therefore, a different approach was chosen for the region close to the glottis. Adatrao and Sciacchitano [36] proposed a background removal technique based on an anisotropic diffusion equation specifically for moving solid objects in PIV images. The results of both background removal techniques for one exemplary image are shown in Figure 4. It can be seen that the original image contains strong reflections at the vocal folds in the left part of the image. Also, sensor noise is visible in the rightmost quarter of the image. The POD-based background removal enables an almost complete removal of the sensor noise while removing most of the light reflections from the vocal folds. However, some artifacts are created around the vocal folds, masking the particle images, e.g., between the vocal folds. In contrast, the anisotropic diffusion-based background removal shows a sharp boundary around the vocal folds, improving the visibility of the particles between them. However, in this case, the sensor noise close to the right image boundary still remains present, albeit with reduced intensity. Therefore, to obtain the best result, the anisotropic diffusion approach is only used in the leftmost part of the images, while the POD-based approach is applied in the remaining part. Looking at Figure 4D, it appears as if there were less particles in the part close to the vocal folds visible than in the remaining image. This is a result of the light reflections from the vocal folds completely masking some particles in vicinity to the vocal folds. The PIV evaluation algorithm was still able to find enough particles in this region to obtain reliable velocity information, however.

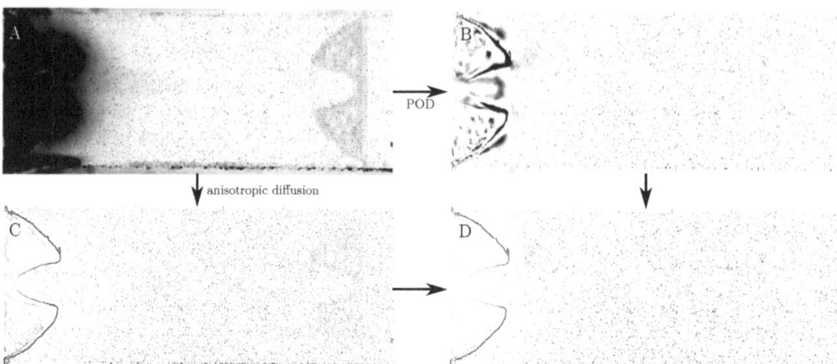

Figure 4. The different background removal techniques. The original images (**A**) are processed by POD (**B**) and anisotropic diffusion (**C**). The results are combined into one image (**D**), where the anisotropic diffusion image is used at the glottis, while the POD image is used in the other regions. The images were inverted to enhance visibility.

Velocity vectors were extracted from the image pairs with the help of the commercial software PIVview2C 3.6.23 (PIVtec GmbH, Göttingen, Germany). For this purpose, a grid of 74×56 correlation windows with an overlap of 50% was defined in the region of interest, leading to a spatial resolution of $\Delta x \times \Delta y = 0.62\,\text{mm} \times 0.31\,\text{mm}$. Outliers were detected via the universal outlier detection by Westerweel and Scarano [37] and interpolated with the information from the surrounding velocity vectors.

2.3. Aeroacoustic Source Computation

One important aspect of understanding the human voice production process is to evaluate the aeroacoustic sources, e.g., Lighthill's source term for low Mach number isentropic turbulent flows

$$T(\boldsymbol{x}, t) \approx \nabla \cdot \nabla \cdot (\rho_0 \boldsymbol{uu}) \tag{1}$$

with the velocities $\boldsymbol{u}$ measured via PIV and the ambient density ρ_0 [23]. This distributed source term is aggregated in a summed source strength based on Lighthill's analogy [24] by neglecting the retarded time effects in this acoustically compact 2D region of interest (ROI)

$$\phi(t) = \frac{1}{4\pi c_0^2 (x_1 - x_0)(y_1 - y_0)} \int_{(x_0, y_0)}^{(x_1, y_1)} T(\boldsymbol{x}, t) \frac{\pi y \mathrm{d}x \mathrm{d}y}{r} = \frac{1}{4 c_0^2 N_x N_y} \sum_i T(\boldsymbol{x}_i, t) \mid y_i \mid . \tag{2}$$

The coordinate locations x_0, x_1, y_0, y_1 are the bounding coordinates of the ROI, respectively, c_0 the isentropic speed of sound, r the direction of a virtual observer point at 1 m distance. It is assumed that the jet is rotationally symmetric around the rotation axis in x-direction, pointing in the flow direction and being centered in the middle of the vocal folds. From this equation, the root-mean-squared value is computed

$$\Phi = \sqrt{\frac{1}{t_1 - t_0} \int_{t_0}^{t_1} (\phi(t))^2 \mathrm{d}t} \,, \tag{3}$$

being a measure of the ability to generate aerodynamic sound. The equation is applied to the measured 2D mid-section, where the velocity's principal direction is recorded. In addition, the aerodynamic input energy is quantified by

$$P = \Delta p + \frac{1}{2} \rho_0 U^2 \,, \tag{4}$$

using the subglottal pressure difference to the ambient pressure Δp. With the input energy, the efficiency of the aeroacoustic sound generation yields

$$\eta = \frac{\Phi c_0^2 (x_1 - x_0)(y_1 - y_0)}{P}.$$

(5)

2.4. Acoustic Characterization of the Vocal Tract

The acoustic properties of the vocal tract were determined using a transmission line model [38]. In this model, the acoustic pressure p_{ac} and volume velocity u_{ac} at the vocal tract input, i.e., the glottis, can be related to the same quantities at the output via chain matrix multiplication:

$$\begin{pmatrix} p_{ac,out} \\ u_{ac,out} \end{pmatrix} = K_{tract} \begin{pmatrix} p_{ac,in} \\ u_{ac,in} \end{pmatrix} = \begin{pmatrix} A_{tract} & B_{tract} \\ C_{tract} & D_{tract} \end{pmatrix} \begin{pmatrix} p_{ac,in} \\ u_{ac,in} \end{pmatrix}$$

(6)

Here, the 2×2-matrix K_{tract} is built from chain multiplication of a series of 2×2-matrices K_i, each representing one part of the vocal tract with constant cross-section. These matrices K_i were computed with the equations derived by Sondhi and Schroeter [38]. In the case of our simplified vocal tract, there are three different cross-sections present: the rectangular section right above the glottis, the circular section of the first tube and the circular section of the second tube. The vocal tract input impedance $Z_{in} = p_{ac,in} / u_{ac,in}$ can be obtained from Equation (6):

$$Z_{in} = \frac{D_{tract} Z_{out} - B_{tract}}{A_{tract} - C_{tract} Z_{out}}$$

(7)

The maxima of the frequency-dependent Z_{in} thereby correspond to the vocal tract resonance frequencies. The transmission line model was implemented following the description given by Story et al. [39]. As the vocal tract walls were fabricated from aluminum and glass, they were modeled as rigid walls. The radiation impedance at the open end was approximated as a vibrating piston in an infinite baffle [40].

3. Results and Discussion

3.1. General Results

As already stated in Section 2.1, the experiment's volume flow rate $\dot{V}$ was set to the minimal flow rate necessary to induce oscillation with contact between the vocal folds. Table 1 lists the resulting $\dot{V}$ for all nine configurations measured. It can be seen that $\dot{V}$ stays roughly constant for $L \leq 340\,\text{mm}$ and decreases monotonically for larger L. The same behavior can be observed in the transglottal pressure $P_{trans} = P_{sub} - P_{supra}$, where P_{sub} and P_{supra} are the mean pressure values measured by the pressure probes in the sub- and supraglottal channel, respectively. The decrease in $\dot{V}$ with increasing L is similar to what Fulcher et al. [41] found, where they used an analytical surface wave model in combination with validation experiments to predict the phonation threshold pressure as a function of the vocal tract length. They related the decreased threshold pressure to an increase in vocal tract inertance due to the increased length.

The oscillation frequency of the vocal folds f_o, as extracted via discrete Fourier transformation from the PIV-measurements also shows a stationary behavior in the range $L \leq 340\,\text{mm}$. It stays within the range $150\,\text{Hz} < f_o < 153\,\text{Hz}$ for these lengths. At $L = 400\,\text{mm}$, f_o jumps to $225.7\,\text{Hz}$, which is 1.5-fold of $150.5\,\text{Hz}$. Increasing the length further leads to a jump back to $\sim 150\,\text{Hz}$ and then a decrease in f_o down to $119.0\,\text{Hz}$. An explanation for this behavior can be found looking at the relationship between f_o and the first vocal tract resonance frequency f_{R1} as computed via the transmission line model. For this purpose, Figure 5 shows the vocal tract input impedance Z_{in} as a function of the frequency f and L. The frequency of the maxima in Z_{in} (indicated by the color yellow) correspond to the vocal tract resonance frequencies f_{Ri}. As expected, f_{Ri} decrease with an increase in L. On top of the contour of Z_{in}, the values of f_o for the different chosen vocal tract lengths from Table 1 are displayed.

Table 1. The main measurement quantities of the PIV-measurements. L denotes the length of the supraglottal channel, $\dot{V}$ the flow rate, P_{trans} the transglottal pressure difference, f_o the oscillation frequency of the vocal folds and f_{R1} the first resonance frequency of the vocal tract as computed via transmission line model.

L in mm	$\dot{V}$ in l/min	P_{trans} in Pa	f_o in Hz	f_{R1} in Hz
200	124	4208	151.3	527
240	120	4100	152.9	433
300	123	4188	151.3	337
340	120	4122	150.5	293
400	107	3791	225.7	244
500	99	3617	158.6	190
600	58	2659	158.6	156
700	55	2527	136.8	132
800	46	2266	119.0	114

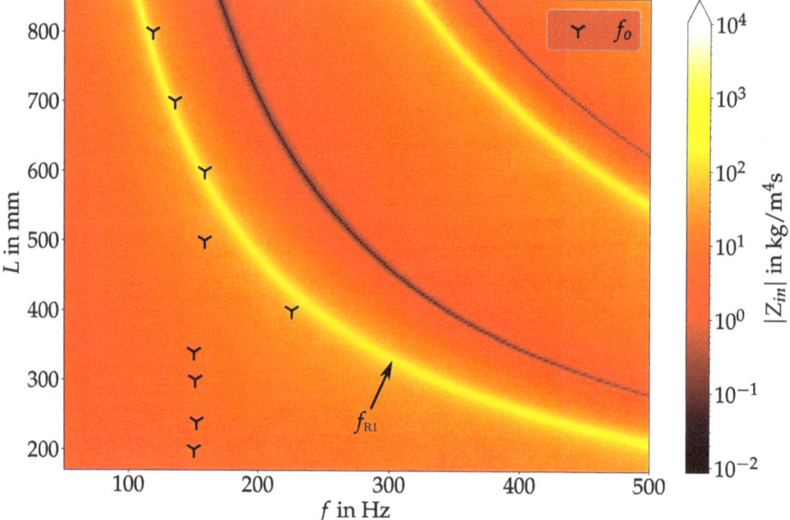

Figure 5. The vocal tract input impedance Z_{in} (computed via transmission line model) shown as color map is a function of frequency f and vocal tract length L. The location of the vocal tract resonance frequency f_{R1} shows as a bright yellow line in the plot. Superimposed are the oscillation frequencies f_o at the individual measurements with different lengths of the supraglottal channel.

Here, it can be seen that for $L < 400$ mm and $L = 500$ mm, f_o and f_{R1} are not in vicinity to each other. Therefore, f_o is approximately constant in this region, with the exception of small variations due to small changes in the experimental conditions as, e.g., a slight variation in $\dot{V}$. For $L \geq 600$ mm, f_{R1} starts falling below 160 Hz and therefore lies in vicinity of the "uninfluenced" value of f_o. This leads to a decrease in f_o with further increasing length in this length range. As a consequence, f_o "jumps" onto f_{R1} in this range and the vocal fold oscillation is coupled to the acoustic standing wave in the vocal tract. This is in accordance to the experimental data observed by Migimatsu et al. [18] in their experimental study with the M5 model. In their work, a much larger increase in L up to ~1 m led to an f_o-jump back to the original uninfluenced value due to f_{R1} being not in the vicinity of the uninfluenced f_o anymore, leading to the domination of the vocal folds' natural mechanical eigenmode. With our experimental setup, this could also be expected; however, this was not investigated. Zhang et al. also observed a locking of f_0 onto the supraglottal resonances [16]. Similarly, Zhang et al. furthermore showed, that also a locking of f_0 onto the resonance frequencies of the subglottal channel can occur, when they studied the influence of the subglottal resonances onto the vocal fold oscillation [16,29]. A special behavior happens at $L = 400\,\text{mm}$. Here, despite the uninfluenced f_0 lying still considerably below f_{R1}, a f_0 jump close to f_{R1} takes place. As the oscillation at $L = 500\,\text{mm}$ falls back to similar f_0 as at the smaller L, this gives the indication that there is a different behavior present at $L = 400\,\text{mm}$ that occurs due to a special combination of the eigenmodes of the vocal folds and the acoustic resonance frequency of the vocal tract. To visualize this, high-speed camera videos have been recorded for three different L: 200 mm, 400 mm and 700 mm, respectively. These lengths were chosen, as they are representative of the three different states of vocal fold oscillation we were able to identify: independent f_0 and f_{R1}, a jump of f_0 to a higher frequency, and a shift of f_0 to lower frequencies. The relationship between f_0 and f_{R1} for these lengths can be seen in Figure 5. Snapshots of one oscillation time period T for each of the chosen lengths are shown in Figure 6. The top row here corresponds to the baseline case, where $f_0 \ll f_{R1}$. In this case, the vocal folds oscillate with a clearly visible convergent-divergent transglottal angle devolution, with a convergent glottal duct shape in the opening-phase and a divergent duct shape in the closing-phase. Similarly, at $L = 700\,\text{mm}$, the same behavior can be seen, albeit with a smaller opening area due to the reduced $\dot{V}$ in this case. In contrast, the oscillation at $L = 400\,\text{mm}$ looks considerably different compared to the other two cases. Here, the change between convergent and divergent shape change in the glottal duct does not directly correlate with opening and closing motion of the vocal folds as described by Titze [42] for aerodynamically driven vocal folds. Therefore, combined with the f_0 jump to the vocal tract resonance frequency and a completely changed oscillatory behavior, this suggests that there is some kind of acoustic coupled motion of the vocal folds present in this case. As this only happens at $L = 400\,\text{mm}$, it is reasonable to assume that an eigenmode of the vocal fold model at a frequency of about 225 Hz is present in this case, that is excited by the acoustic standing wave of the vocal tract. This eigenmode is in the standard case not dominant compared to the 150 Hz-mode and therefore not visible with the other vocal tract lengths.

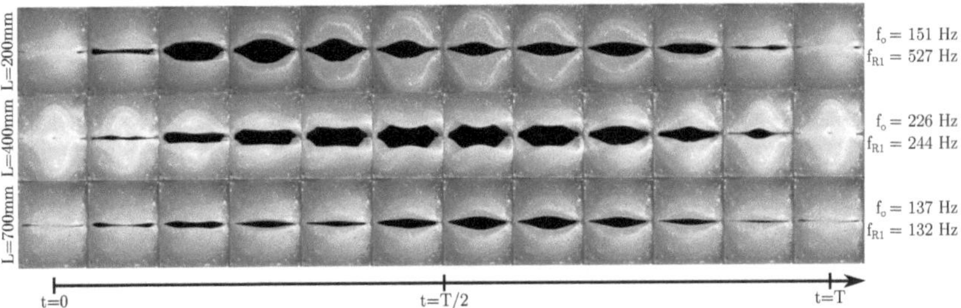

Figure 6. High-speed-camera recordings of the vocal fold oscillation for the three different vocal tract lengths of $L = 200\,\text{mm}$, $L = 400\,\text{mm}$ and $L = 700\,\text{mm}$, respectively. Every row shows the recording for one length over one oscillation period T. Due to the different f_0-values, the actual time steps between two images are different in each row. The values for f_0 and f_{R1} for all three cases are displayed to the right of their respective image series.

3.2. Supraglottal Aerodynamics

PIV measurements have been performed for all configurations of Table 1. Again, the three cases of $L = 200\,\text{mm}$, $L = 400\,\text{mm}$ and $L = 700\,\text{mm}$ are analyzed in more detail as they are representative of the different possible oscillatory behaviors of the vocal folds. Velocity fields of one oscillation cycle for each length are shown in Figures 7–9. Additionally, mean velocity fields for all three configurations are displayed in Figure 10. Looking at Figure 7, one can see that the basic characteristic of the flow is an oscillating jet synchronized to the opening and closing of the vocal folds (displayed in gray on top of the velocity contours). The jet is deflected during the closing phase to the top vocal tract wall, forming a large vortex in the supraglottal channel in the closed phase of the vocal folds. Depending on the cycle, also deflection of the jet downwards to the lower vocal tract wall can happen, leading to a vortex in the closed phase that is rotating in the other direction. This deflection and vortex formation is well known in human phonation, and has been studied extensively in the past (e.g., [21,22,43–46]). If there is approximately 50% of the cycles having an upwards deflection and 50% with a downwards deflection, this leads to a rather symmetric averaged velocity profile, as it can be seen in Figure 10 (top).

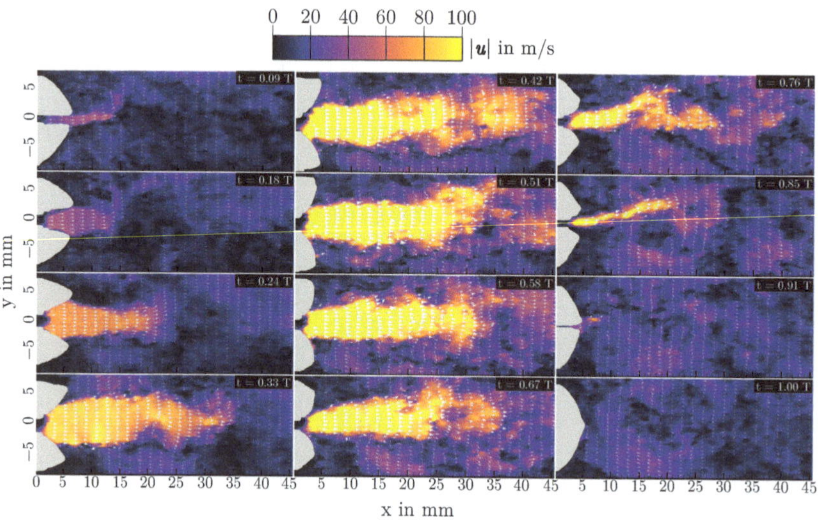

Figure 7. Instantaneous flow fields for 12 different time steps at a vocal tract length of $L = 200\,\text{mm}$. The time steps of the snapshots are shown in their respective top right corner.

Looking at Figure 8, it appears that the basic characteristics of the flow are unchanged from $L = 200\,\text{mm}$ to $L = 400\,\text{mm}$. There is still an oscillating jet flow, which is deflected to one of the vocal tract walls, leading to a large supraglottal vortex occurring during the closing and closed phase. Qualitatively, the acoustic driving of the vocal folds therefore does not appear to change the flow field in the middle plane of the vocal tract significantly. From an aerodynamic point of view, there are some changes, however related to the elongation of the vocal tract. In this case, the supraglottal jet is always deflected towards the lower vocal tract wall, leading to an asymmetry in the mean velocity field shown in Figure 10 (middle). Here, the supraglottal vortex is stabilized by the longer vocal tract. With a shorter vocal tract, the vortex is convected out of the vocal tract by the starting jet in the opening phase of the vocal folds. This leads to a new flow situation, where the jet deflection direction can be changed from one oscillation cycle to the next one. With a longer vocal tract, the vortex is just convected downstream inside the channel, thereby interacting with the jet starting from the vocal folds and deflecting it towards its side of positive x-velocities. Therefore, the direction of jet deflection in this case is dependent on the initial deflection at the beginning of the phonation. Kniesburges et al. observed a similar behavior when changing the supraglottal channel height (y-direction) [22]. Here, an increase in the channel height also led to a stabilized supraglottal vortex that interacted with the glottal jet flow. The jet deflection direction can also change from one phonation process to the next, as it can be seen when comparing the mean velocity profiles of $L = 400\,\text{mm}$ and $L = 700\,\text{mm}$ in Figure 10.

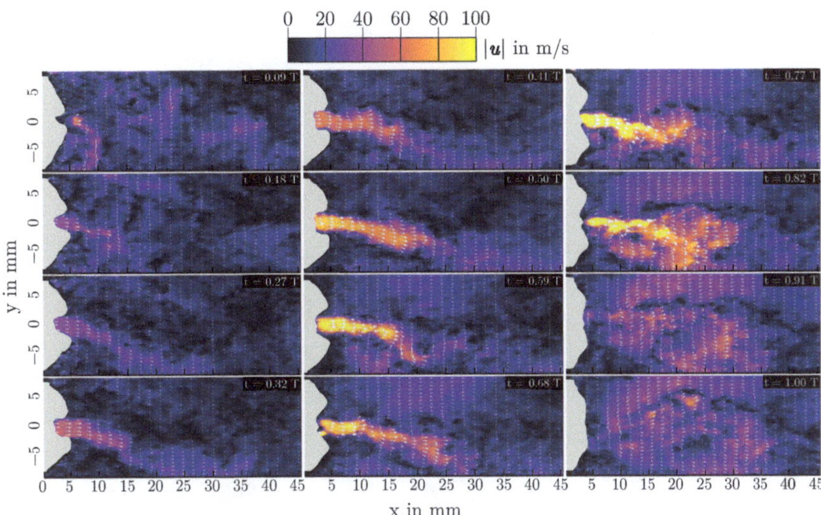

Figure 8. Instantaneous flow fields for 12 different time steps at a vocal tract length of $L = 400\,$mm. The time steps of the snapshots are shown in their respective top right corner.

In the $L = 700\,$mm case, the jet is deflected upwards instead of downwards, also visible in the instationary velocity fields of Figure 9. Comparing the three configurations shown, the differences in the velocity magnitudes are notable, resulting from the different volume flow rates needed for vocal fold oscillation with contact. Generally, the peak flow velocities in this setup are higher than what is found in vivo [1], resulting from the large mean transglottal pressure needed for the single-layer synthetic vocal folds to oscillate with contact.

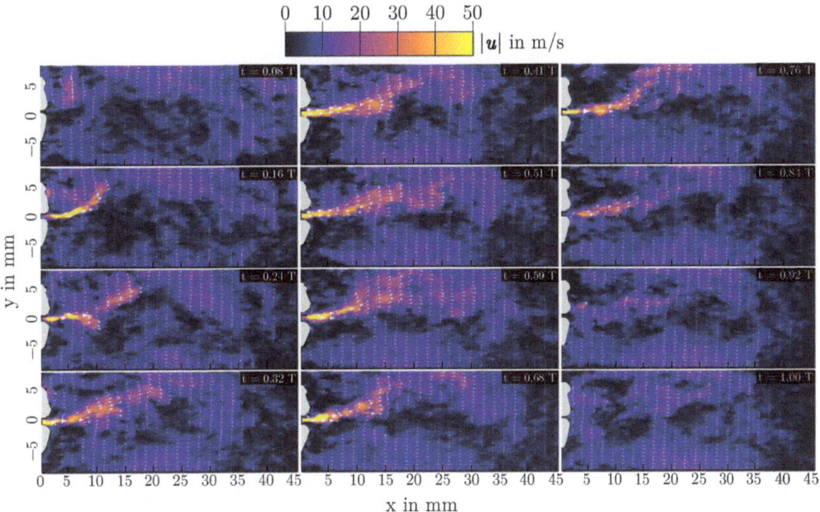

Figure 9. Instantaneous flow fields for 12 different time steps at a vocal tract length of $L = 700\,$mm. The time steps of the snapshots are shown in their respective top right corner. Note the changed color map limits compared to Figures 7 and 8 for improved visibility.

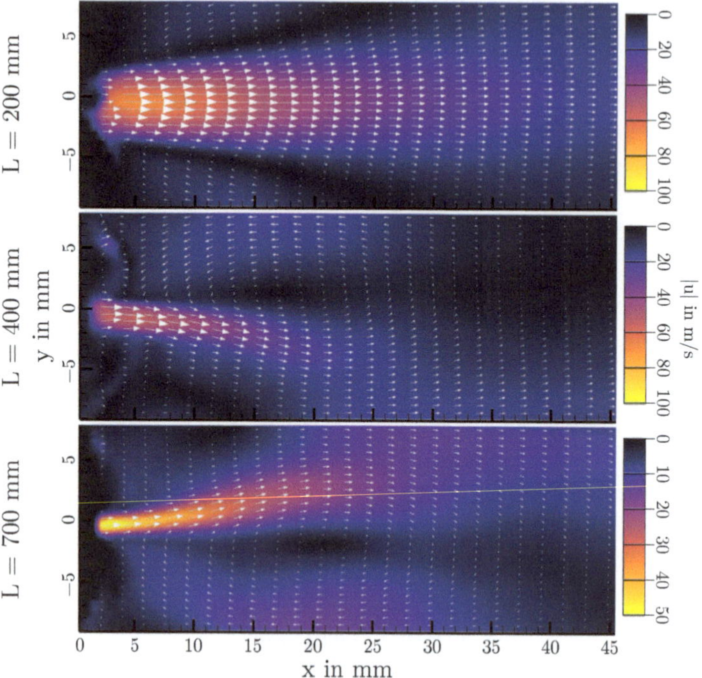

Figure 10. Mean velocity profiles for 200 mm, 400 mm and 700 mm.

More quantitative differences between the three cases can be found by looking at the velocity fields in the frequency domain. Figure 11 shows the power spectral density (PSD) of the velocity magnitude averaged over the whole domain. All three spectra show the same qualitative trend of a general noise level decreasing with increasing frequency and strong harmonic peaks at their respective f_o and higher harmonics. The f_o-shift according to the acoustic resonances as shown in Figure 5 and Table 1 is also observable here. Generally, the decreased velocity magnitudes with increasing L lead to a lower harmonic intensity as well as a lower noise level in the spectra. In the case of $L = 400$ mm, there are also sub-harmonic peaks visible at $1/3f_o, 2/3f_o, 4/3f_o, 5/3f_o$, and so on (with a fundamental frequency of $f_o = 225$ Hz). In this case, the mode at 225 Hz is the strongest, while the 150 Hz mode is still visible in the spectrum. As can be seen in the spectra, the subharmonic peak at $2/3f_o$ coincides perfectly with the f_o-peak at $L = 200$ mm. Therefore, this 150 Hz mode, as well as the peak at ∼75 Hz, can be interpreted as subharmonic peaks. Similarly, Titze [8] found the occurrence of subharmonic peaks at crossings of f_o and f_{R1} in a computational model studying the interaction of supraglottal acoustics and vocal fold oscillation. Kniesburges et al. [22] also observed the appearance of subharmonic peaks in the supraglottal aerodynamic pressure, as well as far field acoustic pressure in a synthetic larynx model. They attributed the subharmonic peaks to small changes in the supraglottal jet location from one oscillation cycle to the next one due to the supraglottal vortex changing direction from cycle to cycle. This, however, is not the same mechanism as apparent in the present study; if the change in rotational direction of the supraglottal vortex was the reason for the subharmonic peaks in our spectra, they would need to occur, especially in the case of $L = 200$ mm, as here we have a symmetric mean velocity field (see Figure 10), indicating a 50:50 distribution of upwards and downwards deflection. In the case of $L = 400$ mm, the supraglottal vortex is more stable, resulting in a 100 % downwards deflection of the jet. This suggests that the subharmonic peaks are not produced by the supraglottal jet location in our case.

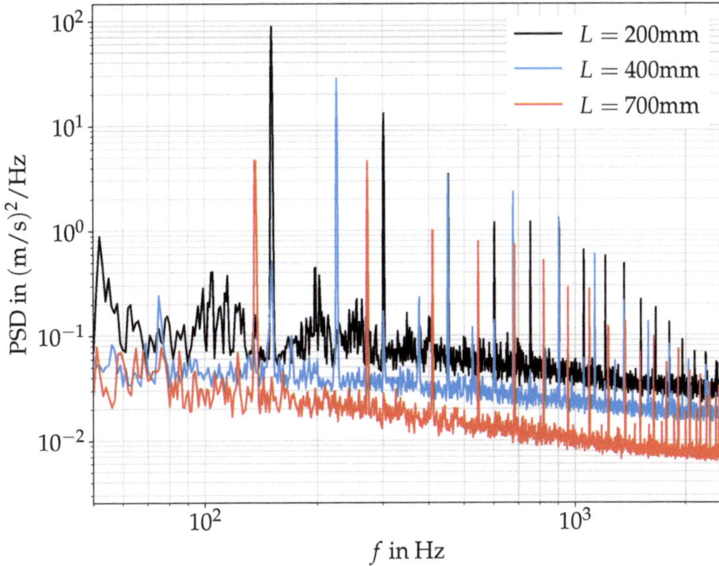

Figure 11. Averaged power spectral density of the flow field obtained by PIV for the three different vocal tract lengths of $L = 200$ mm, $L = 400$ mm and $L = 700$ mm, respectively.

3.3. Aeroacoustic Sources

To investigate the efficiency of the phonation process in the different cases, an aeroacoustic source term computation has been performed on the PIV measurements. For this, the Lighthill analogy was chosen. Root mean square (RMS) values Φ and an aeroacoustic efficiency η were computed as described by Equations (3) and (5). They are shown in Figure 12A and Figure 12B, respectively. It can be seen that Φ decreases with increasing length. This can be expected, as the aeroacoustic source intensity is dependent on the volume flow $\dot{V}$, which generally decreases with increasing length of the supraglottal channel. A similar trend can also be seen in the total subglottal pressure P shown in Figure 12C, which also shows a decrease with increasing length of the channel. Furthermore, the aeroacoustic efficiency η shows a slight decrease with increasing length up to a length of $L = 500$ mm. For larger L, η is rather constant.

Lighthill [24,47] showed that the efficiency of sound generation in free turbulent flows without influence of solid walls in the flow domain scales with the fifth power of the Mach number. To compare our results to this scaling law, Figure 12B also shows a theoretical computation of the aeroacoustic efficiency η_{theor}, which makes use of this fifth power law. The case with $L = 200$ mm is chosen as the baseline case. The proportionality constant of the power law is chosen so that $\eta = \eta_{theor}$ at $L = 200$ mm. For the other cases, the value for η_{theor} is then scaled with the fifth power of the corresponding bulk Mach number. For the channel lengths $L \leq 340$ mm, this law shows a reasonable agreement with the measurement data. It starts deviating from the data for larger L, and shows a strong difference for $L \geq 600$ mm. From Figure 5 we know that this is also the length region where f_o is close to f_{R1}. This suggests that the acoustic resonance frequency of the vocal tract increases the aeroacoustic source intensity strongly by more than one order of magnitude. It also enhances the vocal fold oscillation, as the total subglottal pressure also needed for stable oscillation shows a strong drop by approximately 900 Pa. Overall, the aeroacoustic efficiency η is with approximately 1 % higher than what Lighthill reported for free turbulent flows, which might be explainable by the assumptions we had to make due to the missing information in the third spatial dimension. The assumption of a rotational symmetry of the jet flow leads to a strong correlation in the circumferential direction, which increases the aeroacoustic efficiency. Another reason is that Lighthill did not take the existence of stationary or moving walls in vicinity to the flow field into account. Such walls are known to greatly increase the efficiency of sound production [48,49]. Despite these uncertainties, the η-values found are still useful for a relative comparison between the configurations.

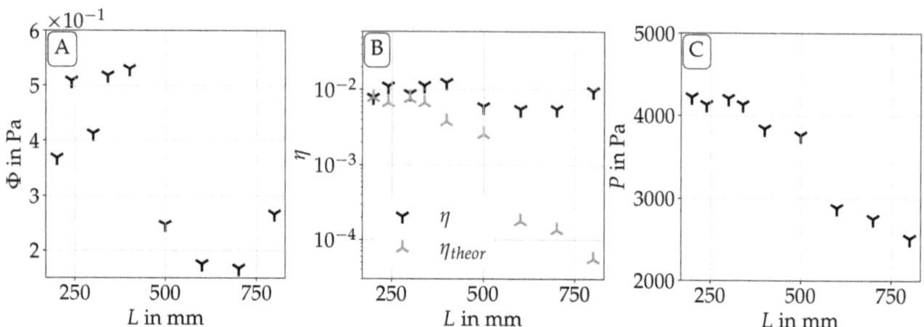

Figure 12. RMS aeroacoustic source level Φ (**A**), aeroacoustic efficiency η (**B**) and the total subglottal pressure P (**C**) of all vocal tract lengths L investigated.

3.4. Acoustic Radiation

Figure 13 shows the acoustic spectra as measured by the microphone for the three main L. The difference in the f_o values is apparent. Furthermore, for $L = 400$ mm strong subharmonic peaks are visible. These subharmonic peaks can be directly related to the aerodynamic flow field, as they are at the same frequencies as in the aerodynamic spectrum of Figure 11. Also notable is the overall much lower noise level for $L = 700$ mm. In contrast to the PIV-related spectra, the peak height at f_o is, however, at a very similar level between the three measurements. This is the case due to the transfer function of the acoustic pressure of the vocal tract. As f_o is very close to a resonance frequency of the vocal tract in the $L = 400$ mm and $L = 700$ mm cases, the transfer function of the vocal tract is very large for those frequencies, leading to an amplification compared to the other frequencies. To obtain a better insight into the changes in the acoustic radiation with changing L, several parameters have been computed from the microphone data. Figure 14A shows the overall sound pressure level (SPL) as a function of L. For most L, the SPL is in a range between 81.8 dB and 84 dB. The one outlier is found for $L = 400$ mm. Here, the SPL rises to almost 95 dB. In this case, two amplifying characteristics coincide: related to the large P and $\dot{V}$ values, the aeroacoustic source intensity and efficiency are very high (see Figure 12A,B). Additionally, f_o is close to f_{R1}, leading to an amplification of the harmonic content in the acoustics. Therefore, a strong increase in the SPL can be expected. This also leads to a high vocal efficiency (VE, calculated following [50]), as seen in Figure 14B. Generally, the cases where f_o is close to f_{R1} show a higher vocal efficiency than the rest of the cases, while also showing higher values for the SNR [51] and the HNR [52], as computed by the GAT. For the case of $L = 400$ mm, the difference in SNR and HNR compared to the other L are, however, much lower than in SPL and VE. This could be related to the strong subharmonic content in the acoustic spectrum, which leads to an erroneous noise content estimation. Also, the CPP [53,54] computed by GAT that is displayed in Figure 14D is very low for this case for the same reason. However, the CPP generally increases with increasing L, which indicates a decrease in noise, in contrast to tonal sound components. An outlier can be found for $L = 600$ mm. Here, some high-intensity, low-frequency noise happened to disturb the acoustic signal at acquisition time, leading to a large low-frequency noise content. This led to a strong decrease in HNR for this length. As this noise increased the overall SPL, the vocal efficiency also supposedly increased here. Generally, the SNR values found for the cases without acoustic backcoupling (small L) lie in the range typical for a healthy voice [51]. With increasing backcoupling, the SNR increases even more, leading to an improved voice quality in these cases. The HNR values reported are in the range of values found for ex vivo studies in the literature [55,56]. They are, however, at the lower end of what is normally found in vivo [57–59]. In our case, this might be attributed to the high volume flow rates needed for the synthetic larynx model to oscillate, leading to increased turbulence broadband noise generation. The high volume flow rate is also responsible for the fact that the VE values are also rather low for all L compared to in vivo data [50].

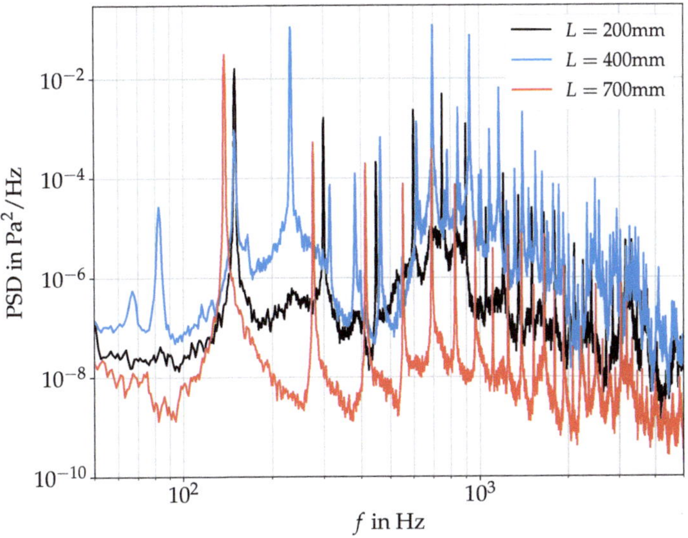

Figure 13. Acoustic spectra for the three different vocal tract lengths of $L = 200\,\text{mm}$, $L = 400\,\text{mm}$ and $L = 700\,\text{mm}$, respectively.

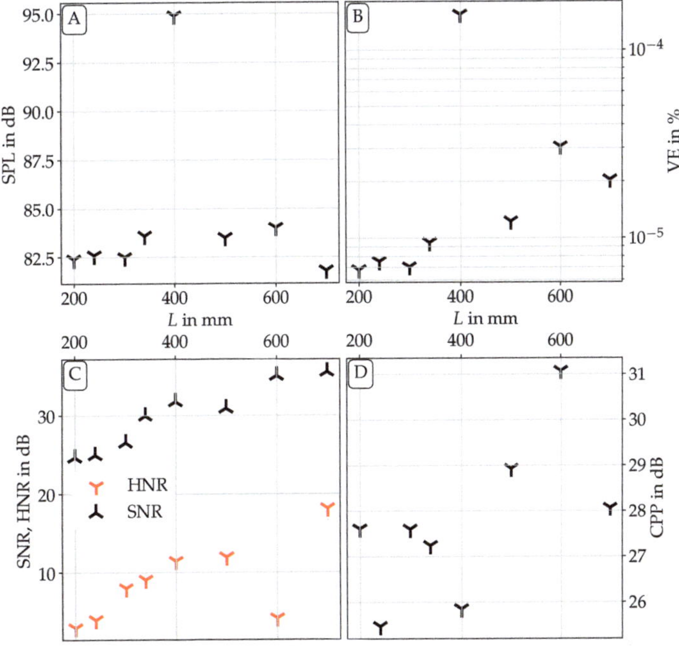

Figure 14. Sound pressure level (**A**), vocal efficiency (**B**), signal-to-noise ratio, harmonics to noise ratio (both **C**) and cepstral peak prominence (**D**) for all vocal tract lengths L investigated.

4. Conclusions

Acoustic back-coupling in the human phonation process has been investigated using a synthetic larynx model and PIV measurements. The vocal tract length was changed systematically in the range $L \in [200, 800]$ mm to vary the relation between fundamental frequency of vocal fold oscillation f_o and lowest resonance frequency of the supraglottal channel f_{R1}. The measurements showed that in the vicinity to each other, f_o is tuned to f_{R1}. Decreasing f_{R1} by increasing L led to a decrease in f_o as well. Increasing f_{R1} to a value higher than the uninfluenced f_o generally did not increase f_o. One exception was the case for $L = 400$ mm. Here, the vocal folds changed their vibration mode, triggered by the acoustic standing waves in the vocal tract having a frequency similar to the eigenfrequency of this mode. The acoustic resonance frequency of the vocal tract did not change the overall characteristics of the supraglottal aerodynamics. However, a changed vocal fold oscillation frequency naturally also led to the change in the dominant frequency in the pulsatile flow field. Looking at the aeroacoustic sources revealed that matching of f_o and f_{R1} resulted in a more than tenfold increase in aeroacoustic efficiency. This also led to an overall increased vocal efficiency, as well as increased SNR, HNR and CPP of the acoustic radiation. This indicates that, at this configuration, a person phonates with higher quality and efficiency. For the case of the professional female singing voice, where f_o and f_{R1} matching can occur at frequencies in the range of approx. 500 Hz, this also means that the singer can phonate longer when f_o and f_{R1} match. This matching is also facilitated by the automatic tuning of f_o to f_{R1} we saw in vicinity.

The elongation of the vocal tract is a simplified approach to be able to study the phonation behavior for different f_o-f_{R1} configurations. In reality, f_o-f_{R1} matching only occurs when f_o is close to lowest resonance frequency of the vocal tract being in the range of 500 Hz, as stated above, which predominantly occurs in children and female singing voice [15]. To increase realism, in future works, more advanced synthetic vocal fold models could be used that show f_o values in this range. Then, more anatomically realistic vocal tract shapes, e.g., from MRI-scans [60] could also be applied, leading to an overall more realistic configuration. Furthermore, the application of tomographic PIV or Lagrangian particle tracking methods could enhance the accuracy of aeroacoustic source term computation. In our case, the overall high aeroacoustic efficiency could be attributed to the rotational symmetry we assumed along the channel axis to obtain some information for the missing third spatial dimension. Tomographic methods would render this assumption unnecessary, improving the accuracy of our evaluation method.

Author Contributions: Conceptualization, C.N. and S.K.; Investigation, C.N. and B.T.; Methodology, C.N. and S.S.; Project Administration, C.N.; Resources, S.K., B.T. and S.B.; Software C.N. and S.S.; Writing—original draft, C.N., S.K. and S.S.; Writing—review and editing, C.N., S.K. and S.S.; Visualization, C.N.; Supervision S.B. and S.K.; Funding acquisition, S.B. All authors have read and agreed to the published version of the manuscript.

Funding: This work is funded by the German Research Foundation (DFG) through the project "Tracing the mechanisms that generate tonal content in voiced speech". Project number: 446965891.

Institutional Review Board Statement: Not applicable.

Informed Consent Statement: Not applicable.

Data Availability Statement: The data presented in this study are not publicly available due to ongoing research in this field.

Conflicts of Interest: The authors declare no conflict of interest. The funders had no role in the design of the study; in the collection, analyses, or interpretation of data; in the writing of the manuscript; or in the decision to publish the results.

Abbreviations

The following abbreviations are used in this manuscript:

CPP	Cepstral Peak Prominence
FSAI	Fluid–Structure–Acoustic Interaction
GAT	Glottis Analysis Tools
HNR	Harmonics-to-Noise Ratio
PIV	Particle Image Velocimetry
POD	Proper Orthogonal Decomposition
PSD	Power Spectral Density
RMS	Root Mean Square
ROI	Region Of Interest
SNR	Signal-to-Noise Ratio
SPL	Sound Pressure Level
VE	Vocal Efficiency

References

1. Mittal, R.; Erath, B.D.; Plesniak, M.W. Fluid Dynamics of Human Phonation and Speech. *Annu. Rev. Fluid Mech.* **2013**, *45*, 437–467. [CrossRef]
2. Bodaghi, D.; Xue, Q.; Zheng, X.; Thomson, S. Effect of Subglottic Stenosis on Vocal Fold Vibration and Voice Production Using Fluid–Structure–Acoustics Interaction Simulation. *Appl. Sci.* **2021**, *11*, 1221. [CrossRef]
3. Döllinger, M.; Zhang, Z.; Schoder, S.; Šidlof, P.; Tur, B.; Kniesburges, S. Overview on state-of-the-art numerical modeling of the phonation process. *Acta Acust.* **2023**, *7*, 25. [CrossRef]
4. Schoder, S.; Maurerlehner, P.; Wurzinger, A.; Hauser, A.; Falk, S.; Kniesburges, S.; Döllinger, M.; Kaltenbacher, M. Aeroacoustic Sound Source Characterization of the Human Voice Production-Perturbed Convective Wave Equation. *Appl. Sci.* **2021**, *11*, 2614. [CrossRef]
5. Lodermeyer, A.; Bagheri, E.; Kniesburges, S.; Näger, C.; Probst, J.; Döllinger, M.; Becker, S. The mechanisms of harmonic sound generation during phonation: A multi-modal measurement-based approach. *J. Acoust. Soc. Am.* **2021**, *150*, 3485–3499. [CrossRef]
6. Titze, I.R.; Story, B.H. Acoustic interactions of the voice source with the lower vocal tract. *J. Acoust. Soc. Am.* **1997**, *101*, 2234–2243. [CrossRef]
7. Fant, G. *Acoustic Theory of Speech Production*; De Gruyter Mouton: The Hague, Netherlands, 1971. [CrossRef]
8. Titze, I.R. Nonlinear source–filter coupling in phonation: Theory. *J. Acoust. Soc. Am.* **2008**, *123*, 2733–2749. [CrossRef]
9. Howe, M.S.; McGowan, R.S. Sound generated by aerodynamic sources near a deformable body, with application to voiced speech. *J. Fluid Mech.* **2007**, *592*, 367–392. [CrossRef]
10. McGowan, R.S.; Howe, M.S. Source-tract interaction with prescribed vocal fold motion. *J. Acoust. Soc. Am.* **2012**, *131*, 2999–3016. [CrossRef]
11. Zañartu, M.; Mongeau, L.; Wodicka, G.R. Influence of acoustic loading on an effective single mass model of the vocal folds. *J. Acoust. Soc. Am.* **2007**, *121*, 1119–1129. [CrossRef]
12. Lucero, J.C.; Lourenço, K.G.; Hermant, N.; Hirtum, A.V.; Pelorson, X. Effect of source–tract acoustical coupling on the oscillation onset of the vocal folds. *J. Acoust. Soc. Am.* **2012**, *132*, 403–411. [CrossRef]
13. Erath, B.D.; Peterson, S.D.; Weiland, K.S.; Plesniak, M.W.; Zañartu, M. An acoustic source model for asymmetric intraglottal flow with application to reduced-order models of the vocal folds. *PLoS ONE* **2019**, *14*, e0219914. [CrossRef]
14. Wade, L.; Hanna, N.; Smith, J.; Wolfe, J. The role of vocal tract and subglottal resonances in producing vocal instabilities. *J. Acoust. Soc. Am.* **2017**, *141*, 1546–1559. [CrossRef]
15. Echternach, M.; Herbst, C.T.; Köberlein, M.; Story, B.; Döllinger, M.; Gellrich, D. Are source-filter interactions detectable in classical singing during vowel glides? *J. Acoust. Soc. Am.* **2021**, *149*, 4565–4578. [CrossRef]
16. Zhang, Z.; Neubauer, J.; Berry, D.A. Influence of vocal fold stiffness and acoustic loading on flow-induced vibration of a single-layer vocal fold model. *J. Sound Vib.* **2009**, *322*, 299–313. [CrossRef]
17. Smith, B.L.; Nemcek, S.P.; Swinarski, K.A.; Jiang, J.J. Nonlinear Source-Filter Coupling Due to the Addition of a Simplified Vocal Tract Model for Excised Larynx Experiments. *J. Voice* **2013**, *27*, 261–266. [CrossRef] [PubMed]
18. Migimatsu, K.; Tokuda, I.T. Experimental study on nonlinear source–filter interaction using synthetic vocal fold models. *J. Acoust. Soc. Am.* **2019**, *146*, 983–997. [CrossRef] [PubMed]
19. Oren, L.; Khosla, S.; Gutmark, E. Intraglottal geometry and velocity measurements in canine larynges. *J. Acoust. Soc. Am.* **2014**, *135*, 380–388. [CrossRef] [PubMed]
20. Oren, L.; Khosla, S.; Gutmark, E. Intraglottal pressure distribution computed from empirical velocity data in canine larynx. *J. Biomech.* **2014**, *47*, 1287–1293. [CrossRef] [PubMed]
21. Lodermeyer, A.; Becker, S.; Döllinger, M.; Kniesburges, S. Phase-locked flow field analysis in a synthetic human larynx model. *Exp. Fluids* **2015**, *56*, 77. [CrossRef]

22. Kniesburges, S.; Lodermeyer, A.; Becker, S.; Traxdorf, M.; Döllinger, M. The mechanisms of subharmonic tone generation in a synthetic larynx model. *J. Acoust. Soc. Am.* **2016**, *139*, 3182–3192. [CrossRef] [PubMed]
23. Lodermeyer, A.; Tautz, M.; Becker, S.; Döllinger, M.; Birk, V.; Kniesburges, S. Aeroacoustic analysis of the human phonation process based on a hybrid acoustic PIV approach. *Exp. Fluids* **2018**, *59*, 13. [CrossRef]
24. Lighthill, M. On sound generated aerodynamically I. General theory. *Proc. Roy. Soc. Lond.* **1952**, *211*, 564–587. [CrossRef]
25. Kaltenbacher, M.; Hüppe, A.; Reppenhagen, A.; Zenger, F.; Becker, S. Computational Aeroacoustics for Rotating Systems with Application to an Axial Fan. *AIAA J.* **2017**, *55*, 3831–3838. [CrossRef]
26. de Luzan, C.F.; Oren, L.; Maddox, A.; Gutmark, E.; Khosla, S.M. Volume velocity in a canine larynx model using time-resolved tomographic particle image velocimetry. *Exp. Fluids* **2020**, *61*, 63. [CrossRef]
27. Scherer, R.C.; Shinwari, D.; Witt, K.J.D.; Zhang, C.; Kucinschi, B.R.; Afjeh, A.A. Intraglottal pressure profiles for a symmetric and oblique glottis with a divergence angle of 10 degrees. *J. Acoust. Soc. Am.* **2001**, *109*, 1616–1630. [CrossRef]
28. Thomson, S.L.; Mongeau, L.; Frankel, S.H. Aerodynamic transfer of energy to the vocal folds. *J. Acoust. Soc. Am.* **2005**, *118*, 1689–1700. [CrossRef]
29. Zhang, Z.; Neubauer, J.; Berry, D.A. The influence of subglottal acoustics on laboratory models of phonation. *J. Acoust. Soc. Am.* **2006**, *120*, 1558–1569. [CrossRef]
30. Durst, F.; Heim, U.; Ünsal, B.; Kullik, G. Mass flow rate control system for time-dependent laminar and turbulent flow investigations. *Meas. Sci. Technol.* **2003**, *14*, 893–902. [CrossRef]
31. Kist, A.M.; Gómez, P.; Dubrovskiy, D.; Schlegel, P.; Kunduk, M.; Echternach, M.; Patel, R.; Semmler, M.; Bohr, C.; Dürr, S.; et al. A Deep Learning Enhanced Novel Software Tool for Laryngeal Dynamics Analysis. *J. Speech Lang. Hear. Res.* **2021**, *64*, 1889–1903. [CrossRef]
32. Maryn, Y.; Verguts, M.; Demarsin, H.; van Dinther, J.; Gomez, P.; Schlegel, P.; Döllinger, M. Intersegmenter Variability in High-Speed Laryngoscopy-Based Glottal Area Waveform Measures. *Laryngoscope* **2020**, *130*, E654–E661. [CrossRef]
33. Raffel, M.; Willert, C.E.; Scarano, F.; Kähler, C.J.; Wereley, S.T.; Kompenhans, J. *Particle Image Velocimetry*; Springer International Publishing: Berlin/Heidelberg, Germany, 2018. [CrossRef]
34. Samimy, M.; Lele, S.K. Motion of particles with inertia in a compressible free shear layer. *Phys. Fluids A Fluid Dyn.* **1991**, *3*, 1915–1923. [CrossRef]
35. Mendez, M.; Raiola, M.; Masullo, A.; Discetti, S.; Ianiro, A.; Theunissen, R.; Buchlin, J.M. POD-based background removal for particle image velocimetry. *Exp. Therm. Fluid Sci.* **2017**, *80*, 181–192. [CrossRef]
36. Adatrao, S.; Sciacchitano, A. Elimination of unsteady background reflections in PIV images by anisotropic diffusion. *Meas. Sci. Technol.* **2019**, *30*, 035204. [CrossRef]
37. Westerweel, J.; Scarano, F. Universal outlier detection for PIV data. *Exp. Fluids* **2005**, *39*, 1096–1100. [CrossRef]
38. Sondhi, M.; Schroeter, J. A hybrid time-frequency domain articulatory speech synthesizer. *IEEE Trans. Acoust. Speech Signal Process.* **1987**, *35*, 955–967. [CrossRef]
39. Story, B.H.; Laukkanen, A.M.; Titze, I.R. Acoustic impedance of an artificially lengthened and constricted vocal tract. *J. Voice* **2000**, *14*, 455–469. [CrossRef]
40. Flanagan, J.L. *Speech Analysis, Synthesis and Perception*; Springer: Berlin/Heidelberg, Germany, 1972; p. 444.
41. Fulcher, L.; Lodermeyer, A.; Kähler, G.; Becker, S.; Kniesburges, S. Geometry of the Vocal Tract and Properties of Phonation near Threshold: Calculations and Measurements. *Appl. Sci.* **2019**, *9*, 2755. [CrossRef]
42. Titze, I. *Principles of Voice Production*; Prentice Hall: Hoboken, NJ, USA, 1994.
43. Neubauer, J.; Zhang, Z.; Miraghaie, R.; Berry, D.A. Coherent structures of the near field flow in a self-oscillating physical model of the vocal folds. *J. Acoust. Soc. Am.* **2007**, *121*, 1102–1118. [CrossRef] [PubMed]
44. Erath, B.D.; Plesniak, M.W. The occurrence of the Coanda effect in pulsatile flow through static models of the human vocal folds. *J. Acoust. Soc. Am.* **2006**, *120*, 1000–1011. [CrossRef]
45. Erath, B.D.; Plesniak, M.W. An investigation of asymmetric flow features in a scaled-up driven model of the human vocal folds. *Exp. Fluids* **2010**, *49*, 131–146. [CrossRef]
46. Erath, B.D.; Plesniak, M.W. Impact of wall rotation on supraglottal jet stability in voiced speech. *J. Acoust. Soc. Am.* **2011**, *129*, EL64–EL70. [CrossRef] [PubMed]
47. Lighthill, M.J. On sound generated aerodynamically II. Turbulence as a source of sound. *Proc. R. Soc. London. Ser. A. Math. Phys. Sci.* **1954**, *222*, 1–32. [CrossRef]
48. Howe, M.S. *Acoustics of Fluid-Structure Interactions*; Cambridge University Press: Cambridge, UK, 1998. [CrossRef]
49. Howe, M.; McGowan, R. Aerodynamic sound of a body in arbitrary, deformable motion, with application to phonation. *J. Sound Vib.* **2013**, *332*, 3909–3923. [CrossRef] [PubMed]
50. Titze, I.R.; Maxfield, L.; Palaparthi, A. An Oral Pressure Conversion Ratio as a Predictor of Vocal Efficiency. *J. Voice* **2016**, *30*, 398–406. [CrossRef] [PubMed]
51. Qi, Y.; Hillman, R.E.; Milstein, C. The estimation of signal-to-noise ratio in continuous speech for disordered voices. *J. Acoust. Soc. Am.* **1999**, *105*, 2532–2535. [CrossRef] [PubMed]
52. Yumoto, E.; Sasaki, Y.; Okamura, H. Harmonics-to-Noise Ratio and Psychophysical Measurement of the Degree of Hoarseness. *J. Speech Lang. Hear. Res.* **1984**, *27*, 2–6. [CrossRef]

53. Hillenbrand, J.; Cleveland, R.A.; Erickson, R.L. Acoustic Correlates of Breathy Vocal Quality. *J. Speech, Lang. Hear. Res.* **1994**, *37*, 769–778. [CrossRef]
54. Hillenbrand, J.; Houde, R.A. Acoustic Correlates of Breathy Vocal Quality: Dysphonic Voices and Continuous Speech. *J. Speech Lang. Hear. Res.* **1996**, *39*, 311–321. [CrossRef]
55. Semmler, M.; Berry, D.A.; Schützenberger, A.; Döllinger, M. Fluid-structure-acoustic interactions in an ex vivo porcine phonation model. *J. Acoust. Soc. Am.* **2021**, *149*, 1657–1673. [CrossRef]
56. Peters, G.; Jakubaß, B.; Weidenfeller, K.; Kniesburges, S.; Böhringer, D.; Wendler, O.; Mueller, S.K.; Gostian, A.O.; Berry, D.A.; Döllinger, M.; et al. Synthetic mucus for an ex vivo phonation setup: Creation, application, and effect on excised porcine larynges. *J. Acoust. Soc. Am.* **2022**, *152*, 3245–3259. [CrossRef] [PubMed]
57. Gorris, C.; Maccarini, A.R.; Vanoni, F.; Poggioli, M.; Vaschetto, R.; Garzaro, M.; Valletti, P.A. Acoustic Analysis of Normal Voice Patterns in Italian Adults by Using Praat. *J. Voice* **2020**, *34*, 961.e9–961.e18. [CrossRef] [PubMed]
58. Gojayev, E.K.; Büyükatalay, Z.Ç.; Akyüz, T.; Rehan, M.; Dursun, G. The Effect of Masks and Respirators on Acoustic Voice Analysis during the COVID-19 Pandemic. *J. Voice* **2021**. [CrossRef] [PubMed]
59. Nguyen, D.D.; Madill, C. Auditory-perceptual Parameters as Predictors of Voice Acoustic Measures. *J. Voice* **2023**. [CrossRef]
60. Story, B.H.; Titze, I.R.; Hoffman, E.A. Vocal tract area functions from magnetic resonance imaging. *J. Acoust. Soc. Am.* **1996**, *100*, 537–554. [CrossRef]

Article

Effect of Ligament Fibers on Dynamics of Synthetic, Self-Oscillating Vocal Folds in a Biomimetic Larynx Model

Bogac Tur [1,*], Lucia Gühring [1], Olaf Wendler [1], Samuel Schlicht [2], Dietmar Drummer [2] and Stefan Kniesburges [1]

[1] Division of Phoniatrics and Pediatric Audiology, Department of Otorhinolaryngology, Head and Neck Surgery, University Hospital Erlangen, Medical School, Friedrich-Alexander-Universität Erlangen-Nürnberg, Waldstrasse 1, 91054 Erlangen, Germany

[2] Institute of Polymer Technology, Friedrich-Alexander-Universität Erlangen-Nürnberg, Am Weichselgarten 10, 91058 Erlangen, Germany

* Correspondence: bogac.tur@uk-erlangen.de

Abstract: Synthetic silicone larynx models are essential for understanding the biomechanics of physiological and pathological vocal fold vibrations. The aim of this study is to investigate the effects of artificial ligament fibers on vocal fold vibrations in a synthetic larynx model, which is capable of replicating physiological laryngeal functions such as elongation, abduction, and adduction. A multi-layer silicone model with different mechanical properties for the musculus vocalis and the lamina propria consisting of ligament and mucosa was used. Ligament fibers of various diameters and break resistances were cast into the vocal folds and tested at different tension levels. An electromechanical setup was developed to mimic laryngeal physiology. The measurements included high-speed video recordings of vocal fold vibrations, subglottal pressure and acoustic. For the evaluation of the vibration characteristics, all measured values were evaluated and compared with parameters from ex and in vivo studies. The fundamental frequency of the synthetic larynx model was found to be approximately 200–520 Hz depending on integrated fiber types and tension levels. This range of the fundamental frequency corresponds to the reproduction of a female normal and singing voice range. The investigated voice parameters from vocal fold vibration, acoustics, and subglottal pressure were within normal value ranges from ex and in vivo studies. The integration of ligament fibers leads to an increase in the fundamental frequency with increasing airflow, while the tensioning of the ligament fibers remains constant. In addition, a tension increase in the fibers also generates a rise in the fundamental frequency delivering the physiological expectation of the dynamic behavior of vocal folds.

Keywords: synthetic vocal fold models; integrated fibers; biomimetic larynx model; physiological vocal fold dynamics

Citation: Tur, B.; Gühring, L.; Wendler, O.; Schlicht, S.; Drummer, D.; Kniesburges, S. Effect of Ligament Fibers on Dynamics of Synthetic, Self-Oscillating Vocal Folds in a Biomimetic Larynx Model. *Bioengineering* **2023**, *10*, 1130. https://doi.org/10.3390/bioengineering10101130

Academic Editor: Chenzhong Li

Received: 8 August 2023
Revised: 13 September 2023
Accepted: 25 September 2023
Published: 26 September 2023

1. Introduction

Human phonation is a multifaceted physical process influenced by a myriad of factors. Humans have the remarkable ability to manipulate their larynx to affect their voice. For instance, the fundamental frequency F_0 can be increased by tensing the vocal folds, a process in which the pre-phonatory vocal fold posturing plays a pivotal role.

The impact of these vocal fold postures is particularly observable in voice pathologies, such as atrophy [1–4] or paresis [5]. These studies underscore the critical influence of glottal closure on human phonation [6–8]. Even in physiological female and child phonation, interestingly, incomplete glottis closure postures occur without pathological voice characteristics [7,9,10].

To investigate the fundamental influence of vocal fold postures on dynamics, ex vivo experiments are highly suitable [11–13]. However, these experiments face challenges such as the scarcity of human larynges and their individuality, which makes it sometimes difficult to formulate universally valid statements. Furthermore, from an experimental perspective, these larynges also need to be processed quickly to prevent degeneration.

For these reasons, synthetic vocal fold models made of silicone have been developed to better investigate the human phonation process [14]. These synthetic models offer the advantage of parametrizable geometry and material properties, and they can be reproducibly fabricated. Artificial larynx models are used to measure the vocal folds collision forces [15–21], characterize supraglottal aerodynamics [22], serve as validation models in developing advanced laryngoscopic techniques [23–25], estimate energy transfer in the vocal folds [26,27], investigate acoustic interaction [28], and study the asymmetric behavior of vocal folds [29,30]. Many research papers report an incomplete glottal closure in synthetic, self-oscillating vocal fold models [31–33]. Reasons for incomplete glottis closure include the isotropic properties of the silicone used for the vocal folds. In contrast, physiological vocal fold vibrations tend to complete glottal closure under various laryngeal postures [1]. In past studies [32], attempts were made by applying vertical restraint to the lateral half of the superior surface of the synthetic vocal folds. It was reported that this led to complete glottal closure due to increased stiffness in the anterior–posterior direction through elongation of the vocal folds [34]. Hirano and Kakita [35] showed that the integration of a muscle layer as well as collagen and elastin fibers in the lamina propria achieved transverse isotropy. Murray and Thomson [36] reproduced glottis closure with a combination of an epithelium layer and an extremely soft cover. In addition, a fiber was included into the ligament, which reinforced the effect of stiffness in the anterior–posterior direction. To represent the physical properties of the collagen and elastin fibers, the study by Shaw et al. [37] was carried out. They cast acrylic and polyester fibers into the surface layer to achieve non-linear stress–strain characteristics of the vocal folds. Xuan and Zhang [38] focus in their study on achieving glottal closure with an epithelium layer and embedded fiber in a single-layer silicone model.

The few models summarized above focused on specific and restricted phonatoric conditions with regard to pre-phonatoric posturing or fiber tension separately. Therefore, the aim of the study presented here is to introduce a synthetic larynx model that is able to reproduce both, different types of pre-phonatoric posturing (elongation and ad-/abduction) combined with different tension levels of the embedded fibers in the ligament layer of the vocal folds model. In this model, fibers are cast into a multi-layer synthetic vocal fold model based on the M5 geometry by Scherer et al. [39] in the anterior–posterior direction. These fibers are tensioned under mechanical force to investigate the effects on the F_0 and the subglottal pressure P_{sub} with an initially pre-phonatory closed glottis. Additionally, this study also examines typical voice parameters from high-speed videos and acoustic measurements, to contextualize the results within the physiology of vocal fold dynamics.

2. Materials and Methods

2.1. Synthetic Larynx Model

The synthetic larynx model includes a vocal fold model made of silicone with the M5 geometry by Scherer et al. [39], which is cast into a silicone ring as shown in Figure 1a. This ring incorporates additional manipulators at various positions, designed to mimic the typical laryngeal functions of ad-/abduction and elongation during pre-phonatory posturing of the vocal folds. The silicone vocal folds are structured in different layers capturing the musculus vocalis and the lamina propria, comprising the ligament and mucosa as shown in Figure 1b,c. Characteristic biomechanical properties have been selected for each layer during the fabrication process [36,40]. Furthermore, fibers are integrated into the ligament to mimic the fibrous structure in the human vocal fold ligament. Figure 1d provides detailed geometric parameters in a mid-coronal cut. Two types of platinum-catalyzed two component silicone rubber (Smooth-On, Inc., Macungie, PA, USA) with specific proportions of silicone thinner are employed to fabricate the larynx model: `Ecoflex 00-30`, `Dragon Skin 10 Slow` and `Silicone Thinner`.

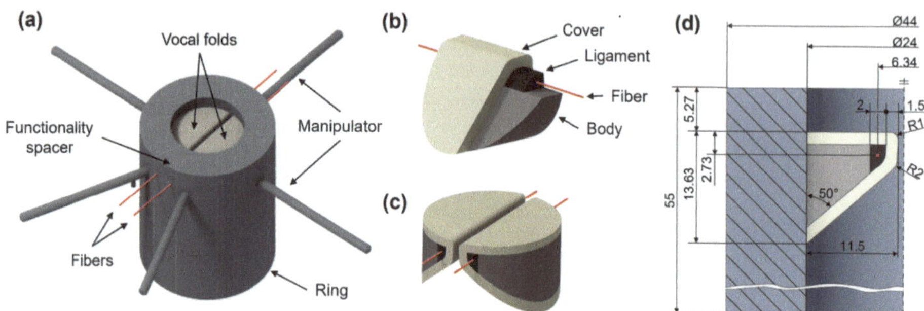

Figure 1. (**a**) 3D CAD model of the entire larynx model, with the manipulators, the vocal folds based on the M5 model, the ring and the fibers (**b**) View of the individual layers in the vocal fold model, with the body, the ligament, the fiber and the cover (**c**) 3D perspective view of the vocal folds as they are arranged in the larynx model (**d**) Mid-coronal cut showing the exact dimensions in mm of the entire model and the position of the fibers.

The silicone ring embodies a balance between rigidity and flexibility essential for laryngeal functions. For this reason, the ring was made of `Dragon Skin 10 Slow`. For all other elements of the model, `Ecoflex 00-30` was cast with a certain amount of thinner. Table 1 shows the different proportions associated with the constituents of the entire larynx model.

Table 1. Material properties of the silicone compounds used in various larynx elements, detailing the specific mixing ratios and corresponding Young's modulus [41].

Larynx Elements	Silicone	Mixing Ratio Part A:Part B:Thinner	Young's Modulus in kPa
Ring	Dragon Skin 10 Slow	1:1:0	151
Functionality spacer	Ecoflex 00-30	1:1:2.5	6 *
Body	Ecoflex 00-30	1:1:2	8.2
Ligament	Ecoflex 00-30	1:1:0	60
Cover	Ecoflex 00-30	1:1:4	2.5

* The Young's modulus was estimated using cubic spline interpolation, leveraging available empirical data points for varying proportions of thinner in the mixture.

In the present study, a total of six different larynx models (MLM1-MLM6) each with a different fiber type were examined to evaluate the resulting vibratory responses of the vocal folds (see Table 2). The diameters of these fibers ranged from 0.108 mm to 0.3 mm, with corresponding break resistances from 1.18 kg to 7.9 kg. The thinnest of these tested fibers is composed of polyvinylidene fluoride (PVDF), while the remainder are made from polyamide 6.6 (PA6.6). In our study, we determined the mechanical properties of the storage modulus E', loss modulus E'', and the loss tangent $\tan \delta$ (the ratio of E'' and E'). The dynamic characterization of the applied pre-conditioned fibers is conducted using a solid analyzer of type `RSA-G2` (TA Instruments, New Castle, DE, USA). To compensate for moisture-dependent effects of polyamide filaments, all samples are prepared under normal conditions (23 °C, 50% humidity) for 30 days. Applied characterizations include frequency sweeps, ranging from 0.5 Hz–50 Hz at a constant strain of 0.1%, and amplitude sweeps at a constant frequency of 1 Hz, with oscillation strains ranging from 0.01 to 1%. All measurements are conducted at a controlled temperature of 25 °C.

Table 2. Overview of the various fiber-based models MLM1-MLM6 including their material composition, diameter, break resistance, and associated dynamic mechanical properties E′-Modulus, E″-Modulus and tan δ, using frequency sweeps from 0.5 Hz–50 Hz at a constant strain of 0.1%.

Model	Fiber					
	Material	Diameter in mm	Break Resistance in kg	E′-Modulus * in GPa	E″-Modulus * in GPa	tan δ *
MLM1 **	PVDF	0.108	1.18	2.91 (0.07)	0.08 (0.01)	0.03 (0.00)
MLM2	PA 6.6	0.12	1.4	2.39 (0.04)	0.16 (0.00)	0.07 (0.00)
MLM3	PA 6.6	0.125	1.82	-	-	-
MLM4	PA 6.6	0.18	2.5	1.19 (0.03)	0.12 (0.00)	0.10 (0.00)
MLM5	PA 6.6	0.25	5.3	1.82 (0.38)	0.16 (0.03)	0.09 (0.00)
MLM6	PA 6.6	0.3	7.9	1.71 (0.10)	0.12 (0.01)	0.07 (0.00)

* The values in parentheses show the standard deviation. ** MLM is the abbreviation for **Multi-Layer-Model**.

The casting process of the synthetic larynx model involves several steps, as displayed in Figure 2a,b. Initially, the silicone ring is cast, incorporating a mold for the functionality spacer. Within this process, the five manipulators are embedded within the silicone ring to control the laryngeal functions, as mentioned above. The silicone ring is then demolded and inserted into a mechanical frame that allows for precise positioning of a 0.9 mm thick cannula through the ring, which serves as a guide for threading the fibers. The fibers are then secured, and the cannula is removed, resulting in the fibers being accurately positioned within the ring. This process is replicated for both the left and right vocal fold.

Subsequent steps involve casting the various layers of the vocal folds using different molds: Initially, the body is cast, followed by the ligament, and finally, the cover. Figure 2c–e shows the molding process for the multi-layer vocal fold model. All parts of the model are cured for at least eight hours at a temperature of 40 °C and cooled down to room temperature. To avoid air deposits, the pre-cast liquid silicone mixtures have been degassed under vacuum for at least 10 min before each casting step. At the end of the final step, the surfaces of the vocal folds are dusted with talcum powder to prevent them from adhering to each other after the mold is completely removed. The molds and the manipulators are made of `Grey Resin` (Formlabs, Somerville, MA, USA) printed with the `Formlabs Form 2` stereolithography 3D printer (Formlabs, Somerville, MA, USA).

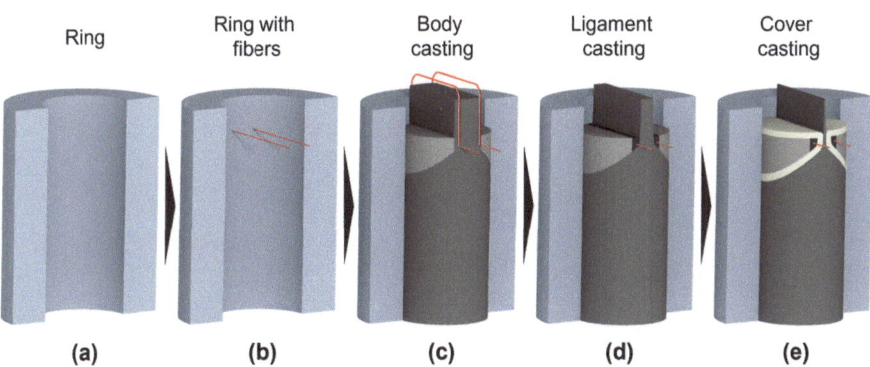

Figure 2. The casting process of the multi-layer vocal folds in the artificial larynx model is shown. (**a**) Ring in the cross-section, (**b**) Ring in the cross-section with integrated fibers, (**c**) Casting of the body, (**d**) Casting of the ligament and (**e**) Casting of the cover.

2.2. Biomimetic Functionality

The biomimetic functionality of the synthetic laryngeal model is realized by including an electromechanical control system to reproduce typical laryngeal dynamics such as elongation, adduction, and abduction. This system is facilitated by manipulators that transmit the linear and rotational motion from the electromotors to the silicone ring displaying the laryngeal cartilages (thyroid and arytenoid) and therefore to the vocal folds.

The apparatus incorporates a total of seven motors, each serving a distinct function that perform the linear and rotational motion as displayed in Figure 3. Three of these 8MT173 Motorized Translation Stages (Standa Ltd., Vilnius, Lithuania) are responsible for linear movement to elongate the vocal folds indicated by the red arrows. Two 8MR151 Motorized Rotation Stages (Standa Ltd., Vilnius, Lithuania) are designated for rotational movement to realize the ad-/abduction motion supported by the remaining two Motorzied XY Scanning Stages (Standa Ltd., Vilnius, Lithuania). The ad-/abduction motion is thereby shown with blue and green arrows in Figure 3.

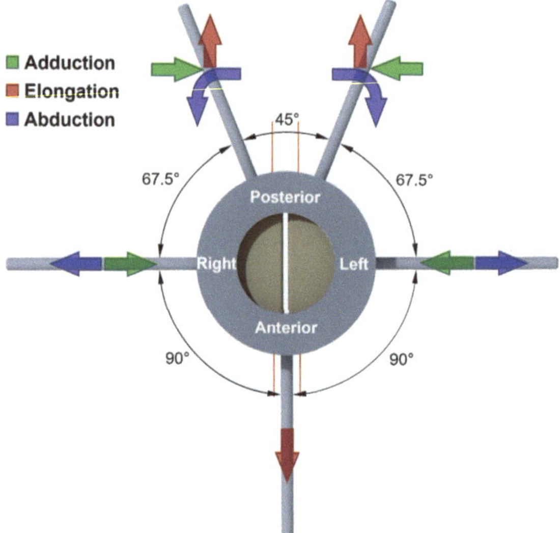

Figure 3. The top view of the artificial larynx model and the manipulators with the angular arrangement. In addition, the laryngeal dynamical motions, i.e., elongation and ad-/abduction, are indicated by colored arrows.

The control of these motorized stages is achieved through an 8SMC5-USB Stepper & DC Motor Controller (Standa Ltd., Vilnius, Lithuania). The motors are controlled by Python scripts which are integrated within a globally working LabVIEW (National Instruments, Austin, TX, USA) measurement script.

2.3. Fiber Guidance and Tensioning System

Another laryngeal function integrated in this model is the application of a pretension in the ligament layer of the vocal folds. This is realized by the inclusion of elastic fibers as described above. These fibers are embedded within the ligament layer in an unloaded state. To control the free fiber ends, the fibers are guided through a customized guidance management system designed to minimize frictional loss and to ensure parallel alignment. This is realized by bearing pulleys to reduce friction at redirection units for the fibers.

The fibers are secured to a 7T67-25 Stable Steel Translation Stage (Standa Ltd., Vilnius, Lithuania) capable of linear movement, with a travel range of 25 mm. This setup allows for precise control over the tension of the ligament. By the integration of highly stiff fibers, a non-linear tension can be introduced by the elongation of the vocal folds and the fibers qualitatively similar to human ligament tissue [42–44].

2.4. Measurement Setup and Data Acquisition

The investigation employed a multimodal measurement setup acquiring different physical parameters from the model. The setup used in this study is a modified version of the setup introduced by Birk [45] for ex vivo larynx models [46–48]. The synthetic larynx model was mounted on an artificial trachea with a diameter of 24 mm. Airflow, which induces the oscillations of the vocal folds, was regulated in standard liter per minute (SLM) by a `1579 A/B` mass flow controller (MKS, Andover, MA, USA) and a 4000B digital power supply (MKS, Andover, MA, USA). The subglottal pressure signal was measured using an `XCS-93-5PSISG` pressure sensor (Kulite Semiconductor Products, Inc., Leonia, NJ, USA) connected to a `PXIe-4330` bridge module (National Instruments, Austin, TX, USA), with a sampling frequency of 44.1 kHz for 1 s. The pressure sensor is located approx. 130 mm below the glottal level.

The glottal region and the dynamics of vocal fold vibration were examined using a `Phantom V2511` digital high-speed camera (Vision Research, Wayne, NJ, USA). The frame rate was set to 4000 frames/s (fps) with a picture resolution of 768×768 pixels. A Canon `EF 180 mm f/3.5L` macro lens (Canon, Inc., Tokyo, Japan) was mounted on the camera to display the vocal folds on the camera chip. The recording duration amounted 600 ms. The start of the high-speed recording was triggered by a `PXIe-6356` multifunctional module (National Instruments, Austin, TX, USA).

In the supraglottal region, acoustic signals were sampled with two 4189 1/2-inch free-field microphones (Brüel & Kjær, Nærum, Denmark) at a distance of at least 30 cm from the model, applying a sampling frequency of 44.1 kHz for 1 s duration. Care was taken that the microphones were not exposed to the airflow coming from the model. Both microphones were connected to a `Nexus 2690` microphone conditioning amplifier (Brüel & Kjær, Nærum, Denmark). The analog voltage signals were sampled and A/D converted by a `PXIe-4492` sound and vibration module (National Instruments, Austin, TX, USA.

The pressure (aerodynamic and acoustic) signals were synchronously sampled using a `LabVIEW` script. All measurements (microphone, pressure sensor and camera) were simultaneously started.

2.5. Data Processing and Analysis

The computation of parameters as F_0 and mean P_{sub} was performed with `MATLAB` R2021b (The MathWorks, Inc., Natick, MA, USA). The spectral analysis is based on Welch's method [49,50], employing a hamming window with a window length of 0.37 s, to yield the power spectral density of the pressure signals [51]. Thereby, the post-processing of the subglottal pressure was conducted relative to the atmospheric pressure.

Glottal dynamic parameters were derived from the high-speed imaging videos of the vocal fold oscillations using our in-house software package `Glottis Analysis Tool 2020 (GAT)` [52,53]. This tool facilitates the segmentation of the glottis area between the vocal folds to obtain the glottal area waveform (GAW), from which characteristic parameters describing the vocal fold dynamics during phonation are calculated. The chosen parameters for glottal dynamics describe the dynamical behavior regarding the glottal closure, i.e., Glottis gap index (GGI) and Closing quotient (ClQ), the vibration periodicity, i.e., Amplitude periodicity (AP) and Time periodicity (TP), and symmetry of the vocal folds, i.e., Phase asymmetry index (PAI) and Amplitude symmetry index (ASI). For the calculation of these parameters, 20 consecutive cycles were used, which is the minimum number of cycles that provide stable parameters in high-speed video recordings [54,55]. An expanded discourse and relevant literature concerning these parameters can be found in Table 3a.

GAT was also used for the physiological parameter evaluation of acoustic and subglottal pressure. Regularity parameters, Jitter (jitt) and Shimmer (shim), and sound quality parameters, Harmonics-to-noise ratio (HNR), Normalized noise energy (NNE) and Cepstral peak prominence (CPP), were considered. For these analyses, 100 cycles were used [56,57]. Table 3b furnishes additional data and bibliographic references pertinent to these parameters [58].

Table 3. Overview of computed (**a**) glottal dynamic parameters and (**b**) acoustic parameters.

Parameter	Abbreviation/Unit	Description/Range	Reference
(a) Glottal dynamic parameters			
Glottis gap index	GGI/AU	0: full closure/[0, 1]	[59]
Closing quotient	ClQ/AU	1: completely open/[0, 1]	[60]
Amplitude periodicity	AP/AU	1: periodic/[0, 1]	[61]
Time periodicity	TP/AU	1: periodic/[0, 1]	[61]
Phase asymmetry index	PAI/AU	0: symmetric/[0, 1]	[61]
Amplitude symmetry index	ASI/AU	1: symmetric/[0, 1]	[62]
(b) Acoustic parameters			
Harmonics-to-noise ratio	HNR/dB	Higher is better *	[63]
Normalized noise energy	NNE/dB	Smaller is better *	[64]
Cepstral peak prominance	CPP/dB	Higher is better *	[65]
Shimmer	$shim/\%$	Smaller is better *	[66]
Jitter	$jitt/\%$	Smaller is better *	[66]

* The aphorism "Smaller/Higher is better" pertains to the realm of healthy modal phonation, and is to be construed in the context of efficiency, regularity, harmonic richness, and noise levels.

2.6. Measuring Protocol

In the course of our investigation, we fabricated six distinct synthetic larynx models, each representing a different fiber type as previously specified in Table 2. The procedure for data collection and measurement was equal for all models. Each model was placed on the artificial trachea, with the manipulators mounted to the motors. Subsequently, the fibers were secured to the linear stage via the fiber guidance and tensioning system. The translation stage, with a travel range of 0–25 mm, was initially set at 5 mm representing the initial tension level in just tight-free condition.

We followed a systematic protocol to collect data from the synthetic larynx models. The process, which was repeated for each model, consisted of the following steps:

1. Relax the translation stage to 0 mm to ensure the fibers are in a completely tension-free state with no force applied.
2. Adduct the vocal folds until complete glottal closure is achieved.
3. Take a reference measurement of the subglottal pressure with the flow completely switched off.
4. Manually increase the flow rate until the oscillation onset flow is identified, which is the point at which the synthetic vocal folds begin to oscillate stably.
5. Record the first measurement at the onset of oscillation.
6. Iteratively increase the flow rate by increments of 10 SLM until reaching the maximum flow rate of 200 SLM.
7. Switch off the flow.
8. Increase the fiber tension by elongating the fiber by 5 mm.
9. Establish complete glottis closure if not already closed.
10. Repeat the procedure starting from identifying the onset flow (step 4).

This procedure was repeated, with each measurement starting from the onset flow after adjusting the fiber tension and ensuring complete glottis closure. Thereby, the maximum elongation of the fiber was 25 mm, which is the maximum tension level. Using this procedure, each larynx model was tested at six tension levels, for each at least 2 flow rate levels.

3. Results and Discussion

3.1. General Phonation Parameters

Within the study, we performed a total of $N = 213$ measurements with six larynx models. The results are presented in Figure 4, with Figure 4a depicting the F_0 in Hz and Figure 4b illustrating the P_{sub} in Pa as a function of flow in SLM for all models and all tension levels of the fibers. Table 4 provides a comprehensive summary of the general phonation parameters for all models, as well as the parameters at onset.

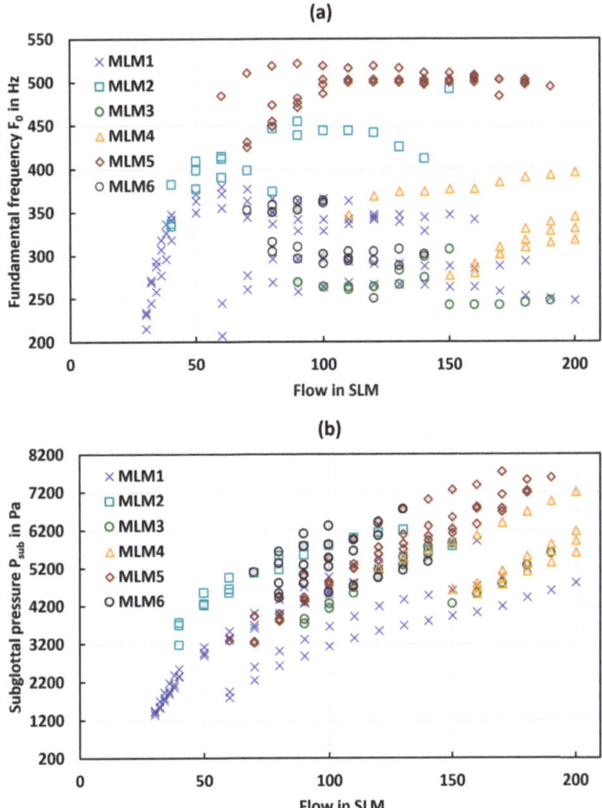

Figure 4. The figure presents the measured parameters (**a**) the F_0 in *Hz* and (**b**) the P_{sub} in *Pa* as a function of flow in *SLM*. for different tension levels of the fiber.

A noteworthy observation from the data is the wide range of F_0 values produced by models MLM1 and MLM2, as detailed in Table 4a. In contrast, MLM3–MLM6 show a much smaller F_0 range than MLM1 and MLM2 with the highest frequencies for MLM5.

The mechanical properties of the individual layers of the synthetic multi-layer vocal fold models, as shown in Table 1, are within physiological ranges of related tissues [43,67–73]. These properties contribute to the generation of F_0 values during vocal fold oscillation, which align with the physiological range of human phonation [42,74]. While F_0 for males typically lies between 100 and 220 Hz [75], females show a slightly higher F_0 range [76]. The synthetic models presented in this study also demonstrate higher F_0 values beyond normative human phonation, up to the ranges of professional female singers [77]. In comparison to other models, i.e., ex vivo porcine and ovine [45,47,48] studies as well as other synthetic models [16,18,20,36,38,40], our synthetic model vibrates at significantly higher F_0 values. Thus, this model enables us to reliably study the phonation process at these high frequencies in the range of professional female singers as mentioned above.

According to Figure 4b, all models exhibited an increase in P_{sub} with increasing flow. Furthermore, a trend was observed indicating that P_{sub} increases with the diameter of the fibers. While MLM1 displayed the largest range in P_{sub}, the other models demonstrated higher overall P_{sub} values.

Measuring the P_{sub} in vivo presents significant challenges, leading to a scarcity of publications that can be used to reference the physiological range of the mean P_{sub}. Despite this, Holmberg et al. [60] report a P_{sub} for males and females between 580–680 Pa and 600–700 Pa, respectively. In contrast, Sundberg et al. [75,78] report P_{sub} values between 800–2200 Pa, whereas Baken and Orlikoff [79] estimate the P_{sub} values to range from 353–1941 Pa. In our study, the measured P_{sub} values of the synthetic models were predominantly higher than these reported physiological ranges. MLM1, however, lied in these ranges, which will be discussed in more detail in Section 3.5. These higher values of the P_{sub} in synthetic models matched well with many previous studies [31,51,80]. When compared to ex vivo measurements [45–48,81], the P_{sub} in our models is also higher. However, the P_{sub} increase with increasing flow is similarly reproduced [56,82]. The reason for a higher P_{sub} in the presented models could be the flow resistance by the integrated fibers. While the onset flow rates shown in Table 4 observed in the presented model do not fall within the physiological range typical of normal phonation, it is consistent with other studies that applied synthetic as well as ex vivo larynx models [36,38,83].

Table 4. Range of phonation parameters grouped for the data at the onset flow rate only and the all data with higher flow rate. The parameter N corresponds to the number of measurements conducted for the respective model. The standard deviation is indicated in parentheses besides the mean value.

	MLM1 (N = 76)	MLM2 (N = 20)	MLM3 (N = 18)	MLM4 (N = 26)	MLM5 (N = 48)	MLM6 (N = 25)
(a) Phonation onset						
mean F_0 in Hz	226.63 (13.55)	371.44 (31.71)	260.19 (12.68)	298.77 (28.42)	453.54 (25.92)	328.82 (22.38)
mean P_{sub} in Pa	1593.66 (235.96)	4375.36 (906.12)	3944.00 (223.64)	4661.83 (100.76)	3444.03 (318.67)	4789.36 (325.45)
mean flow rate in SLM	42 (14.69)	66 (38.78)	110 (28.28)	145 (20.61)	70 (7.07)	82 (9.79)
(b) All recordings						
mean F_0 in Hz	306.42 (43.66)	411.28 (38.46)	264.82 (17.65)	337.18 (36.87)	496.32 (20.55)	318.36 (30.43)
max F_0 in Hz	379.52	492.57	306.84	395.67	522.18	363.37
min F_0 in Hz	207.25	333.76	242.24	277.23	425.28	250.32
mean P_{sub} in Pa	3595.90 (1281.65)	5026.83 (863.09)	4872.53 (642.95)	5549.97 (708.14)	5702.86 (1245.18)	5405.57 (614.18)
max P_{sub} in Pa	5918.14	6206.71	5851.08	7200.75	7737.04	6745.46
min P_{sub} in Pa	1350.85	3172.69	3733.03	4515.16	3221.27	4417.77

3.2. Glottal Dynamic Parameters

The glottal dynamic parameters are presented as boxplots in Figure 5, with a focus on the glottal gap (GGI, ClQ), periodicity (AP, TP), and symmetry (ASI, PAI) of the vocal folds oscillations.

Glottal gap parameters: Except for MLM3 and MLM6, all models exhibit a high variability in the GGI, with MLM3 and MLM4 showing notably high GGI values compared to the rest, see Figure 5a. An observable trend is the increase in the median GGI with increasing fiber diameter, with MLM3 displaying the highest median as an exception. The GGI describes the ratio between the minimum and maximum glottal area and can take values between 0 and 1 (see Table 3a). A GGI value close to 0 indicates large vibrational changes as well as complete glottal closure, whereas values approaching 1 describe less change in the glottal area [59,84].

The GGI ranges of all fabricated synthetic models measured in this study are notably higher compared to those reported in ex vivo [46–48] and in vivo [7,85–88] studies. However, the GGI for the MLM1 comprises the entire range starting with GGI = 0 representing regular phonation up to GGI = 0.78 representing a severe glottis closure insufficiency [31,89,90].

In terms of the ClQ, which describes the closing time of the glottis per cycle, all models, except for notable outliers for MLM1 and MLM5, fall within the expected range for physiological vocal fold vibration, as seen in Figure 5b. These ClQ values align well with those reported in ex vivo studies involving porcine [45,48], and ovines [47] vocal folds. Furthermore, in vivo studies suggest that the observed ClQ values are representative of healthy phonation [86].

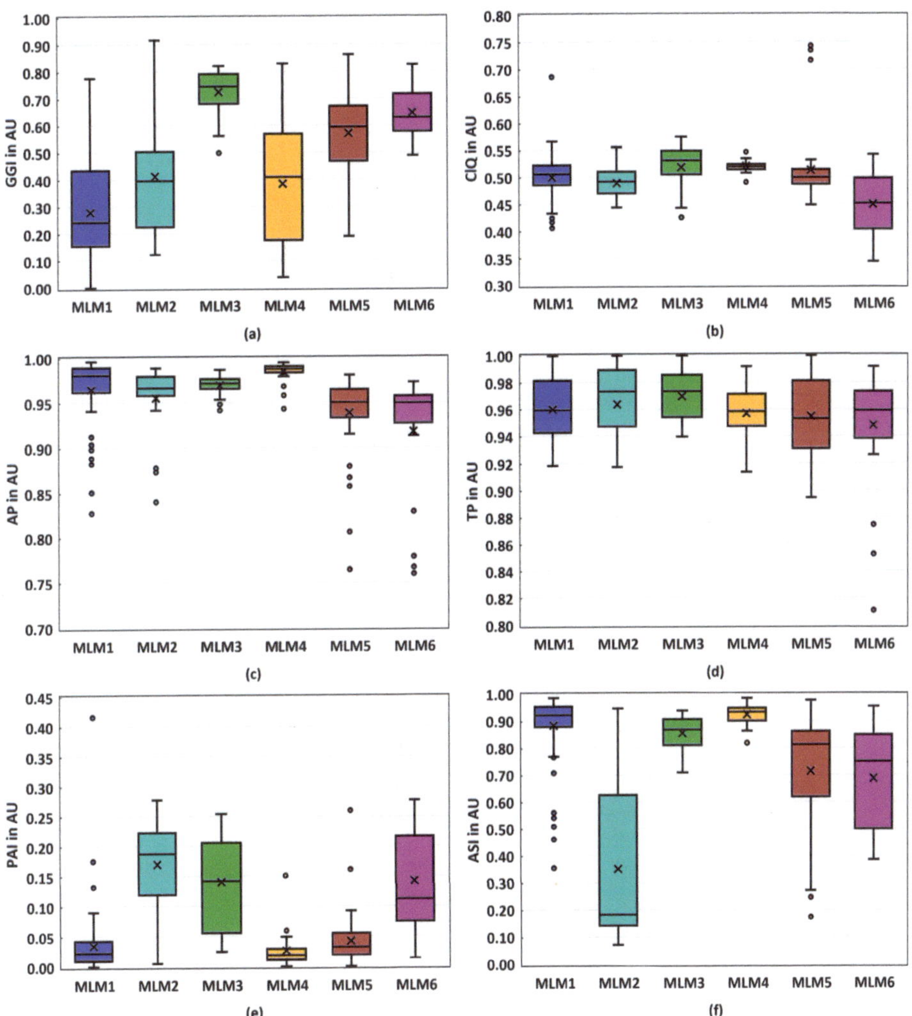

Figure 5. Illustrates boxplots for the glottal dynamics parameters of the models MLM1-MLM6: (**a**) Glottis gap index (GGI) (**b**) Closing quotient (ClQ) (**c**) Amplitude periodicity (AP) (**d**) Time periodicity (TP) (**e**) Phase asymmetry index (PAI) (**f**) Amplitude symmetry index (ASI).

Periodicity parameters:

The AP for all models show a median above 0.95, with relatively low variability across all models, see Figure 5c. However, some outliers can be observed. Similarly, for TP, all models also record a median above 0.95, albeit with higher variability than AP, see Figure 5d. Apart from MLM6, no outliers were measured. The synthetic vocal fold models indicate a high degree of periodicity in the amplitude of vocal fold vibration. This is consistent with the expectation for physiological vocal fold vibration [47,48].

Symmetry parameters:

MLM1, MLM4, and MLM5 exhibit a very low PAI with minimal variability, shown in Figure 5e. The remaining models display both a higher PAI and significantly greater variability. The ASI for MLM2 shows both the highest variability and the lowest median. While MLM5 and MLM6 are not as low as MLM1–MLM3, MLM4 shows the highest ASI with minimal variability. Apart from MLM1, the remaining models show little or even no outliers, see Figure 5f. The symmetry of vocal fold vibration is a critical aspect of voice production, and asymmetric characteristics of the vocal fold motion can be indicative of vocal pathologies [91,92]. In terms of the PAI, models MLM1, MLM4, and MLM5 exhibit very low values with minimal variability, suggesting a high degree of symmetry of vocal fold vibration. This is in line with the other reports of ex vivo [47,48,86] and in vivo studies [88]. The remaining models display both a higher PAI and significantly greater variability, indicating less symmetry in their vocal fold vibration. Conversely, the high ASI and minimal variability of MLM4 indicate a high degree of symmetry in the amplitude of vocal fold vibration. These findings underscore the importance of considering both the phase and the amplitude symmetry in the evaluation of synthetic vocal fold models.

3.3. Acoustic Parameters

In this section, we present the acoustic parameters. Interestingly, an observable trend is that the median of the HNR decreases with increasing fiber diameter, with the exception of the MLM4 model. It shows the highest dispersion in the MLM1 model, accompanied by a relatively high median value, see Figure 6a. However, the highest HNR is produced by the MLM4 model, which shows minimal dispersion.

Compared to the literature, the synthetic models exhibit HNR values that are similarly high to those reported in ex vivo studies [46–48].

The NNE presents a different pattern. With the exception of the MLM4 and MLM6, an increase in fiber diameter corresponds to an increase in the NNE value, which is shown in Figure 6b. Therein, the MLM1 and MLM4 models exhibited the lowest median values, with MLM1 showing a higher variability. In this context as well, the values fall within a physiological range when compared to ex vivo studies [47].

All models display relatively high CPP values, as shown in Figure 6c, with MLM2 producing the highest median. Figure 6d,e show the computed values for the Jitter and Shimmer parameter. Except for the MLM4 model, all other models demonstrate similar behavior of these two parameters. An observable trend is that the median values increase with the fiber diameter. The lowest values of both Jitter and Shimmer are again shown by the models MLM1 and MLM4, respectively. In comparison to previous studies, the values for CPP, jitt, and shim observed in our synthetic models closely align with those reported in ex vivo experiments [46–48].

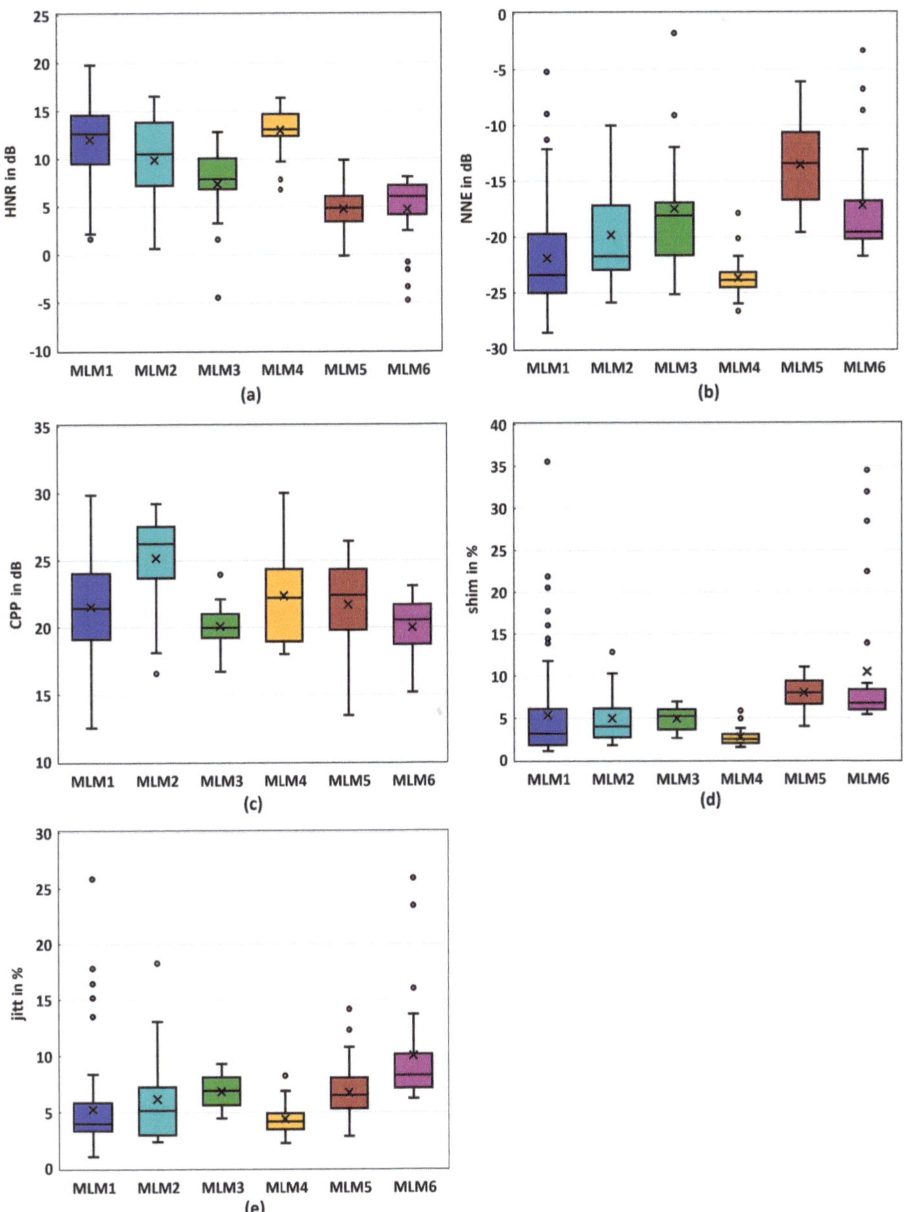

Figure 6. Boxplots for the acoustic parameters of the models MLM1-MLM6: (**a**) Harmonics to noise ratio (HNR) (**b**) Normalized noise energy (NNE) (**c**) Ceptral peak prominance (CCP) (**d**) Shimmer (shim) (**e**) Jitter (jitt).

3.4. Evaluation of Reproduction Characteristics of the Larynx Model

Based on the results presented above, the models were evaluated with regard to their ability of reproducing physiological parameters by using a scoring system. This will facilitate the identification of the most suitable model for physiological vocal folds oscillations to illustrate the impact of fiber tension on F_0 and P_{sub}. For each parameter, the six models are rated from the best (score 6) to the worst (score 1) with regard to physiological range. After each model has been scored for each parameter, the total score of each model was calculated by summing up the individual scores of the single parameters for the respective model.

The scoring system takes into account several factors: the number of measurements per model, with more measurements getting a higher score, as shown in Table 4; the range of the fundamental frequency with a higher score for a larger frequency range F_0 (see Table 3); the glottal flow resistance R_B [93] with a larger score for a higher resistance; the glottal dynamic parameters as seen in Table 3a, and the acoustic parameters shown in Table 3b both being closer to their optimum.

The resulting total scores are displayed in Table 5 for each model with MLM1 and MLM4 exhibiting the highest scores, which indicate that the calculated parameters largely fall within the range of physiological vocal fold vibrations. Given that the ranges for the fundamental frequency of both models largely overlap (see Figure 4 and Table 3), the subsequent analysis will focus solely on the MLM1 model with the thinnest fiber. All other results demonstrating the influence of fiber tension are provided in the Supplementary Material.

Table 5. Comparative evaluation of six larynx models based on physiological parameter reproduction scores (++: 6, +: 5, o+: 4, o-: 3, -: 2, - -: 1).

	MLM1	MLM2	MLM3	MLM4	MLM5	MLM6
N	++	-	- -	o+	+	o-
F_0 range	++	+	- -	o+	-	o-
R_B	o-	++	-	- -	o+	+
GGI	++	o+	- -	+	o-	-
ClQ	++	+	o-	-	o+	- -
AP	o+	o-	+	++	-	- -
TP	o+	+	++	o-	-	- -
PAI	+	- -	o-	++	o+	-
ASI	+	- -	o+	++	o-	-
HNR	+	o+	o-	++	-	- -
NNE	+	o+	o-	++	- -	-
CPP	o+	++	-	+	o-	- -
shim	+	o+	o-	++	- -	-
jitt	++	o+	-	+	o-	- -
Score	70	54	39	65	39	27

3.5. Influence of Fiber Tension

As described in Section 2.3, the fibers in the models were elongated from 0 mm–20 mm in 5 mm increments. Figure 7 shows the effect of the fiber tension on F_0 and P_{sub} for MLM1. As expected, F_0 increases with increasing flow for each level of tension. Furthermore, it can be observed that for a flow rate of less than 40 SLM, the frequency increases linearly until it reaches a saturation point at a flow greater than 40 SLM. This effect of the linear increase in F_0 is particularly noticeable for the tension levels of 10 mm, 15 mm, and 20 mm.

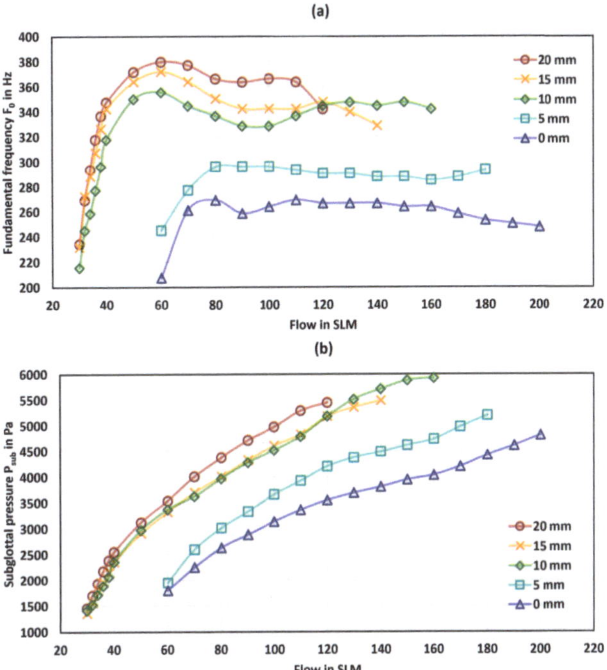

Figure 7. The figure shows the influence of different fiber tension levels for the MLM1 model. Illustrates (**a**) the F_0 in Hz and (**b**) the P_{sub} in Pa as a function of flow in SLM.

The P_{sub} also exhibits the effect of nearly linear increasing with the flow rate. Similarly, F_0 and P_{sub} also increase with increasing fiber tension at a constant flow rate. The increase in F_0 and P_{sub} is highest between the stages of 5 mm and 10 mm. Xuan and Zhang [38] reported an increase in the F_0 when a fiber was integrated into the body of a synthetic vocal fold model. This model was based on a one-layer isotropic vocal fold model, which served as the baseline for their study [26,32,34]. Interestingly, when the fiber was integrated into the cover of the model, there was no significant change in the F_0. However, the P_{sub} exhibited a more intriguing behavior. While the integration of the fiber into the body led to a decrease in P_{sub}, it increased when the fiber was embedded in the cover. It should be noted that in their study, the fibers were not subjected to any tension. In another study, Murray and Thomson [36] investigated the influence of fiber tension. They employed a multi-layer model and integrated the fibers into the ligament. By applying tension to the fibers using weights at each end, they achieved an increase in F_0 of approximately 10%.

The integration of fibers into the present synthetic model produced similar effects as reported in other studies about incorporated fibers in synthetic vocal fold models [36,38]. In contrast to these studies, the fibers can be elongated variably, allowing for the control of frequency at specific flow stages. As previously mentioned, the impact of fiber elongation is physiologically relevant, particularly in the range where the F_0 increases linearly before reaching a saturation point, which was also reported elsewhere [94].

4. Conclusions

This study presents the design, construction, and evaluation of a synthetic larynx model, incorporating various functional aspects of human laryngeal dynamics. The model is fabricated using two types of platinum-catalyzed silicone rubber, with integrated manipulators and fibers, to mimic the biomechanical properties of the human larynx and vocal folds. These models are systematically tested with varying fiber types and their elongation, providing a comprehensive understanding of the impact of these variables on vocal fold vibration.

Our analysis demonstrates that these synthetic larynx models are capable of replicating the key features of physiological vocal fold oscillation, as indicated by a wide range of glottal dynamic and acoustic parameters. Despite some disparities, the models exhibit behavior largely in line with the expected physiological up to pathological ranges. Notably, the fundamental frequency F_0 and subglottal pressure P_{sub} values generated by the synthetic models fall within the physiological ranges reported in in vivo and ex vivo studies, indicating their potential for reproducing the characteristics of human phonation.

It is found that the larynx model presents a wide range of F_0 values depending on the different fiber types, indicating the crucial role of fiber characteristics within the ligament to control the vibratory responses of the vocal folds. Furthermore, our results show that an increase in fiber diameter corresponds to an increase in P_{sub}, indicating an increase in flow resistance due to the greater stiffness of thicker fibers. Further studies will be dedicated to elucidating the correlations between the fiber characteristics and the observed outcomes.

The models' ability to replicate the periodicity, symmetry, and glottal gap parameters of vocal fold vibration further corroborates their physiological similarity. However, it is also observed that certain configurations of the synthetic models may not fully replicate typical dynamics.

We also evaluate the influence of fiber tension on the F_0 and P_{sub}, revealing that F_0 increases linearly with increasing flow until reaching a saturation point. This observation shows the high relevance of including fibers into the ligament in synthetic larynx models to accurately reproduce human vocal fold dynamics.

The development and evaluation of these synthetic larynx models offer a significant contribution to the field of voice research. Furthermore, this is the first synthetic larynx model that provides dynamic phonation characteristics of professional female singers. Additionally, the model is able to analyze the regular as well as irregular phonation by varying the pre-phonatory settings of the vocal folds in an asymmetric manner.

Supplementary Materials: The following supporting information can be downloaded at: https://www.mdpi.com/article/10.3390/bioengineering10101130/s1, Figures S1–S5: The figures show the influence of different fiber tension levels for the models MLM2-MLM6. Illustrates (a) the F_0 in Hz and (b) the P_{sub} in Pa as a function of flow in SLM.

Author Contributions: Conceptualization, B.T. and S.K.; methodology, B.T., O.W. and S.K.; validation, B.T., S.S. and S.K.; formal analysis, B.T., L.G. and S.S.; investigation, B.T. and S.K.; resources, S.K., O.W. and D.D.; data curation, B.T.; writing—original draft preparation, B.T.; writing—review and editing, S.K., S.S., O.W. and D.D.; visualization, B.T.; supervision, S.K. and D.D.; project administration, S.K.; funding acquisition, S.K. and D.D. All authors have read and agreed to the published version of the manuscript.

Funding: This research received no external funding.

Data Availability Statement: The data presented in this study are not publicly available due to ongoing research in this field.

Acknowledgments: We acknowledge financial support by Deutsche Forschungsgemeinschaft and Friedrich-Alexander-Universität Erlangen-Nürnberg within the funding programme "Open Access Publication Funding".

Conflicts of Interest: The authors declare no conflict of interest.

References

1. Van Den Berg, J.; Tan, T.S. Results of experiments with human larynxes. *Pract. Oto-Rhino-Laryngol.* **1959**, *21*, 425–450. [CrossRef] [PubMed]
2. Choi, J.; Son, Y.I.; So, Y.K.; Byun, H.; Lee, E.K.; Yun, Y.S. Posterior glottic gap and age as factors predicting voice outcome of injection laryngoplasty in patients with unilateral vocal fold paralysis. *J. Laryngol. Otol.* **2012**, *126*, 260–266. [CrossRef] [PubMed]

3. Morrison, M.D.; Rammage, L.A.; Belisle, G.M.; Pullan, C.B.; Nichol, H. Muscular tension dysphonia. *J. Otolaryngol.* **1983**, *12*, 302–306. [PubMed]
4. Nguyen, D.D.; Kenny, D.T.; Tran, N.D.; Livesey, J.R. Muscle tension dysphonia in Vietnamese female teachers. *J. Voice Off. J. Voice Found.* **2009**, *23*, 195–208. [CrossRef] [PubMed]
5. Bhatt, J.; Verma, S. Management of glottal insufficiency. *Otorinolaringologia* **2014**, *64*, 101–107.
6. Stevens, K.N. *Acoustic Phonetics*, 1. Paperback ed.; Number 30 in Current Studies in Linguistics Series; MIT Press: Cambridge, MA, USA, 2000.
7. Patel, R.R.; Dixon, A.; Richmond, A.; Donohue, K.D. Pediatric high speed digital imaging of vocal fold vibration: A normative pilot study of glottal closure and phase closure characteristics. *Int. J. Pediatr. Otorhinolaryngol.* **2012**, *76*, 954–959. [CrossRef]
8. Fritzen, B.; Hammarberg, B.; Gauffin, J.; Karlsson, I.; Sundberg, J. Breathiness and insufficient vocal fold closure. *J. Phon.* **1986**, *14*, 549–553. [CrossRef]
9. Cielo, C.A.; Schwarz, K.; Finger, L.S.; Lima, J.M.; Christmann, M.K. Glottal Closure in Women with No Voice Complaints or Laryngeal Disorders. *Int. Arch. Otorhinolaryngol.* **2019**, *23*, e384–e388. [CrossRef]
10. Schneider, B.; Bigenzahn, W. Influence of glottal closure configuration on vocal efficacy in young normal-speaking women. *J. Voice* **2003**, *17*, 468–480. [CrossRef]
11. Berry, D.A.; Montequin, D.W.; Tayama, N. High-speed digital imaging of the medial surface of the vocal folds. *J. Acoust. Soc. Am.* **2001**, *110*, 2539–2547. [CrossRef]
12. Döllinger, M.; Tayama, N.; Berry, D.A. Empirical Eigenfunctions and medial surface dynamics of a human vocal fold. *Methods Inf. Med.* **2005**, *44*, 384–391. [PubMed]
13. Boessenecker, A.; Berry, D.A.; Lohscheller, J.; Eysholdt, U.; Doellinger, M. Mucosal wave properties of a human vocal fold. *Acta Acust. United Acust.* **2007**, *93*, 815–823.
14. Kniesburges, S.; Thomson, S.L.; Barney, A.; Triep, M.; Sidlof, P.; Horáček, J.; Brücker, C.; Becker, S. In vitro experimental investigation of voice production. *Curr. Bioinform.* **2011**, *6*, 305–322. [CrossRef]
15. Spencer, M.; Siegmund, T.; Mongeau, L. Determination of superior surface strains and stresses, and vocal fold contact pressure in a synthetic larynx model using digital image correlation. *J. Acoust. Soc. Am.* **2008**, *123*, 1089–1103. [CrossRef]
16. Motie-Shirazi, M.; Zañartu, M.; Peterson, S.D.; Mehta, D.D.; Hillman, R.E.; Erath, B.D. Collision Pressure and Dissipated Power Dose in a Self-Oscillating Silicone Vocal Fold Model With a Posterior Glottal Opening. *J. Speech Lang. Hear. Res.* **2022**, *65*, 2829–2845. [CrossRef] [PubMed]
17. Chen, L.J.; Mongeau, L. Verification of two minimally invasive methods for the estimation of the contact pressure in human vocal folds during phonation. *J. Acoust. Soc. Am.* **2011**, *130*, 1618–1627. [CrossRef]
18. Horáček, J.; Bula, V.; Radolf, V.; Šidlof, P. Impact stress in a self-oscillating model of human vocal folds. *Adv. Vib. Eng.* **2016**, *13*.
19. Motie-Shirazi, M.; Zañartu, M.; Peterson, S.D.; Mehta, D.D.; Kobler, J.B.; Hillman, R.E.; Erath, B.D. Toward Development of a Vocal Fold Contact Pressure Probe: Sensor Characterization and Validation Using Synthetic Vocal Fold Models. *Appl. Sci.* **2019**, *9*, 3002. [CrossRef]
20. Motie-Shirazi, M.; Zañartu, M.; Peterson, S.D.; Erath, B.D. Vocal fold dynamics in a synthetic self-oscillating model: Contact pressure and dissipated-energy dose. *J. Acoust. Soc. Am.* **2021**, *150*, 478–489. [CrossRef]
21. Weiss, S.; Sutor, A.; Rupitsch, S.J.; Kniesburges, S.; Doellinger, M.; Lerch, R. Development of a small film sensor for the estimation of the contact pressure of artificial vocal folds. *J. Acoust. Soc. Am.* **2013**, *133*, 3618. [CrossRef]
22. Neubauer, J.; Zhang, Z.; Miraghaie, R.; Berry, D.A. Coherent structures of the near field flow in a self-oscillating physical model of the vocal folds. *J. Acoust. Soc. Am.* **2007**, *121*, 1102–1118. [CrossRef] [PubMed]
23. Popolo, P.S.; Titze, I.R. Qualification of a quantitative laryngeal imaging system using videostroboscopy and videokymography. *Ann. Otol. Rhinol. Laryngol.* **2008**, *117*, 404–412. [CrossRef] [PubMed]
24. Semmler, M.; Kniesburges, S.; Birk, V.; Ziethe, A.; Patel, R.; Dollinger, M. 3D Reconstruction of Human Laryngeal Dynamics Based on Endoscopic High-Speed Recordings. *IEEE Trans. Med. Imaging* **2016**, *35*, 1615–1624. [CrossRef] [PubMed]
25. Semmler, M.; Kniesburges, S.; Parchent, J.; Jakubaß, B.; Zimmermann, M.; Bohr, C.; Schützenberger, A.; Döllinger, M. Endoscopic Laser-Based 3D Imaging for Functional Voice Diagnostics. *Appl. Sci.* **2017**, *7*, 600. [CrossRef]
26. Thomson, S.L.; Mongeau, L.; Frankel, S.H. Aerodynamic transfer of energy to the vocal folds. *J. Acoust. Soc. Am.* **2005**, *118*, 1689–1700. [CrossRef] [PubMed]
27. Motie-Shirazi, M.; Zañartu, M.; Peterson, S.D.; Erath, B.D. Vocal fold dynamics in a synthetic self-oscillating model: Intraglottal aerodynamic pressure and energy. *J. Acoust. Soc. Am.* **2021**, *150*, 1332–1345. [CrossRef]
28. Migimatsu, K.; Tokuda, I.T. Experimental study on nonlinear source-filter interaction using synthetic vocal fold models. *J. Acoust. Soc. Am.* **2019**, *146*, 983–997. [CrossRef]
29. Zhang, Z.; Hieu Luu, T. Asymmetric vibration in a two-layer vocal fold model with left-right stiffness asymmetry: Experiment and simulation. *J. Acoust. Soc. Am.* **2012**, *132*, 1626–1635. [CrossRef]
30. Zhang, Z.; Kreiman, J.; Gerratt, B.R.; Garellek, M. Acoustic and perceptual effects of changes in body layer stiffness in symmetric and asymmetric vocal fold models. *J. Acoust. Soc. Am.* **2013**, *133*, 453–462. [CrossRef]
31. Kniesburges, S.; Lodermeyer, A.; Semmler, M.; Schulz, Y.K.; Schützenberger, A.; Becker, S. Analysis of the tonal sound generation during phonation with and without glottis closure. *J. Acoust. Soc. Am.* **2020**, *147*, 3285–3293. [CrossRef]

32. Zhang, Z.; Neubauer, J.; Berry, D.A. The influence of subglottal acoustics on laboratory models of phonation. *J. Acoust. Soc. Am.* **2006**, *120*, 1558–1569. [CrossRef] [PubMed]

33. Zhang, Z.; Neubauer, J.; Berry, D.A. Influence of vocal fold stiffness and acoustic loading on flow-induced vibration of a single-layer vocal fold model. *J. Sound Vib.* **2009**, *322*, 299–313. [CrossRef]

34. Zhang, Z. Restraining mechanisms in regulating glottal closure during phonation. *J. Acoust. Soc. Am.* **2011**, *130*, 4010–4019. [CrossRef] [PubMed]

35. Hirano, M. Cover-body theory of vocal cord vibration. *Speech Sci.* **1985**, 1–46.

36. Murray, P.R.; Thomson, S.L. Vibratory responses of synthetic, self-oscillating vocal fold models. *J. Acoust. Soc. Am.* **2012**, *132*, 3428–3438. [CrossRef]

37. Shaw, S.M.; Thomson, S.L.; Dromey, C.; Smith, S. Frequency response of synthetic vocal fold models with linear and nonlinear material properties. *J. Speech Lang. Hear. Res.* **2012**, *55*, 1395–1406. [CrossRef]

38. Xuan, Y.; Zhang, Z. Influence of Embedded Fibers and an Epithelium Layer on the Glottal Closure Pattern in a Physical Vocal Fold Model. *J. Speech Lang. Hear. Res.* **2014**, *57*, 416–425. [CrossRef]

39. Scherer, R.C.; Shinwari, D.; De Witt, K.J.; Zhang, C.; Kucinschi, B.R.; Afjeh, A.A. Intraglottal pressure profiles for a symmetric and oblique glottis with a divergence angle of 10 degrees. *J. Acoust. Soc. Am.* **2001**, *109*, 1616–1630. [CrossRef]

40. Murray, P.R.; Thomson, S.L.; Smith, M.E. A Synthetic, Self-Oscillating Vocal Fold Model Platform for Studying Augmentation Injection. *J. Voice* **2014**, *28*, 133–143. [CrossRef]

41. Kniesburges, S.; Birk, V.; Lodermeyer, A.; Schützenberger, A.; Bohr, C.; Becker, S. Effect of the ventricular folds in a synthetic larynx model. *J. Biomech.* **2007**, *55*, 128–133. [CrossRef]

42. Titze, I.R. *Principles of Voice Production*; Prentice Hall: Upper Saddle River, NJ, USA, 1994.

43. Min, Y.B.; Titze, I.R.; Alipour-Haghighi, F. Stress-Strain Response of the Human Vocal Ligament. *Ann. Otol. Rhinol. Laryngol.* **1995**, *104*, 563–569. [CrossRef] [PubMed]

44. Titze, I.; Alipour, F. *The Myoelastic-Aerodynamic Theory of Phonation*; National Center for Voice and Speech: Salt Lake City, UT, USA, 2006.

45. Birk, V.; Döllinger, M.; Sutor, A.; Berry, D.A.; Gedeon, D.; Traxdorf, M.; Wendler, O.; Bohr, C.; Kniesburges, S. Automated setup for ex vivo larynx experiments. *J. Acoust. Soc. Am.* **2017**, *141*, 1349–1359. [CrossRef] [PubMed]

46. Semmler, M.; Berry, D.A.; Schützenberger, A.; Döllinger, M. Fluid–structure-acoustic interactions in an ex vivo porcine phonation model. *J. Acoust. Soc. Am.* **2021**, *149*, 1657–1673. [CrossRef] [PubMed]

47. Jakubaß, B.; Peters, G.; Kniesburges, S.; Semmler, M.; Kirsch, A.; Gerstenberger, C.; Gugatschka, M.; Döllinger, M. Effect of functional electric stimulation on phonation in an ex vivo aged ovine model. *J. Acoust. Soc. Am.* **2023**, *153*, 2803. [CrossRef] [PubMed]

48. Peters, G.; Jakubaß, B.; Weidenfeller, K.; Kniesburges, S.; Böhringer, D.; Wendler, O.; Mueller, S.K.; Gostian, A.O.; Berry, D.A.; Döllinger, M.; et al. Synthetic mucus for an ex vivo phonation setup: Creation, application, and effect on excised porcine larynges. *J. Acoust. Soc. Am.* **2022**, *152*, 3245–3259. [CrossRef] [PubMed]

49. Hayes, M.H. *Statistical Digital Signal Processing and Modeling*; Wiley: Hoboken, NJ, USA, 1996.

50. Stoica, P.G.; Moses, R.; Stoica, P.; Moses, R.L. *Spectral Analysis of Signals*; Pearson: Prentice Hall: Upper Saddle River, NJ, USA, 2005.

51. Kniesburges, S.; Lodermeyer, A.; Becker, S.; Traxdorf, M.; Döllinger, M. The mechanisms of subharmonic tone generation in a synthetic larynx model. *J. Acoust. Soc. Am.* **2016**, *139*, 3182–3192. [CrossRef] [PubMed]

52. Kist, A.M.; Gómez, P.; Dubrovskiy, D.; Schlegel, P.; Kunduk, M.; Echternach, M.; Patel, R.; Semmler, M.; Bohr, C.; Dürr, S.; et al. A Deep Learning Enhanced Novel Software Tool for Laryngeal Dynamics Analysis. *J. Speech Lang. Hear. Res.* **2021**, *64*, 1889–1903. [CrossRef]

53. Maryn, Y.; Verguts, M.; Demarsin, H.; van Dinther, J.; Gomez, P.; Schlegel, P.; Döllinger, M. Intersegmenter Variability in High-Speed Laryngoscopy-Based Glottal Area Waveform Measures. *Laryngoscope* **2020**, *130*, E654–E661. [CrossRef]

54. Schlegel, P.; Semmler, M.; Kunduk, M.; Döllinger, M.; Bohr, C.; Schützenberger, A. Influence of Analyzed Sequence Length on Parameters in Laryngeal High-Speed Videoendoscopy. *Appl. Sci.* **2018**, *8*, 2666. [CrossRef]

55. Karnell, M.P.; Hall, K.D.; Landahl, K.L. Comparison of fundamental frequency and perturbation measurements among three analysis systems. *J. Voice* **1995**, *9*, 383–393. [CrossRef]

56. Alipour, F.; Scherer, R.C.; Finnegan, E. Pressure-flow relationships during phonation as a function of adduction. *J. Voice Off. J. Voice Found.* **1997**, *11*, 187–194. [CrossRef]

57. Mehta, D.D.; Zañartu, M.; Quatieri, T.F.; Deliyski, D.D.; Hillman, R.E. Investigating acoustic correlates of human vocal fold vibratory phase asymmetry through modeling and laryngeal high-speed videoendoscopy. *J. Acoust. Soc. Am.* **2011**, *130*, 3999–4009. [CrossRef] [PubMed]

58. Titze, I.R.; National Center for Voice and Speech; Workshop on Acoustic Voice Analysis. *Workshop on Acoustic Voice Analysis: Summary Statement*; National Center for Voice and Speech: Salt Lake City, UT, USA, 1995.

59. Patel, R.; Dubrovskiy, D.; Döllinger, M. Characterizing vibratory kinematics in children and adults with high-speed digital imaging. *J. Speech Lang. Hear. Res.* **2014**, *57*, S674–S686. [CrossRef] [PubMed]

60. Holmberg, E.B.; Hillman, R.E.; Perkell, J.S. Glottal airflow and transglottal air pressure measurements for male and female speakers in soft, normal, and loud voice. *J. Acoust. Soc. Am.* **1988**, *84*, 511–529. [CrossRef] [PubMed]

61. Qiu, Q.; Schutte, H.; Gu, L.; Yu, Q. An Automatic Method to Quantify the Vibration Properties of Human Vocal Folds via Videokymography. *Folia Phoniatr. Logop.* **2003**, *55*, 128–136. [CrossRef]

62. Wang, S.G.; Park, H.J.; Lee, B.J.; Lee, S.M.; Ko, B.; Lee, S.M.; Park, Y.M. A new videokymography system for evaluation of the vibration pattern of entire vocal folds. *Auris Nasus Larynx* **2016**, *43*, 315–321. [CrossRef]

63. Yumoto, E.; Gould, W.J.; Baer, T. Harmonics-to-noise ratio as an index of the degree of hoarseness. *J. Acoust. Soc. Am.* **1982**, *71*, 1544–1549. [CrossRef]

64. Kasuya, H.; Endo, Y.; Saliu, S. Novel acoustic measurements of jitter and shimmer characteristics from pathological voice. In Proceedings of the 3rd European Conference on Speech Communication and Technology (Eurospeech 1993), Berlin, Germany, 21–23 September 1993; pp. 1973–1976. [CrossRef]

65. Hillenbrand, J.; Houde, R.A. Acoustic correlates of breathy vocal quality: Dysphonic voices and continuous speech. *J. Speech Lang. Hear. Res.* **1996**, *39*, 311–321. [CrossRef] [PubMed]

66. Bielamowicz, S.; Kreiman, J.; Gerratt, B.R.; Dauer, M.S.; Berke, G.S. Comparison of voice analysis systems for perturbation measurement. *J. Speech Lang. Hear. Res.* **1996**, *39*, 126–134. [CrossRef]

67. Chan, R.W.; Rodriguez, M.L. A simple-shear rheometer for linear viscoelastic characterization of vocal fold tissues at phonatory frequencies. *J. Acoust. Soc. Am.* **2008**, *124*, 1207–1219. [CrossRef]

68. Alipour, F.; Vigmostad, S. Measurement of Vocal Folds Elastic Properties for Continuum Modeling. *J. Voice* **2012**, *26*, 816.e21–816.e29. [CrossRef] [PubMed]

69. Chhetri, D.K.; Zhang, Z.; Neubauer, J. Measurement of Young's Modulus of Vocal Folds by Indentation. *J. Voice* **2011**, *25*, 1–7. [CrossRef] [PubMed]

70. Oren, L.; Dembinski, D.; Gutmark, E.; Khosla, S. Characterization of the Vocal Fold Vertical Stiffness in a Canine Model. *J. Voice* **2014**, *28*, 297–304. [CrossRef] [PubMed]

71. Dion, G.R.; Coelho, P.G.; Teng, S.; Janal, M.N.; Amin, M.R.; Branski, R.C. Dynamic nanomechanical analysis of the vocal fold structure in excised larynges: Nanomechanical Analysis of the VF Structure. *Laryngoscope* **2017**, *127*, E225–E230. [CrossRef] [PubMed]

72. Chan, R.W.; Titze, I.R. Viscoelastic shear properties of human vocal fold mucosa: Measurement methodology and empirical results. *J. Acoust. Soc. Am.* **1999**, *106*, 2008–2021. [CrossRef] [PubMed]

73. Chan, R.W.; Fu, M.; Young, L.; Tirunagari, N. Relative Contributions of Collagen and Elastin to Elasticity of the Vocal Fold Under Tension. *Ann. Biomed. Eng.* **2007**, *35*, 1471–1483. [CrossRef]

74. Zemlin, W.R. *Speech and Hearing Science: Anatomy and Physiology*, 3rd ed.; Prentice-Hall: Englewood Cliffs, NJ, USA, 1988.

75. Sundberg, J.; Fahlstedt, E.; Morell, A. Effects on the glottal voice source of vocal loudness variation in untrained female and male voices. *J. Acoust. Soc. Am.* **2005**, *117*, 879–885. [CrossRef]

76. Jones, S.M.; Garrett, C. *Laryngeal Endoscopy and Voice Therapy*; OCLC: 956917938; Compton: Oxford, UK, 2016.

77. Echternach, M.; Sundberg, J.; Baumann, T.; Markl, M.; Richter, B. Vocal tract area functions and formant frequencies in opera tenors' modal and falsetto registers. *J. Acoust. Soc. Am.* **2011**, *129*, 3955–3963. [CrossRef]

78. Sundberg, J.; Titze, I.; Scherer, R. Phonatory control in male singing: A study of the effects of subglottal pressure, fundamental frequency, and mode of phonation on the voice source. *J. Voice* **1993**, *7*, 15–29. [CrossRef]

79. Baken, R.J.; Orlikoff, R.F. *Clinical Measurement of Speech and Voice*, 2nd ed.; Delmar Cengage Learning: Clifton Park, NJ, USA, 2010.

80. Fulcher, L.; Lodermeyer, A.; Kähler, G.; Becker, S.; Kniesburges, S. Geometry of the Vocal Tract and Properties of Phonation near Threshold: Calculations and Measurements. *Appl. Sci.* **2019**, *9*, 2755. [CrossRef]

81. Döllinger, M.; Berry, D.A.; Kniesburges, S. Dynamic vocal fold parameters with changing adduction in ex vivo hemilarynx experiments. *J. Acoust. Soc. Am.* **2016**, *139*, 2372–2385. [CrossRef]

82. Alipour, F.; Scherer, R.C. Effects of oscillation of a mechanical hemilarynx model on mean transglottal pressures and flows. *J. Acoust. Soc. Am.* **2001**, *110*, 1562–1569. [CrossRef]

83. Falk, S.; Kniesburges, S.; Schoder, S.; Jakubaß, B.; Maurerlehner, P.; Echternach, M.; Kaltenbacher, M.; Döllinger, M. 3D-FV-FE Aeroacoustic Larynx Model for Investigation of Functional Based Voice Disorders. *Front. Physiol.* **2021**, *12*, 616985. [CrossRef] [PubMed]

84. Kunduk, M.; Doellinger, M.; McWhorter, A.J.; Lohscheller, J. Assessment of the variability of vocal fold dynamics within and between recordings with high-speed imaging and by phonovibrogram. *Laryngoscope* **2010**, *120*, 981–987. [CrossRef] [PubMed]

85. Södersten, M.; Hertegård, S.; Hammarberg, B. Glottal closure, transglottal airflow, and voice quality in healthy middle-aged women. *J. Voice* **1995**, *9*, 182–197. [CrossRef] [PubMed]

86. Semmler, M.; Kniesburges, S.; Pelka, F.; Ensthaler, M.; Wendler, O.; Schützenberger, A. Influence of Reduced Saliva Production on Phonation in Patients With Ectodermal Dysplasia. *J. Voice* **2021**, S089219972100206X. [CrossRef]

87. Schlegel, P.; Kunduk, M.; Stingl, M.; Semmler, M.; Döllinger, M.; Bohr, C.; Schützenberger, A. Influence of spatial camera resolution in high-speed videoendoscopy on laryngeal parameters. *PLoS ONE* **2019**, *14*, e0215168.[CrossRef]

88. Schützenberger, A.; Kunduk, M.; Döllinger, M.; Alexiou, C.; Dubrovskiy, D.; Semmler, M.; Seger, A.; Bohr, C. Laryngeal High-Speed Videoendoscopy: Sensitivity of Objective Parameters towards Recording Frame Rate. *BioMed Res. Int.* **2016**, *2016*, 4575437. [CrossRef]

89. Zhang, Z. Compensation Strategies in Voice Production With Glottal Insufficiency. *J. Voice* **2019**, *33*, 96–102. [CrossRef]

90. Graham, E.; Angadi, V.; Sloggy, J.; Stemple, J. Contribution of Glottic Insufficiency to Perceived Breathiness in Classically Trained Singers. *Med. Probl. Perform. Artist.* **2016**, *31*, 179–184. [CrossRef]
91. Schlegel, P.; Kniesburges, S.; Dürr, S.; Schützenberger, A.; Döllinger, M. Machine learning based identification of relevant parameters for functional voice disorders derived from endoscopic high-speed recordings. *Sci. Rep.* **2020**, *10*, 10517. [CrossRef] [PubMed]
92. Bonilha, H.S.; Deliyski, D.D.; Whiteside, J.P.; Gerlach, T.T. Vocal Fold Phase Asymmetries in Patients With Voice Disorders: A Study Across Visualization Techniques. *Am. J.-Speech-Lang. Pathol.* **2012**, *21*, 3–15. [CrossRef] [PubMed]
93. Van Den Berg, J.; Zantema, J.T.; Doornenbal, P. On the Air Resistance and the Bernoulli Effect of the Human Larynx. *J. Acoust. Soc. Am.* **1957**, *29*, 626–631. [CrossRef]
94. Colton, R.H. Physiological mechanisms of vocal frequency control: The role of tension. *J. Voice* **1988**, *2*, 208–220. [CrossRef]

bioengineering

MDPI

Article

The Effects of Negative Pressure Induced by Flow Separation Vortices on Vocal Fold Dynamics during Voice Production

Weili Jiang [1,2], Xudong Zheng [1,2], Charles Farbos de Luzan [3], Liran Oren [3], Ephraim Gutmark [4] and Qian Xue [1,2,*]

[1] Mechanical Engineering Department, Rochester Institute of Technology, Rochester, NY 14623, USA; wxjeme@rit.edu (W.J.); xxzeme@rit.edu (X.Z.)
[2] Mechanical Engineering Department, University of Maine, Orono, ME 04469, USA
[3] Department of Otolaryngology Head and Neck Surgery, University of Cincinnati, Cincinnati, OH 45267, USA; farboscs@ucmail.uc.edu (C.F.d.L.); orenl@ucmail.uc.edu (L.O.)
[4] Department of Aerospace Engineering and Engineering Mechanics, University of Cincinnati, Cincinnati, OH 45267, USA; gutmarej@ucmail.uc.edu
* Correspondence: qxxeme@rit.edu

Abstract: This study used a two-dimensional flow-structure-interaction computer model to investigate the effects of flow-separation-vortex-induced negative pressure on vocal fold vibration and flow dynamics during vocal fold vibration. The study found that negative pressure induced by flow separation vortices enhances vocal fold vibration by increasing aeroelastic energy transfer during vibration. The result showed that the intraglottal pressure was predominantly negative after flow separation before gradually recovering to zero at the glottis exit. When the negative pressure was removed, the vibration amplitude and flow rate were reduced by up to 20%, and the closing speed, flow skewness quotient, and maximum flow declination rate were reduced by up to 40%. The study provides insights into the complex interactions between flow dynamics, vocal fold vibration, and energy transfer during voice production.

Keywords: flow separation vortices; vocal fold; intraglottal negative pressure

Citation: Jiang, W.; Zheng, X.; Farbos de Luzan, C.; Oren, L.; Gutmark, E.; Xue, Q. The Effects of Negative Pressure Induced by Flow Separation Vortices on Vocal Fold Dynamics during Voice Production. *Bioengineering* **2023**, *10*, 1215. https://doi.org/10.3390/bioengineering10101215

Academic Editor: Wen Wang

Received: 15 August 2023
Revised: 22 September 2023
Accepted: 17 October 2023
Published: 18 October 2023

1. Introduction

The production of the human voice is a complex process characterized by the finely tuned vibration of the true vocal fold pair within the larynx. This mechanism transforms a continuous stream of respiratory air into a pulsating airflow, forming the primary sound source of the voice. The larynx also houses a pair of false vocal folds, which are situated just above the true vocal folds. While the false vocal folds generally have a minimal role in voice production, they have the potential to positively contribute to sound intensity [1] and be engaged in certain types of voice production [2]. One of the important goals in voice production research is to understand the fundamental mechanisms that govern the intricate interactions among glottal aerodynamics, tissue biomechanics, vibratory dynamics, and acoustics. The improved understanding can provide scientific insights into the management of voice health.

Considerable research has focused on the development of intraglottal pressure during the closing phase of vocal fold vibrations. This process holds significance because, during this period, the glottis forms a divergent shape that can cause flow separation and complex pressure forces on the vocal fold surfaces impacting vibration and flow dynamics. For instance, in experiments using excised canine larynges, Oren et el. [3] found a positive correlation between the intraglottal negative pressure and the sound pressure level. Moreover, vortical structures and the associated flow turbulence generated from the flow separation process were found to affect quadrupole sound source, characterized by their broad spectral range and high frequencies [4–6], and vocal fold vibration [3,7–9]. Research has shown that flow separation occurs when the maximum cross-sectional area is one to two times the minimum glottal opening area [10,11]. As airflow passes this point, a circulating flow area forms between the jet and the medial glottal wall. Within this area, the intraglottal pressure is negative (gauge pressure) and it gradually recovers to atmospheric pressure (zero gauge pressure) at the glottal exit. Studies have observed that the lowest intraglottal pressure ranges between −0.5 and −0.2 times the subglottal pressure (gauge pressure) in the absence of the supraglottal tract [9,12–15], and between −0.5 and −1.2 times the subglottal pressure (gauge pressure) when the supraglottal tract is present [16,17].

In an experimental study involving excised canine larynges [7], it was shown that vortices form near the superior aspect of the folds after flow separation. These flow separation vortices (FSV) induce increased negative pressure at the superior aspect of the folds. Later experimental and computational studies also showed that the strength of FSV, and subsequently the negative pressure they augment, are proportional to the magnitude of the glottal divergent angle [3,8]. Farbos de Luzan et al. [18] quantified the intraglottal negative pressure induced by FSV using large eddy simulation by comparing the pressure fields between a divergent channel and a straight channel. The geometric model is static but with a time-varying pressure waveform applied at the inlet of the domain. They suggested that FSV in the divergent section of the true vocal folds was responsible for 136% more pressure reduction during vocal fold closing. Sundström et al. [8] investigated the impact of flow-separation-vortex-induced negative pressure (FSV_{NP}) on the glottal dynamics using flow-structure-interaction (FSI) modeling. They reported that the FSV produced strong negative pressure on the folds, which was correlated with the vortical strength, and that the aerodynamic force induced by the FSV was at times higher than the elastic recoil force in the tissues. The increased negative pressure induced by FSV leads to a hypothesis that FSV can exert an additional pulling force on the vocal fold during closing and increase the flow deceleration rate. Notably, the rapid deceleration of airflow during the closing phase, quantified by MFDR, has been shown to have strong correlations with sound pressure level (SPL), acoustic energy in higher harmonics, and vocal efficiency loudness [19–21].

Much research effort has been dedicated to examining the effect of flow separation on glottal dynamics. Zhang [22] employed a 2D continuum vocal fold model to investigate how the location of flow separation influences the phonation set, revealing effects on threshold pressure, frequency, and vibration patterns. In another study, Pelorson et al. [23] employed a boundary layer theory-based approach to investigate the effect of dynamically moving flow separation, finding that it decreases the fundamental frequency and MFDR compared to fixed flow separation. However, a limitation of these studies is that zero pressure was assumed after flow separation, neglecting the effect of negative pressure.

A few more recent studies have explored the potential effects of FSV_{NP} on vocal fold dynamics. Pirnia et al. [24] experimentally investigated the steady state response of cantilevered plates when subjected to tangential advection of periodic generated vortex rings. The velocity field was measured by a particle image velocimetry (PIV) system while the pressure field and the plate energy were calculated using the Poisson pressure equation and Kirchhoff–Love plate theory, respectively. The results were applied to vocal fold vibration by comparing it to a plate with similar non-dimensioned mass and stiffness parameters. They estimated that the ratio of energy transfer due to vortex loading to total aerodynamic energy transfer was negligible. However, modeling the vocal fold structure as a plate is a great simplification of the geometry and the boundary conditions. Moreover, due to experimental limitations, the modulus of elasticity of the plate was as high as 19.9 kPa, which was much larger than the transverse Young's modulus of the vocal fold reported in the literature [25,26]. Farahani and Zhang [27] employed a 2D computational model of the vocal fold to explore the impact of FSV_{NP} on sound production. They utilized the Bernoulli equation to calculate intraglottal pressures; however, when the glottis was divergent, they introduced a sinusoidal spatial distribution of negative pressure on the vocal fold surface between the flow separation location and the superior edge. The spatial mean value of the applied negative pressure was up to 0.15 times the subglottal pressure, which was estimated from an experimental measurement. They observed a 12.5% increase in MFDR and a 1 dB increase in sound intensity.

The complete understanding of how flow separation and the resulting pressure changes affect voice production remains elusive. In the current study, we aim to quantify the effects of FSV_{NP} on vocal fold vibration by employing an innovative modeling approach. Specifically, we coupled a three-mass vocal fold model with a Navier–Stokes equation-based flow model to simulate fully coupled FSI during vocal fold vibration. Leveraging the Navier–Stokes equation, our FSI model provides high-resolution dynamic pressure solutions on vocal fold surfaces throughout the vibration cycle. To isolate the effect of FSV_{NP} during glottis closing, we created a comparative model that mirrors the original FSI setup, with the only exception of removing the negative pressure loading on the vocal folds when the glottis is in divergent shape. This approach differs from the method in Farahani and Zhang [27], where the treatment of flow separation location, intraglottal negative pressure values and spatial destitution of the pressure was simplified. In our approach, the removed negative pressure in the comparative model is accurately calculated from the Navier–Stokes equations and includes precise spatial and temporal variations. By comparing the simulation results from the original FSI model and the comparative model, we aim to examine the effect of FSV_{NP} on comprehensive vocal parameters, including glottal flow rate, vocal fold vibratory dynamics, and aerodynamic energy transfer. We hypothesize that FSV_{NP} can enhance vocal fold vibration by increasing aeroelastic energy transfer during vibration.

2. Materials and Methods

2.1. Computational Methodology

The aerodynamics of the glottal flow is numerically simulated using a hydrodynamic/acoustic splitting method [28,29]. The method was developed for simulating low-Mach number flow dynamics to avoid the high computational cost of full compressible flow simulations. Our previous work verified this method for simulating glottal aerodynamics and vocal tract acoustics [30]. The results showed that the splitting method showed good agreement with the compressible flow simulation for low-Mach number internal flow problems with both velocity and pressure boundary conditions. In this method, the flow variables are divided into perturbed and incompressible components: $v = v' + V$ and $p = p' + P$, where v, v', and V represent the total, the perturbed, and the incompressible components of the velocity, respectively. p, p', and P represent the pressure's total, perturbed, and incompressible components. The incompressible components are calculated from the unsteady, viscous, incompressible Navier–Stokes equations:

$$\frac{\partial V_i}{\partial x_i} = 0$$

$$\frac{\partial V_i}{\partial t} + \frac{\partial V_i V_j}{\partial x_j} = -\frac{1}{\rho}\frac{\partial P}{\partial x_i} + v\frac{\partial^2 V_i}{\partial x_j \partial x_i}$$

where ρ and v are the flow density and the kinematic viscosity of the incompressible flow, respectively.

The perturbed components are calculated from the linearized perturbed compressible equations (LPCE):

$$\frac{\partial v'_j}{\partial t} + \frac{\partial v'_i V_i}{\partial x_j} + \frac{1}{\rho}\frac{\partial p'}{\partial x_j} = 0$$

$$\frac{\partial p'}{\partial t} + V_i\frac{\partial p'}{\partial x_i} + \gamma P\frac{\partial v'_i}{\partial x_i} + v'_i\frac{\partial P}{\partial x_i} = -\left(\frac{\partial P}{\partial t} + V_i\frac{\partial P}{\partial x_i}\right)$$

where γ is the ratio of the specific heats. When solving the LPCE, the values of the incompressible variables are obtained from the incompressible N-S equation calculation. To resolve the moving geometries, a sharp-interface immersed boundary method based on the ghost-cell approach is employed for treating boundary conditions. The incompressible flow solver is described in more detail in [31], while the compressible flow solver is detailed in [28].

Figure 1a illustrates the schematic of the three-mass model of the vocal fold. In this model, the body-cover structure is represented by three lumped masses connected through springs and dampers [32]. The model only considers the lateral motion of the vocal fold. The equations of the motion of the three masses are

$$m_u\ddot{x}_u = -k_u\left[(x_u - x_b) + \eta_u(x_u - x_b)^3\right] - d_u(\dot{x}_u - \dot{x}_b) + k_c(x_l - x_u) + Fex_u$$

$$m_l\ddot{x}_l = -k_l\left[(x_l - x_b) + \eta_l(x_l - x_b)^3\right] - d_l(\dot{x}_l - \dot{x}_b) - k_c(x_l - x_u) + Fex_l$$

$$m_b\ddot{x}_b = -k_b\left[x_b + \eta_b x_b^3\right] - d_b\dot{x}_b + k_u\left[x_u - x_b + \eta_u(x_u - x_b)^3\right] + d_u(\dot{x}_u - \dot{x}_b) + k_l\left[x_l - x_b + \eta_l(x_l - x_b)^3\right] + d_l(\dot{x}_l - \dot{x}_b)$$

where the subscripts u and l represent the upper and lower portions of the cover layer, respectively, and b represents the body layer. x_s, $\dot{x}_s$, $\ddot{x}_s$ ($s = u, l, b$) represent the displacement, velocity, and acceleration of the masses. m_s, k_s, d_s ($s = u, l, b$) are the mass, stiffness, and damping coefficients. η_s ($s = u, l, b$) are the nonlinear spring coefficients, and were set to 100 in our simulations. k_c is the stiffness of the spring connecting the upper and lower masses. Fex_s ($s = u, l$) represent the external force applied on the masses in the lateral direction. d_s were calculated as $d_s = 2\zeta_s\sqrt{m_s k_s}$, where $\zeta_u = \zeta_l = 0.4$ and $\zeta_b = 0.2$ [32].

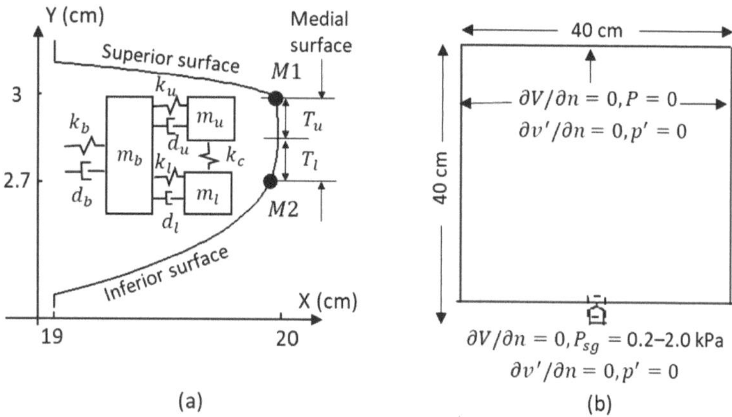

Figure 1. Simulation setup. (**a**) The schematic of the three-mass model of the vocal fold. (**b**) Flow domain and boundary condition. The dashed square highlights the glottal region.

The vocal fold profile was reconstructed from CT scans of the larynx of a 30-year-old male (Figure 1a) [30]. The profile was represented by one hundred marker points, through which the three-mass model and flow model were coupled. As depicted in Figure 1a, $M1$ (y = 3 cm) and $M2$ (y = 2.7 cm) are the two marker points representing the locations of the upper and lower masses, respectively. At each time step, the velocity and displacement of $M1$ and $M2$ are updated from the values of $\dot{x}_u$ and $\dot{x}_l$ in the three-mass model. The velocity and displacement of other marker points are updated using linear interpolations. The flow solver is marched by one time step with the updated vocal fold surface velocity and displacement. Then, the pressure loading on the upper (from y = 2.85 cm to y = 3 cm) and lower (from y = 2.7 cm to y = 2.85 cm) halves of the medial vocal fold surface are integrated to obtain the external forces (Fex_u and Fex_l) acting on the upper and lower masses. Finally, with the updated external forces, the three-mass model is marched by one time step.

In the simulation group where FSV_{NP} was eliminated, an artificial treatment was introduced in the coupling process. Whenever the divergent angle of the two vocal folds surpassed 5 degrees, any negative pressures on the vocal fold surfaces were set to zero when computing the external forces (Fex_u and Fex_l). This process eliminated the influence of negative pressures on the vocal fold dynamics due to flow separation. Nonetheless, the remaining steps of the coupling process remained the same as described earlier.

2.2. Simulation Setup

The simulation setup generally follows our previous work [30] metero and a comprehensive validation of the simulation setup was provided in [30]. The parameters of the three-mass model are provided in Table 1. Most of these values are from Story and Titze [32], except for the upper mass, which was increased from 0.01 g to 0.04 g. Previous simulations [30] showed that when the upper and lower masses were of equal value (0.01 g), the model generated small glottal angles (12 degrees between the two vocal folds) that are not conducive to generating flow separations. In addition, excised canine larynges experiments measured maximum divergent glottal angles of 37 and 51 degrees at subglottal pressures of 1.2 and 1.8 kPa, respectively [33]. Based on these observations, we increased the upper mass to achieve larger glottal angles for generating flow separations. With this adjustment, the model generates the maximum divergent angle of 48 and 60 degrees at subglottal pressures of 1.2 and 1.8 kPa, respectively.

Table 1. Parameters of the three-mass model of the vocal folds.

	m_u (g)	m_l (g)	m_b (g)	k_u (N/m)	k_l (N/m)	k_b (N/m)	k_c (N/m)	T_u (cm)	T_l (cm)
Value	0.04	0.01	0.05	5	7	100	1	0.15	0.15

Figure 1b depicts the airway configuration. The glottal region is highlighted using a dashed square. Downstream of the glottis, an open environment is represented by a large box measuring 37 cm × 40 cm. The configuration was discretized using 256 × 256 non-uniform Cartesian grids. The vocal fold region ($19 \leq X \leq 21$ cm, $2.17 \leq Y \leq 3.17$ cm) had the highest grid resolution, with 128 × 98 non-uniform Cartesian grids. The grid density in the vocal fold region was the same as that used in [30], who performed a grid-independent study on a similar configuration.

In the incompressible flow solver, a subglottal pressure was imposed at the glottis inlet, with values ranging from 0.2 to 2 kPa. A zero gauge pressure was applied at the exit of the far-field domain. Non-slip, non-penetration boundary conditions were imposed on the wall of the vocal tract and vocal fold. The incompressible air was assumed to have a density of 1.145 kg/m^3 and a kinematic viscosity of 1.65×10^{-5} m^2/s, corresponding to 35 °C. The LPCE solver used a hard-wall boundary condition for the vocal fold walls ($\frac{\partial p'}{\partial n} = 0$, $v' \cdot \hat{n} = 0$, where $\hat{n}$ denotes the face normal). A buffer zone was incorporated to eliminate acoustic reflections at the inlet of the subglottal tract and the exit of the far-field domain. During vibrations, the vocal fold contact was modeled by enforcing a 0.16 mm minimum gap between the folds until the aerodynamic force was sufficient to separate them. This gap size was approximately 6% of the maximum gap observed at 1.2 kPa subglottal pressure.

3. Results and Discussion

3.1. Flow Dynamics

A series of FSI simulations were conducted by varying the subglottal pressure from 0.2 to 2 kPa in increments of 0.2 kPa. These cases are referred to as baseline cases. Sustained vibrations were observed when the subglottal pressure was above 0.4 kPa. The resulting fundamental frequency ranged from 101 to 144 Hz, glottal opening ranged from 1.2 to 3.5 mm, maximal divergent angle ranged from 12 to 62 degrees, open quotient mostly ranged from 0.51 to 0.53, and skewness quotient was around 2.5. The glottal opening was determined by measuring the minimum distance between the upper and lower masses. The glottal angle (α) was defined as the angle between the two vocal fold medial surfaces (Figure 2c) and was calculated using the positions of the $M1$ and $M2$ markers points:

$$\alpha = \mathrm{atan}\left(\frac{dx_{M1} - dx_{M2}}{dy_{M1,\,M2}}\right)$$

where dx_{M1} and dx_{M2} are the lateral distances between the pair of $M1$ and $M2$, respectively, and $dy_{M1,\,M2}$ is the vertical distance between $M1$ and $M2$, which is the same for both sides of the vocal folds. The open quotient was defined as a ratio of the open glottis duration (T_o) to the vibration period (T). The skewness quotient was the duration ratio of the flow acceleration (T_1) to flow deceleration (T_2). T, T_o, T_1 and T_2 are denoted in Figure 2a in one of the vibration cycles. Unless otherwise noted, the data reported in the results section averaged over four consecutive sustained cycles. The fundamental frequency, open quotient, and skewness quotient fell within the expected physiological range [34]. For the glottal opening and the maximal divergent angle, the value under high subglottal pressure is higher than what is typically measured or employed in other studies. For instance, the maximal divergent angle measured in an excised canine larynx was reported to be 51 degrees at 1.8 kPa subglottal pressure [33]. The divergent angle of up to 40 degrees was typically employed in studies involving synthetic vocal folds [9,13] or simulations [4,6,35]. A glottal width of 3 mm was utilized in the numerical study of intraglottal vortices [35]. However, it is worth noting that the subglottal pressures in those referenced studies were also lower. For instance, in [13], subglottal pressure ranged from 0.3 to 1.48 kPa; in [9], [6], [4], and [35], it was 0.5 kPa, 0.35 kPa, 0.59 kPa, and 0.3 kPa, respectively. The higher subglottal pressure employed in our study could potentially have contributed to the large glottal angle and opening. A larger divergent angle has been shown to shift the flow separation location upstream [10,22,36]. However, as reported in [36], when the divergent angle was 40 degrees, the flow separation location was close to the glottis entrance, with the reverse flow just past the separation points. This flow separation pattern is similar to what we observed in our current study (Figure 2). Therefore, we consider the glottal opening and divergent angle we observed to be within the reasonable range, and though the divergent angle may exceed 40 degrees under high-pressure conditions in our study, it might affect the flow separation location only in a very limited range near the glottis entrance.

The baseline cases displayed consistent vibratory dynamics and flow patterns. To illustrate this, we present the 1.2 kPa subglottal pressure case data, which is representative of all cases, in Figure 2. Figure 2a depicts the time history of the flow rate, demonstrating sustained vibration. Figure 2b–d shows the phase-averaged data of flow rate and glottal opening, glottal angle, and spatial-average flow pressure on the vocal fold medial surface, respectively. Phase time is normalized to 0–1, with 0 indicating the beginning of the glottal opening. This figure focuses on the duration of the opening and closing phases (from 0 to 0.55). The flow rate was calculated at the glottal exit. A positive value of the glottal angle indicates a convergent shape, and a negative value indicates a divergent glottis. The average flow pressure on the vocal fold medial surface was calculated by averaging the external forces on the upper and lower masses and dividing it by the thickness of the vocal fold medial surface. The lines depict the phase-averaged values, while the width of the shaded region shows the standard deviation. The two vertical dashed lines indicate the moments of maximum glottis opening and maximum flow rate, and the horizontal dashed lines indicate zero glottal angle and zero vocal fold surface pressure, respectively.

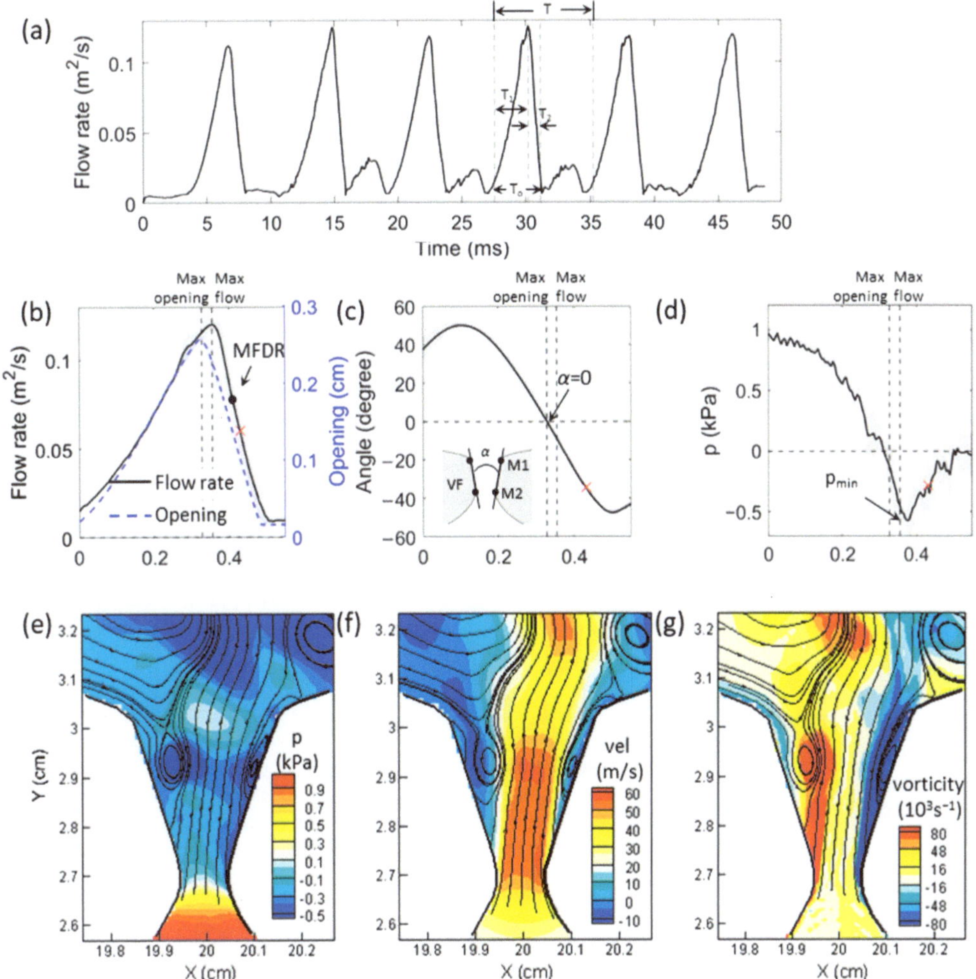

Figure 2. Glottal flow parameters and flow field contours in the baseline configuration at 1.2 kPa subglottal pressure. (**a**) Time history of the glottal flow. Phase averaged (**b**) flow rate and glottal opening, (**c**) glottal angle, and (**d**) mean pressure of the medial surface of the vocal folds, respectively. MFDR instant is denoted in (**b**). (**e–g**) Show intraglottal pressure, vertical velocity, and vorticity field, respectively, with flow streamlines superimposed. The phase time corresponding to (**e–g**) is denoted as a cross 'x' in (**b–d**). In (**a**), T is the period of the vibration; T_o is the duration of open glottis; T_1 and T_2 correspond to the time duration of flow acceleration and deceleration, respectively. See text for a detailed description of the figures.

The results show that during the initial phase of the glottal opening, a large positive (convergent) angle was present due to the earlier opening of the inferior edge. This is attributed to the propagation of a mucosal wave on the vocal fold surface. As the flow rate increased, the convergent angle also increased until it reached its maximum. The surface pressure decreased rapidly during the entire opening phase. At the point of maximum glottal opening, the angle reduced to zero, and the surface pressure also dropped to nearly zero. As the glottis began to close, the angle became negative, indicating a divergent glottal shape. The divergent angle continued to increase during closing and reached its maximum towards the end of the closing phase. The surface pressure was negative throughout the closing phase, dropping to its lowest point (-0.57 kPa when Phase $= 0.37$) shortly before the MFDR occurred. There was a small phase difference between the maximum glottis opening and maximum flow rate, likely due to the small inertance in the downstream box. The most negative intraglottal pressure corresponds to 48% subglottal pressure, which is in line with observations from previous studies (-0.5 to -0.2 times subglottal pressure) [9,12–15]. The mean negative pressure during the closing phase was approximately -0.33 kPa, equivalent to 28% of the subglottal pressure.

In Figure 2e–g, the contours of flow pressure, the vertical component of flow velocity, and vorticity inside the glottis at a representative time instant during glottal closing (Phase $= 0.43$) are shown. The phase time corresponding to Figure 2e–g is denoted as cross in Figure 2b–d, which locates around the mid-closing phase. Additionally, flow streamlines are superimposed to illustrate the separated flow patterns. A recirculation zone is formed between the glottal jet and each medial wall, characterized by flow entering from above the glottis and forming FSV. The locations of the lowest pressure and the strongest vorticity correspond to the FSV location. The calculated vorticity exhibits a magnitude level similar to that reported in [8].

Overall, the baseline simulations demonstrate flow and vibration dynamics that are in line with physiological observations and quantities. During the glottis closing, a recirculation zone is formed between the glottal jet and each vocal fold wall. This leads to the generation of local negative pressures and influences the adjacent wall pressures. Specifically, for the representative subglottal pressure of 1.2 kPa, the resultant wall pressure on the medial surface of the vocal fold during the closing phase reaches a minimum value of -0.57 kPa and a mean value of -0.33 kPa, corresponding to 48% and 28% of the subglottal pressure, respectively.

3.2. Effects of Eliminating FSV_{NP}

In this section, we compare the baseline results with a series of simulations under the same conditions, except that the negative pressures applied on the medial wall during the closing phase are set to be atmospheric. This process (referred to as (-) FSV_{NP}) aims to eliminate the FSV_{NP} contribution to the mechanism of the fold vibrations.

Figure 3a–c compares the displacement of the masses (M1 and M2) and glottal opening with their corresponding time derivatives shown in Figure 3d–f for the representative subglottal pressure of 1.2 kPa. Due to left-right symmetry, only one vocal fold was studied (Figure 3a,b,d,e). The glottal opening (Figure 3c) was determined as the minimum distance between the two sides of the masses. Figure 3f is the derivative of the glottal opening, representing the speed of the vocal fold displacement. In the closing phase, it is referred to as the closing speed. Overall, the (-) FSV_{NP} case demonstrates smaller vibration amplitudes and closing speeds than the baseline case. Quantitatively, the vibration amplitude of the lower and upper masses in the (-) FSV_{NP} case was 3.8% and 6.8% lower than those in the baseline case, respectively. The maximum glottis opening in the (-) FSV_{NP} case was 8.8% lower than the baseline. The maximum closing speed of the upper and lower mass in the (-) FSV_{NP} case was 26.3% and 33.3% smaller than those in the baseline case. For the upper mass, due to the phase delay between the two masses, the maximum closing speed occurred in the closed phase (thus not shown). The maximum glottal closing speed in the (-) FSV_{NP} case was 30.5% smaller than in the baseline case. These results suggest that the FSV_{NP} played an essential role in promoting vocal fold vibration and facilitating glottis closing, likely because the negative pressures pull the vocal fold to close, resulting in additional energy transfer to the vocal fold.

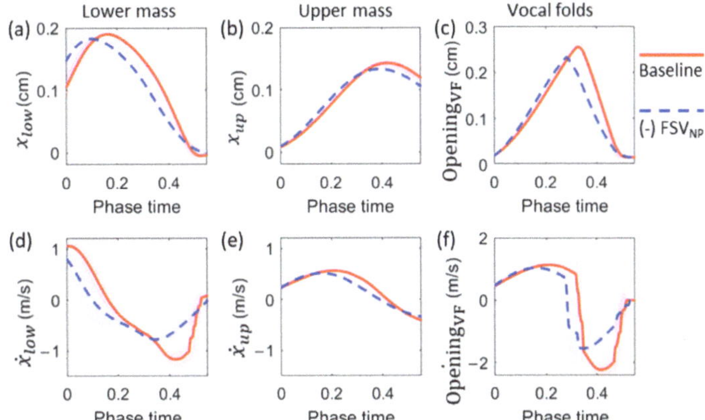

Figure 3. Comparison of the vibration parameters between the baseline and (-) FSV$_{NP}$ cases at a representative subglottal pressure of 1.2 kPa. The lines show the phase-averaged values, and the shaded area indicates the standard deviation. (**a,b**) Displacement of the lower mass and upper mass, respectively. (**c**) Glottal opening. (**d,e**) The velocity of the lower mass and upper mass, respectively. (**f**) Time-derivative of the glottal opening.

Figure 4 compares the phase-averaged flow rate between the baseline and (-) FSV$_{NP}$ cases at 1.2 kPa subglottal pressure. The (-) FSV$_{NP}$ case showed a lower flow rate and slower deceleration, leading to smaller MFDR and flow skewness values. Quantitatively, the MFDR in the (-) FSV$_{NP}$ case was 80 m^2/s^2, 34% lower than the value in the baseline case (120.7 m^2/s^2). The flow skewness quotient was 1.64, 41% lower than the value in the baseline case (2.8). These observations are consistent with the reduced glottal opening and closing speed in the (-) FSV$_{NP}$ case.

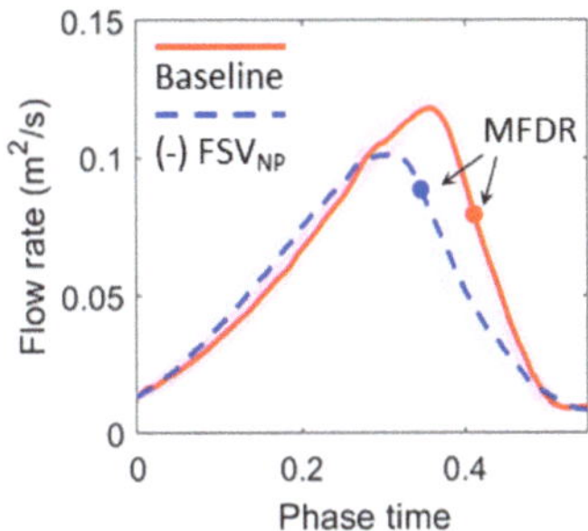

Figure 4. Comparison of the flow rate between the baseline and (-) FSV$_{NP}$ cases at a representative subglottal pressure of 1.2 kPa. The line plot shows the phase-averaged value, and the shaded area indicates the standard deviation. MFDR instants are denoted.

To investigate the aeroelastic energy transfer, we computed the power transfer from the airflow to the vocal fold and the total flow work over the vocal fold medial surface during the opening and closing of the fold vibrations. The power transfer was computed by multiplying the flow pressure by the normal component of the velocity vectors and integrating over the vocal fold medial surface. The flow work was then obtained by integrating the power values over time. The overall energy balance did not consider viscous force because their contribution is about two orders of magnitude smaller than the aerodynamic pressure [37]. Figure 5 compares the time history of power transfer and flow work when the glottis was open between the (-) FSV$_{NP}$ and baseline cases. The power plot shows the phase-averaged values as curves, with the shades representing the standard deviation. The work plot shows data from one vibration cycle to avoid the accumulation effect due to the integration over time. In the baseline case, a distinct peak in power transfer was observed around the time when the maximum FSV_{NP} (Figure 2d) and maximum glottis closing speed (Figure 3f) occurred. The flow work plot shows that in the baseline case, the flow work continuously increased during most of the closing phase, while, in the (-) FSV_{NP} case, it remained almost constant during glottis closing. These results support the notion that FSV_{NP} generated additional energy transfer from the airflow to the fold during the closing phase.

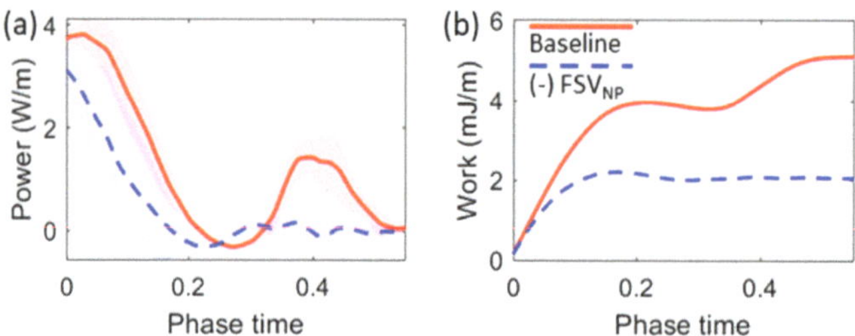

Figure 5. Comparison of (**a**) power transfer from the airflow to the vocal fold and (**b**) the total flow work over the medial wall during opening and closing phases between the baseline and (-) FSV_{NP} cases. The line plot in (**a**) shows the phase averaged value while the width of the shaded area indicates the standard deviation. In (**b**), the work is shown for one vibration cycle to avoid the accumulation effect due to the integration over time.

In Figure 6, we compare key vibration and flow parameters between the baseline and (-) FSV_{NP} cases over the range of simulated subglottal pressures. The parameters examined include the maximum opening of the upper and lower masses (a,b), maximum closing speed of the lower mass (c), maximum glottis opening (d), maximum glottis closing speed (e), maximum flow rate (f), flow skewness quotient (g), MFDR (h), and energy transfer in one vibration cycle (i). The percentage difference between the two cases is represented by the bars in the figures. The results indicate that removing FSV_{NP} had a more pronounced effect when the subglottal pressure was above 1 kPa. These effects were consistent across this pressure range, causing reduced vibration amplitude, closing speed, flow rate, flow skewness, MFDR, and energy transfer. Quantitatively, the effects on vibration amplitude and flow rate were up to 20%, while those on closing speed, flow skewness quotient, MFDR, and energy transfer were more significant, reaching up to 40%. The most notable effects were observed in the 1–1.2 kPa subglottal pressure range. No consistent effects were observed below a subglottal pressure of 1 kPa. Moreover, some relative errors may appear large at low subglottal pressures due to small values in the baseline case (e.g., Figure 6b,c,e at a subglottal pressure of 0.4 kPa).

In our study, removing FSV_{NP} at a subglottal pressure of 1.2 kPa, corresponding to approximately 28% of the subglottal pressure, resulted in a 34% reduction in MFDR compared to the baseline. Farahani and Zhang [27] reported similar findings, showing that applying additional negative pressures downstream of the flow separation with a mean value of approximately 15% of subglottal pressure resulted in a 12.5% increase in MFDR. They estimated that a 12.5% increase in MFDR resulted in a 1 dB change in SPL. Using the same method, we estimated that the 34% decrease in MFDR would result in a 2.4 dB change in SPL.

These results suggested that FSV_{NP} enhances vocal fold vibrations and sound intensity during voice production by promoting greater energy transfer from the airflow to the vocal fold. It is also worth noting that when the FSV_{NP} was removed, complete glottis closure was never achieved at the lowest subglottal pressure of 0.4 kPa, whereas it was achieved in the baseline case. This observation also suggests the importance of FSV_{NP} in facilitating stronger vibrations.

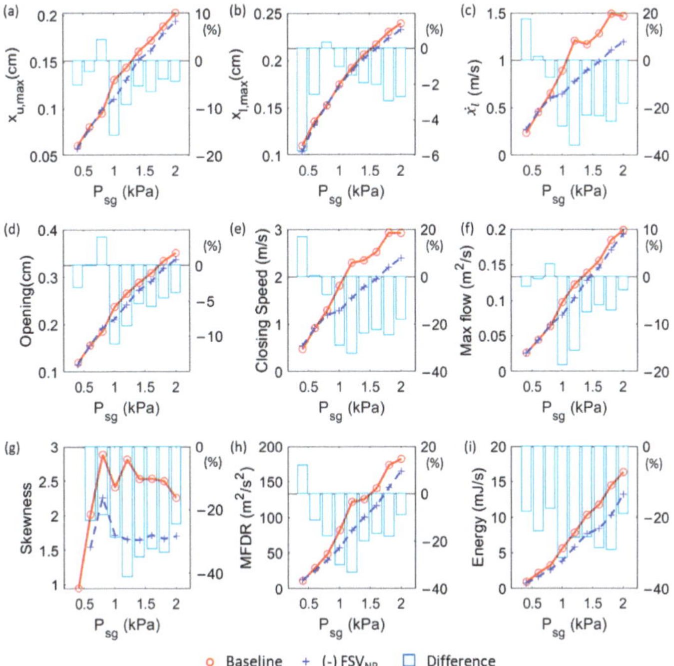

Figure 6. Comparison of important vibration and flow parameters between the baseline and (-) FSV$_{NP}$ cases over the range of simulated subglottal pressures. (**a**) Maximum opening of the upper mass. (**b**) Maximum opening of the lower mass. (**c**) The maximum closing speed of the lower mass. (**d**) Glottal opening. (**e**) Vocal fold closing speed. (**f**) The maximum glottal flow. (**g**) Skewness quotient. (**h**) Maximum flow declination rate (MFDR). (**i**) Total energy transfer from the airflow to vocal fold tissue during one vibration cycle.

4. Conclusions

This study employed a two-dimensional FSI computer model to investigate the effects of FSV$_{NP}$ on vocal fold vibration and flow dynamics during vocal fold vibration. The numerical model integrated a Navier–Stokes equation-based incompressible flow model, a linearized perturbed compressible equation-based acoustic model, and a three-mass vocal fold model. The baseline simulations predicted flow and vibration dynamics that are consistent with physiological observations and quantities. Specifically, FSV formed between the medial wall and the separated glottal flow resulting in an intraglottal pressure that was predominantly negative after the location of flow separation and before gradually recovering to zero at the glottis exit. In a representative subglottal pressure of 1.2 kPa, the mean negative pressure in the closing phase was approximately -0.33 kPa, equivalent to 28% of the subglottal pressure. The maximum FSV$_{NP}$ occurred near the time instant of MFDR and corresponded to the location of the FSV.

To isolate the effects of FSV$_{NP}$, a comparison simulation group was conducted in which negative pressures arising after flow separation were removed in FSI. Notably, the results showed consistent effects above a subglottal pressure of 1 kPa, where removing FSV$_{NP}$ resulted in reduced vibration amplitude, flow rate, closing speed, flow skewness quotient, and MFDR. Quantitatively, the vibration amplitude and flow rate were reduced by up to 20%, and the closing speed, flow skewness quotient, and MFDR were reduced by up to 40%. In energy transfer analysis, the FSV$_{NP}$ generated an additional energy transfer peak from the airflow to the vocal fold during glottis closing, thereby increasing overall energy transfer over a cycle. Quantitatively, removing FSV$_{NP}$ resulted in an energy loss of up to 32%. These findings suggest that FSV_{NP} enhances vocal fold vibration by increasing the aeroelastic energy transfer during vibration. This result emphasizes the significant role of the FSV on the vocal fold dynamics.

It is important to acknowledge the limitations of this study. Firstly, though two-dimensional assumption is very frequently employed in the simulations of the vocal fold vibration and laryngeal flow [38], previous studies have shown that the negative pressure from a model with a constant glottal shape in the anterior–posterior direction could be 16% lower than a model with the glottal opening gradually decreasing to zero towards the two ends [12]. Additionally, the pressure variation in the latter model was largest around the mid-coronal plane. It decreased towards the two ends, with intraglottal pressure always being positive at the two ends [15]. In the current study, only mid-coronal plane parameters were considered by employing a two-dimensional model, which might have overestimated the mean intraglottal negative pressure. Therefore, caution should be taken when generalizing these results to three-dimensional scenarios.

The second limitation is that the three-mass model used in the study has a larger upper mass than the lower mass to achieve the desired glottal divergent angle and flow separation during closing, which is opposite to what is usually seen in in two-mass models, where the lower mass is greater to mimic the characteristics of the body layer [23,39]. Nevertheless, it was not the first time that a larger upper mass was used in a three-mass model [40]. Ultimately, our simulation results indicated that our model parameters predicted reasonable dynamics of vocal fold vibrations, with dynamic parameters that agree with physiological quantities.

Overall, this study contributes to a deeper understanding of the FSV and the associated FSV_{NP}'s influence on vocal fold vibration and flow dynamics, which may have implications for improving voice production and treating voice disorders.

Author Contributions: Conceptualization, W.J., Q.X., X.Z., C.F.d.L., L.O. and E.G.; methodology, W.J., Q.X., X.Z., C.F.d.L., L.O. and E.G.; software, W.J., Q.X. and X.Z.; formal analysis, W.J., Q.X., X.Z., C.F.d.L., L.O. and E.G.; investigation, W.J., Q.X. and X.Z.; writing—original draft preparation, W.J., Q.X. and X.Z.; writing—review and editing, W.J., Q.X., X.Z., C.F.d.L., L.O. and E.G.; supervision, Q.X. and X.Z.; project administration, Q.X. and L.O.; funding acquisition, L.O. and Q.X. All authors have read and agreed to the published version of the manuscript.

Funding: The research was funded by NIH Grant No. R01DC009435 from the National Institute on Deafness and Other Communication Disorders (NIDCD). This work used Expanse at the San Diego Supercomputer Center (SDSC) through allocation CTS180004 from the Extreme Science and Engineering Discovery Environment (XSEDE) [41], which was supported by National Science Foundation grant number #1548562.

Institutional Review Board Statement: Not applicable.

Informed Consent Statement: Not applicable.

Data Availability Statement: The data presented in this study are available upon request.

Acknowledgments: We would like to acknowledge the significant contributions of Sid Khosla, who unexpectedly passed away during the course of this research. Khosla played an important role in the early stage of this research and provided a wealth of invaluable insights that shaped our work. We are deeply grateful for his contributions.

Conflicts of Interest: The authors declare no conflict of interest.

References

1. Alipour, F.; Jaiswal, S.; Finnegan, E. Aerodynamic and Acoustic Effects of False Vocal Folds and Epiglottis in Excised Larynx Models. *Ann. Otol. Rhinol. Laryngol.* **2007**, *116*, 135–144. [CrossRef] [PubMed]
2. Saraniti, C.; Speciale, R.; Santangelo, M.; Massaro, N.; Maniaci, A.; Gallina, S.; Serra, A.; Cocuzza, S. Functional Outcomes after Supracricoid Modified Partial Laryngectomy. *J. Biol. Regul. Homeost. Agents* **2019**, *33*, 1903–1907.
3. Oren, L.; Khosla, S.; Gutmark, E. Medial Surface Dynamics as a Function of Subglottal Pressure in a Canine Larynx Model. *J. Voice* **2019**, *35*, 69–76. [CrossRef] [PubMed]

4. Schickhofer, L.; Mihaescu, M. Analysis of the Aerodynamic Sound of Speech through Static Vocal Tract Models of Various Glottal Shapes. *J. Biomech.* **2020**, *99*, 109484. [CrossRef]
5. Zhang, Z. Mechanics of Human Voice Production and Control. *J. Acoust. Soc. Am.* **2016**, *140*, 2614–2635. [CrossRef] [PubMed]
6. Lasota, M.; Šidlof, P.; Maurerlehner, P.; Kaltenbacher, M.; Schoder, S. Anisotropic Minimum Dissipation Subgrid-Scale Model in Hybrid Aeroacoustic Simulations of Human Phonation. *J. Acoust. Soc. Am.* **2023**, *153*, 1052–1063. [CrossRef]
7. Oren, L.; Khosla, S.; Gutmark, E. Intraglottal Geometry and Velocity Measurements in Canine Larynges. *J. Acoust. Soc. Am.* **2014**, *135*, 380–388. [CrossRef]
8. Sundström, E.; Oren, L.; Farbos de Luzan, C.; Gutmark, E.; Khosla, S. Fluid-Structure Interaction Analysis of Aerodynamic and Elasticity Forces during Vocal Fold Vibration. *J. Voice* **2022**. *Online ahead of print.* [CrossRef]
9. Li, S.; Scherer, R.C.; Wan, M.X.; Wang, S.P.; Song, B. Intraglottal Pressure: A Comparison between Male and Female Larynxes. *J. Voice* **2019**, *34*, 813–822. [CrossRef]
10. Alipour, F.; Scherer, R.C. Flow Separation in a Computational Oscillating Vocal Fold Model. *J. Acoust. Soc. Am.* **2004**, *116*, 1710–1719. [CrossRef]
11. Pelorson, X.; Hirschberg, A.; Wijnands, A.; Bailliet, H. Description of the Flow through In-Vitro Models of the Glottis during Phonation. *Acta Acust.* **1995**, *3*, 191–202.
12. Scherer, R.C.; Torkaman, S.; Kucinschi, B.R.; Afjeh, A.A. Intraglottal Pressures in a Three-Dimensional Model with a Non-Rectangular Glottal Shape. *J. Acoust. Soc. Am.* **2010**, *128*, 828–838. [CrossRef] [PubMed]
13. Scherer, R.C.; Shinwari, D.; De Witt, K.J.; Zhang, C.; Kucinschi, B.R.; Afjeh, A.A. Intraglottal Pressure Profiles for a Symmetric and Oblique Glottis with a Divergence Angle of 10 Degrees. *J. Acoust. Soc. Am.* **2001**, *109*, 1616–1630. [CrossRef] [PubMed]
14. Jiang, J.J.; Titze, I.R. Measurement of Vocal Fold Intraglottal Pressure and Impact Stress. *J. Voice* **1994**, *8*, 132–144. [CrossRef] [PubMed]
15. Alipour, F.; Scherer, R.C. Dynamic Glottal Pressures in an Excised Hemilarynx Model. *J. Voice* **2000**, *14*, 443–454. [CrossRef] [PubMed]
16. DeJonckere, P.H.; Lebacq, J.; Titze, I.R. Dynamics of the Driving Force During the Normal Vocal Fold Vibration Cycle. *J. Voice* **2017**, *31*, 649–661. [CrossRef]
17. Motie-Shirazi, M.; Zañartu, M.; Peterson, S.D.; Erath, B.D. Vocal Fold Dynamics in a Synthetic Self-Oscillating Model: Intraglottal Aerodynamic Pressure and Energy. *J. Acoust. Soc. Am.* **2021**, *150*, 1332–1345. [CrossRef]
18. Farbos de Luzan, C.; Oren, L.; Gutmark, E.; Khosla, S.M. Quantification of the Intraglottal Pressure Induced by Flow Separation Vortices Using Large Eddy Simulation. *J. Voice* **2021**, *35*, 822–831. [CrossRef] [PubMed]
19. Kreiman, J.; Gerratt, B.R. Perceptual Interaction of the Harmonic Source and Noise in Voice. *J. Acoust. Soc. Am.* **2012**, *131*, 492–500. [CrossRef]
20. Sapienza, C.M.; Stathopoulos, E.T. Comparison of Maximum Flow Declination Rate: Children versus Adults. *J. Voice* **1994**, *8*, 240–247. [CrossRef]
21. Gauffin, J.; Sundberg, J. Spectral Correlates of Glottal Voice Source Waveform Characteristics. *J. Speech Lang. Heart Res.* **1989**, *32*, 556–565. [CrossRef] [PubMed]
22. Zhang, Z. Influence of Flow Separation Location on Phonation Onset. *J. Acoust. Soc. Am.* **2008**, *124*, 1689–1694. [CrossRef] [PubMed]
23. Pelorson, X.; Hirschberg, A.; van Hassel, R.R.; Wijnands, A.P.J.; Auregan, Y. Theoretical and Experimental Study of Quasisteady-Flow Separation within the Glottis during Phonation. Application to a Modified Two-Mass Model. *J. Acoust. Soc. Am.* **1994**, *96*, 3416–3431. [CrossRef]
24. Pirnia, A.; Browning, E.A.; Peterson, S.D.; Erath, B.D. Discrete and Periodic Vortex Loading on a Flexible Plate; Application to Energy Harvesting and Voiced Speech Production. *J. Sound Vib.* **2018**, *433*, 476–492. [CrossRef]
25. Alipour, F.; Vigmostad, S. Measurement of Vocal Folds Elastic Properties for Continuum Modeling. *J. Voice* **2012**, *26*, 816.e21–816.e29. [CrossRef]
26. Chhetri, D.K.; Zhang, Z.; Neubauer, J. Measurement of Young's Modulus of Vocal Folds by Indentation. *J. Voice* **2011**, *25*, 1–7. [CrossRef]
27. Farahani, M.H.; Zhang, Z. A Computational Study of the Effect of Intraglottal Vortex-Induced Negative Pressure on Vocal Fold Vibration. *J. Acoust. Soc. Am.* **2014**, *136*, EL369–EL375. [CrossRef] [PubMed]
28. Seo, J.H.; Mittal, R. A High-Order Immersed Boundary Method for Acoustic Wave Scattering and Low-Mach Number Flow-Induced Sound in Complex Geometries. *J. Comput. Phys.* **2011**, *230*, 1000–1019. [CrossRef]
29. Seo, J.H.; Moon, Y.J. Linearized Perturbed Compressible Equations for Low Mach Number Aeroacoustics. *J. Comput. Phys.* **2006**, *218*, 702–719. [CrossRef]
30. Bodaghi, D.; Jiang, W.; Xue, Q.; Zheng, X. Effect of Supraglottal Acoustics on Fluid-Structure Interaction during Human Voice Production. *J. Biomech. Eng.* **2021**, *143*, 041010. [CrossRef]
31. Mittal, R.; Dong, H.; Bozkurttas, M.; Najjar, F.M.M.; Vargas, A.; von Loebbecke, A. A Versatile Sharp Interface Immersed Boundary Method for Incompressible Flows with Complex Boundaries. *J. Comput. Phys.* **2008**, *227*, 4825–4852. [CrossRef] [PubMed]
32. Story, B.H.; Titze, I.R. Voice Simulation with a Body-Cover Model of the Vocal Folds. *J. Acoust. Soc. Am.* **1995**, *97*, 1249–1260. [CrossRef] [PubMed]

33. Jiang, W.; Farbos De Luzan, C.; Wang, X.; Oren, L.; Khosla, S.M.; Xue, Q.; Zheng, X. Computational Modeling of Voice Production Using Excised Canine Larynx. *J. Biomech. Eng.* **2022**, *144*, 021003. [CrossRef] [PubMed]
34. Titze, I.R. *Principles of Voice Production*; National Center for Voice and Speech: Iowa City, IA, USA, 2000.
35. Mihaescu, M.; Khosla, S.M.; Murugappan, S.; Gutmark, E.J. Unsteady Laryngeal Airflow Simulations of the Intra-Glottal Vortical Structures. *J. Acoust. Soc. Am.* **2010**, *127*, 435–444. [CrossRef]
36. Li, S.; Scherer, R.C.; Wan, M.; Wang, S.; Wu, H. The Effect of Glottal Angle on Intraglottal Pressure. *J. Acoust. Soc. Am.* **2006**, *119*, 539–548. [CrossRef] [PubMed]
37. Thomson, S.L.; Mongeau, L.; Frankel, S.H. Aerodynamic Transfer of Energy to the Vocal Folds. *J. Acoust. Soc. Am.* **2005**, *118*, 1689. [CrossRef]
38. Döllinger, M.; Zhang, Z.; Schoder, S.; Šidlof, P.; Tur, B.; Kniesburges, S. Overview on State-of-the-Art Numerical Modeling of the Phonation Process. *Acta Acust.* **2023**, *7*, 25. [CrossRef]
39. Ishizaka, K.; Flanagan, J.L. Synthesis of Voiced Sounds From a Two-Mass Model of the Vocal Cords. *Bell Syst. Technol. J.* **1972**, *51*, 1233–1268. [CrossRef]
40. Story, B.H. An Overview of the Physiology, Physics and Modeling of the Sound Source for Vowels. *Acoust. Sci. Technol.* **2002**, *23*, 195–206. [CrossRef]
41. Towns, J.; Cockerill, T.; Dahan, M.; Foster, I.; Gaither, K.; Grimshaw, A.; Hazlewood, V.; Lathrop, S.; Lifka, D.; Peterson, G.D.; et al. XSEDE: Accelerating Scientific Discovery. *Comput. Sci. Eng.* **2014**, *16*, 62–74. [CrossRef]

bioengineering

Article

Time-Dependent Fluid-Structure Interaction Simulations of a Simplified Human Soft Palate

Peng Li *, Marco Laudato * and Mihai Mihaescu

Department of Engineering Mechanics, FLOW, KTH Royal Institute of Technology, 10044 Stockholm, Sweden; mihaescu@kth.se
* Correspondence: penl@kth.se (P.L.); laudato@kth.se (M.L.)

Abstract: Obstructive Sleep Apnea Syndrome (OSAS) is a common sleep-related disorder. It is characterized by recurrent partial or total collapse of pharyngeal upper airway accompanied by induced vibrations of the soft tissues (e.g., soft palate). The knowledge of the tissue behavior subject to a particular airflow is relevant for realistic clinic applications. However, in-vivo measurements are usually impractical. The goal of the present study is to develop a 3D fluid-structure interaction model for the human uvulopalatal system relevant to OSA based on simplified geometries under physiological conditions. Numerical simulations are performed to assess the influence of the different breathing conditions on the vibrational dynamics of the flexible structure. Meanwhile, the fluid patterns are investigated for the coupled fluid-structure system as well. Increasing the respiratory flow rate is shown to induce larger structural deformation. Vortex shedding induced resonance is not observed due to the large discrepancy between the flow oscillatory frequency and the natural frequency of the structure. The large deformation for symmetric breathing case under intensive respiration is mainly because of the positive feedback from the pressure differences on the top and the bottom surfaces of the structure.

Keywords: obstructive sleep apnea; fluid-structure interaction; soft palate; 3D simplified model

Citation: Li, P.; Laudato, M.; Mihaescu, M. Time-Dependent Fluid-Structure Interaction Simulations of a Simplified Human Soft Palate. *Bioengineering* **2023**, *10*, 1313. https://doi.org/10.3390/bioengineering10111313

Academic Editor: Iris Little

Received: 14 October 2023
Revised: 8 November 2023
Accepted: 11 November 2023
Published: 14 November 2023

1. Introduction

Obstructive Sleep Apnea (OSA) is a common, complex and highly prevalent sleep-disordered breathing condition [1], which is characterized by partial or complete cessation of airflow during sleep owing to upper airway collapse [2]. Estimates show that more than nearly 1 billion individuals are affected by OSA worldwide [3]. In the middle-aged labor force, approximately 4% of men and 2% of women are likely to fulfill the minimal diagnostic standards for the OSA syndrome [4]. Usually, patients with OSA have loud snoring [5] and are suffering from a range of harmful sequelae, including daytime fatigue, morning headache [6], fatigue-related accidents and risk of systemic hypertension and heart failure [7]. Even though not every snoring person has sleep apnea, Huang et al. [8] estimates that 10% of habitual snorers are at a high risk of complete airway collapse which leads to OSA.

The typical phenomenology of OSA concerns the soft palate. The human palate forms a separation between the nasal and oral cavities. It is comprised of two distinct parts, i.e., hard palate and soft palate [9]. The hard palate is immobile, consisted of bone, while the soft palate is a movable soft tissue, suspended from the posterior border of the hard palate. The respiratory airflow can create instabilities and vibrations in the soft palate, resulting in snoring sound. In the case of sleep apnea, the airway is blocked due to posterior collapse of the soft palate. Another common pattern of OSA is related to pharyngeal collapse [10]. At the level of the oropharynx, the collapse is lateral, most commonly due to the presence of large tonsils. In the hypopharynx or the tongue area, the tongue base collapses antero-posteriorly and squeezes the epiglottis in between to cause complete obstruction.

Compliant or elastic structures subjected to fluid flow can exhibit strong oscillation, resulting from the interaction between the flow and the structural components. The interplay between the fluid and structural components is usually termed as Fluid Structure Interaction (FSI). FSI happens in a wide range of systems, from the flapping of bee wings, the flag flutter, the aircraft wing flutter, to the vibration of human soft palate. FSI-based modeling of OSA has gained a remarkable momentum in the recent scientific literature [11–13]. The main focus of such FSI analyses was the dynamics of the airflow and the soft tissues. In most of the investigations, 3D accurate, imaged models are reconstructed from Magnetic Resonance Imaging (MRI) and Computer Tomography (CT) scanning. Huang et al. [14] developed a 3D finite element model from CT image data. The model included realistic anatomical details of the airway surrounding structures such as the skull, neck, cartilage and soft tissues. FSI simulations were performed both for normal and OSA respiratory processes. The simulation results demonstrate a positive feedback of the upper airway collapse through the interaction of the anterior airway wall and the pressure gradient on the cross-section area during OSA. In [13], Pirnar et al. reestablished the pharyngeal geometric model from CT images of a patient, where the soft palate with uvula was considered as an isolated structure. They found that airway occlusion could occur at the region of the velopharynx without considering gravity during inspiratory process. Chen et al. [15] used a reconstructed 3D model of the upper airway to analyze the impact of the uvula flapping on the airway vibration. From the simulation results, they demonstrated that the flapping frequencies of the uvula tend to affect more significantly on the airway vibration compared with the flapping amplitude. However, in their work, the uvula was considered to be a rigid body with certain flapping frequencies. And the airflow was assumed to be incompressible.

Due to high computational cost and geometric complexity of the 3D anatomically accurate model, several studies were conducted on a simplified model of the upper airway. The main advantage of employing simpler geometries is that the fundamental physical behaviour of the system clearly emerges and can be more easily controlled [16]. Moreover, as numerical models require validation, it is much simpler to find in literature high-quality experimental data sets to validate the numerical results [17]. Tetlow & Lucey [18] modelled the soft palate as a cantilevered flexible plate placed within a unsteady, viscous, incompressible channel flow to mimic the mechanical process for the dynamics of the soft palate in OSA. They revealed that constant inlet velocity conditions can over-estimate the energy exchange between the flow and the structure. They also showed that offsetting the position of the plate within the channel can cause a larger deformation of the plate. The main limitation of the study was that the model was 2D and the plate was treated as a classical thin plate, which means the transverse shear stress was out of consideration. Viscous effects on the plate were also ignored. In order to have a better comprehension of the OSA mechanisms, Khalili et al. [19] developed a new 2D model of the upper airway. The soft palate was modeled as a flexible plate, placed in a compressible, viscous, laminar channel flow. The flow patterns were analyzed when both the nasal and oral inlets were open and when one of them was closed. Besides, an acoustic analysis was performed to the acoustic wave propagation induced by the palatal flutter. However, the 3D effects of the geometry were not considered. Geometry dimensions, as well as material properties (e.g., Young's modulus, Poisson's ratio) were not in physiological condition. Based on a 2D FSI model, Cisonni et al. [20] developed a 2D FSI model, based on a flexible, tapered cantilever axially mounted in a channel flow, to represent the uvuloplatal system in snoring conditions. The effects of material inhomogeneity on the vibrational dynamics were investigated by varying the ratio of the uvula thickness to the hard palate thickness. They demonstrated that higher degree of tapering would decrease the fluid speed required to initiate the structural vibration instability. In their work, the flow was treated incompressible in laminar regime. The flexible palate was described by the one-dimensional Kirchoff–Love beam, where the friction force was not incorporated.

The large part of the studies on the fundamental dynamical behaviour of the uvulopalatal system under OSA conditions is limited by either low dimensional studies or by strict assumptions on the airflow. In this paper, these limitations are tackled by the implementation of a simplified 3D soft palate with uvula under the effects of a compressible, viscous unsteady airflow. The flow is confined in a cylindrical tube, representing the pharyngeal airway. The effects of different inlet boundary conditions of the flow on the oscillatory behavior of the uvulopalatal system are investigated by means of a two-way FSI coupling approach. Understanding the dynamics of the uvula is preparatory to future aeroacoustic simulations that will be used for assessment of sound generation into relation with OSA.

The paper is structured as follows. In Section 2, the model development for the fluid and solid domains are described. In Section 3, the results of the fluid simulations, with the solid structure fixed, are presented. The eigen modes of the 3D solid structure are analysed by employing a modal analysis. The results of 2-way FSI simulations are finally described. An in-depth discussion of the results and conclusions with possible future developments can be found in Section 4.

2. Methods

In this work, a fully two-way FSI coupling [21] is employed. Within each inner time iteration, the results of the Computational Fluid Dynamics (CFD) analysis i.e., velocity, pressure, and wall shear stress at the interfaces are used to calculate the external loads imposed by the fluid on the soft structure. The consequent displacement of the solid structure is computed and transferred back to the CFD solver by morphing the fluid mesh. The geometry under analysis in this study is a simplified 3D geometry of the pharyngeal airway with a elastic structure resembling the soft palate described in Section 2.1. The implementation of the computation model used for the fluid and solid domains is reported in Section 2.2 and Section 2.3, respectively. Convergence study and numerical validation of the grid are presented at the end of this section.

2.1. Simplified 3D Model for Palatal System

The goal of the numerical study is to analyse a simplified 3D model of the palatal system to clarify the main physical mechanisms underlying the palatal snoring. Since the palatal system has very complex geometry and in-vivo measurements are usually not practical, with a simplified geometry it is easier to analyse the effects of vibrations of the soft palate. Moreover, simplified systems can be more easily adapted to entail the wide varying spectrum of geometries of the soft palates reported for human beings. The mean and standard deviation of the human palatal system's geometrical dimensions and mechanical properties are reported in Table 1 and in Table 2, respectively. It is to be noted that the value of the Poisson's ratio is connected to the material characteristics of the soft palate, which has been experimentally determined. The geometry employed in this work refers to the mean values of these tables.

Table 1. Anatomical dimensions of the soft palate.

	Values from Literature	Values Employed in This Study
Pharyngeal radius, D (cm)	3.8 [22]	3.8
Soft palate length, l_{sp} (cm)	4.7 ± 0.7 [23]	4.7 (=1.24D)
Palate thickness, h (cm)	0.74 ± 0.14 [24]	0.74 (=0.19D)
Uvula length, l_u (cm)	1.8 ± 0.1 [25]	1.8 (=0.47D)
Uvula width, w_u (cm)	0.8 ± 0.04 [25]	0.8 (=0.21D)

Table 2. Mechanical properties of the soft palate.

	Values from Literature	Values Employed in This Study
Young's modulus, E (Pa)	585–1410 [26]	600
Poisson's ratio, ν_{sp} (-)	0.22–0.54 [26]	0.45
Density, ρ_{sp} (kg/m^3)	1022–1116 [26]	1060

Figure 1 shows the model geometry where the pharynx is modeled as a tube with circular cross-section. The palate system is placed in the middle of the tube, separating the nasal (upper) and oral (lower) cavities. The soft palate is attached to the back of the hard palate. The inlets are placed $5D$ upstream of the soft palate, while the outlet $10D$ downstream, to avoid spurious effects from the boundaries. The total length of the tube is $16.24D$. The inlets boundary conditions can be tuned with different flow rates which correspond to different breathing conditions. In *nose-only* breathing condition, $\dot{Q}_1 \neq 0, \dot{Q}_2 = 0$. In *symmetric* breathing condition, $\dot{Q}_1 = \dot{Q}_2$. A total of 6 different breathing conditions, spanning from *symmetric* to *nose-only*, are considered in this work. In the schematic presented in Figure 1, one point probe is placed on the tip of the soft palate and is used to monitor the solid displacement. The two monitoring points a and b are symmetrically placed along the vertical direction, in correspondence of the tip of the soft palate. The point probe c is located $3D$ downstream of the soft part for measuring the fluctuation in the wake region.

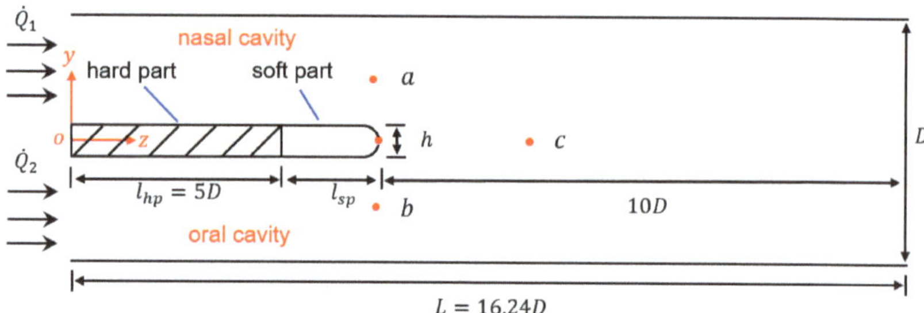

Figure 1. Sketch of the fluid domain with dimensions and the applied inflow boundary conditions shown.

The oropharyngeal anatomy [27] shows the uvula is a drop-shaped muscular tissue attached to the posterior edge of the soft palate. The soft palate is modelled as a rectangular thick plate in Figure 2, smoothly attached by the quasi-cylindrical uvula at the middle of the side edge. All the other sides are considered clamped.

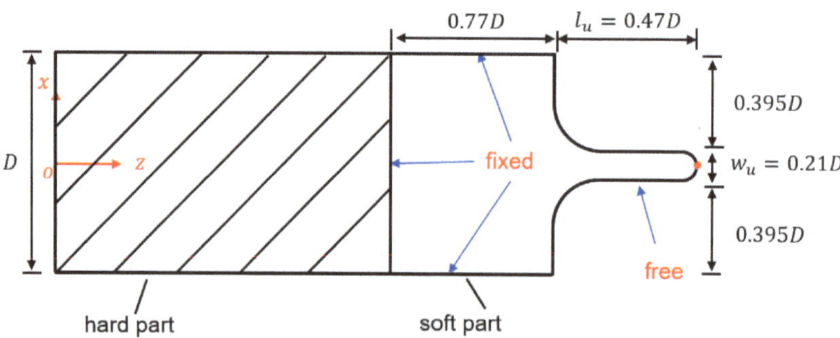

Figure 2. Model geometry for the soft palate with dimensions and the associated boundary conditions.

2.2. Fluid Domain

In the present study, the unsteady compressible 3D Navier-Stokes equations are solved to describe the fluid flow around the human palatal system. The corresponding numerical model is implemented in the commercial software Star-CCM+ (v. 17.06.008) by Siemens. The working fluid is air and it is modeled as an ideal, Newtonian, and isothermal fluid. Large Eddy Simulation (LES) is used to produce high-fidelity simulations of the instantaneous time behaviour of the system by employing less computational resources of a direct numerical simulation. Concerning the human upper respiratory tract, LES have been already successfully employed in several studies, among which the first ever LES applied to OSA was conducted by Mihaescu et al. (2011) [28]. Chen & Gutmark [29] show that LES can give better prediction of the root mean square (RMS) than the Reynolds stress model (RSM). The sub-grid scale model used in this study is the Wall-Adapting Local Eddy-viscosity (WALE) by Nicoud & Ducros [30], due to its flexibility in handling wall bounded flows and the transition regime with a reasonable computational cost. The model is closed by adopting the $k - \omega$ Shear Stress Transport (SST) [31] turbulence model. Since the Mach number of the involved flow in this study is way below 0.3, a pressure-based segregated solver is implemented, so that the conservation equations of mass and momentum are solved in a sequential manner. Despite the low Mach number, the fluid is modeled as compressible. This is motivated by the perspective to employ the current numerical model to directly assess the sound generation due to the structural vibrations [32]. This can be an additional information to be used for quantifying obstructive airway disorder. The spatial mesh grid spacing δ is tuned such that δ is larger than the Kolmogorov scale but smaller than the Taylor micro-scale. For the time integration procedure, a second-order implicit method is applied and the time step size Δt is chosen to satisfy the Courant-Friedrichs-Levy (CFL) criterion, i.e., $CFL < 1.0$.

Flow rate boundary conditions are applied to the two inlets, while the outlet is set as zero pressure boundary. No-slip boundary conditions are imposed on the walls and at the FSI interface. In [33], experimental measurements show that in a human nasal cavity the respiration rates corresponding to calm, medium and intensive breathing are equivalent to 180 mL/s, 560 mL/s and 1100 mL/s, respectively. Table 3 shows the flow rate conditions accordingly employed in this work, where $\dot{Q}$ is the total flow rate, defined as the convex sum of the nasal cavity ($\dot{Q}_1$) and the oral cavity ($\dot{Q}_2$). Consequently, $\dot{Q}_1$ and $\dot{Q}_2$ can be defined in terms of the openness factor α as:

$$\dot{Q}_1 = \alpha \dot{Q} \tag{1}$$

$$\dot{Q}_2 = (1 - \alpha)\dot{Q}, \tag{2}$$

where α ranges from 0.5 to 1.0. The case $\alpha = 0.5$ corresponds to symmetric breathing scenario (mouth-nose breathing), while $\alpha = 1.0$ refers to nose-only breathing with mouth closed. In this work, the FSI behaviour of the system is investigated by considering, for each flow rate in Table 3, 6 different values of the openness factor α.

Table 3. Various respiratory rates in a real human nose [33].

Breathing Condition	Calm	Medium	Intensive
Flow rate, $\dot{Q}$ (L/s)	0.36	1.12	2.2

2.3. Solid Domain

The solid domain is modeled as an isotropic linear elastic material with Young's modulus E, Poisson's ratio ν_{sp} and density ρ_{sp} (see Table 2 for the corresponding numerical values). In the model, gravity is not considered. Finite Element Method (FEM) is applied to account for the structural deformation as response to the shear stress and pressure imposed by the fluid on the fluid-structure interface.

2.4. Mesh Sensitivity Study

The determination of an efficient mesh for both the solid and fluid domains is achieved by means of a mesh sensitivity study. Unstructured polyhedral mesh is applied to the fluid domain. The viscous sub-layer is resolved by adopting 8 prism layers on the walls. Unstructured tetrahedral mesh has been employed for the solid domain. To prevent self-locking i.e., a numerical underestimation of the displacement which can be caused by linear tetrahedral finite element, 10-node quadratic tetrahedral finite elements are employed.

Figure 3a shows the topology of the fluid computational domain. The region surrounding the soft structure, where the most interesting dynamical phenomena happen, is discretized with a finer mesh which is able to resolve the FSI behaviour of the system with high fidelity. A local refinement is applied in the wake region, as is shown in Figure 3a, in order to well resolve the flow characteristics. The mesh for the solid soft structure is shown in Figure 3b. The mesh convergence study is conducted in sequential steps. The first step consists in keeping the solid structure fixed and applying three meshes of 0.6 M, 0.9 M, 1.2 M cells to the flow domain. The inflow boundary condition is set as $\dot{Q} = 0.36\,\text{L/s}$, $\alpha = 1.0$, which corresponds to nose-only calm respiratory. Convergence is assessed by considering time-averaged axial velocity profiles on the line probes shown in Figure 3a. In the second step, the most efficient mesh is implemented for the fluid domain, and three meshes of 0.053 M, 0.3 M, 0.5 M cells are considered for the solid domain which is now allowed to deform. The time histories of the displacement magnitude on the tip point of the structure are monitored and compared to identify the most efficient solid mesh.

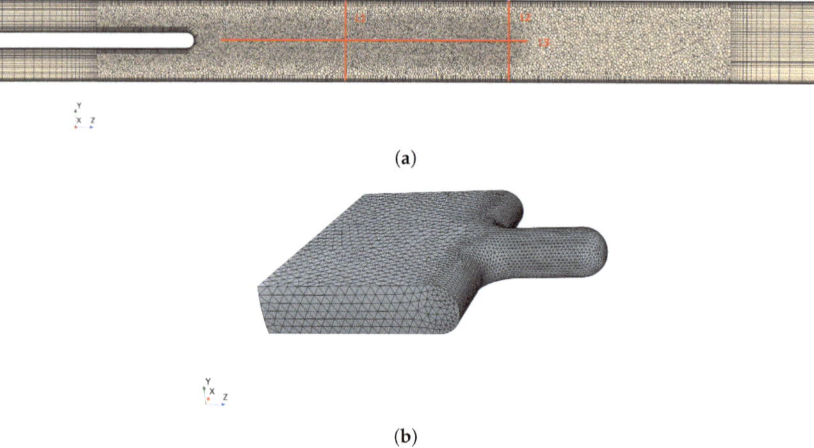

(a)

(b)

Figure 3. (**a**) A cut plane ($x = 0$) of the fluid mesh, showing the locations of the velocity line probes, with L_1, L_2 cross the flow and L_3 in the axial direction. In the same region it is possible to see the mesh refinement applied in the area downstream the structure. (**b**) Mesh of the solid domain.

As shown in Figure 4a–c, the aforementioned 3 fluid meshes with increasing resolution (noted in the figures as *coarse*, *medium*, and *fine* mesh) are applied and reach good convergence in CFD simulations. The solid grid independence study with three solid meshes is also conducted and reaches a similar result in FSI simulations, as shown in Figure 4d. The grid convergence study shows that 0.9 M fluid cells and 0.3 M solid cells are adequate for the FSI simulations.

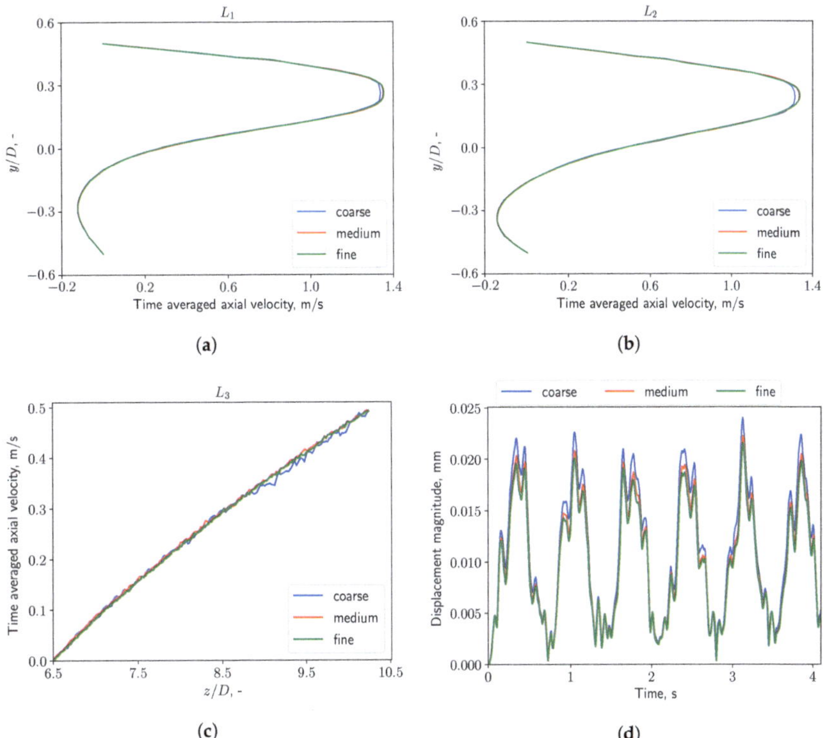

Figure 4. (**a–c**): Time-averaged axial velocity evaluated at line probes L_1, L_2 and L_3 when the solid structure is fixed. y/D, z/D: the non-dimensional y and z coordinate, normalized by the tube diameter D , respectively. (**d**): Time history of the displacement magnitude of the uvula tip.

3. Results

In this section, the results of the numerical simulation are presented and discussed. Section 3.1 shows the flow patterns and characteristics by analyzing the pressure fluctuation spectra in the wake region. The influence of different breathing patterns, as well as different inlet opening conditions are investigated (see Table 3). In Section 3.2, a modal analysis of the solid structure is presented to understand the vibration characteristics of the structure. When the structure is allowed to move, the solid and the fluid come to interact. The corresponding results are shown in Section 3.3.

3.1. Analysis of Fluid Domain

The axial component of the velocity fields are presented in Figure 5. The flow of the calm breathing case ($\dot{Q} = 0.36$ L/s) presents symmetric wake behind the body and no pronounced fluctuations are seen. When the boundary conditions are switched to medium breathing scenario ($\dot{Q} = 1.12$ L/s), the wake starts to oscillate due to the growth of the fluctuations in the flow. When the inlet flow rate is increased to 2.2 L/s (intensive breathing), the wake is characterised by a vortex shedding from the body representing the palate system.

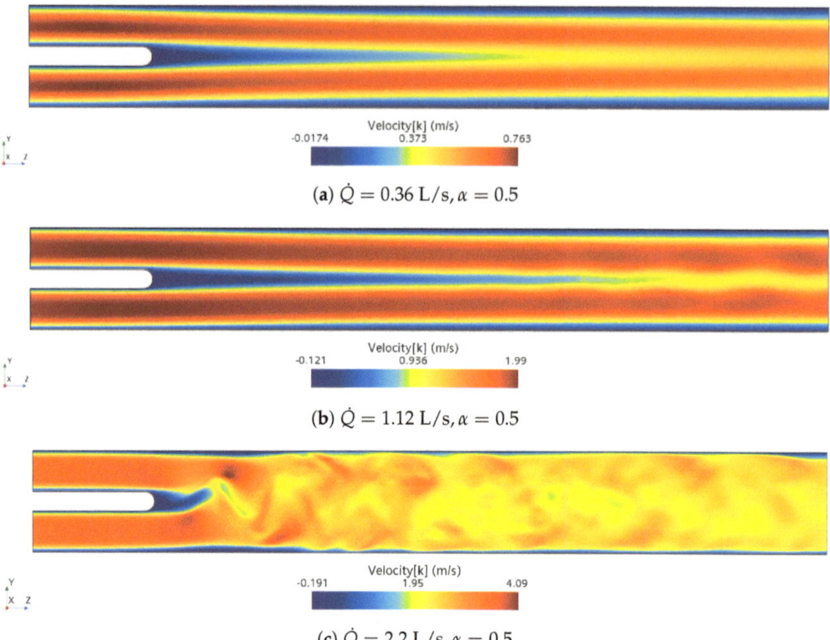

(**a**) $\dot{Q} = 0.36 \, \mathrm{L/s}, \alpha = 0.5$

(**b**) $\dot{Q} = 1.12 \, \mathrm{L/s}, \alpha = 0.5$

(**c**) $\dot{Q} = 2.2 \, \mathrm{L/s}, \alpha = 0.5$

Figure 5. Instantaneous axial velocity fields on the $x = 0$ cut plane for three breathing flow rates with symmetric inlet boundary conditions.

Figure 6 displays the mean axial flow velocity when nose-only breathing ($\alpha = 1.0$) occurs. In this condition, the flow coming from the nasal inlet (i.e., the upper inlet) is dominant over the oral one (lower inlet). The flow mixes downstream the body and recirculation bubbles downstream and under the body appear as described by the streamlines. Concerning the flow unsteadiness, the oscillating frequency of the flow is evaluated by analyzing the spectrum of the pressure fluctuations at the point probe located downstream the uvula shown in Figure 1. The Strouhal number is defined as:

$$St = \frac{fh}{u_{mix}}, \tag{3}$$

with f the oscillating frequency in the wake, h the height of the solid structure, and u_{mix} the mixed surface-averaged velocity of the two inlets. And u_{mix} is defined in the following form:

$$u_{mix} = \sqrt{\alpha \left(\frac{\dot{Q}_1}{S}\right)^2 + (1 - \alpha)\left(\frac{\dot{Q}_2}{S}\right)^2}, \tag{4}$$

where S is the cross section area of one of the inlets.

In Figure 7a, one can see when in restful breathing condition ($\dot{Q} = 0.36 \, \mathrm{L/s}$), the oscillation of the flow is negligible. When the respiratory rate is increased to the medium condition ($\dot{Q} = 1.12 \, \mathrm{L/s}$), the flow oscillates with a frequency ranging from 40 Hz to 56 Hz. For intensive breathing condition ($\dot{Q} = 2.2 \, \mathrm{L/s}$), the frequency varies between 73 Hz and 130 Hz, depending on the openness of the inlets (i.e., the α parameter). Figure 7b shows that the Strouhal numbers St for both medium and intensive breathing conditions vary in the vicinity of 0.2, which is expected by the presented flow characteristics. The predicted frequencies in this study lie within the palatal snoring frequency range 21–323 Hz, which was evaluated from snoring acoustic measurements conducted by Brietzke & Mair (2006) [34]. A study by Agrawal et al. (2002) [35] revealed that patients with palatal snoring had a median peak frequency at 137 Hz.

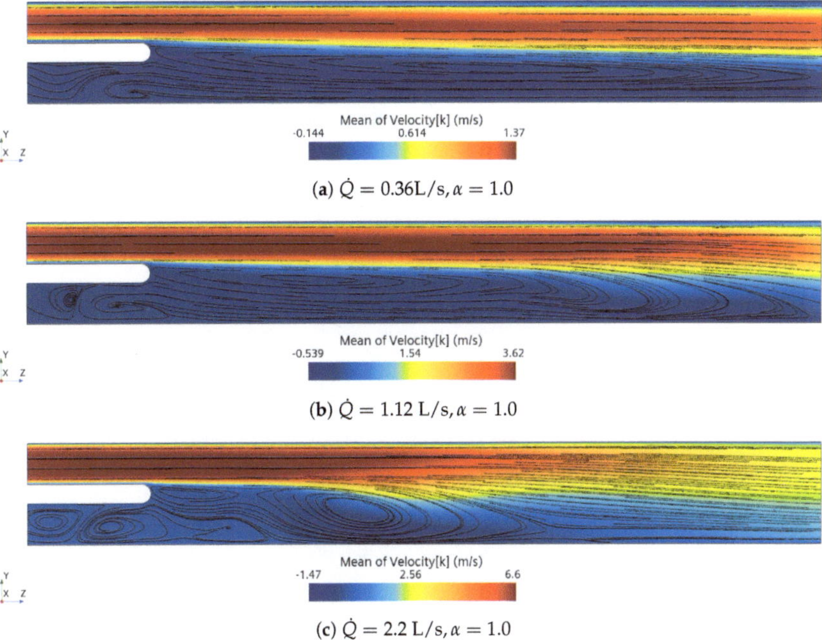

(**a**) $\dot{Q} = 0.36\,L/s, \alpha = 1.0$

(**b**) $\dot{Q} = 1.12\,L/s, \alpha = 1.0$

(**c**) $\dot{Q} = 2.2\,L/s, \alpha = 1.0$

Figure 6. Axial velocity mean fields on the yz cut plane for three respiratory flow rates with nose-only breathing.

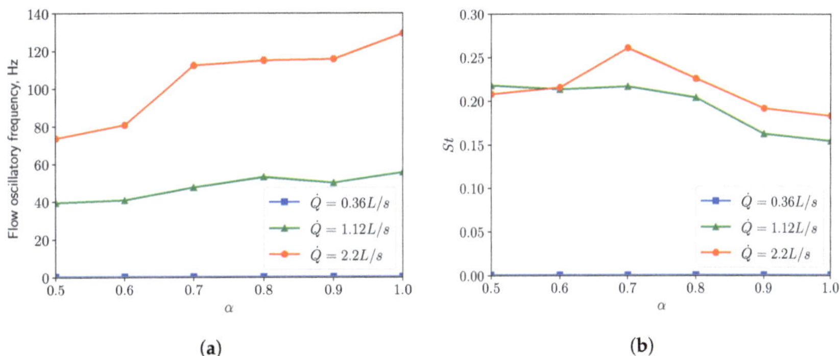

(**a**)

(**b**)

Figure 7. (**a**) shows the flow oscillatory frequency in the wake against the openness factor α for three different respiratory rates, with (**b**) displaying the corresponding Strouhal number.

3.2. Modal Analysis of Solid Structure

The purpose of modal analysis is to determine the eigenmode shapes and the corresponding natural frequencies of the solid structure during free vibration. The distribution of energy in the set of eigenmodes is not constant and often only few modes are needed to represent the motion of a structure with an acceptable approximation. In the present study, the modal analysis is carried out by employing the commercial FEM solver COMSOL Multiphysics (v6.1). The same clamped boundary conditions are applied to the three side edges, while the other edge is set free, as is shown in Figure 2.

To determine which modes are significant for the representation of the response of the structure when subjected to external loads, the effective mass associated with each mode is calculated. It measures the amount of the total mass participating in that mode in a given excitation direction. A mode with large effective mass is a significant contributor to the structural response in a given excitation direction. Figure 8 shows the ratio of effective mass to the total mass of the solid η in percentage plotted against the first 50 modes for the three main directions. A threshold value of $\eta > 5\%$ has been selected to define the most significant modes. Moreover, as will be discussed in Section 3.3, the vertical displacement (y direction) is dominant under most of the boundary conditions. Consequently, the analysis will focus only on the modes concerning vertical displacement (red dots in Figure 8). Modes 1, 3, 6, and 14 are the most significant modes for the vertical deformation, as can be seen in Table 4. Their mode shapes are presented in Figure 9. Mode 1 is a simply bending mode, while the other three are higher-order modes. Mode 14, the lateral (in the x-direction) part of the soft palate undergoes larger deformation compared with the uvula's tip, whereas the other three modes have larger deformation on the tip point.

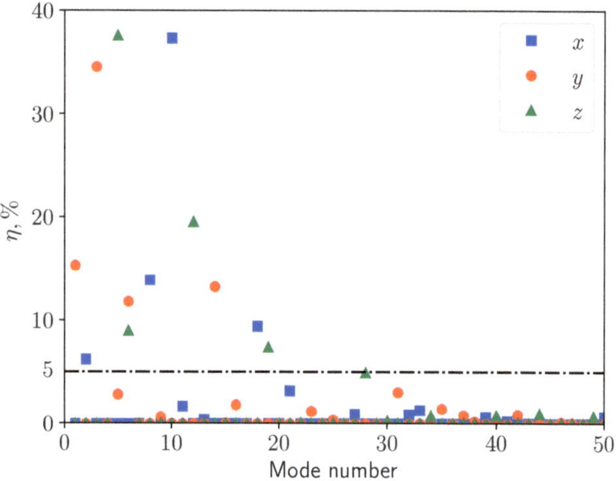

Figure 8. Effective mass ratio.

Table 4. The first 15 modes of the solid structure with the effective mass ratio in y direction.

Mode	Natural Frequency (Hz)	Effective Mass Ratio, η (%)
1	1.428	15.32
2	2.623	7.22×10^{-8}
3	3.979	34.56
4	6.122	2.07×10^{-5}
5	6.726	2.83
6	6.762	11.81
7	8.515	3.95×10^{-10}
8	8.712	5.75×10^{-8}
9	10.061	0.64
10	11.086	4.20×10^{-8}
11	11.090	5.47×10^{-6}
12	11.943	9.74×10^{-9}
13	13.348	1.00×10^{-3}
14	13.384	13.26
15	14.221	0.05

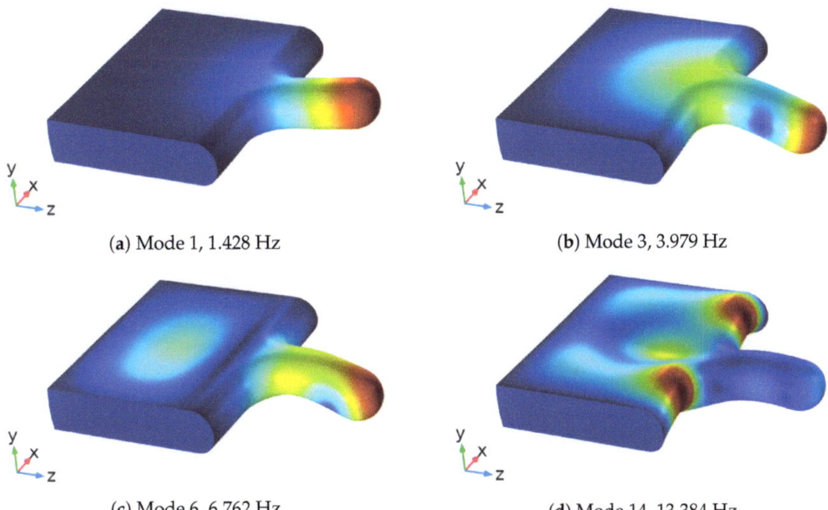

(**a**) Mode 1, 1.428 Hz

(**b**) Mode 3, 3.979 Hz

(**c**) Mode 6, 6.762 Hz

(**d**) Mode 14, 13.384 Hz

Figure 9. The 4 most significant modes for the vertical deformation of the structure. Color represents the displacement magnitude.

3.3. FSI Simulations

In fully coupled FSI scenario, the soft palate starts to vibrate due to the interaction with the flow. The pressure and the wall shear stress imposed by the fluid are driving the motion of the soft palate. By following the same boundary conditions scheme of the previous section, the simulation's goal is to investigate the effects of the three breathing regimes on the motion of the structure. In particular, the displacement on the tip point of the structure (see Figure 2) is monitored. Figure 10 shows the maximum displacement magnitude of the tip point. The displacement magnitude is defined as

$$\xi = \sqrt{\xi_x^2 + \xi_y^2 + \xi_z^2}, \tag{5}$$

with ξ_x, ξ_y, ξ_z the displacement in x, y and z direction, respectively. The maximum displacement magnitude indicates the largest deformation that the structure can have and it is employed as a kinematic representative of the dynamics of the solid.

Figure 10 shows that as $\dot{Q}$ grows, the maximum displacement magnitude $|\xi|_{max}$ increases. This can be expected since the fluid will exert larger pressure and wall shear stress on the structure, leading to larger deformation. A similar pattern is displayed in Figure 11, which shows that higher respiratory rate results in larger pressure difference between the top and bottom surfaces of the uvula, which causes to larger deformation. For the cases of calm ($\dot{Q} = 0.36$ L/s) and medium ($\dot{Q} = 1.12$ L/s) breathing, the structure undergoes larger displacement as the openness factor α increases. One possible explanation is that the flow from the nasal inlet is dominant over the flow coming from the oral inlet. Consequently, the pressure difference acting on the structure increases, as is shown in Figure 11, causing larger deformation in the structure. However, in the case of intensive respiration, $|\xi|_{max}$ and α are not in a monotonically increasing relation anymore. The structure undergoes large deformation when the inlet boundary conditions are symmetric ($\alpha = 0.5$). The displacement decreases quickly as α increases from 0.5 to 0.6. For larger values of α, the displacement increases again almost in a monotonic fashion. When α reaches 0.9, a quick increment in the deformation emerges. This behavior is probably because more fluctuations and small structures in the flow are generated and provide more energy to the vibration of the soft palate. In the next subsections, a careful analysis of the behaviour of the soft palate under different breathing conditions is provided. In particular, to have a better understanding of the underlying physical mechanisms, the displacement spectra of the structure will be analysed.

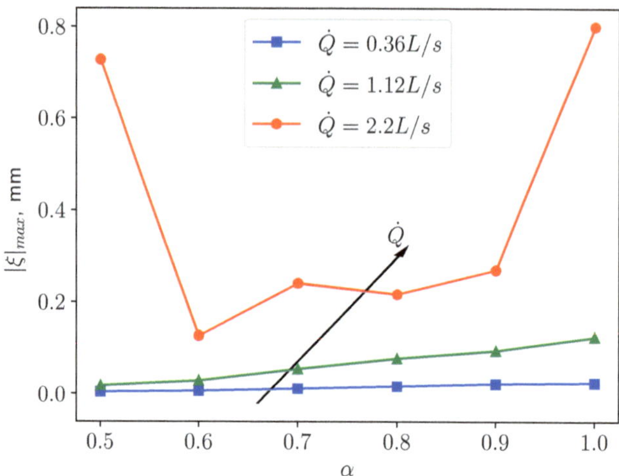

Figure 10. The maximum tip displacement magnitude of the structure under different respiratory rates.

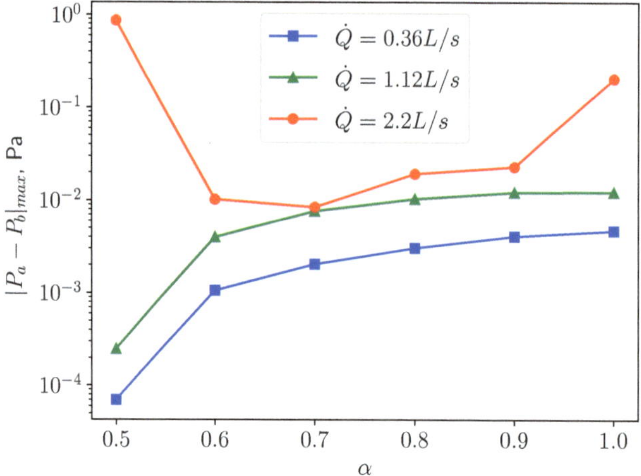

Figure 11. The maximum pressure difference between point probes *a* and *b* (see Figure 1) located above the top and below the bottom surfaces of the uvula, respectively.

3.3.1. Calm Breathing: $\dot{Q} = 0.36$ L/s

For the calm respiratory scenario, the time histories of the displacement components of the uvula's tip are shown in Figure 12, for different values of α. It can be seen that higher values of α correspond to a larger vertical displacement (ζ_y). The amplitude of the span-wise displacement (ζ_x) is almost constant w.r.t. α, while the axial displacement (ζ_z) amplitude slightly increases. Moreover, ζ_y is dominant over the other two components when the inlet boundary conditions are not symmetric, i.e., $\alpha \neq 0.5$. The pressure difference between the top and the bottom surfaces of the solid structure grows, causing larger vertical deformation. For this reason, the following analysis concerns only the vertical displacement.

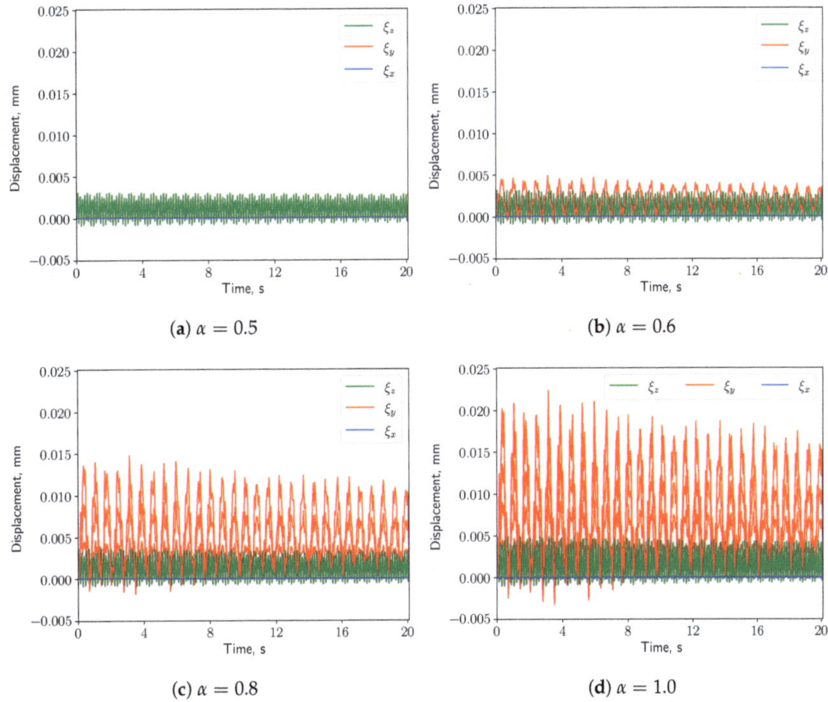

Figure 12. (**a**–**d**) show the displacement components at the structure tip when $\dot{Q} = 0.36$ L/s.

Figure 13 presents the corresponding spectra of the vertical displacement (ξ_y), with the Power Spectrum Density (PSD) plotted versus the frequency. The first two dominant frequencies are compared with the natural frequencies of the solid structure in Figure 14, with f_1 and f_5 corresponding to the first and the fifth eigen-frequencies of the solid domain (see Table 4). From the modal analysis in Section 3.2, the larger contribution to the vertical deformation comes from the first mode. The power level of the first peak increases as α increases, as can be seen in the spectrum. With higher openness factor α, the larger difference in pressure between the two sides of the solid induces larger deformation in the soft palate in the vertical direction.

Figure 15 shows the time history of the pressure at the probe points *a* and *b* shown in Figure 1 on the top and bottom of the uvula. Both pressure signals are in phase, indicating that the effect of the vortex shedding are negligible under such low respiratory rate. Since the fluid flow is not oscillating, no resonance effects between the solid and the fluid are expected (see Figure 7a).

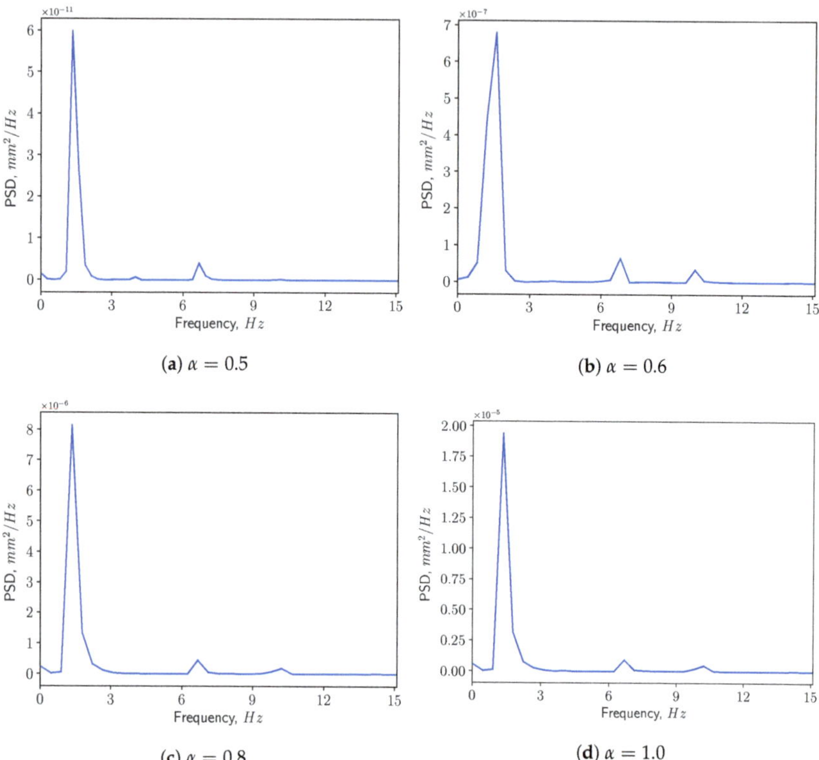

(a) $\alpha = 0.5$

(b) $\alpha = 0.6$

(c) $\alpha = 0.8$

(d) $\alpha = 1.0$

Figure 13. Power spectrum density of the vertical displacement (ζ_y) at the tip in calm breathing ($\dot{Q} = 0.36$ L/s).

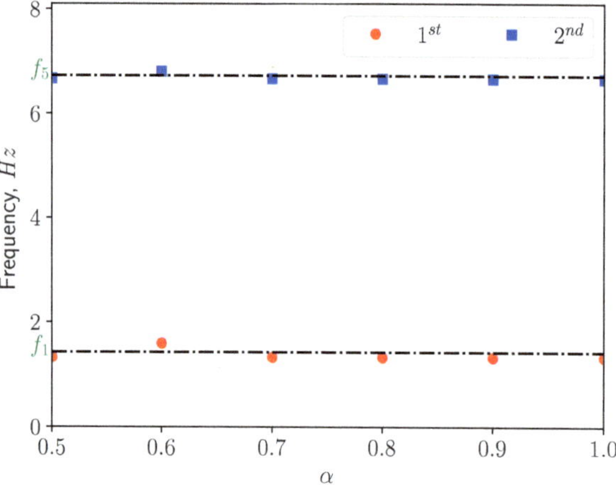

Figure 14. The first two dominant frequencies of the power spectrum of the vertical displacement (ζ_y) in calm breathing ($\dot{Q} = 0.36$ L/s).

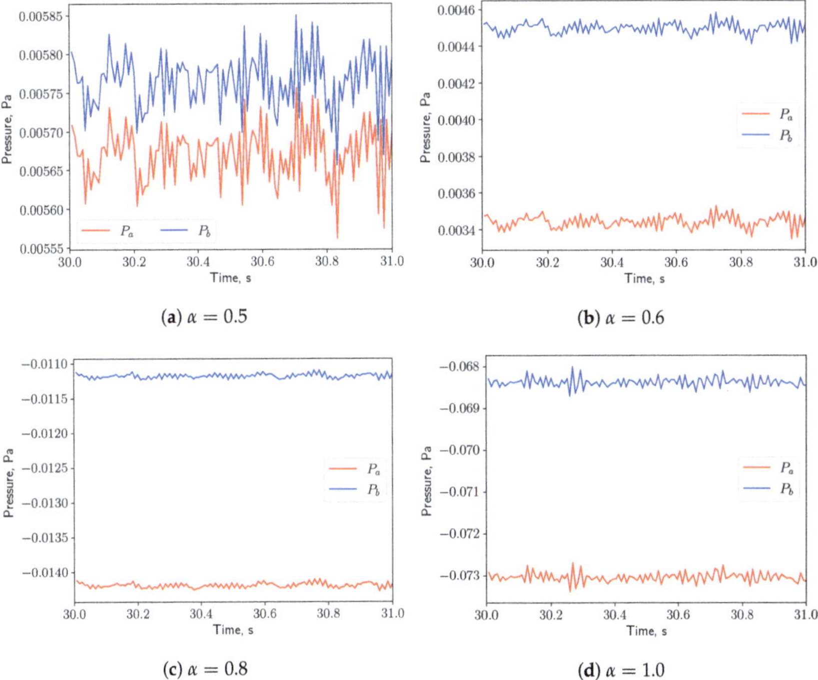

(**a**) $\alpha = 0.5$

(**b**) $\alpha = 0.6$

(**c**) $\alpha = 0.8$

(**d**) $\alpha = 1.0$

Figure 15. (**a**–**d**) show the pressure signals on the top and bottom surface of the uvula in calm breathing ($\dot{Q} = 0.36$ L/s).

3.3.2. Medium Breathing: $\dot{Q} = 1.12$ L/s

When the medium respiratory rate boundary conditions are employed, the solid structure undergoes larger deformation compared with the calm breathing cases. Remarkably, also the span-wise and axial components of the displacement shown in Figure 16 increase as α grows. This is mainly because the pressure gradients in vertical and span-wise directions as well as the drag force acting on the solid structure are becoming larger, compared with the calm breathing cases. When $\alpha = 0.5$, due to the symmetry of the inlet boundary conditions and of the geometry, the small pressure differences in x & y directions induce small deformation in these two directions w.r.t. the vertical direction. The axial deformation is dominant, which is mainly driven by the drag force imposed by the fluid. As long as the inlet boundary conditions are no more symmetric, i.e., $\alpha \neq 0.5$, the vertical deformation starts to increase rapidly w.r.t. the other ones.

The power spectrum of the solid displacement ζ_y is computed and presented in Figure 17. The first two dominant frequencies are shown in Figure 18. The frequency corresponding to the first peak coincides with the first eigen-frequency of the structure. This means that under this conditions, the first eigen-mode of the structure is excited. This can be confirmed by Figure 18, where f_2, f_5 and f_9 are the natural frequencies of the eigen-modes 2, 5, and 9. From Table 4, one can see the contributions from modes 2, 5, and 9 are less than 5%. Additionally, it can be noted that the power level of the first peak increases as α grows since the amplitude of ζ_y is larger.

The pressures at the probe points shown in Figure 19 oscillate in phase for all the values of α, indicating that the effects of the vortex shedding are negligible under this breathing scenarios. Resonance is not expected, since the natural frequency of the solid (1.428 Hz) is not matching from the flow oscillatory frequency (see Figure 7a).

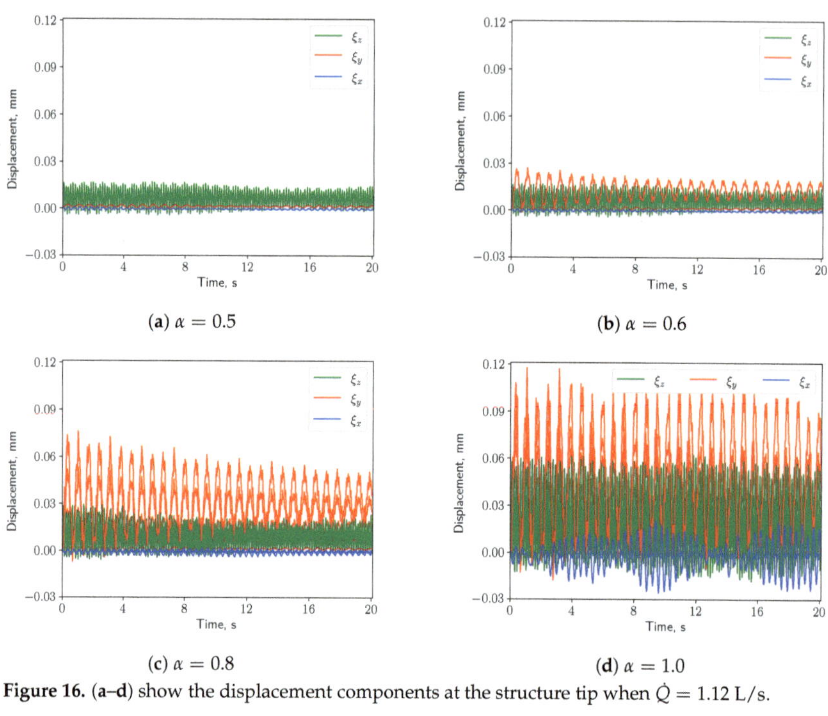

Figure 16. (**a**–**d**) show the displacement components at the structure tip when $\dot{Q} = 1.12$ L/s.

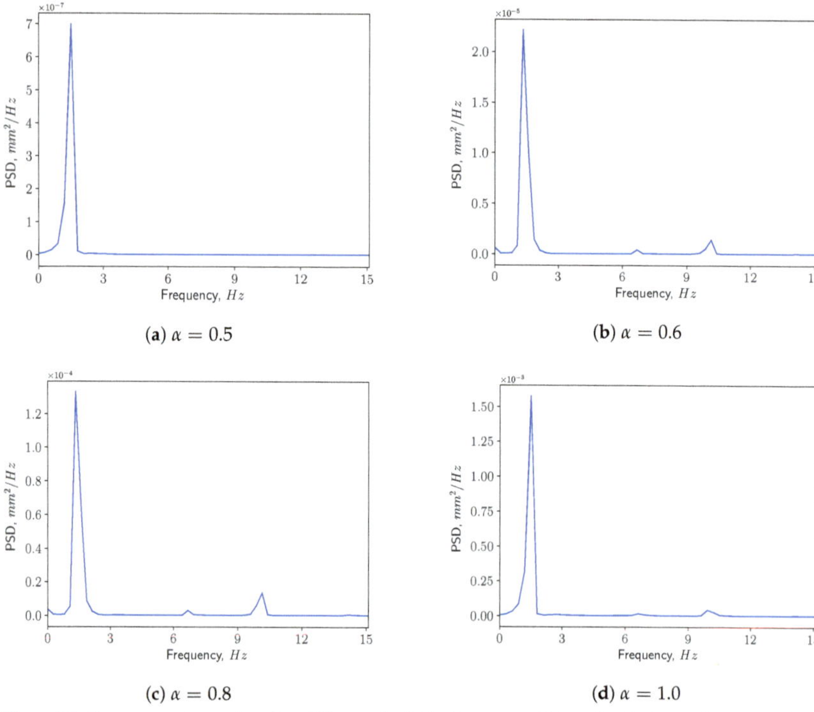

Figure 17. Power spectrum density of the vertical displacement (ξ_y) at the tip in medium breathing ($\dot{Q} = 1.12$ L/s).

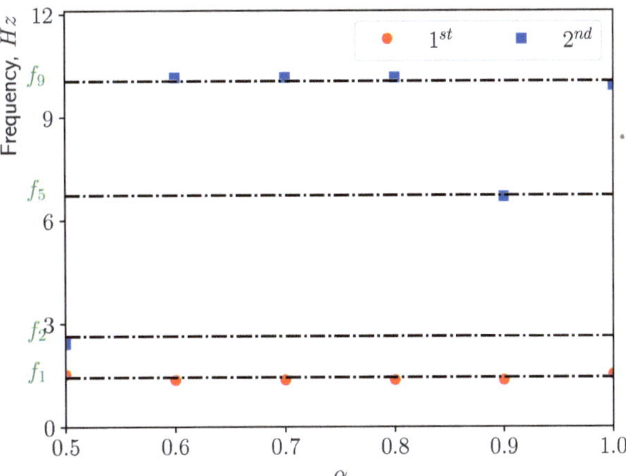

Figure 18. The first two dominant frequencies of the power spectrum of the vertical displacement (ζ_y) in medium breathing ($\dot{Q} = 1.12\ \text{L/s}$).

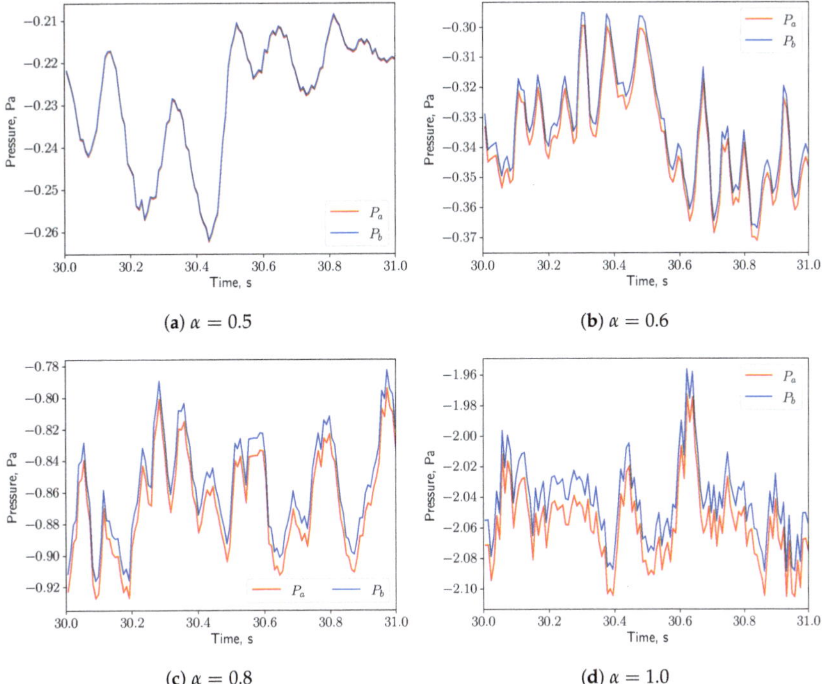

(a) $\alpha = 0.5$

(b) $\alpha = 0.6$

(c) $\alpha = 0.8$

(d) $\alpha = 1.0$

Figure 19. (a–d) show the pressure signals on the top and bottom surface of the uvula in medium breathing ($\dot{Q} = 1.12\ \text{L/s}$).

3.3.3. Intensive Breathing: $\dot{Q} = 2.2\ \text{L/s}$

When intensive breathing rate is implemented at the inlet boundaries, the deformation of the structure is larger (see Figure 20). Interestingly, under symmetric condition ($\alpha = 0.5$) the structure undergoes comparable deformation as the case of nose-only breathing ($\alpha = 1.0$). The displacements decrease as α grows from 0.5 to 0.6, and then increase again for larger values of α. Such behavior is remarkably different than the calm and medium breathing cases. Figure 5 shows a vortex which is shed from the soft palate, whereas in the other two cases no pronounced vortex shedding appears. The large deformation in Figure 20a is connected to the vortex shedding, which induces a phase shift in the pressure signal on both sides of the soft palate.

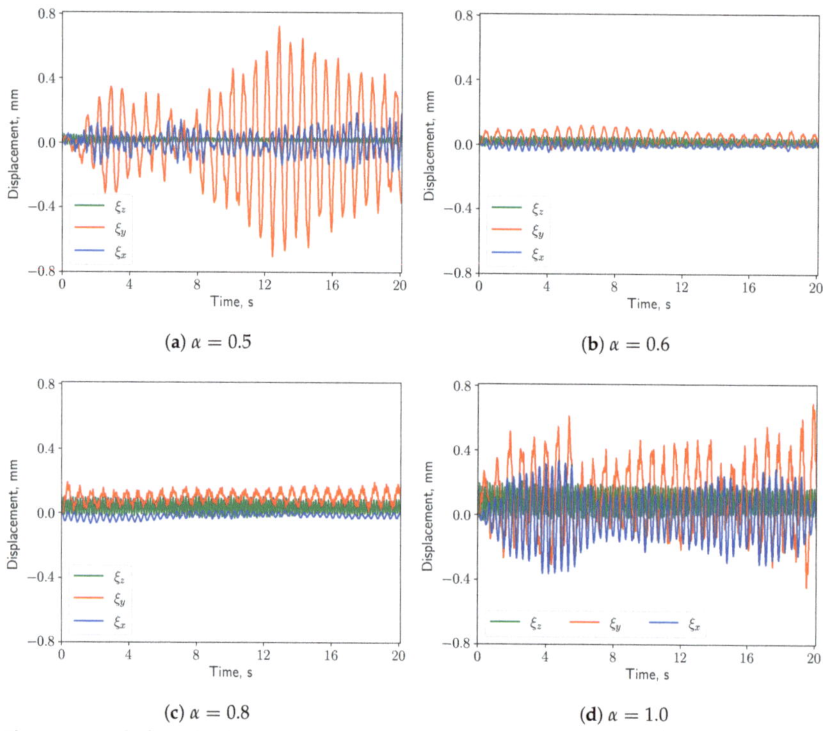

Figure 20. (**a**–**d**) show the displacement components at the structure tip when $\dot{Q} = 2.2\,\text{L/s}$.

Figure 21 shows the power spectrum density of the vertical displacement of the solid structure. The first two dominant frequencies for each case are presented in Figure 22, comparing with the eigen-frequencies of the structure. In the spectra, one can see that for most of the cases the first dominant frequency still coincides with the first natural frequency of the solid structure. This means that the first eigen-mode of the structure plays significant role in the vertical bending of the soft palate. For the case with $\alpha = 0.9$, the first dominant frequency is close to the 9th eigen-frequency of the solid. From Table 4, it is possible to remark that the effective mass ratio associated to this mode is less than 1%. Consequently, it is possible to neglect the contribution of this mode on the displacement.

In Figure 23, for symmetric breathing condition ($\alpha = 0.5$), the time histories of the pressure at the probe points on the top and bottom of the uvula are presented. It is possible to remark that the pressure signals have a 180 deg phase shift, which is associated to the vortex shedding. High pressure and low pressure zones appear alternately on the both sides of the solid structure. This gives a positive feedback on the structure and eventually leads to larger deformation. When α is increased to 0.6, the flow is no more in a symmetric condition and the pressures on the top and bottom surfaces of the structure oscillate in phase. No positive feedback from the flow is added to the solid structure and the displacement decreases. Even though the vortex shedding process is present in the symmetric case and the vibration of the solid structure is intensified, no resonance effects are expected, as the natural frequency of the solid (1.428 Hz) does not match with the vortex shedding frequency (see Figure 7a).

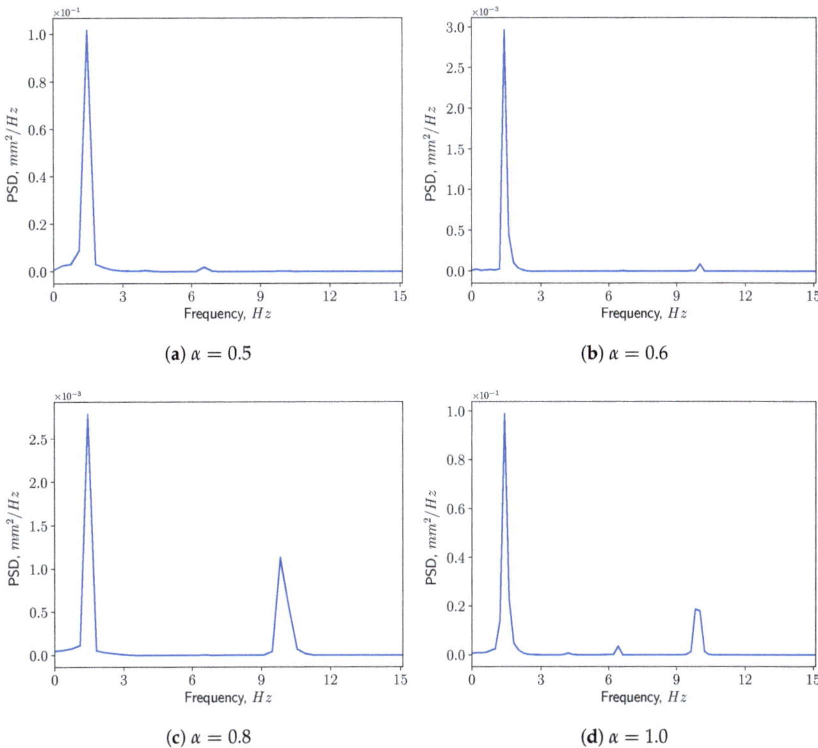

(**a**) $\alpha = 0.5$

(**b**) $\alpha = 0.6$

(**c**) $\alpha = 0.8$

(**d**) $\alpha = 1.0$

Figure 21. Power spectrum density of the vertical displacement (ξ_y) at the tip in intensive breathing ($\dot{Q} = 2.2\,\text{L/s}$).

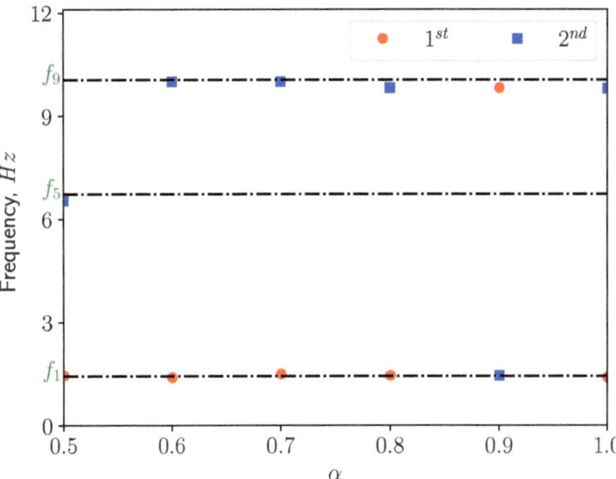

Figure 22. The first two dominant frequencies of the power spectrum of the vertical displacement (ξ_y) in intensive breathing ($\dot{Q} = 2.2\,\text{L/s}$).

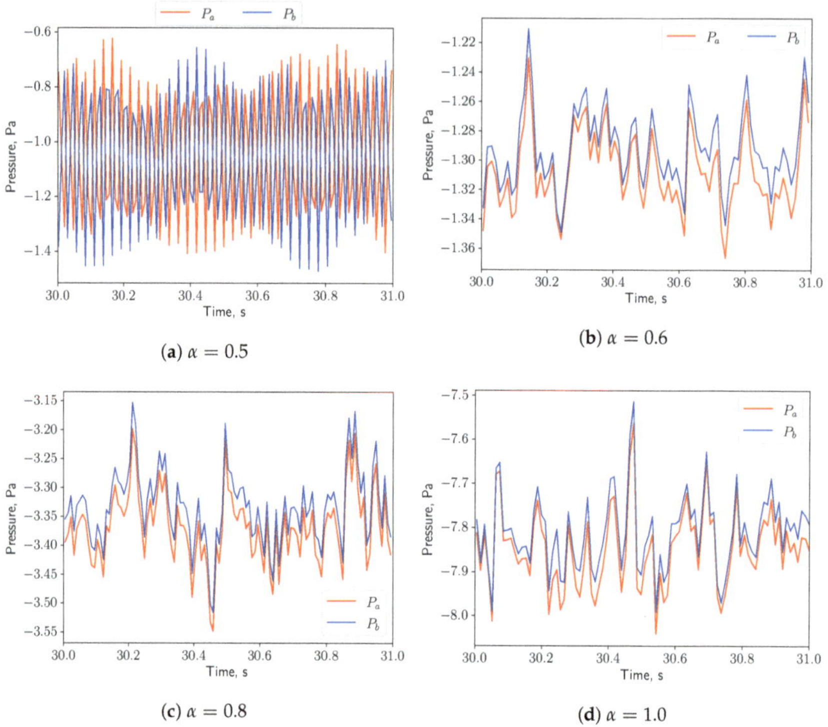

(a) $\alpha = 0.5$

(b) $\alpha = 0.6$

(c) $\alpha = 0.8$

(d) $\alpha = 1.0$

Figure 23. (**a**–**d**) show the pressure signals on the top and bottom surface of the uvula in intensive breathing ($\dot{Q} = 2.2$ L/s).

4. Discussion & Conclusions

A simplified 3D airway model including the soft palate in physiological boundary and geometric conditions is developed to study the interaction among the uvulopalatal system and the airflow by employing two-way coupling FSI analysis. The vibrational response of the soft structure is found to be considerably dependent on the investigated parameters i.e., the inlet flow rate $\dot{Q}$ and the inlet openness factor α.

The present study's simplified 3D airway model makes it possible to investigate the uvulopalatal vibrational characteristics and the airflow changes. The FSI analysis shows that larger structural deformations are present for increasing inlet respiratory rate $\dot{Q}$. In the cases of calm and medium breathing scenarios, the structure undergoes larger displacements for larger values of the openness factor α. In intensive respiration, the structure undergoes large deformations for $\alpha = 0.5$. This is mainly due to the positive feedback from the vortex shedding of the flow. However, the resonance between the structure and the fluid does not occur since the natural frequency and the flow oscillation frequency do not match. The effects of the axial velocity on the components of the structural displacement can be investigated by analyzing the results from the analysis of the volume flow rate $\dot{Q}$. When $\dot{Q}$ is increasing, as can be seen from Figures 12, 16 and 20, y displacement (vertical) and z displacement (axial) grow accordingly.

As is shown in Figure 20a, from a phenomenological standpoint, the vortex-induced vibration is not associated to a linear resonance. The 3D FSI description provided in this work, is able to link the more complex vortex shedding phenomenology to the motion of the soft palate. The 3D description allows for an in-depth description of the coupling between the airflow and the soft palate. Such understanding is crucial to determine the behaviour of the soft palate under pathological conditions. As the factor α increases, the pressures on the top and bottom surfaces of the structure are oscillating in phase due to the non-symmetric boundary conditions. For $0.6 < \alpha < 0.9$, the deformation increases almost monotonically. There is a sharp increment when α varies from 0.9 to 1.0. This is probably because more fluctuations and small structures in the flow are generated and provide more energy to the solid structural vibration.

The solver used in this paper has been previously validated using available experimental data for different simulated scenarios. The chosen numerical flow solver setup has been validated in a previous study by comparing the predicted surface pressures resulting from the compressible flow solution through a vocal folds model against corresponding experimental pressure wall measurements [32]. In a different study which considered changes in the convergent-divergent vocal tract model geometry (using different trans-glottal angles), the compressible flow solver has been validated by comparing the pressure predictions to corresponding available experimental mid-line pressure data [36]. Concerning the abilities of the current solver to predict deformations, the buckling critical pressure of a collapsible tube as a function of cross-sectional area—the "tube law"—has been obtained and validated against experimental data for different geometrical parameters associated with the flexible tube [16]. In all these cases, an overall good agreement between the pressure values predicted by the solver and the experimental pressure measurements has been reached.

One of the limitations of this study is that gravity and damping effects are not included in the modelling scheme. The effect of these assumptions on the structural response will be object of a future investigation. Another limitation is that the solid structure is assumed to be linear elastic. Figure 10 shows that the largest deformation under physiological conditions can be as high as 11% compared with the height of the structure. A parametric study comparing different material laws, such as Neo-Hookean elasticity and the Ogden hyperelasticity, is to be conducted. In this work, the variability and uncertainty of the geometry and material properties of the uvulopalatal system is not investigated. A further study considering these uncertainty effects [37] is planned.

An interesting perspective to generalize the results of this work, is to consider unsteadiness in the boundary conditions, to include in the picture the natural time dependence of the human respiratory rate. Another future work is to study the effects of the pharyngeal airway occlusion on the dynamics of the uvulopalatal system.

Author Contributions: P.L. conceived the simulations and the post-processing techniques. M.L. collaborated in the implementation of the simulations and data processing. M.M. initiated the project and collaborated in the analysis of the results. All authors have read and agreed to the published version of the manuscript.

Funding: The project is funded by Swedish Research Council through Grant VR 2020-04857. M.L. is financed by KTH Engineering Mechanics in the thematic areas Biomechanics, Health and Biotechnology (BHB) and by the Swedish Research Council Grant VR 2022-03032. M.M. is funded by the Swedish Research Council Grant VR 2020-04857.

Institutional Review Board Statement: Not applicable.

Informed Consent Statement: Not applicable.

Data Availability Statement: Simulation and post-processed data can be provided by the corresponding author under reasonable request.

Acknowledgments: The authors acknowledge the high performance computing (HPC) resources provided by the National Academic Infrastructure for Supercomputing in Sweden (NAISS, Project NAISS 2023/1-19) at PDC Centre for High-Performance Computing (PDC-HPC) and at National Supercomputer Center (NSC). The authors acknowledge PRACE for awarding access to the Fenix Infrastructure resources at CINECA, which are partially funded from the European Union's Horizon 2020 research and innovation programme through the ICEI project under the grant agreement No. 800858.

Conflicts of Interest: The authors declare that there are no conflicts of interest regarding the publication of this article.

Abbreviations

The following abbreviations are used in this manuscript:

OSA	Obstructive Sleep Apnea
OSAS	Obstructive Sleep Apnea Syndrome
FSI	Fluid Structure Interaction
CFD	Computational Fluid Dynamics
FEM	Finite Element Method
PSD	Power Spectrum Density
MRI	Magnetic Resonance Imaging
CT	Computer Tomography

References

1. Peppard, P.E.; Young, T.; Barnet, J.H.; Palta, M.; Hagen, E.W.; Hla, K.M. Increased prevalence of sleep-disordered breathing in adults. *Am. J. Epidemiol.* **2013**, *177*, 1006–1014. [CrossRef] [PubMed]
2. Romero, E.; Krakow, B.; Haynes, P.; Ulibarri, V. Nocturia and snoring: Predictive symptoms for obstructive sleep apnea. *Sleep Breath.* **2010**, *14*, 337–343. [CrossRef] [PubMed]
3. Lechat, B.; Naik, G.; Reynolds, A.; Aishah, A.; Scott, H.; Loffler, K.A.; Vakulin, A.; Escourrou, P.; McEvoy, R.D.; Adams, R.J.; et al. Multinight prevalence, variability, and diagnostic misclassification of obstructive sleep apnea. *Am. J. Respir. Crit. Care Med.* **2022**, *205*, 563–569. [CrossRef] [PubMed]
4. Young, T.; Palta, M.; Dempsey, J.; Skatrud, J.; Weber, S.; Badr, S. The occurrence of sleep-disordered breathing among middle-aged adults. *N. Engl. J. Med.* **1993**, *328*, 1230–1235. [CrossRef]
5. Patel, J.A.; Ray, B.J.; Fernandez-Salvador, C.; Gouveia, C.; Zaghi, S.; Camacho, M. Neuromuscular function of the soft palate and uvula in snoring and obstructive sleep apnea: A systematic review. *Am. J. Otolaryngol.* **2018**, *39*, 327–337. [CrossRef] [PubMed]
6. Osman, A.M.; Carter, S.G.; Carberry, J.C.; Eckert, D.J. Obstructive sleep apnea: Current perspectives. *Nat. Sci. Sleep* **2018**, *10*, 21–34. [CrossRef]
7. Malhotra, A.; White, D.P. Obstructive sleep apnoea. *Lancet* **2002**, *360*, 237–245. [CrossRef]
8. Huang, L.; Quinn, S.J.; Ellis, P.D.; Williams, J.E.F. Biomechanics of snoring. *Endeavour* **1995**, *19*, 96–100. [CrossRef]
9. Blausen Medical. Medical gallery of Blausen medical. *Wiki J. Med.* **2014**, *1*, 1–79.
10. Ambekar, A.A.; Saksena, S.G.; Bapat, J.S.; Butani, M.T. Correlation of bedside airway screening tests with airway obstruction during drug-induced sleep endoscopy. *Asian J. Anesthesiol.* **2019**, *57*, 117–124.
11. Zhao, M.; Barber, T.; Cistulli, P.A.; Sutherl, K.; Rosengarten, G. Simulation of upper airway occlusion without and with mandibular advancement in obstructive sleep apnea using fluid-structure interaction. *J. Biomech.* **2013**, *46*, 2586–2592. [CrossRef] [PubMed]
12. Mylavarapu, G.; Gutmark, E.; Shott, S.; Fleck, R.; Mahmoud, M.; McConnell, K.; Szczesniak, R.; Hossain, M.M.; Huang, G.; Tadesse, D.G.; et al. Predicting critical closing pressure in children with obstructive sleep apnea using fluid-structure interaction. *J. Appl. Physiol.* **2021**, *131*, 1629–1639. [CrossRef] [PubMed]
13. Pirnar, J.; Dolenc-Grošelj, L.; Fajdiga, I.; Žun, I. Computational fluid-structure interaction simulation of airflow in the human upper airway. *J. Biomech.* **2015**, *48*, 3685–3691. [CrossRef] [PubMed]
14. Huang, R.; Li, X.; Rong, Q. Control mechanism for the upper airway collapse in patients with obstructive sleep apnea syndrome: A finite element study. *Sci. China Life Sci.* **2013**, *56*, 366–372. [CrossRef]
15. Chen, Y.; Feng, X.; Shi, X.Q.; Cai, W.; Li, B.; Zhao, Y. Computational fluid–structure interaction analysis of flapping uvula on aerodynamics and pharyngeal vibration in a pediatric airway. *Sci. Rep.* **2023**, *13*, 2013. [CrossRef]
16. Laudato, M.; Mosca, R.; Mihaescu, M. Buckling critical pressures in collapsible tubes relevant for biomedical flows. *Sci. Rep.* **2023**, *13*, 9298. [CrossRef]
17. Gregory, A.L.; Agarwal, A.; Lasenby, J. An experimental investigation to model wheezing in lungs. *R. Soc. Open Sci.* **2021**, *8*, 201951. [CrossRef]
18. Tetlow, G.A.; Lucey, A.D. Motions of an offset plate in Viscous channel flow: A model for flutter of the soft palate. *World Congr. Med. Phys. Biomed. Eng.* **2006**, *14*, 3457–3460.
19. Khalili, M.; Larsson, M.; Müller, B. Interaction between a simplified soft palate and compressible viscous flow. *J. Fluids Struct.* **2016**, *67*, 85–105. [CrossRef]
20. Cisonni, J.; Lucey, A.D.; Elliott, N.S.J. Tapered-Cantilever Based Fluid-Structure Interaction Modelling of the Human Soft Palate. In *Fluid-Structure-Sound Interactions and Control: Proceedings of the 5th Symposium on Fluid-Structure-Sound Interactions and Control*; Springer: Singapore, 2021; Volume 5, pp. 45–49.
21. Benra, F.K.; Dohmen, H.J.; Pei, J.; Schuster, S.; Wan, B. A comparison of one-way and two-way coupling methods for numerical analysis of fluid-structure interactions. *J. Appl. Math.* **2011**, *2011*, 853560. [CrossRef]
22. Alkhader, M.; Alrashdan, M.S.; Abdo, N.; Abbas, R. Usefulness of Hard Palate Measurements in Predicting Airway Dimensions in Patients Referred for Cone Beam CT. *Open Dent. J.* **2021**, *15*, 505–511. [CrossRef]

23. Rodenstein, D.O.; Dooms, G.; Thomas, Y.; Liistro, G.; Stanescu, D.C.; Culee, C.; Aubert-Tulkens, G. Pharyngeal shape and dimensions in healthy subjects, snorers, and patients with obstructive sleep apnoea. *Thorax* **1990**, *45*, 722–727. [CrossRef] [PubMed]

24. Abu Allhaija, E.S.; Al-Khateeb, S.N. Uvulo-glosso-pharyngeal dimensions in different anteroposterior skeletal patterns. *Angle Orthod.* **2005**, *75*, 1012–1018. [PubMed]

25. Stauffer, J.L.; Buick, M.K.; Bixler, E.O.; Sharkey, F.E.; Abt, A.B.; Manders, E.K.; Kales, A.; Cadieux, R.J.; Barry, J.D.; Zwillich, C.W. Morphology of the uvula in obstructive sleep apnea1-3. *Am. Rev. Respir. Dis.* **1989**, *140*, 724. [CrossRef] [PubMed]

26. Birch, M.J.; Srodon, P.D. Biomechanical properties of the human soft palate. *Cleft-Palate-Craniofacial J.* **2009**, *46*, 268–274. [CrossRef]

27. Uvulectomy, If It Sticks Out, Cut It Off? Available online: http://www.circumstitions.com/uvulectomy.html (accessed on 1 July 2023).

28. Mihaescu, M.; Mylavarapu, G.; Gutmark, E.J.; Powell, N.B. Large eddy simulation of the pharyngeal airflow associated with obstructive sleep apnea syndrome at pre and post-surgical treatment. *J. Biomech.* **2011**, *44*, 2221–2228. [CrossRef]

29. Chen, J.; Gutmark, E. Numerical investigation of airflow in an idealized human extra-thoracic airway: A comparison study. *Biomech. Model. Mechanobiol.* **2014**, *13*, 205–214. [CrossRef]

30. Nicoud, F.; Ducros, F. Subgrid-scale stress modelling based on the square of the velocity gradient tensor. *Flow Turbul. Combust.* **1999**, *62*, 183–200. [CrossRef]

31. Menter, F.R. Two-equation eddy-viscosity turbulence models for engineering applications. *AIAA J.* **1994**, *3*, 1598–1605. [CrossRef]

32. Schickhofer, L.; Malinen, J.; Mihaescu, M. Compressible flow simulations of voiced speech using rigid vocal tract geometries acquired by MRI. *J. Acoust. Soc. Am.* **2019**, *145*, 2049–2061. [CrossRef]

33. Hahn, I.; Scherer, P.W.; Mozell, M.M. Velocity profiles measured for airflow through a large-scale model of the human nasal cavity. *J. Appl. Physiol.* **1993**, *75*, 2273–2287. [CrossRef] [PubMed]

34. Brietzke, S.E.; Mair, E.A. Acoustical analysis of snoring: Can the probability of success be predicted? *Otolaryngol.—Head Neck Surg.* **2006**, *135*, 417–420. [CrossRef] [PubMed]

35. Agrawal, S.; Stone, P.; Mcguinness, K.; Morris, J.; Camilleri, A.E. Sound frequency analysis and the site of snoring in natural and induced sleep. *Clin. Otolaryngol. Allied Sci.* **2002**, *27*, 162–166. [CrossRef]

36. Schickhofer, L.; Mihaescu, M. Analysis of the aerodynamic sound of speech through static vocal tract models of various glottal shapes. *J. Biomech.* **2020**, *99*, 109484. [CrossRef] [PubMed]

37. Hamdia, K.M.; Ghasemi, H. Quantifying the uncertainties in modeling soft composites via a multiscale approach. *Int. J. Solids Struct.* **2022**, *256*, 111959. [CrossRef]

 *bioengineering*

Article

Vibro-Acoustic Platelet Activation: An Additive Mechanism of Prothrombosis with Applicability to Snoring and Obstructive Sleep Apnea

Daniel E. Palomares [1,2], Phat L. Tran [2,3], Catherine Jerman [3], Moe Momayez [2,4], Pierre Deymier [2,5], Jawaad Sheriff [6], Danny Bluestein [6], Sairam Parthasarathy [2,3,7] and Marvin J. Slepian [1,2,3,6,*]

1 Department of Biomedical Engineering, University of Arizona, Tucson, AZ 85724, USA; dpalomares@arizona.edu
2 Arizona Center for Accelerated Biomedical Innovation, University of Arizona, Tucson, AZ 85724, USA; lephat.tran@gmail.com (P.L.T.); mmomayez@arizona.edu (M.M.); deymier@arizona.edu (P.D.); spartha1@arizona.edu (S.P.)
3 Department of Medicine, University of Arizona, Tucson, AZ 85724, USA; cfj@arizona.edu
4 Department of Mining & Geological Engineering, University of Arizona, Tucson, AZ 85724, USA
5 Department of Materials Science & Engineering, University of Arizona, Tucson, AZ 85724, USA
6 Department of Biomedical Engineering, Stony Brook University, Stony Brook, NY 11794, USA; jawaad.sheriff@stonybrook.edu (J.S.); danny.bluestein@stonybrook.edu (D.B.)
7 Health Sciences Center for Sleep and Circadian Sciences, University of Arizona, Tucson, AZ 85724, USA
* Correspondence: slepian@arizona.edu; Tel.: +1-520-626-4314; Fax: +1-520-626-7625

 **check for updates**

Citation: Palomares, D.E.; Tran, P.L.; Jerman, C.; Momayez, M.; Deymier, P.; Sheriff, J.; Bluestein, D.; Parthasarathy, S.; Slepian, M.J. Vibro-Acoustic Platelet Activation: An Additive Mechanism of Prothrombosis with Applicability to Snoring and Obstructive Sleep Apnea. *Bioengineering* **2023**, *10*, 1414. https://doi.org/10.3390/bioengineering10121414

Academic Editors: Rossana Madrid and Andrea Cataldo

Received: 1 October 2023
Revised: 28 November 2023
Accepted: 8 December 2023
Published: 12 December 2023

Abstract: Introduction: Obstructive sleep apnea (OSA) and loud snoring are conditions with increased cardiovascular risk and notably an association with stroke. Central in stroke are thrombosis and thromboembolism, all related to and initiaing with platelet activation. Platelet activation in OSA has been felt to be driven by biochemical and inflammatory means, including intermittent catecholamine exposure and transient hypoxia. We hypothesized that snore-associated acoustic vibration (SAAV) is an activator of platelets that synergizes with catecholamines and hypoxia to further amplify platelet activation. **Methods:** Gel-filtered human platelets were exposed to snoring utilizing a designed vibro-acoustic exposure device, varying the time and intensity of exposure and frequency content. Platelet activation was assessed via thrombin generation using the Platelet Activity State assay and scanning electron microscopy. Comparative activation induced by epinephrine and hypoxia were assessed individually as well as additively with SAAV, as well as the inhibitory effect of aspirin. **Results:** We demonstrate that snore-associated acoustic vibration is an independent activator of platelets, which is dependent upon the dose of exposure, i.e., intensity x time. In snoring, acoustic vibrations associated with low-frequency sound content (200 Hz) are more activating than those associated with high frequencies (900 Hz) (53.05% vs. 22.08%, $p = 0.001$). Furthermore, SAAV is additive to both catecholamines and hypoxia-mediated activation, inducing synergistic activation. Finally, aspirin, a known inhibitor of platelet activation, has no significant effect in limiting SAAV platelet activation. **Conclusion:** Snore-associated acoustic vibration is a mechanical means of platelet activation, which may drive prothrombosis and thrombotic risk clinically observed in loud snoring and OSA.

Keywords: platelet activation; snoring; vibration; obstructive sleep apnea; fluid–structure interactions; shear stress; aspirin

1. Introduction

Obstructive sleep apnea (OSA) is an increasingly prevalent, clinically significant condition affecting 2% to 14% of community screened populations with an associated heavy economic burden [1–4]. Sleep-related changes in muscle tone in OSA lead to intermittent upper airway collapse, increased airway resistance, breathing pauses, and recurrent episodes of hypoxia and hypercapnia during sleep [2,5–13]. These reductions (hypopneas) and cessations (apneas) of inspiratory flow provoke arousal from sleep with excitation of the sympathetic nervous system with catecholamine release [2,5,7,8,10,11,13–16]. Although loud snoring, early-morning headache, and daytime sleepiness are common presenting complaints, the significance of OSA relates to the association and causal relationship to a wide range of disease states burdened with significant morbidity and mortality. In particular, OSA is associated with atrial fibrillation, congestive heart failure, hypertension, coronary artery disease, diabetes mellitus, and stroke [2,3,9,13–15,17].

One of the most devastating consequences of OSA is stroke. OSA significantly increases the risk of stroke independent of other cardiovascular and cerebrovascular risk factors, including hypertension [2,5,8,11,12,15,17,18]. Central in OSA stroke is thrombosis. OSA-mediated stroke is largely ischemic, caused by underlying intra-arterial thrombosis or thromboemboli related to an associated prothrombotic state [8,9,11,13,15,18–20]. Pro-thrombosis in OSA is thought to be driven by sympathetic activation, catecholamine surge, and acute blood pressure changes during apneic episodes, resulting in oxidative stress, endothelial dysfunction, and inflammation [10,14,18,21–24]. Additionally, platelet dysfunction, including increased spontaneous activation and aggregation in patients with OSA, has been previously reported [11,15,19,21,25–30]. Hypoxia is known to facilitate prothrombosis via enhanced platelet activation, aggregation, and adhesion and may play a role in the association of OSA and stroke [21,26,28–31]. Despite these described biochemical and cellular mechanisms, our current understanding of the mechanisms of hypercoagulability and platelet dysfunction in OSA remains incomplete.

Physical forces such as shear stress and pressure are known drivers of platelet activation independent of biochemical agonists [28–30]. The cell membrane serves as the first point of contact between a cell and its environment, sensing external cues and transducing these inputs internally, resulting in a phenotypic response. In addition, cells possess a wide range of mechanoreceptors that transduce and transmit extracellular physical force intracellularly via the process of mechanotransduction [32–34]. Vibration is an additional physical force that may be transduced by cells [35]. Vibration of anatomic structures in the pharyngeal airway causes snoring [5,36–38]. OSA is strongly associated with snoring, specifically loud, vibratory snoring in the low-frequency range [39,40]. Snore-mediated vibrations, over a range of acoustic frequencies, may be transmitted through retropharyngeal tissues, including the carotids and other vasculature, representing a potential source of external physical force sufficient to activate platelets independent of other triggers of platelet activation such as hypoxia or sympathetic activation [36,37,41–43]. Via this transmission, fluid–structure interactions further transmit forces to flowing blood and contain platelets moving through both the carotid artery and jugular veins, which may affect platelet activation [5,16,19,44]. We explore this potential here.

In the present study, we hypothesized that snore-associated acoustic vibration (SAAV) is an activator of platelets that will synergize with catecholamines and hypoxia to further amplify platelet activation. To test this hypothesis, we first exposed platelets to snore-associated acoustic vibration using a designed vibro-acoustic exposure device, examining platelet activation and the dose dependency of snore vibration as an activator. As a second step, we explored the synergistic effects of hypoxia and/or exogenous epinephrine with SAAV on platelet activation. Finally, to limit SAAV-mediated platelet activation, we examined the effect of exogenous aspirin on limiting snore-mediated platelet activation.

2. Methods

2.1. Fabrication of Vibro-Acoustic Exposure Device

A vibro-acoustic exposure device (VAED) was fabricated to allow in vitro exposure of platelets to snoring recorded from human volunteers in a sleep lab or to other sounds/tones. Two configurations of the VAED were designed for differing vibro-acoustic exposure: contact and non-contact (Figure 1). The contact design was formed using a midrange speaker, Petri dishes, and securing tape. A commercial midrange speaker, 13.35 cm in diameter (J.W. Speakers, Germantown, WI, USA), with emitting frequencies between 100 and 8000 Hz was selected. A crystal polystyrene Petri dish (145 mm, Greiner Bio-One, Kremsmünster, Austria) was mounted on the midrange speaker using electrical tape (Scotch, 3M, Vinyl). The polystyrene dish was firmly secured in intimate contact across the face of the speaker, allowing transmitted sound to induce vibration in the affixed Petri dish platform. Smaller 35 mm diameter Petri dishes (Greiner Bio-One, Kremsmünster, Austria) containing desired platelet samples were then similarly taped onto the large Petri dish surface to perform an experiment. In the non-contact design, the speaker was suspended in a frame 1 cm above the 145 mm diameter Petri dish, which was secured to a firm base. Similar to the contact design, smaller 35 mm diameter Petri dishes (Greiner Bio-One, Kremsmünster, Austria) containing desired platelet samples were then identically affixed within the large Petri dish surface to perform an experiment.

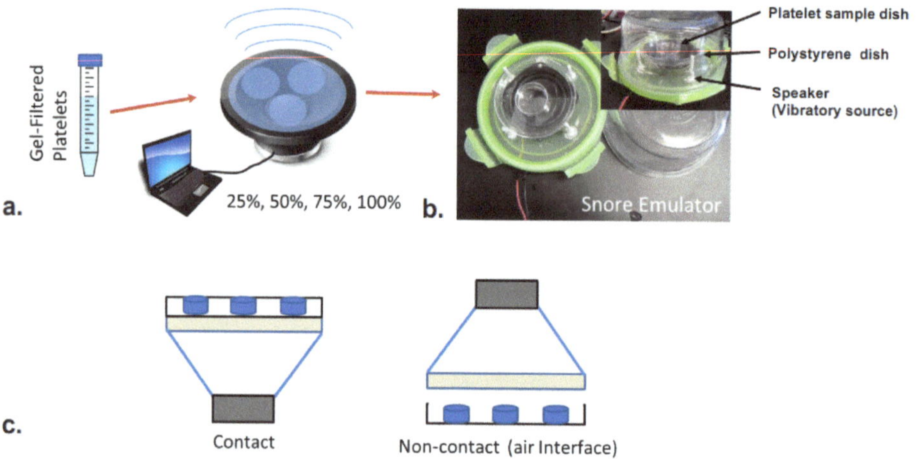

Figure 1. Vibro-Acoustic Exposure Device (VAED). (**a**) Gel-filtered platelets were exposed to snoring in either a contact or non-contact configuration. Snore intensity was varied via computer-controlled digital input over a range (25–100%). (**b**) In the contact mode, the large Petri dish holding samples is in intimate contact with and supported by the speaker membrane. In contact mode, samples are exposed to both sound waves emitted by the speaker as well as vibrations associated with the motion of the speaker membrane. (**c**) In non-contact mode, samples are only subjected to acoustic waves propagating from the speaker toward the Petri dish through an air interface.

For experiments, the VAED was actuated via a digital input signal derived from the snore recording—either as a composite of frequencies (whole snore) or a defined narrow bandwidth of frequency contained within snoring (200 Hz. and 900 Hz.). To measure vibration induced by the VAED, a laser vibrometer (VibroOne, Polytec GmbH) was utilized to obtain the amplitude of the vibro-acoustic wave emitted at different levels of amplification.

2.2. Platelet Preparation

Whole blood was collected from consenting healthy adult volunteers of both sexes who had not taken aspirin or ibuprofen for two weeks in accordance with a University of Arizona IRB-approved protocol (protocol #1810013264A002). Whole blood was centrifuged to obtain platelet-rich plasma (PRP), which was filtered through a column of Sepharose 2B beads (Sigma-Aldrich, St. Louis, MO, USA) to collect gel-filtered platelets (GFP) [45]. GFP was diluted to a count of 20,000/µL in HEPES-modified Tyrode's buffer, with 3 mM $CaCl_2$ added 10 min prior to experiments.

2.3. Platelet Activation State

The platelet activation state (PAS) is a well-characterized means of measuring mechanically mediated platelet activation and was utilized for all studies herein [46,47]. The PAS assay records the rate of thrombin generation, utilizing acetylated prothrombin while incubating with factor Xa over 10 min at 37 °C at a final platelet count of 5000/µL. To ensure linear kinetics, the feedback action of generated thrombin is blocked via the use of acetylated prothrombin [47]. For all experiments, PAS values were normalized to a sonicated sample of platelets using Equation (1). For sonication, platelet samples were activated using a Branson Sonifier 150, at 10 W for 10 s with a microprobe (Branson, Branson, MO, USA). The sonicated platelets act as a positive control for mechanically activating a platelet. For all protocols outlined herein, serial 25 µL aliquots of GFP samples were analyzed via this assay.

$$Normalized\ PAS = \frac{Experimental\ PAS}{Sonicated\ PAS} \tag{1}$$

2.4. Effect of Snoring on Platelet Activation

Gel-filtered platelets were added to the smaller sample Petri dishes and were then exposed to a snore recording (whole snore composite) for 40 min via the VAED, with unexposed GFP in Petri dishes serving as control. All samples were incubated at 37 °C. At serial time points of 0, 10, 20, 30, and 40 min, serial GFP aliquots were collected from both vibration and non-vibration groups, and platelet activation was assessed via the PAS assay.

2.5. Scanning Electron Microscopy

Scanning electron microscopy (SEM) was used to visualize morphological changes in platelets exposed to snore vibration. GFP (20,000 platelets/µL) was exposed to snore vibration using the VAED as above, with non-vibration controls, and samples were collected at times 0, 20, and 40 min. Aliquots, 30 µL, of each collected sample were placed on glass coverslips (13 mm diameter) and immediately fixed with the addition of 30 µL of 2% *v/v* glutaraldehyde in HEPES-modified platelet buffer for 30 min. Coverslips were rinsed with sequential dilutions of glutaraldehyde in ddH$_2$O, then dehydrated with a series of increasing ethanol concentrations reaching 100%. Samples were gold-coated using an Anatech Hummer 6.6 (Anatech, Sparks, NV, USA) sputter system. SEM images were acquired with a FEI Inspec-S SEM (FEI Company, Hillsboro, OR, USA) at 1–10,000×.

2.6. Effect of Vibro-Acoustic Stimulation on Platelet Activation

For frequency band-specific studies, 200 Hz and 900 Hz were selected as the exposure frequencies based on the identified peak frequency regions (Figure 2b) of recorded snoring. The open-source software Audacity® (version 2.2.2) generated each frequency as a continuous sinusoidal wave for 40 min using an amplitude of 1AU (Arbitrary Units), and this was utilized for activation of the VAED. Each frequency was tested individually to determine its activation capability. For a given experiment, GFP samples (in duplicate) of 1.5 mL of GFP (20,000 platelets/µL) were placed in the smaller Petri dishes and then secured to the vibration face of the VAED (contact experiments) or placed on a base 1 cm. below the suspended speaker (non-contact experiments). For all experiments, resting, non-exposed GFP, and GFP exposed to sonication served as non-exposure and max activation controls, respectively. All experiments were conducted over 40 min in a humidified incubator at 37 °C. For all sample conditions, serial aliquots of GFP were collected at 10-min intervals and analyzed via the PAS assay.

2.7. Effect of Epinephrine on Platelets Pre-Exposed to Snore

Gel-filtered platelets in small Petri dishes were exposed to snore-associated vibration (whole snore) as above for 40 min, with unexposed GFP in small Petri dishes serving as control. Following 10 min of exposure or non-exposure, 10 mM epinephrine (epi) was added to samples of both groups, i.e., vibration-exposed and non-exposed platelets. As a control for the epinephrine group, HEPES-modified Tyrode's buffer was added to a set of exposed and no-exposed GFP in Petri dishes as well. All samples were then incubated at 37 °C for 40 min, and serial aliquots were removed at 10-minute intervals and analyzed via the PAS assay.

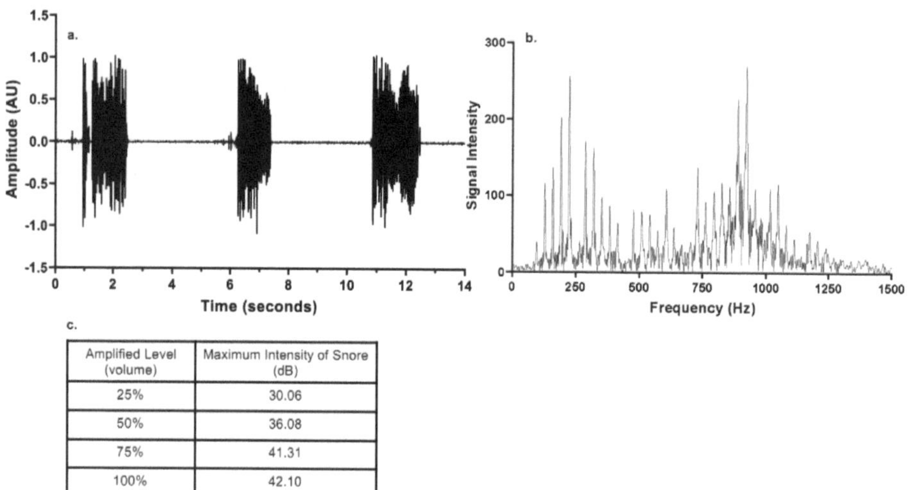

Figure 2. (**a**) Typical OSA snore-note periodic active and silent intervals. (**b**) Frequency spectrum of snore. (**c**) Maximum intensity of snore as a function of amplified level.

2.8. Effect of Hypoxia on Platelets Exposed to Snore

GFP was incubated for 30 min at 37 °C with overflowing CO_2 in an incubator to induce a hypoxic state. To maintain hypoxia throughout the vibration experiment, the VAED was placed in a chamber that was pumped with 100% CO_2. The pressure inside the chamber was kept at less than 2 psi. After 30 min of pre-incubation, the VAED was turned on, and GFP was exposed to snore-associated vibration (whole snore recording) at 100% acoustic intensity for 40 min as outlined above. Samples were collected at serial time points, and platelet activation was assessed via the PAS assay.

2.9. Effect of Aspirin on Snore-Mediated Platelet Activation

Aspirin (ASA, Sigma, St. Louis, MO, USA) was dissolved in sodium bicarbonate solution (324 mg ASA, 965 mg citric acid, and 1744 mg sodium hydrogen carbonate in 50 mL double-distilled H_2O) and diluted to 125 µM final concentration. To verify the inhibitory efficacy of the ASA preparation, GFP was treated with ASA at a final concentration of 25 µM for 10 min at 37 °C. ASA-treated GFP was then exposed to 25 µM arachidonic acid (AA), with ASA-treated GFP and untreated GFP (no ASA, no AA) serving as control. All samples were incubated for 10 min at 37 °C, and the platelet activation state was determined. Separately, ASA-treated GFP, 37 °C × 10 min) was exposed to snore-associated vibration at 100% acoustic intensity for 40 min. as outlined above, with a non-vibration group as controls. In addition, GFP-only samples, GFP treated with epinephrine, and GFP exposed to hypoxia were prepared as outlined above. These specimens were then incubated with ASA at a final concentration of 25 µM at 37 °C for 10 min and similarly exposed to snore-associated vibration (whole snore recording) on the VAED over a 40 min period, with non-exposed samples as control. Samples were collected from all groups at serial time points, and platelet activation was assessed via the PAS assay.

2.10. Optoacoustic Vibration Patterns Generated by Snore Frequency Components

The VAED was utilized to obtain the fluid vibration patterns generated by snore frequencies. Smaller Petri dishes were secured onto the larger Petri dish vibration surface of the VAED with the cover off to allow contained fluid visualization. Water (3 mL) was placed in the secured smaller dishes, and the VAED was activated with defined frequencies at 100% output—as utilized for the activation experiments above, for 10 s as a high-resolution camera phone (4K@30/60 fps, 1080p@30/60/120/240 fps; gyro-EIS, OIS, Pixel 6 pro, Google, Mountain View, CA, USA) recorded the vibrational patterns. Frames of the videos generated were assembled as a composite for comparison.

2.11. Statistical Analysis

For all experiments, a two-tailed Welch's t-test was used to compare the means of non-exposed to exposed platelets, with $p < 0.05$ indicating a significant difference.

3. Results

3.1. Mechanical Characterization of Vibro-Acoustic Exposure Device

Snoring was recorded in the sleep laboratory of the University of Arizona from OSA patients with a high-fidelity audio system. A typical snore recording (amplitude vs. time) is shown in Figure 2a for 14 s, normalized for depiction over a range of a -1 to 1 scale (AU, arbitrary units). A wide frequency range content was detected for recorded snoring, ranging from 100 to 1300 Hz (Figure 2b). The snore recording could be readily amplified and played back as an input signal for the actuation of the VAED. The characterization of the amplitude of the vibro-acoustic wave emitted at different levels of amplification by a laser vibrometer is shown in Figure 2b. The frequencies from 200 to 300 Hz and 850 to 950 Hz displayed a higher intensity and decibel content. Powering the VAED over a range of amplification from 0–100% in 25% increments revealed a range of emitted intensities (Figure 2c). This enabled quantification of the intensity of the reproduced snore and its induced vibration as a function of amplified level (volume).

3.2. Effect of Contact versus Noncontact on Snore-Mediated Platelet Activation

Exposure of platelets to snoring, either as a composite recording (whole snore) or for the selected frequencies tested, resulted in platelet activation only in the contact mode, i.e., only for platelets subjected to direct transmission of vibration to the fluid platelet sample (Figure 3). Platelets exposed to identical snore frequencies and intensities in a noncontact mode, with an interposed air gap with no detectable fluid vibration, did not demonstrate activation over the tested 40-min period. There was no difference in platelet activation for the non-contact exposure group from that of a baseline resting control over 40 min as well. As a decision outcome of this first study, all subsequent experiments herein were conducted in contact mode.

3.3. Effect of Time of Exposure on Snore-Mediated Platelet Activation

In the contact VAED mode, platelets were exposed to 42.10 dB, i.e., 100% maximum intensity of snore, for 40 min. Snore exposure was compared to non-snore-exposed resting platelets. After 10 min of snore vibration exposure, a significant difference was already detectable versus resting platelets (Figure 3). Over the entire duration of the study exposure period, the snore-exposed samples were found to have a higher net platelet activity state than the resting sample at 20 min, with values of 0.96% and 0.61%, respectively (p-value = 0.003) and at 30 min with values at 1.20% and 0.69% (p-value = 0.001), respectively. The 40-min time point for snore-exposed samples had the highest increase in platelet activity state versus other time points and continued to be significantly different from the resting non-exposed samples with values of 1.55% and 0.72% (p-value = 0.001), respectively.

3.4. Effect of Snore Vibro-Acoustic Exposure on Platelet Morphology

Continued exposure of platelets to snore-associated acoustic vibration led to progressive changes in platelet morphology (Figure 4). Notably, with time, an increase in the degree of platelet activation was detectable morphologically. In the absence of vibro-acoustic exposure, platelets were noted to retain an unactivated discoid morphology. In contrast, following 20 min of snore vibration exposure, platelets began to extend pseudopods. At 40 min, more pronounced pseudopod extensions were noted along with platelet spreading, indicative of greater activation.

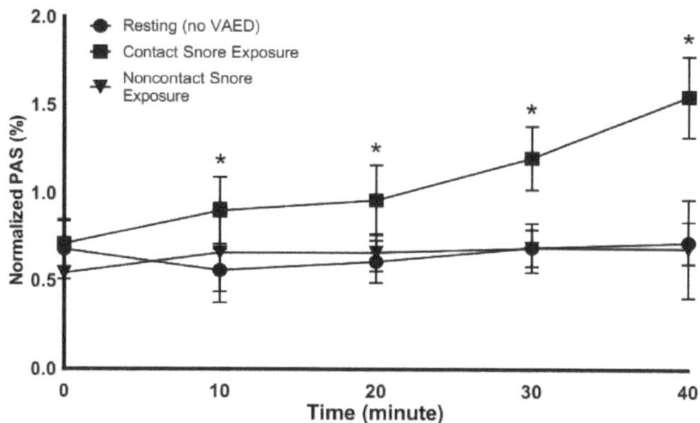

Figure 3. Platelet activity state (PAS) in response to whole snore exposure via the vibro-acoustic exposure device. Snore-exposed platelets increased activation with time of exposure, with significant difference versus non-contact and resting platelets detectable starting at 10 min of exposure (* $p < 0.05$).

With Vibro-acoustic exposure

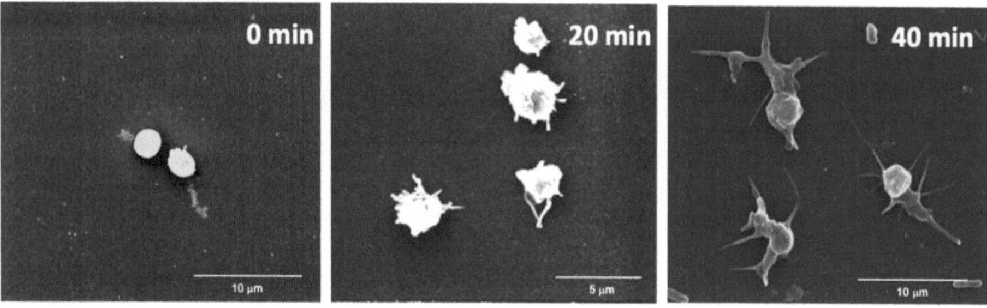

Without Vibro-acoustic exposure

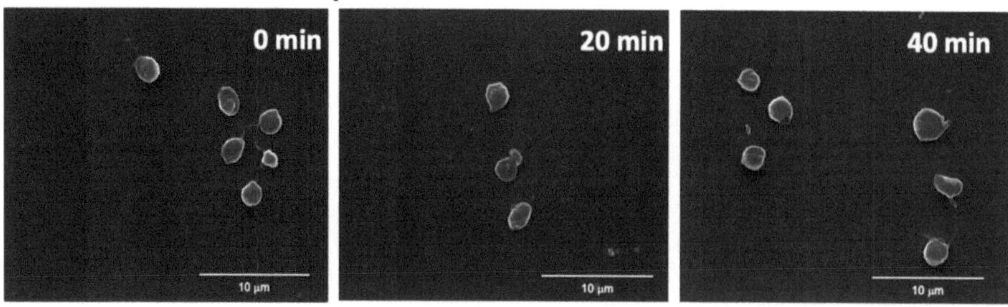

Figure 4. Effect of snore-associated vibration on platelet morphology. Top row—SEM Images of platelets exposed to whole snore-associated vibration for 0, 20, and 40 min. Bottom row—SEM of control platelets without snore-associated vibration exposure at 0, 20, and 40 min.

3.5. Effect of Varying Snore Intensity on Platelet Activation

Exposure of platelets to varying intensities of snore-associated vibration (whole snore recording) revealed varying degrees of activation. Over 30 min, a linear, progressive increase in platelet activation was noted, correlating with increasing intensity of vibration exposure (Figure 5). Progressing from 0 to 100% exposure in 25% increments resulted in an increase of activation from 1.2%, to 1.4%, to 1.5%, to 1.7%, and 2.1%, respectively. The increase observed at 100% was significant in comparison to the baseline 0% control ($p = 0.047$).

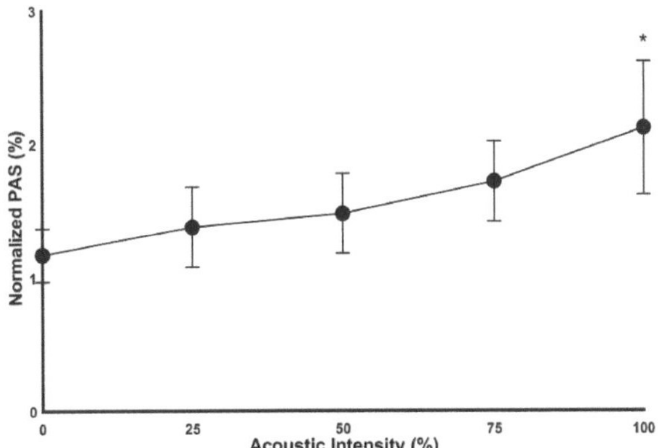

Figure 5. Platelet activation as a function of the intensity of vibro-acoustic exposure. All exposures for 30 min. (100% exposure, * $p < 0.05$ vs. non-exposure control).

3.6. Effect of Differing Dominant Frequencies on Snore-Mediated Platelet Activation

Frequency spectrum analysis of snore recordings identified two high-energy frequency ranges: 150 to 250 Hz and 850 to 950 Hz (Figure 2). The median frequencies from the two ranges, i.e., 200 and 900 Hz, were used as representative of the dominant frequencies in snoring to determine the effect of frequency on platelet activity. The two frequencies were used at the same vibro-acoustic intensity corresponding to the amplified level of the 42.10 dB snore. Platelet activation state relative to the tested frequencies is shown in Figure 6. The frequency of 200 Hz had a greater effect on increasing platelet activation state than 900 Hz, achieving 53.05% versus 22.08%, respectively ($p = 0.001$). The controls for both were 15.92% and 13.78%, respectively.

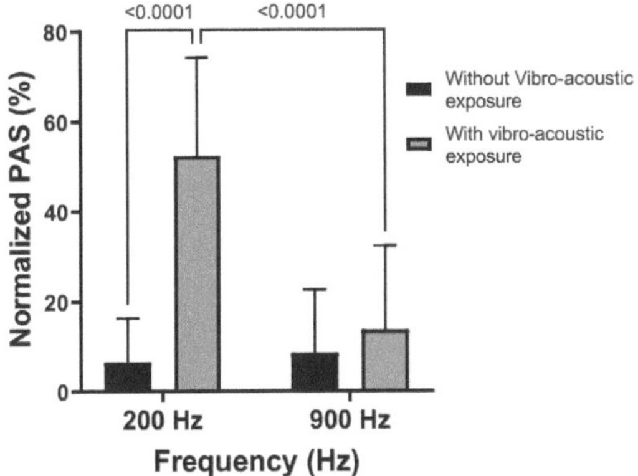

Figure 6. Effect of defined frequency exposure on platelet activation. All samples were exposed to continuous (40 min) frequency-mediated vibration via the VAED. (200 Hz and 900 Hz tested vs. non-vibration control).

3.7. Effect of Epinephrine versus Epinephrine + Snore-Associated Vibration on Platelet Activation

Epinephrine alone was observed to activate platelets (Figure 7a). In comparison to snore vibration, epinephrine led to a platelet activity state of 2.79% ± 0.02, which was significant compared to 0.68% ± 0.01 for untreated control (p-value = 0.001). In comparison, snore vibration alone led to a PAS of 1.55% ± 0.01, which was also significant relative to the control (p = 0.001). Notably, the combination of epinephrine + snore vibration led to a much greater increase in platelet activation (7.06% + 0.08) than either epinephrine or snore alone (p = 0.001). The overall increase was > 10-fold relative to control, 2.5-fold > epi alone, and 4.5-fold > snore alone.

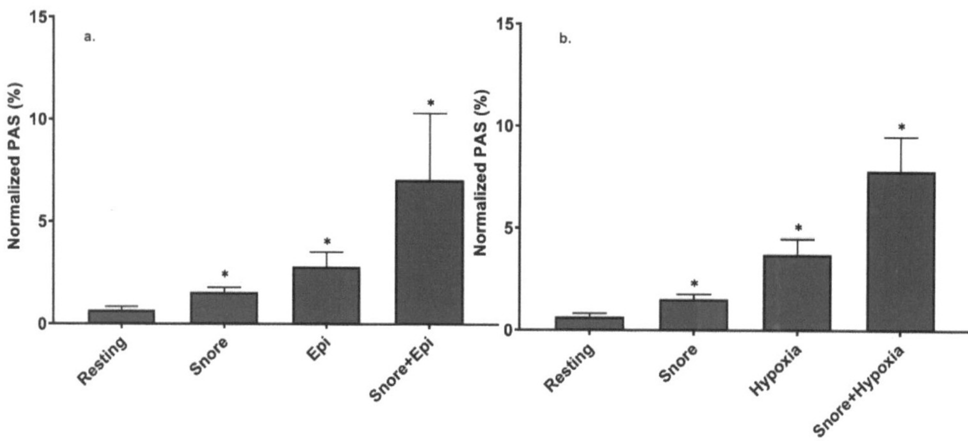

Figure 7. Comparative versus the additive effect of epinephrine and hypoxia with snore-associated vibration-mediated platelet activation. (**a**) snore exposure vs. epinephrine vs. combination of snore + epinephrine. (**b**) snore vs. hypoxia vs. combination of snore + hypoxia. (All agonists were significantly different from resting, * p <0.05).

3.8. Effect of Hypoxia versus Hypoxia + Snore-Associated Vibration on Platelet Activation

Platelet exposure to hypoxia, under the conditions employed, resulted in platelet activation (Figure 7b). In comparison to snore vibration, hypoxia led to a PAS of 3.73% ±0.03, which was significant compared to 0.68% ±0.01 for untreated control (p-value = 0.001). In comparison, snore vibration alone led to a PAS of 1.55% ± 0.01, which was also significant relative to the control (p = 0.001). Like epinephrine, the combination of hypoxia + snore-associated vibration led to a much greater increase in platelet activation state, i.e., 7.83% ±0.08, than either hypoxia or snore alone (p = 0.001).

3.9. Combined Effects of Epinephrine and Hypoxia with Snore Vibration Platelet Activation

The combination of snore-associated vibration with epinephrine and hypoxia led to an even greater degree of platelet activation than either epinephrine or hypoxia alone (Figure 8). Combining all three agonists led to a net synergy of activation. Compared to baseline control activation of 0.68%, the triple combination led to a net 14.55% ± 0.11. activation versus 7.06% for snore + epinephrine and 7.83% for snore + hypoxia.

3.10. Effect of Aspirin on Snore and Combinational Agonist-Mediated Platelet Activation

Aspirin did not limit snore vibration-mediated platelet activation (Figure 9). The concentration of aspirin tested was observed to have no inhibitory effect on snore vibration-mediated platelet activation (Figure 9). In fact, a minor degree of activation was noted for aspirin-treated platelets exposed to snore vibration (25 μM), i.e., 3.54% versus 0.68% for snore alone. Aspirin had no effect in limiting platelet activation resulting from synergistic activation induced by snore vibration + epinephrine + hypoxia. As an overall control for this series of experiments, the aspirin preparation tested was found to be inhibitory of arachidonic acid-mediated platelet activation, with arachidonic acid resulting in 13.73% activation versus ASA + AA in 3.71%, $p = 0.036$.

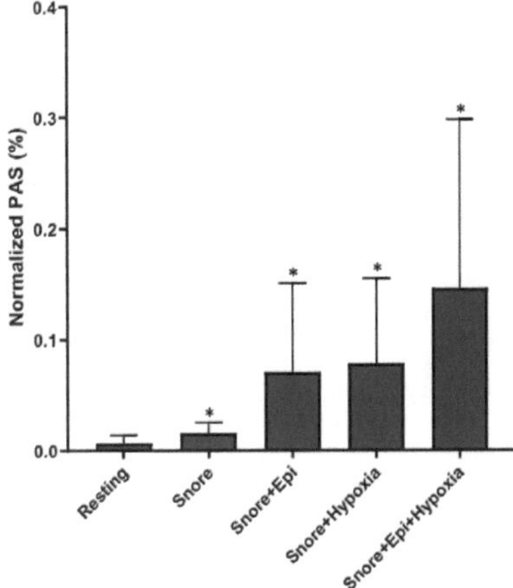

Figure 8. Additive/Synergistic effects of exogenous agonists. Snore vs. snore + epi vs. snore + hypoxia vs. combination of snore + epi + hypoxia. (All combinations of agonists are significantly different from resting, * $p < 0.05$).

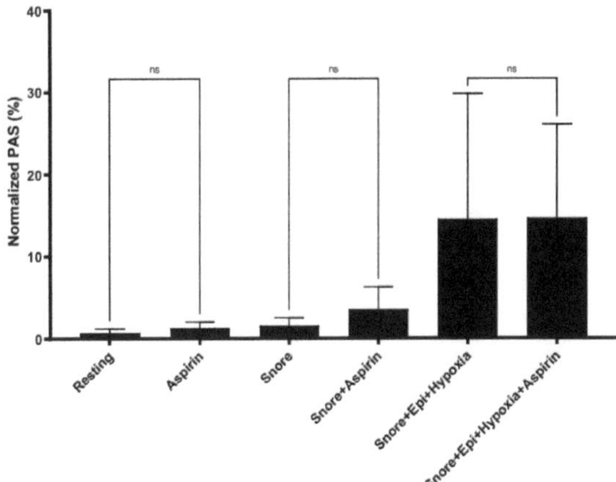

Figure 9. Effect of aspirin as a means of limiting snoring and combinational agonist-mediated platelet activation. Aspirin had no effect in limiting activation for each experimental condition versus its non-aspirin control).

3.11. Optoacoustic Vibration Patterns Generated by Snore Frequency Components

A progressive degree of fluid disturbance and vibration was visually detectable and found to be inversely correlated with the frequency of exposure (Figure 10). At higher frequencies, shimmering optoacoustic patterns were noted, with limited disturbance to bulk fluid. As frequency decreased, increasing disturbance with fluid movement was detectable, with fluid ejection and splash observed at <500 Hz. Low-frequency bandwidths induced noticeably greater fluid oscillation than high-frequency bandwidth exposure.

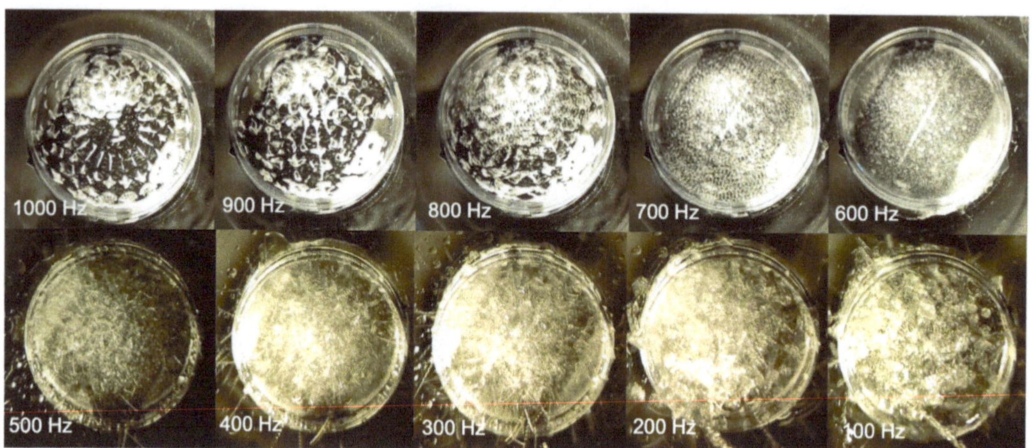

Figure 10. Vibro-acoustic effect of acoustic frequency on fluid dispersion and ejection. (1000–100 Hz).

4. Discussion

Obstructive sleep apnea is a highly prevalent disease and is associated with significant cardiovascular morbidity and mortality. Of particular concern is the association of OSA with stroke. Relatedly, snoring alone has also been found to result in an increased risk of stroke [11,12,15,19]. Central to stroke pathophysiology are thrombosis and thromboembolic phenomena, all based on and initiating with platelet activation [12,19]. In the present study, we determined that the vibrational component of snoring, common to both intense snoring and snoring associated with OSA, is a direct "mechanical" agonist of platelets, resulting in platelet activation. Vibration-mediated activation was found to be associated with exposure time and intensity, i.e., dose-dependent, with low-frequency components being the most activating. Furthermore, our results suggest that snore-associated vibration may act synergistically with both catecholamines and hypoxia—both biochemical responses associated with OSA—to induce an even greater level of platelet activation than either agonist alone. Aspirin, a classical antiplatelet agent, was found to have no efficacy in limiting snore vibration-mediated platelet activation alone or in the setting of synergistic epinephrine and/or hypoxia agonists.

4.1. Vibro-Acoustic Stimulation of Biological Cells

Sound applied to an aqueous fluid is a pressure wave propagating within the medium at speeds on the order of 1500 m/s. At the frequency of audible sound, such as that associated with snoring, the wavelength is on the order of several meters. Any microscale heterogeneity, such as platelets in the fluid, will essentially experience a constant pressure field, leading to negligible pressure forces. As such, the absence of a pressure gradient or any other sudden change in the corresponding force field experienced by the platelet will not result in activation. Platelet activation requires forces acting on length scales on the order of the platelet diameter (2–4 μm). Long-wavelength audible acoustic waves cannot produce so-called acoustic cavities that may be able to concentrate mechanical energy on a scale smaller than the wavelength. Furthermore, low-intensity snore acoustic waves cannot form cavitation, which is a known activator of cells [48].

The loss of acoustic momentum that results from attenuation or dissipation of the sound field in a viscous fluid may result in a nonlinear time-independent fluid motion, i.e., stationary vortices, known as acoustic streaming [36,37,49]. The scale of any acoustic streaming happens on the scale of the wavelength, therefore eliminating that mechanism as a possible activator of platelets.

In addition to acoustic waves, we have considered the effects of vibration on platelet activation. As shown in Figure 10, the vibrations of the fluid receptacle create ripples on the surface of the fluid, which are known as Faraday waves [49]. Faraday waves are nonlinear standing waves occurring when the vibration frequency exceeds a critical value, the fluid surface becomes unstable, and protrusions/ripples form on the surface. The amplitude of the Faraday waves scales to the intensity of the vibro-acoustic exposure device. Fluid flow associated with gradients in velocity resulting from these instabilities may potentially lead to shear forces on microscale platelets. Our study demonstrates that exposure to snoring via a contact configuration of the VAED led to platelet activation via vibration as opposed to audible acoustic waves. Disturbed fluid motion, with associated non-physiological shear forces, is a known activator of platelets [37,50–53]. With the VAED operating in the contact configuration, stronger mechanical vibrations and vibro-acoustic waves are likely to be operative on platelets versus acoustic waves alone. Furthermore, with a significant impedance mismatch between air and sample fluid in the noncontact mode of operation, one expects that platelets will be subjected to primarily the effect of the low-intensity vibro-acoustic waves, with air inducing greater damping of high versus low-frequency vibro-acoustic waves [54–56]. This is consistent with the absence of platelet activation observed for the noncontact studies.

4.2. Vibration Effects on Stresses Acting on Flowing Platelets

Vibration-related shear has been described as an activator of flowing cells [44]. OSA and snoring vibrations that are transmitted to adjoining carotid arteries and jugular veins induce flow disturbances via fluid–structure interaction that is transmitted to the vessel walls, where these vibrations are translated into random flow disturbances characterized by instantaneous acceleration forces acting on the flowing blood. Out-of-phase accelerations acting on a cell nucleus or its intracellular constituents appear to play a role in the cellular response to vibrations [44]. High-frequency flow disturbances such as those induced by OSA and snoring may generate both laminar viscous shear and turbulent stresses. The latter, known as Reynolds stresses, are random and can easily increase the total stresses acting on flowing platelets by an order of magnitude [57,58]. Platelets respond and activate earlier by such a combination of stresses. The turbulent Reynolds stresses contain high-frequency random vibrations that act directly on the platelet membrane and its receptors. Those are then transduced to the intracellular components via the cytoskeleton and cytoplasm of the platelet and may initiate earlier activation [34,59,60].

4.3. Effects of Vibro-Acoustic Dose on Platelet Activation

Our studies demonstrate that with increasing length of snore-associated exposure, greater platelet activation occurs (Figure 3). This concept of dose, i.e., intensity × time, is well established for the mechanical activation of platelets [45,61,62]. In the case of shear stress, this phenomenon is referred to as net stress accumulation. In prior work, our group has shown that for a given intensity of stress, the degree of platelet activation increases with the repetitiveness (frequency) of stress exposure [63,64]. Furthermore, even with cessation of exposure, platelets are sensitized and will continue to activate proportionally to the magnitude of the initial shear stress dose [45]. This progressive increase in platelet activation is seen here as well with snore-associated acoustic vibration exposure.

The importance of time of exposure, dose, and post-stress sensitization is reflected in our studies examining whole snore exposure versus selective frequency exposure. For whole composite snore exposure, a PAS of 1.55% was achieved at 40 min. By contrast, at 200 Hz, a PAS of 53.05%, and at 900 Hz, a PAS of 22.08% was achieved. This difference in activation may be explained by understanding the actual period of vibration exposure induced with each snore regime. For whole snore—i.e., actual snoring—sound and vibration events only occur intermittently and episodically. In considering a series of repetitive snores as in actual snoring, the duty cycle of sound exposure represents 20–25% of the time interval between snoring events, with 70–80% of the time being silent, free of vibrational and mechanical exposure. By contrast, for the single-frequency studies, sound and vibration were produced continuously without interruption, leading to 40 min of intense mechanical exposure. This longer duration of persistent vibration is reflected in the greater degree of platelet activation observed.

At 42.10 dB for 40 min, platelets are adequately stimulated to activate. For reference, 42.10 dB corresponds to a pressure of 0.025 dynes/cm^2 [40,64]. Obstructive sleep apnea is designated as moderate or severe when the snoring episode has a duration of 30 min or more [14]. It has been suggested that exposure to a high vibration intensity over a short period of time may activate platelets in a way similar to that of a lower intensity over a longer time [65]. This may explain the possibility of significant platelet activation even at low vibration intensities if exposure exceeds 40 min.

4.4. Effect of Frequency on Platelet Activation

Examination of the frequency spectrum of typical snore recordings revealed two high-energy content regions centering around 200 Hz and 900 Hz. Of these two frequencies, 200 Hz led to greater platelet activation compared with 900 Hz, despite exposure to the same (100%) intensity of vibration. These results suggest that the low range frequency leads to greater oscillation and fluid disturbance, directly imparting activating forces on the platelets. This is consistent with the visibly detectable fluid disturbances observed (Figure 10). We have previously shown that flowing platelets are sensitive to specific frequencies in their shear stress trajectories through gaps in prosthetic heart valves, with higher thrombin generation observed in the 82–94 Hz band despite a lower intensity, 0.56 dB, than other frequency regions [66]. Furthermore, short exposures of platelets to low oscillatory frequencies trigger significant thrombin generation compared to constant low-shear stress, and thrombin generation continues to increase even after cessation of oscillatory flow [58,64]. This suggests a differential stress-biochemical coupling effect, where the frequency components are sensed and mechano-transduced by the platelets, triggering downstream biochemical events, e.g., receptor activation, cytoskeletal changes, and granule release. This differential stress-biochemical coupling mechanism is observed in other cell types, including vascular endothelial cells, fibroblasts, and bone mesenchymal stem cells [67–69].

The midrange speaker utilized in these studies has an operating frequency range of 100 to 1500 Hz, which cannot explain, based on mechanical intensity response to voltage stimulus, the difference in activation for the two tested frequencies. The variation in activation as a function of frequency is likely due to the difference in response of the fluid and contained platelets to vibrations generated by these defined frequencies.

4.5. Combined Effects of Vibration and Biochemical Agonists on Platelet Activation

Patients suffering from severe OSA tend to have intermittent spikes in blood catecholamine levels associated with startle and awakening after intermittent apneic episodes. Similarly, these patients have transient episodes of relative hypoxemia [11,12,15,16,21]. Our results demonstrate and confirm that both epinephrine exposure and hypoxia induce platelet activation. Notably, our results extend insight into potential mechanisms operative in OSA. The finding that snore-associated vibration activates platelets and that this activation is further amplified in the setting of epinephrine and hypoxia, or conversely that snore-associated vibration amplifies the basal activation observed with these known platelet agonists, is suggestive of additional mechanisms that may be operative and underlie the demonstrated increased thrombosis and stroke risk of OSA and loud snoring. Furthermore, these mechanisms may also contribute to an overall increased cardiovascular risk [11,12,15,16,19,21]. Underlying this synergy of agonists may be alteration of platelet membrane porogenicity induced via snore-associated vibration, enhanced exposure of agonist receptors resulting as a result of vibrational change, or facilitated presentation and mixing of biochemical agonists, or enhanced activation initiation via hypoxia-mediated mitochondrial and internal energy changes. These potential mechanisms underlying the observed synergies are plausible, though they remain to be experimentally defined.

4.6. Combined Effect of Aspirin, Agonist, and Snore-Associated Vibration on Platelet Activation

Aspirin is the prototypic antiplatelet agent, demonstrated to reduce prothrombotic cardiovascular risk [70]. It is well established that the mode of action of aspirin is via inhibition of arachidonate and thromboxane A2 pathways within the platelet [71,72]. Over the past few years, it has been demonstrated that ASA has had limited efficacy clinically in limiting platelet activation in settings of elevated shear stress and flow disturbances, e.g., as in the setting of implanted mechanical circulatory support devices, in fact, driving excessive bleeding [73]. Furthermore, definitive studies have recently established that ASA has a limited ability to inhibit shear-mediated platelet activation [74–76]. Underlying this lack of effect has been the realization that shear-mediated platelet activation does not involve the classical biochemical agonist receptor pathways but rather mechanisms such as mechano-transduction, mechano-destruction with membrane damage, and mechanosensitive channel activation [34,77]. It is very likely that the observed lack of efficacy of ASA in the setting of snore vibration-mediated platelet activation relates to this similar dichotomy as to the pathway of activation involved, with mechanical means differing from traditional biochemical agonist pathways. Consistent with this have been clinical studies demonstrating the limited efficacy of aspirin in reducing OSA-associated thrombosis and stroke [78,79].

Beyond the lack of inhibitory efficacy of aspirin, we observed a slight trend toward increased platelet activation for aspirin-treated platelets exposed to snoring. Although we caution that these data are limited and in vitro, it has been reported in other settings for aspirin to have the potential for a paradoxical prothrombotic effect [80], with conceivable mechanisms related to variable Cox-1 vs. Cox-2 reactivity [81,82] or variable platelet subpopulation reactivity [83], among other possibilities.

5. Limitations

The vibro-acoustic exposure device, constructed from plastic and fiber materials, has material properties that differ from those of the airway, neck, and blood vessel tissue in patients. As such, the vibration resonance, wave transmission, and fluid–structure interactions may differ from the in vitro testing here versus the in vivo patient situation. The present study utilizes sound recordings typical of patients with OSA. Despite this, patients over a range of body habitus, differing age, and degree of throat and neck tissue laxity may have differing frequencies of snoring and differing coupling characteristics regarding fluid–structure interactions and transmission of sound-mediated vibration to blood and fluid and contained platelets. The finding here may represent potential mechanisms operative for a particular defined spectrum of individuals. The study here utilized platelets in a fixed dish. Future studies are planned for snore vibration exposure to flowing platelet samples.

6. Conclusions

Obstructive sleep apnea and loud snoring both carry an increased risk of stroke. Thrombotic stroke in these conditions has traditionally been hypothesized to be driven by biochemical and inflammatory mediators. In the present study, we demonstrate that snore-associated vibration is a mechanical agonist of platelets that leads to platelet activation. Snore-associated vibration activates platelets in a dose-dependent fashion, with activation continuing to increase with increasing time of exposure. Low frequency, 200 Hz, appears to activate platelets greater than higher frequencies. Furthermore, snore-associated vibration as an agonist of platelet activation is synergistic with known activators and risk factors of thrombosis in OSA and snoring, i.e., catecholamines and hypoxemia. Aspirin, a standard clinical antiplatelet agent, is ineffective in limiting snore vibration-mediated platelet activation. These studies identify a new risk factor and potential mechanism operative in snoring and OSA, i.e., acoustic vibration-mediated platelet activation. Further defining the mechanical-cellular coupling mechanisms operative will afford opportunities for translational advancement to reduce the clinically significant consequences associated with thrombotic and thromboembolic consequences of OSA and snoring.

Author Contributions: Conceptualization, D.E.P., S.P., M.M., P.D., D.B. and M.J.S.; Methodology, D.E.P., P.L.T., M.M., J.S., P.D. and M.J.S.; Validation, D.E.P., M.M., P.D. and M.J.S.; Formal analysis, D.E.P., M.M., P.D., J.S. and M.J.S.; Investigation, D.E.P., P.L.T. and C.J.; Resources, M.J.S.; Data curation, D.E.P. and M.J.S.; Writing—original draft preparation, D.E.P. and M.J.S.; Writing—review and editing, D.E.P., S.P., M.M., J.S., P.D., D.B. and M.J.S.; Visualization, D.E.P.,P.L.T., M.M. and M.J.S.; Supervision, M.M., P.D. and M.J.S.; Project administration, M.J.S.; Funding acquisition, M.J.S. and D.B. All authors have read and agreed to the published version of the manuscript.

Funding: This study was supported by the Arizona Center for Accelerated Biomedical Innovation (ACABI) of the University of Arizona and by the National Institute of Health (NIH) (T32HL007955 to D.P. and 5U01HL131052 to D.B. and M.J.S.).

Institutional Review Board Statement: All human subjects research was carried out in accordance with the University of Arizona Institutional Review Board (IRB) approved protocol (University of Arizona IRB protocol #1810013264A002). Informed consent was obtained from all donors included in the study. No animal studies were carried out by the authors for this article.

Informed Consent Statement: Not applicable.

Data Availability Statement: Data that support the findings of this study are available for the corresponding author upon request.

Acknowledgments: We thank Jubin George, Marcus Hutchinson and Lukas Jensen for technical assistance in the conduct of this study.

Conflicts of Interest: Daniel Palomares, Phat L. Tran, Catherine Jerman, Moe Momayez, Pierre Deymier, and Jawaad Sheriff, report no conflict of interest for this study. Sai Parthasarathy reports personal fees from Jazz Pharmaceuticals, Inc., other from UpToDate, Inc., grants from NIH (R25-HL126140, R33-HL151254; OT2-HL161847; R21-HD109777; C06-OD028307; HL140144; HL138377; OT2-HL156912 and OT2HL158287), grants from PCORI (DI-2018C2-13161, CER-2018C2-13262), grants from Department of Defense (W81XWH20C0051 and W81XWH2110025), grants from Pima County Health Department (CPIMP211275), grants from Arizona Commerce Authority (LTR DTD 021822), grants from Sergey Brin Foundation, grants from Philips, Inc. (0483-06-161311-73077), grants from Sommetrics, Inc., grants from American Academy of Sleep Medicine Foundation (AASMF; 169-SR-17), grants from Regeneron, Inc., outside the submitted work; In addition, Sai Parthasarathy and Marvin Slepian have a patent US20160213879A1 that is licensed to SaiOx, Inc, unrelated to the present work. Danny Bluestein reports grants from NIH (U01 HL131052, U01 EB026414, R41 HL134418), outside the submitted work. Danny Bluestein and Marvin Slepian have U.S. Patents # 9,655,720 and 10,772,722, unrelated to the present work. Danny Bluestein and Marvin Slepian have an equity interest in PolyNova Cardiovascular Inc., unrelated to the present work. Marvin Slepian reports grants from NIH (R25-DK128859, U01 HL131052, U01 EB026414, R41 HL134418), and from Johnson and Johnson, Medidata and Regenesis Biomedical, outside the submitted work: In addition, Marvin Slepian and Sai Parthasarathy have a patent US20160213879A1 that is licensed to SaiOx, Inc, unrelated to the present work. Danny Bluestein and Marvin Slepian have U.S. Patents # 9,655,720 and 10,772,722, unrelated to the present work. Danny Bluestein and Marvin Slepian have an equity interest in PolyNova Cardiovascular Inc., unrelated to the present work.

References

1. Victor, L.D. Obstructive Sleep Apnea. *Am. Fam. Physician* **1999**, *60*, 2279–2286. [PubMed]
2. Parish, J.M.; Somers, V.K. Obstructive sleep apnea and cardiovascular disease. *Mayo Clin. Proc.* **2004**, *79*, 1036–1046. [CrossRef] [PubMed]
3. Semelka, M.; Wilson, J.; Floyd, R.; Health, E.; Hospital, L. Diagnosis and Treatment of Obstructive Sleep Apnea in Adults. *Am. Fam. Physician* **2016**, *94*, 355–360. [PubMed]
4. Alghanim, N.; Comondore, V.R.; Fleetham, J.; Marra, C.A.; Ayas, N.T. The economic impact of obstructive sleep apnea. *Lung* **2008**, *186*, 7–12. [CrossRef] [PubMed]
5. Somers, V.K.; White, D.P.; Amin, R.; Abraham, W.T.; Costa, F.; Culebras, A.; Daniels, S.; Floras, J.S.; Hunt, C.E.; Olson, L.J.; et al. Sleep Apnea and Cardiovascular Disease: An American Heart Association/American College of Cardiology Foundation scientific statement from the American Heart Association Council for High Blood Pressure Research Professional Education Committee, Council on Clinical Cardiology, Stroke Council, and Council on Cardiovascular Nursing. *Circulation* **2008**, *118*, 1080–1111. [CrossRef] [PubMed]

6. Young, T.; Palta, M.; Dempsey, J.; Peppard, P.E.; Nieto, F.J.; Hla, K.M. Burden of Sleep Apnea: Rationale, Design, and Major Findings of the Wisconsin Sleep Cohort Study. *WMJ* **2009**, *108*, 246–249.

7. Friedman, O.; Logan, A.G. The Price of Obstructive Sleep Apnea—Hypopnea: Hypertension and Other Ill Effects. *Am. J. Hypertens.* **2009**, *22*, 474–483. [CrossRef]

8. Palomaki, H. Snoring and the risk of ischemic brain infarction. *Stroke* **1991**, *22*, 1021–1025. [CrossRef]

9. Bai, J.; He, B.; Wang, N.; Chen, Y.; Liu, J.; Wang, H.; Liu, D. Snoring Is Associated with Increased Risk of Stroke: A Cumulative Meta-Analysis. *Front. Neurol.* **2021**, *12*, 574649. [CrossRef]

10. Lavie, L. From Oxidative Stress to Cardiovascular Risk in Obstructive Sleep Apnoea. *Somnologie* **2006**, *10*, 113–119. [CrossRef]

11. Redline, S.; Yenokyan, G.; Gottlieb, D.J.; Shahar, E.; O'Connor, G.T.; Resnick, H.E.; Diener-West, M.; Sanders, M.H.; Wolf, P.A.; Geraghty, E.M.; et al. Obstructive sleep apnea-hypopnea and incident stroke: The sleep heart health study. *Am. J. Respir. Crit. Care Med.* **2010**, *182*, 269–277. [CrossRef]

12. Yaggi, H.K.; Concato, J.; Kernan, W.N.; Lichtman, J.H.; Brass, L.M.; Mohsenin, V. Obstructive Sleep Apnea as a Risk Factor for Stroke and Death. *N. Engl. J. Med.* **2005**, *353*, 2034–2041. [CrossRef] [PubMed]

13. Sahlin, C.; Sandberg, O.; Gustafson, Y.; Bucht, G.; Carlberg, B.; Stenlund, H.; Franklin, K.A. Obstructive sleep apnea is a risk factor for death in patients with stroke: A 10-year follow-up. *Arch. Intern. Med.* **2008**, *168*, 297–301. [CrossRef] [PubMed]

14. Butt, M.; Dwivedi, G.; Khair, O.; Lip, G.Y.H. Obstructive sleep apnea and cardiovascular disease. *Int. J. Cardiol.* **2010**, *139*, 7–16. [CrossRef]

15. Jehan, S.; Farag, M.; Zizi, F.; Pandi-Perumal, S.R.; Chung, A.; Truong, A.; Tello, D.; McFarlane, S.I. Obstructive sleep apnea and stroke HHS Public Access. *Sleep Med. Disord.* **2018**, *2*, 120–125.

16. Palomaki, H.; Partinen, M.; Juvela, S.; Kaste, M. Snoring as a Risk Factor for Sleep-Related Brain Infarction. 1989. Available online: http://ahajournals.org (accessed on 21 November 2023).

17. Bhopal, R.S. Snoring as a risk factor for disease. *Br. Med. J.* **1985**, *291*, 1204. [CrossRef] [PubMed]

18. Gupta, A.; Shukla, G. Obstructive sleep apnea and stroke. *J. Clin. Sleep Med.* **2018**, *14*, 1819. [CrossRef]

19. Lipford, M.C.; Flemming, K.D.; Calvin, A.D.; Mandrekar, J.; Brown, R.D.; Somers, V.K.; Caples, S.M. Associations between cardioembolic stroke and obstructive sleep apnea. *Sleep* **2015**, *38*, 1699–1705. [CrossRef]

20. Peng, Y.-H.; Liao, W.-C.; Chung, W.-S.; Muo, C.-H.; Chu, C.-C.; Liu, C.-J.; Kao, C.-H. Association between obstructive sleep apnea and deep vein thrombosis/pulmonary embolism: A population-based retrospective cohort study. *Thromb. Res.* **2014**, *134*, 340–345. [CrossRef]

21. Eisensehr, I.; Noachtar, S. Haematological aspects of obstructive sleep apnoea. *Sleep Med. Rev.* **2001**, *5*, 207–221. [CrossRef]

22. Eckert, D.J.; Malhotra, A.; Jordan, A.S. Mechanisms of apnea. *Prog Cardiovasc Dis.* **2009**, *51*, 313–323. [CrossRef] [PubMed]

23. Hermann, D.M.; Bassetti, C.L. Role of sleep-disordered breathing and sleep-wake disturbances for stroke and stroke recovery. *Neurology* **2016**, *87*, 1407–1416. [CrossRef] [PubMed]

24. Wolk, R.; Somers, V.K. Sleep and the metabolic syndrome. *Exp. Physiol.* **2007**, *92*, 67–78. [CrossRef] [PubMed]

25. Bokinsky, G.; Miller, M.; Ault, K.; Husband, P.; Mitchell, J. Spontaneous Platelet Activation and Aggregation During Obstructive Sleep Apnea and Its Response to Therapy with Nasal Continuous Positive Airway Pressure. *Chest* **1995**, *108*, 625–630. [CrossRef] [PubMed]

26. Sanner, B.M.; Konermann, M.; Tepel, M.; Groetz, J.; Mummenhoff, C.; Zidek, W. Platelet function in patients with obstructive sleep apnoea syndrome. *Eur. Respir. J.* **2000**, *16*, 648–652. [CrossRef] [PubMed]

27. Geiser, T.; Buck, F.; Beat, J.M.; Bassetti, C.; Haeberli, A.; Gugger, M. In vivo Platelet Activation Is Increased during Sleep in Patients with Obstructive Sleep Apnea Syndrome. *Respiration* **2002**, *69*, 229–234. [CrossRef]

28. Peled, N.; Kassirer, M.; Kramer, M.R.; Rogowski, O.; Shlomi, D.; Fox, B.; Berliner, A.S.; Shitrit, D. Increased erythrocyte adhesiveness and aggregation in obstructive sleep apnea syndrome. *Thromb. Res.* **2008**, *121*, 631–636. [CrossRef]

29. Minoguchi, K.; Yokoe, T.; Tazaki, T.; Minoguchi, H.; Oda, N.; Tanaka, A.; Yamamoto, M.; Ohta, S.; O'Donnell, C.P.; Adachi, M. Silent brain infarction and platelet activation in obstructive sleep apnea. *Am. J. Respir. Crit. Care Med.* **2007**, *175*, 612–617. [CrossRef]

30. Yardim-Akaydin, S.; Caliskan-Can, E.; Firat, H.; Ardic, S.; Simsek, B. Influence of gender on C-reactive protein, fibrinogen, and erythrocyte sedimentation rate in obstructive sleep apnea. *Antiinflamm. Antiallergy Agents Med. Chem.* **2014**, *13*, 56–63. [CrossRef]

31. Tyagi, T.; Ahmad, S.; Gupta, N.; Sahu, A.; Ahmad, Y.; Nair, V.; Chatterjee, T.; Bajaj, N.; Sengupta, S.; Ganju, L.; et al. Altered expression of platelet proteins and calpain activity mediate hypoxia-induced prothrombotic phenotype. *Blood* **2014**, *123*, 1250–1260. [CrossRef]

32. Ingber, D.E. Cellular mechanotransduction: Putting all the pieces together again. *FASEB J.* **2006**, *20*, 811–827. [CrossRef]

33. Wang, N. Review of cellular mechanotransduction. *J. Phys. D Appl. Phys.* **2017**, *50*, 233002. [CrossRef]

34. Slepian, M.J.; Sheriff, J.; Hutchinson, M.; Tran, P.; Bajaj, N.; Garcia, J.G.; Saavedra, S.S.; Bluestein, D. Shear-mediated platelet activation in the free flow: Perspectives on the emerging spectrum of cell mechanobiological mechanisms mediating cardiovascular implant thrombosis. *J. Biomech.* **2017**, *50*, 20–25. [CrossRef]

35. Vibrations in Fluid–Structure Interaction Systems. In *Flow-Induced Vibrations*; Academic Press: Cambridge, MA, USA, 2014; pp. 359–401. [CrossRef]

36. Mihaescu, M.; Murugappan, S.; Kalra, M.; Khosla, S.; Gutmark, E. Large Eddy simulation and Reynolds-Averaged Navier-Stokes modeling of flow in a realistic pharyngeal airway model: An investigation of obstructive sleep apnea. *J. Biomech.* **2008**, *41*, 2279–2288. [CrossRef]

37. Mihaescu, M.; Mylavarapu, G.; Gutmark, E.J.; Powell, N.B. Large Eddy Simulation of the pharyngeal airflow associated with Obstructive Sleep Apnea Syndrome at pre and post-surgical treatment. *J. Biomech.* **2011**, *44*, 2221–2228. [CrossRef] [PubMed]

38. Fleck, R.J.; Ishman, S.L.; Shott, S.R.; Gutmark, E.J.; McConnell, K.B.; Mahmoud, M.; Mylavarapu, G.; Subramaniam, D.R.; Szczesniak, R.; Amin, R.S. Dynamic volume computed tomography imaging of the upper airway in obstructive sleep apnea. *J. Clin. Sleep Med.* **2017**, *13*, 189–196. [CrossRef] [PubMed]

39. Beck, R.; Odeh, M.; Oliven, A.; Gavriely, N. The acoustic properties of snores. *Eur. Respir. J.* **1995**, *8*, 2120–2128. [CrossRef]

40. Pevernagie, D.; Aarts, R.M.; De Meyer, M. The acoustics of snoring. *Sleep Med. Rev.* **2010**, *14*, 131–144. [CrossRef]

41. Hedner, J.A.; Wilcox, I.; Sullivan, C.E. Speculations on the interaction between vascular disease and obstructive sleep apnea. In *Sleep and Breathing*; Saunders, N.A., Sullivan, C., Eds.; Dekker: New York, NY, USA, 1994; pp. 823–846.

42. Amatoury, J.; Howitt, L.; Wheatley, J.R.; Avolio, A.P.; Amis, T.C. Snoring-related energy transmission to the carotid artery in rabbits. *J. Appl. Physiol.* **2006**, *100*, 1547–1553. [CrossRef] [PubMed]

43. Chuang, H.H.; Liu, C.H.; Wang, C.Y.; Lo, Y.L.; Lee, G.S.; Chao, Y.P.; Li, H.-Y.; Kuo, T.B.J.; Yang, C.C.H.; Shyu, L.-Y.; et al. Snoring sound characteristics are associated with common carotid artery profiles in patients with obstructive sleep apnea. *Nat. Sci. Sleep* **2021**, *13*, 1243–1255. [CrossRef]

44. Uzer, G.; Manske, S.L.; Chan, M.E.; Chaing, F.P.; Rubin, C.T.; Frame, M.D.; Judex, S. Separating Fluid Shear Stress from Acceleration during Vibrations in Vitro: Identification of Mechanical Signals Modulating the Cellular Response. *Cell. Mol. Bioeng.* **2012**, *5*, 266–276. [CrossRef] [PubMed]

45. Sheriff, J.; Bluestein, D.; Girdhar, G.; Jesty, J. High-shear stress sensitizes platelets to subsequent low-shear conditions. *Ann. Biomed. Eng.* **2010**, *38*, 1442–1450. [CrossRef] [PubMed]

46. Jesty, J.; Bluestein, D. Acetylated Prothrombin as a Substrate in the Measurement of the Procoagulant Activity of Platelets: Elimination of the Feedback Activation of Platelets by Thrombin. 1999. Available online: http://www.idealibrary.com (accessed on 21 November 2023).

47. Roka-Moiia, Y.; Walk, R.; Palomares, D.E.; Ammann, K.R.; Dimasi, A.; Italiano, J.; Sheriff, J.; Bluestein, D.; Slepian, M.J. Platelet Activation via Shear Stress Exposure Induces a Differentiating Biomarker "Signature" of Activation versus Biochemical Agonists. *Thromb. Haemost.* **2020**, *120*, 776–792. [PubMed]

48. Wu, P.; Wang, X.; Lin, W.; Bai, L. Acoustic characterization of cavitation intensity: A review. *Ultrason. Sonochemistry* **2022**, *82*, 105878. [CrossRef] [PubMed]

49. Patel, U.N.; Rothstein, J.P.; Modarres-Sadeghi, Y. Vortex-induced vibrations of a cylinder in inelastic shear-thinning and shear-thickening fluids. *J. Fluid Mech.* **2022**, *934*, A39. [CrossRef]

50. Ruggeri, Z.M.; Orje, J.N.; Habermann, R.; Federici, A.B.; Reininger, A.J. Activation-Independent Platelet Adhesion and Aggregation under Elevated Shear Stress Running title: Platelet adhesion and aggregation in flowing blood. *Blood* **2006**, *108*, 1903–1911. [CrossRef]

51. Nobili, M.; Sheriff, J.; Morbiducci, U.; Redaelli, A.; Bluestein, D. Platelet activation due to hemodynamic shear stresses: Damage accumulation model and comparison to in vitro measurements. *ASAIO J.* **2008**, *54*, 64–72. [CrossRef]

52. Casa, L.D.C.; Deaton, D.H.; Ku, D.N. Role of high shear rate in thrombosis. *J. Vasc. Surg.* **2015**, *61*, 1068–1080. [CrossRef]

53. Girdhar, G.; Xenos, M.; Alemu, Y.; Chiu, W.-C.; Lynch, B.E.; Jesty, J.; Einav, S.; Slepian, M.J.; Bluestein, D. Device thrombogenicity emulation: A novel method for optimizing mechanical circulatory support device thromboresistance. *PLoS ONE* **2012**, *7*, e32463. [CrossRef]

54. Bass, H.E.; Sutherland, L.C.; Zuckerwar, A.J.; Blackstock, D.T.; Hester, D.M. Atmospheric absorption of sound: Further developments. *J. Acoust. Soc. Am.* **1995**, *97*, 680–683. [CrossRef]

55. Gilbert, K.E.; Bass, H.E. Acoustic Waves. In *Encyclopedia of Atmospheric Sciences*; Judith, A., Curry, J.A., Pyle, J.A., Eds.; Academic Press Inc. Ltd.: London, UK, 2003.

56. Available online: https://www.acs.psu.edu/drussell/Demos/Absorption/Absorption.html (accessed on 21 November 2023).

57. Einav, S.; Bluestein, D. Dynamics of blood flow and platelet transport in pathological vessels. *Ann. N. Y. Acad. Sci.* **2004**, *1015*, 351–366. [CrossRef] [PubMed]

58. Soares, J.S.; Sheriff, J.; Bluestein, D. A novel mathematical model of activation and sensitization of platelets subjected to dynamic stress histories. *Biomech. Model. Mechanobiol.* **2013**, *12*, 1127–1141. [CrossRef] [PubMed]

59. Zhang, P.; Gao, C.; Zhang, N.; Slepian, M.J.; Deng, Y.; Bluestein, D. Multiscale Particle-Based Modeling of flowing platelets in blood plasma using dissipative particle dynamics and coarse grained molecular dynamics. *Cel. Mol. Bioeng.* **2014**, *7*, 552–574. [CrossRef] [PubMed]

60. Zhang, P.; Zhang, L.; Slepian, M.J.; Deng, Y.; Bluestein, D. A multiscale biomechanical model of platelets: Correlating with in-vitro results. *J. Biomech.* **2017**, *50*, 26–33. [CrossRef]

61. Hellums, J.D. Whitaker lecture: Biorheology in thrombosis research. *Ann. Biomed. Eng.* **1994**, *22*, 445–455. [CrossRef] [PubMed]

62. Bluestein, D.; Niu, L.; Schoephoerster, R.T.; Dewanjee, M.K. Fluid mechanics of arterial stenosis: Relationship to the development of mural thrombus. *Ann. Biomed. Eng.* **1997**, *25*, 344–356. [CrossRef] [PubMed]

63. Sheriff, J.; Soares, J.S.; Xenos, M.; Jesty, J.; Slepian, M.J.; Bluestein, D. Evaluation of shear-induced platelet activation models under constant and dynamic shear stress loading conditions relevant to devices. *Ann. Biomed. Eng.* **2013**, *41*, 1279–1296. [CrossRef]
64. Erbe, C.; Thomas, J.A. Introduction to Acoustic Terminology and Signal Processing. *Explor. Anim. Behav. Through Sound* **2022**, *1*, 111–152. [CrossRef]
65. Maimon, N.; Hanly, P.J. Does Snoring Intensity Correlate with the Severity of Obstructive Sleep Apnea? *JCSM* **2010**, *6*, 5. [CrossRef]
66. Consolo, F.; Sheriff, J.; Gorla, S.; Magri, N.; Bluestein, D.; Pappalardo, F.; Slepian, M.; Redaelli, A. High Frequency Components of Hemodynamic Shear Stress Profiles are a Major Determinant of Shear-Mediated Platelet Activation in Therapeutic Blood Recirculating Devices. *Nat. Sci. Rep.* **2017**, *7*, 4994. [CrossRef]
67. Feaver, R.E.; Gelfand, B.D.; Blackman, B.R. Human haemodynamic frequency harmonics regulate the inflammatory phenotype of vascular endothelial cells. *Nat. Commun.* **2013**, *4*, 1525. [CrossRef]
68. Sei, Y.J.; Ahn, S.I.; Virtue, T.; Kim, T.; Kim, Y. Detection of frequency-dependent endothelial response to oscillatory shear stress using a microfluidic transcellular monitor. *Sci. Rep.* **2017**, *7*, 10019. [CrossRef]
69. Lei, X.; Liu, B.; Wu, H.; Wu, X.; Wang, X.L.; Song, Y.; Zhang, S.-S.; Li, J.-Q.; Bi, L.; Pei, G.X. The effect of fluid shear stress on fibroblasts and stem cells on plane and groove topographies. *Cell Adhes. Migr.* **2020**, *14*, 12–23. [CrossRef] [PubMed]
70. Awtry, E.H.; Loscalzo, J. Aspirin. *Circulation* **2000**, *101*, 1206–1218. Available online: http://circ.ahajournals.org/content/101/10/1206.abstract (accessed on 21 November 2023). [CrossRef]
71. Vane, J.R.; Botting, R.M. The mechanism of action of aspirin. *Thromb. Res.* **2003**, *110*, 255–258. [CrossRef]
72. Mekaj, Y.H.; Daci, F.T.; Mekaj, A.Y. New insights into the mechanisms of action of aspirin and its use in the prevention and treatment of arterial and venous thromboembolism. *Ther. Clin. Risk Manag.* **2015**, *11*, 1449–1456. [CrossRef] [PubMed]
73. Mehra, M.R.; Crandall, D.L.; Gustafsson, F.; Jorde, U.P.; Katz, J.N.; Netuka, I.; Uriel, N.; Connors, J.M.; Sood, P.; Heatley, G.; et al. Aspirin and left ventricular assist devices: Rationale and design for the international randomized, placebo-controlled, non-inferiority ARIES HM3 trial. *Eur. J. Heart Fail.* **2021**, *23*, 1226–1237. [CrossRef]
74. Valerio, L.; Tran, P.L.; Sheriff, J.; Brengle, W.; Ghosh, R.; Chiu, W.C.; Redaelli, A.; Fiore, G.B.; Pappalardo, F.; Bluestein, D.; et al. Aspirin has limited ability to modulate shear-mediated platelet activation associated with elevated shear stress of ventricular assist devices. *Thromb. Res.* **2016**, *140*, 110–117. [CrossRef] [PubMed]
75. Valerio, L.; Sheriff, J.; Tran, P.L.; Brengle, W.; Redaelli, A.; Fiore, G.B.; Pappalardo, F.; Bluestein, D.; Slepian, M.J. Routine clinical anti-platelet agents have limited efficacy in modulating hypershear-mediated platelet activation associated with mechanical circulatory support. *Thromb. Res.* **2018**, *163*, 162–171. [CrossRef]
76. Sheriff, J.; Girdhar, G.; Chiu, W.-C.; Jesty, J.; Slepian, M.J.; Bluestein, D. Comparative efficacy of in vitro and in vivo metabolized aspirin in the DeBakey ventricular assist device. *J. Thromb. Thrombolysis* **2014**, *37*, 499–506. [CrossRef]
77. Roka-Moiia, Y.; Ammann, K.R.; Miller-Gutierrez, S.; Sweedo, A.; Palomares, D.; Italiano, J.; Sheriff, J.; Bluestein, D.; Slepian, M.J. Shear-mediated platelet activation in the free flow II: Evolving mechanobiological mechanisms reveal an identifiable signature of activation and a bi-directional platelet dyscrasia with thrombotic and bleeding features. *J. Biomech.* **2021**, *123*, 110415. [CrossRef]
78. Scinico, M.; Sostin, O.V.; Agarwal, R.; Kapoor, A.D.; Petrini, J.R.; Mendez, J.L. A Pilot Study of Aspirin Resistance in Obstructive Sleep Apnea Patients. *Clin. Investig. Med.* **2021**, *44*, 55–63. [CrossRef] [PubMed]
79. Li, N.; Wen, W.; Cai, X.; Zhu, Q.; Hu, J.; Heizhati, M.; Yuan, Y.; Gan, L.; Dang, Y.; Yang, W.; et al. The Use of Aspirin Increases the Risk of Major Adverse Cardiac and Cerebrovascular Events in Hypertensive Patients with Obstructive Sleep Apnea for the Primary Prevention of Cardiovascular Disease: A Real-World Cohort Study. *J. Clin. Med.* **2022**, *11*, 7066. [CrossRef] [PubMed]
80. Christian, D.; Omar, A.; Vanessa, D.; Francisco, X.E. Paradoxical Thrombotic Effects of Aspirin: Experimental Study on 1000 Animals. *Cardiovasc. Hematol. Disord. Drug Targets* **2010**, *10*, 103–110.
81. Lei, J.; Zhou, Y.; Xie, D.; Zhang, Y. Mechanistic insights into a classic wonder drug-aspirin. *J. Am. Chem. Soc.* **2015**, *137*, 70–73. [CrossRef] [PubMed]
82. Christiansen, M.; Grove, E.L.; Hvas, A.M. Contemporary Clinical Use of Aspirin: Mechanisms of Action, Current Concepts, Unresolved Questions, and Future Perspectives. *Semin. Thromb. Hemost.* **2021**, *47*, 800–814. [CrossRef] [PubMed]
83. Södergren, A.L.; Ramström, S. Platelet subpopulations remain despite strong dual agonist stimulation and can be characterized using a novel six-colour flow cytometry protocol. *Sci. Rep.* **2018**, *8*, 1441. [CrossRef]

bioengineering

Review

Computational Rhinology: Unraveling Discrepancies between In Silico and In Vivo Nasal Airflow Assessments for Enhanced Clinical Decision Support

Sverre Gullikstad Johnsen 

SINTEF, NO-7465 Trondheim, Norway; sverre.g.johnsen@sintef.no

Abstract: Computational rhinology is a specialized branch of biomechanics leveraging engineering techniques for mathematical modelling and simulation to complement the medical field of rhinology. Computational rhinology has already contributed significantly to advancing our understanding of the nasal function, including airflow patterns, mucosal cooling, particle deposition, and drug delivery, and is foreseen as a crucial element in, e.g., the development of virtual surgery as a clinical, patient-specific decision support tool. The current paper delves into the field of computational rhinology from a nasal airflow perspective, highlighting the use of computational fluid dynamics to enhance diagnostics and treatment of breathing disorders. This paper consists of three distinct parts—an introduction to and review of the field of computational rhinology, a review of the published literature on in vitro and in silico studies of nasal airflow, and the presentation and analysis of previously unpublished high-fidelity CFD simulation data of in silico rhinomanometry. While the two first parts of this paper summarize the current status and challenges in the application of computational tools in rhinology, the last part addresses the gross disagreement commonly observed when comparing in silico and in vivo rhinomanometry results. It is concluded that this discrepancy cannot readily be explained by CFD model deficiencies caused by poor choice of turbulence model, insufficient spatial or temporal resolution, or neglecting transient effects. Hence, alternative explanations such as nasal cavity compliance or drag effects due to nasal hair should be investigated.

Keywords: computational rhinology; computational fluid dynamics (CFD); large eddy simulation (LES); nasal airflow; nasal resistance; rhinomanometry (RMM); turbulence

check for
updates

Citation: Johnsen, S.G. Computational Rhinology: Unraveling Discrepancies between In Silico and In Vivo Nasal Airflow Assessments for Enhanced Clinical Decision Support. *Bioengineering* **2024**, *11*, 239. https://doi.org/10.3390/bioengineering11030239

Academic Editor: Mohammad Rahimi-Gorji

Received: 9 January 2024
Revised: 9 February 2024
Accepted: 17 February 2024
Published: 28 February 2024

1. Introduction

Computational fluid dynamics (CFD) is an emerging in silico tool in rhinology, leveraging engineering techniques for mathematical modelling of nasal airflow. The interdisciplinary integration of CFD in rhinology is part of computational rhinology, a specialized branch of biomechanics. Computational rhinology has already contributed significantly to advancing our understanding of nasal function, including airflow patterns, mucosal cooling, particle deposition, and drug delivery. Future prospects in computational rhinology encompass the development of virtual surgery as a clinical, patient-specific decision support tool and the refinement of patient selection criteria for treating common nasal airway disorders. These promising advancements may also extend into the broader field of otorhinolaryngology.

The present study derives from a collaborative effort between St. Olavs hospital the University hospital of Trondheim, the Norwegian University of Science and Technology, and the research foundation SINTEF. Our aim is to improve the understanding of obstructive sleep apnea (OSA) by employing engineering tools such as mathematical modelling [1–11]. One notable discovery by the research team underscores the potential of minor anterior nasal surgical intervention (e.g., correcting nasal septum deviation) to alleviate OSA, alone. This was clearly demonstrated in the patient included in **Part III**

(Section 4) of this paper, whose OSA markedly improved following surgical septum deviation correction. Other research groups have reported similar positive outcomes [12]. OSA is caused by repetitive collapses of the pharyngeal walls during sleep, and the impact of surgical nasal cavity modification on the onset of OSA is not yet fully understood. Isolated nasal surgery is thus not generally recommended as the first-line treatment for OSA [13], and there are no objective clinical methods available to identify patients who will benefit from such surgery.

CFD has been proposed as an attractive objective tool for predicting how alterations to the upper airways affect patient-specific airflow. However, its effectiveness in such detailed applications is hindered by the lack of in vivo nasal airflow measurements, which are essential for validating CFD models. In vivo rhinomanometry (RMM) stands as the sole method capable of supplying clinical nasal airflow data. While CFD has demonstrated reliability when compared to in vitro airflow measurements in physical nasal cavity replicas, in silico (CFD-based) RMM has been reported to severely underpredict the nasal resistance measured by in vivo RMM, without adequate explanations given. This apparent paradox was the primary driver for the current study. To empower CFD as a practical, patient-specific clinical decision support tool, it is vital to understand the possible reasons for the gross disagreement observed when comparing in silico and in vivo RMM. While this paper does not conclusively resolve this issue, it presents new evidence and discussions to narrow down the list of possible explanations, offering a solid foundation for future development and utilization of CFD-based simulation tools for improved understanding of (patient-specific) nasal function and as clinical decision support in rhinology.

This paper is written to be accessible to readers at all levels, from beginners to experts, in the disciplines of otorhinolaryngology and CFD. Thus, it aims to bridge the gap between these traditionally distinct scientific fields. This paper not only serves as an introductory guide to computational rhinology but also explores and discusses unresolved controversies, dilemmas, and paradoxes within the field and presents new evidence that will help unravel the current disparity between measurements and simulations in rhinology. This paper includes a comprehensive bibliography, meticulously compiled through a combination of ancestry and descendancy literature review approaches, utilizing internet-based publication databases such as Google Scholar, Researchgate, PubMed, and journal web pages. Additionally, it offers novel results from finely resolved large eddy simulation (LES)-based CFD simulations of active anterior RMM. Key terms and concepts regarding rhinology and CFD are presented in Sections 2.1–2.4, and an overview of nomenclature and abbreviations is provided in the back matter. Raw data from the simulations are available in the Supplementary Data [14].

This paper is structured into three main parts. **Parts I** and **II** can be read independently, but **Part III** relies on the two former parts in the sense that these serve as the scientific background of the in silico study performed.

Part I (Section 2 Computational Rhinology) offers a general overview of computational rhinology, with a particular focus on nasal airway obstruction and subjective and objective clinical measures. It also traces the evolution of CFD as a clinical decision support tool over the last decades. A comprehensive review is given of important research questions whose answers are crucial for the adoption of CFD as a clinical decision support tool.

Part II (Section 3 Overview of Published Literature on in vitro and in silico Nasal Airflow Studies) provides a bibliographic survey, encompassing research involving in vitro measurements in physical nasal cavity replicas and studies focusing on in silico experiments conducted in digital nasal cavity models. These studies are classified by the nature of their inflow boundary conditions (steady state vs. transient) and modeling techniques and the citations are summarized in tabular format.

Part III (Section 4 In silico RMM Simulation Results) presents novel patient-specific findings derived from in silico simulations of active anterior RMM, achieved through high-fidelity, transient LES simulations. This part presents one of the most detailed CFD simulations of nasal airflow to date, including three breathing cycles on each side of the nose within fully resolved simulations. The simulations were designed to elucidate core research queries highlighted in **Parts I** and **II**, specifically focusing on whether the discrepancy between in vivo and in silico RMM can be attributed to CFD model deficiencies such as the following:

- Unresolved transitional or turbulent effects.
- Unresolved spatial phenomena such as vortices and eddies.
- Transient effects like hysteresis, developing boundary layers, and meandering.

Assessment of the finely resolved LES-based CFD simulations indicated that they were highly accurate, providing a reliable benchmark for the present study. They revealed that neither pronounced transitional/turbulent nor transient effects are significant contributors to the observed disparities. In conclusion, a pseudo-steady laminar model with a relatively coarse computational mesh yields near-perfect predictions compared to the transient, fully resolved LES model. While the former model can be run within hours on a desktop computer, the latter required approximately 3 million CPU-hours.

2. Part I—Computational Rhinology

2.1. Rhinology

In humans, the nasal cavity is the primary conduit for lung ventilation, supplying the body with fresh oxygen while expelling carbon dioxide. Additionally, it plays a vital role in the olfactory system by transporting odors to the olfactory sensory system. It is a highly intricate flow channel that is optimized for various functions, including humidification and heating of inhaled air as well as air filtering through particle deposition [15,16].

While nasal breathing is the typical mode of respiration, there are situations when nasal breathing alone may not provide sufficient oxygen saturation in the blood, and oral breathing becomes necessary. Oral breathing may be required in cases of impaired nasal patency, where the nasal passages are partially blocked or restricted, hindering effective airflow. Oral breathing lacks, however, most of the traits of nasal breathing, and it is well known that excessive oral breathing has adverse effects on, e.g., dentofacial development, oral health, and digestive and breathing disorders (e.g., obstructive sleep apnea) [17].

In the medical field, the study of the nasal cavity falls within the discipline of otorhino-laryngology, with a specific focus on the nasal cavity known as rhinology. While rhinology aims to enhance diagnostic and treatment approaches for nasal and sinonasal disorders, its significance extends beyond its primary domain. Otorhinolaryngology, as a whole, benefits from the advancements made in rhinology. Additionally, the knowledge and expertise of rhinologists are essential in the fields of allergy/immunology, pulmonology, sleep medicine, head and neck surgery, and facial plastic and reconstructive surgery. It follows that advancements in understanding nasal physiology and the complexities of nasal airflow can lead to improved diagnostic techniques and innovative treatment strategies across a wide range of disorders within these associated medical fields.

Nasal airway obstruction (NAO), which can be caused by structural abnormalities or deformities such as deviated septum, nasal polyps, turbinate hypertrophy, nasal injuries, or other underlying causes, is an important subtopic of rhinology. It is generally accepted that NAO affects the nasal airflow pattern and correlates with symptoms such as nasal congestion, trouble breathing, sleep-disordered breathing, and others. Mathematical modelling of nasal airflow has been suggested as an important tool to complement objective clinical measurements in the assessment of NAO and evaluation of treatment options [18–20].

2.2. The Relationship between Obstructive Sleep Apnea and Nasal Airway Obstruction

Obstructive sleep apnea (OSA) is a common breathing disorder caused by recurrent, temporary upper airway collapses during sleep [21]. The collapses are primarily attributed to the Venturi effect, triggered by accelerated airflow through constrictions enveloped by soft tissue within the oropharyngeal tract. A pressure difference between the soft tissue and the airway may cause partial (hypopnea) or complete (apnea) respiratory blockages when neuromuscular response is impaired or relaxed. Interruptions in breathing can lead to frequent awakenings or transitions from deeper to lighter stages of sleep, resulting in overall reduced sleep quality and leading to symptoms of daytime sleepiness, fatigue, and reduced cognitive function.

It is widely acknowledged that OSA can have severely negative effect on patients' health and wellbeing, as it is correlated with conditions such as cardiovascular diseases, metabolic syndrome and diabetes, and learning disabilities and cognitive development [22–26]. Moreover, OSA has implications for tasks such as operating motor vehicles due to national regulations and driver's license restrictions [27]. Insurance companies also approach individuals with OSA differently from their healthier counterparts, potentially affecting coverage eligibility [28]. The concern arises that professional truck drivers might avoid OSA screening to circumvent challenges with their driver's licenses or life insurance coverage.

Various treatment options are available for OSA, including continuous positive airway pressure (CPAP), mandibular advancement devices (MAD), and surgical intervention, complemented by lifestyle adjustments and weight management. The severity of OSA is conventionally quantified using the apnea–hypopnea index (AHI), representing the count of obstructive incidents per hour of sleep.

Young et al. [29] conducted a study that revealed a significant association between NAO and sleep-disordered breathing, including conditions such as snoring and OSA. This finding has been supported by several other studies, demonstrating the negative impact of high nasal flow resistance on snoring and OSA [30–33]. Singh et al. [17] pointed out that oral breathing is mainly caused by NAO and that several aspects of oral breathing affect OSA adversely. There is, however, weak correlation between AHI and patients' subjective assessment of nasal symptoms [34]. Hoel et al. [35,36] found that patients with OSA and increased nasal resistance had a higher ratio of hypopneas to apneas.

Scott and Kent [37] advocated the vital role of the nasal cavity in breathing, noting that 90 percent of airflow occurs through the nasal cavity and approximately 60 percent of flow resistance is attributed to the nasal passages. Understanding nasal airflow is therefore crucial for comprehending the development of OSA. They also acknowledged the complexity of surgical interventions, stating that individualized approaches are necessary for optimal outcomes. However, the success rate of septoplasty, a common surgical procedure to alleviate NAO, varies by between 43 and 85 percent based on objective and subjective measures [38].

In the context of CPAP treatment for OSA, Nakata et al. [39] emphasized the significance of addressing nasal resistance, stating that increased nasal resistance can contribute to CPAP failure. They suggested that surgical correction of severe nasal obstruction should be considered to enhance the effectiveness of CPAP therapy.

A recent review found conflicting meta-studies regarding the effect of nasal surgery on AHI and concluded that isolated nasal surgery should not generally be recommended as the first-line treatment for OSA [13]. There are, however, studies that indicate that septoplasty has a pivotal role in combination with inferior turbinate surgery [7,40] or multilevel palate and/or tongue surgery [41], to improve AHI.

To improve the effectiveness of NAO treatment, it is anticipated that patient-specific planning tools, such as mathematical models, can serve as valuable decision support tools [42]. These have the potential to enhance treatment outcomes by providing personalized insights and guidance for surgical interventions and other treatment options.

2.3. Clinical Evaluation of Nasal Patency

Nasal patency refers to the degree or extent to which the nasal passages are open and unobstructed [43]. It is influenced by various factors, including the size and shape of the nasal passages, the condition of the nasal mucosa (lining of the nose), the presence of any anatomical abnormalities or obstructions, and the function of the nasal gateway. Because nasal patency is not a physical quantity that can be measured directly, it is the role of the rhinologist to use various clinical measures to assess the nasal patency.

Objective, clinical measurements to evaluate nasal patency include rhinomanometry (RMM), acoustic rhinometry (AR), and peak nasal inspiratory flow (PNIF) in addition to endoscopy and medical imaging techniques such as computed tomography (CT) or magnetic resonance imaging (MRI) [44–46]. Despite the objective nature of these measurements, their results must be scrutinized by a medical expert to exploit their diagnostic potential.

Subjective measures include self-reporting questionnaires designed to assess patients' subjective experience and perception of their nasal patency and quality of life by marking the perceived level of a specific symptom on a prescribed scale. Popular measures include variations of the Visual Analog Scale (VAS) and Nasal Obstruction Symptom Evaluation (NOSE) scale [47] among others.

Computational fluid dynamics (CFD) is an emerging, objective tool in rhinology. It is based on the mathematical modelling of nasal airflow and has the potential to be an important supplement to current clinical measures to advance the understanding of nasal airflow [18–20].

Many studies have investigated the correlation between the various objective and subjective measures, but the results are not conclusive.

Several studies [44,48–50] indicate that the objective standard measurements performed during clinical exam have limited value in the diagnosis of NAO patients. Subjective sensation of nasal obstruction may be caused by several factors other than objective nasal flow resistance, but unilateral nasal flow resistance correlates better with subjective sensation of nasal patency than bilateral nasal flow resistance [45]. Mozzanica et al. [51] reported significant correlation between subjective sensation of nasal patency, measured with I-NOSE and VAS questionnaires, and RMM data, however, and Hueto et al. [33] highlighted the usefulness of RMM in determining the appropriate pressure settings for CPAP treatment in OSA patients. Preoperative clinical evaluation commonly involves objective measures, and the clinical value of RMM should not be downplayed [52,53].

Zhao and Dalton [18] pointed out that standard rhinometric measurements are poorly correlated with patient-reported subjective symptoms and questioned their clinical value for evaluation of nasal obstruction and subjective evaluation of treatment outcome. Eccles and co-authors [54,55] effectively showed how subjective sensing of nasal patency can be decoupled from objectively measured nasal resistance by application of menthol, which affects the sensory ability of the trigeminal nerve endings. The dissociation between subjective and objective evaluation of nasal patency is evident in empty nose syndrome (ENS), which is a condition that typically occurs after surgical procedures to alleviate NAO and minimize nasal resistance. Paradoxically, some individuals who undergo these surgeries may experience increased sensations of NAO, among other symptoms, even though objective tests show that their nasal passages are open. Impaired trigeminal nerve function has been pointed to as an explanation for ENS [56]. Malik et al. [57] studied how the formation of a middle meatus jet stream is characteristic for ENS patients. Di et al. [58] and Li and co-authors [59,60] studied nasal aerodynamics in ENS patients using CFD.

It can be deduced that the same sensors that cause subjective sensing of nasal patency are responsible for subjective sensing of heat exchange between the nasal airflow and the nasal tissue, and several studies have demonstrated correlations between mucosal cooling and temperature, obtained from CFD simulations, and patients' subjective evaluation of nasal patency [50,61–68].

Obviously, both local mucosal cooling and overall nasal resistance depend on the nasal airflow pattern. Whereas subjective sensation of nasal patency may depend more on the cooling effect, transport and deposition of nasal spray may be more closely correlated with nasal resistance [69]. The fact that the nasal airflow pattern is highly sensitive to local nasal anatomy/geometry suggests that objective tools such as CFD are required to complement standard clinical flow characterization techniques such as RMM and bridge the gap between subjective and objective clinical measurements.

2.3.1. Nasal Flow Resistance

Flow resistance, R, is an intrinsic flow channel-specific attribute that correlates the pressure drop over the channel, ΔP, and the volumetric flowrate, Q;

$$\Delta P = RQ. \tag{1}$$

The flow resistance can vary with the flowrate, and its characteristics can be determined by measuring the pressure drop for known flowrates, or vice versa. For pressure-driven flows such as respiratory airflow, Equation (1) offers insight into how the breathing effort (quantified as pressure drop) must increase to maintain a consistent flowrate when encountering heightened flow resistance. For instance, to achieve an adequate inspiratory flowrate in an obstructed airway, greater intrathoracic negative pressure is required than in an unobstructed airway. This offers a simple explanation for why oral breathing may become the favored mode of respiration in the presence of NAO. In general, heightened flow resistance serves as an indicator of diminished patency.

The nature of (internal) fluid flow heavily depends on the geometry of the flow channel. e.g., for fully developed, steady, laminar, single-phase flow in a straight channel, the flow resistance is inversely proportional to the hydraulic diameter, $D_h \equiv 4A/O$, raised to the fourth power (Hagen–Poiseuille equation);

$$R = \frac{128L\mu}{\pi D_h^4}, \tag{2}$$

where A, O, and L are the cross-sectional area, perimeter, and length of the flow channel, respectively, and μ is the dynamic viscosity of the fluid.

In complex flow channels such as the nasal cavities, slight changes in the geometrical features may cause unpredictable airflow response. e.g., it can be imagined that in parts of the nasal cavity featuring narrow passages such as in the nasal vestibule or the olfactory slit, a slight modification may reduce the perimeter considerably without affecting the cross-sectional area notably. This may severely impact the local hydraulic diameter. Moreover, flow instabilities or recirculation zones triggered by geometrical features (e.g., abrupt changes in flow direction or cross-sectional area) may result in effective cross-sections smaller than the actual cross-section. It is not obvious that the concept of hydraulic diameter and associated standard flow resistance correlations are applicable for nasal airflow [70–73].

A peculiar feature of the nasal passages, known as the nasal cycle, is caused by temporal, asymmetric swelling and deswelling of the nasal mucosa. This effect causes an intermittent variation in nasal resistance that can be observed through the sensation of unilateral nasal obstruction. It is believed that this effect is important for the removal of deposited dust particles, etc. Not all humans have it, however, and the periodicity of the phenomenon is neither the same between different individuals nor constant in the same individual [74,75]. The nasal cycle is expected to affect unilateral nasal resistance measurements.

Nasal cavity compliance may permit local expansion or contraction of the nasal cavity cross-section due to periods of over- and under-pressure occurring during the respiratory cycle. This may cause local pressure dependency in the hydraulic diameter, hence the nasal resistance, which may introduce asymmetry with respect to exhalation/inhalation and temporal effects such as hysteresis in the pressure–flow relationship (Equation (1)).

2.3.2. Rhinomanometry

Rhinomanometry (RMM) is the only technique in clinical use that allows for quantitative assessment of the respiratory function of the nose [76], and it is thus of significant importance for the calibration and validation of patient-specific mathematical models of nasal air flow. The theory and background have been thoroughly covered by Vogt et al. [77]. RMM is a method that measures the pressure drop in the nose as a function of volumetric air flowrate. The resulting pressure–flow curves relate the volumetric flowrate to the pressure drop. The measured volumetric flow and pressure drops form the basis for calculation of the nasal resistance and estimation of representative hydraulic diameters [78]. Figure 1 shows an example of a pressure–flow curve.

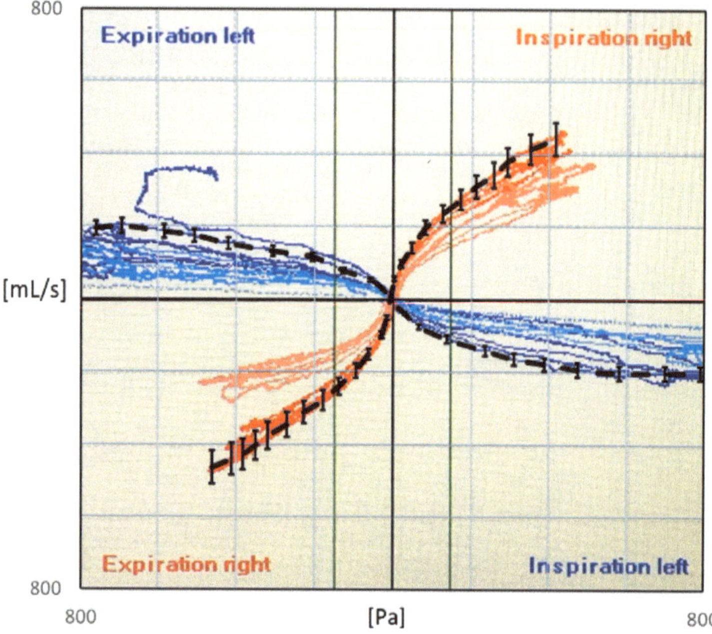

Figure 1. Rhinomanometry output for the current patient. Red corresponds to right side and blue corresponds to left side RMM, and light/dark colors indicate before/after administration of decongestive nasal spray. Black curves with 10% error bars show the selected "measured data" used in the current paper.

Whereas bilateral RMM considers the simultaneous measurement of both nasal passages, unilateral RMM considers only one nasal passage at the time by occluding one nostril while assessing the airflow in the open passage. It follows from the definition (Equation (1)) that the reciprocal bilateral resistance is the sum of the reciprocals of the individual unilateral resistances of the two nasal passages:

$$\frac{1}{R_{bi}} = \frac{1}{R_{uni,left}} + \frac{1}{R_{uni,right}}. \tag{3}$$

In posterior RMM, the pressure is measured directly in or close to the nasopharynx using a pressure probe typically inserted via the oral cavity. In anterior RMM, one nostril is closed, and the pressure probe is inserted into the nasal vestibule behind the occlusion to provide an indirect measurement of choanae pressure. It has been shown that posterior and anterior RMM are equivalent with respect to (unilateral) pressure measurement [79]. Passive RMM is performed by enforcing external nasal airflow. More commonly employed is active RMM, where the patient's own physiological airflow is utilized. The RMM procedure combining active breathing with anterior measurement is denoted as active anterior RMM (AAR).

Because the nasal passages are lined with a mucosal membrane subject to unpredictable temporal variations in swelling, RMM is typically performed twice, before and after decongestion. Application of topical nasal decongestant (e.g., xylometazoline) or physical exercise serves to eliminate the vascular component of nasal obstruction caused by swelling of the turbinates, allowing for quantification of the anatomical component of NAO [80]. The anatomical component is determined solely by the rigid tissue of the nose (e.g., bone and cartilage), and it is thus related to the maximal nasal volume, independent of the nasal cycle [45]. In clinical rhinology, such decongestion tests are performed to quantify the different roles of the skeleton and mucosa in NAO. For the mathematical modelling of airflow in the nasal cavities, it is convenient to disregard the complexities associated with soft tissue and temporal variations in unilateral nasal resistance.

2.3.3. Mathematical Illustration of the Principles of Rhinomanometry and Inherent Hysteresis Using Bernoulli's Equation

The mathematical background of rhinomanometry has been thoroughly covered by others [77]. Here, a brief illustration of the concept of rhinomanometry is provided, based on the mathematical description of simple pipe flow.

In pipe and duct flow engineering, it is common practice to estimate pressure drops by utilizing Bernoulli's equation [81]. For unsteady, fully developed, horizontal, incompressible flow through a straight, rigid duct of constant cross-sectional area, the pressure drop per unit length results from the contributions of a friction term and an unsteady inertial term,

$$\frac{\Delta P}{L} = \frac{8\rho}{\pi^2 D_h^5} \underbrace{\left[\underbrace{f_D Q}_{\text{friction term}} + \underbrace{\frac{\pi D_h^3}{2} \frac{\partial \ln Q}{\partial t}}_{\text{unsteady term}} \right]}_{\text{flow resistance per unit length, } R/L} Q \; , \tag{4}$$

where ρ is the fluid mass density and f_D is the Darcy friction factor,

$$f_D = \begin{cases} f_{D,lam} = 64/\text{Re} & \text{for} \quad \text{Re} \leq 2000 \\ f_{D,turb}(\text{Re}, \varepsilon_r) & \text{for} \quad \text{Re} \geq 4000 \end{cases} \; , \tag{5}$$

where Re is the Reynolds number, which for internal duct flow, can be expressed as

$$\text{Re} = 4Q/\pi v D_h \; , \tag{6}$$

where $v = \mu/\rho$ is the fluid kinematic viscosity and ε_r is the relative wall roughness. Equation (4) reduces to Equation (2) for laminar flow (Re $\leq$ 2000) with $\partial Q/\partial t = 0$. Several alternative formulations exist for the turbulent friction factor, e.g., the Haaland equation [82],

$$\frac{1}{\sqrt{f_{D,turb}}} = -1.8 \log_{10}\left[\left(\frac{\varepsilon_r}{3.7}\right)^{1.11} + \frac{6.9}{\text{Re}} \right]. \tag{7}$$

In laminar flow, the flow is characterized by smooth, locally parallel streamlines, and flow variables behave deterministically. In the case of turbulent flow, however, flow variables behave stochastically. In duct flow engineering, the flow is typically considered laminar for Reynolds numbers below 2000 and turbulent for Reynolds numbers above 4000, but these thresholds may vary depending on the specific flow configuration. In the intermediate range, the flow is transitional, which is a generally poorly understood, complex and dynamic flow state. For transitional flow, the friction factor can be approximated by a smooth, weighted average of the laminar and turbulent friction factors, $f_{D,tran} \approx wt \cdot f_{D,lam} + (1 - wt) \cdot f_{D,turb}$, where the weight, wt, varies smoothly between 1 and 0 in the laminar and turbulent regimes, respectively.

To improve the general understanding of RMM, a simplified model is used to represent nasal airflow, based on a straight, smooth ($\varepsilon_r = 0$) duct of a given hydraulic diameter and fully developed sinusoidal (respiratory) flow,

$$Q = Q_{\max} \sin(\omega t) \; , \tag{8}$$

where t is the time variable, $\omega = 2\pi/\tau$ is the angular frequency, and τ is the period of the respiratory cycle. Employing this model, synthetic RMM pressure–flow curves can be generated to study the impact of the various parameters. In Figure 2, data are shown for $Q_{\max} = 600$ mL/s, $\tau = 5$ s, and $D_h = 10$ mm, to compare the effects of assuming laminar or turbulent flow and to illustrate the effect of the unsteady term. Figure 2a shows that there is a phase shift between the flowrate and pressure drop due to the unsteady term. In Figure 2b, it can be seen that this causes a hysteresis effect such that the pressure–flow curve does not pass through the origin.

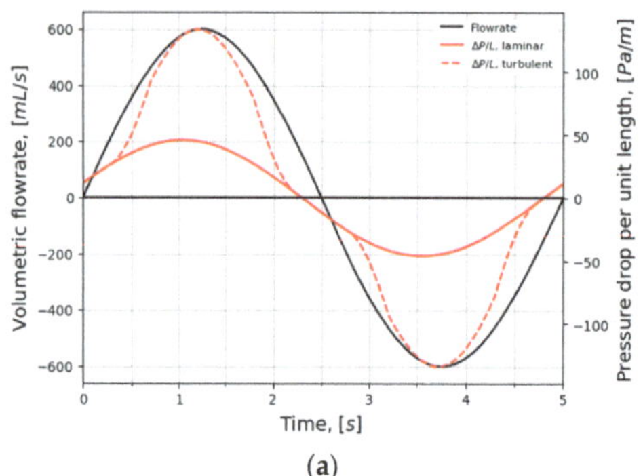

(a)

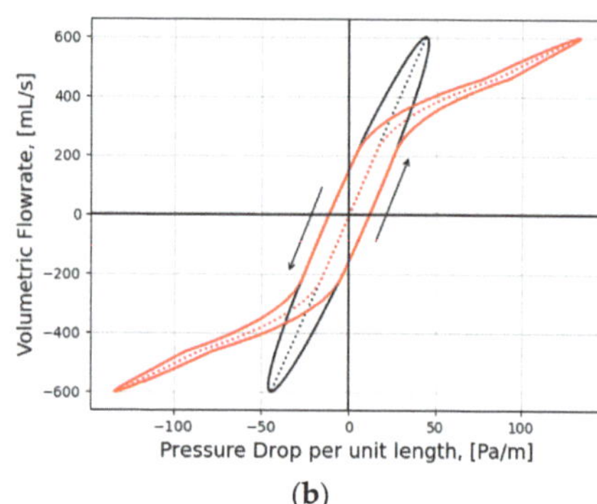

(b)

Figure 2. Synthetic rhinomanometry data obtained using a sinusoidal volumetric flowrate (Equation (8)), with $Q_{\max} = 600$ mL/s and $\tau = 5$ s, and pressure drop per unit length calculated from Bernoulli's equation, Equation (4), for a horizontal, straight, smooth pipe of hydraulic diameter $D_h = 10$ mm. (a) Volumetric flowrate (black) and pressure drop per unit length (red) as functions of time, for laminar (solid curve) and turbulent (dashed curve) flow. (b) Pressure–flow curves for laminar (black) and turbulent (red) flow. Dotted curves neglect the unsteady term in Equation (4). Arrows indicate the evolution in time.

The hysteresis width is found by evaluating Equation (4) at $Q = 0$,

$$W_{\Delta P} = \frac{8\rho L \omega Q_{\max}}{\pi D_h^2} , \tag{9}$$

where Equation (8) was used. It follows that the relative hysteresis width, defined as the width of the pressure–flow curve hysteresis at the level of $Q = 0$ divided by the maximum laminar pressure drop, and can be expressed as

$$W_{\Delta P,rel} = \omega D_h^2 / 16\nu. \tag{10}$$

Figure 3 illustrates how the relative hysteresis width increases for increasing hydraulic diameter while the flow resistance per unit length decreases.

In general, the following can be observed through analysis of Equation (4):

- Steady, laminar pressure–flow curves are straight lines passing through the origin.
- The slope of the pressure–flow curve decreases with increasing flow resistance. That is, laminar pressure–flow curves are steeper than turbulent ones due to the higher friction factor in turbulent flow. Distinct change in slope in in vivo RMM pressure–flow curves may thus be an indication of transition to turbulence.
- The unsteady flow resistance term causes a hysteresis effect, such that the pressure–flow curve becomes a closed loop not passing through the origin.
- The relative hysteresis width is determined by the hydraulic diameter and the period of the respiratory cycle.

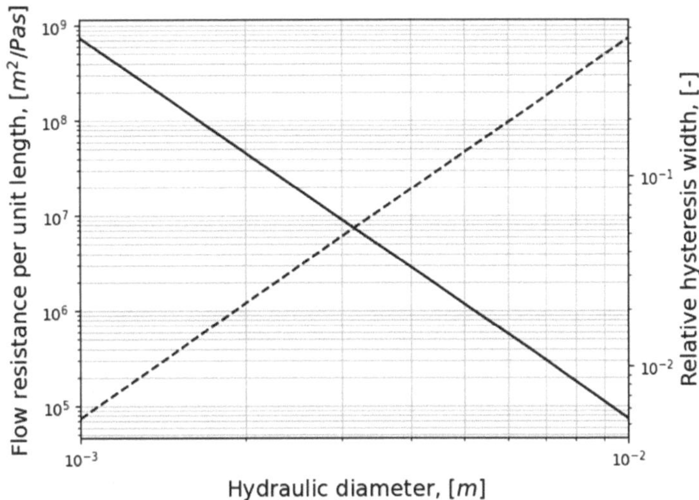

Figure 3. Flow resistance per unit length (Equation (2)) (solid curve) and relative hysteresis width (Equation (10)) (dashed curve) for laminar air flow.

The complex shape of the nasal cavity as well as effects of lateral wall movement may complicate this picture considerably for nasal airflow. Blevins [81] provides an overview of applicable methods to approximate friction factors and pressured drop for channels with noncircular cross-sections, bends, changes in flow area, etc.

2.4. Computational Fluid Dynamics in Rhinology

Computational fluid dynamics (CFD) combines numerical mathematics, computational sciences, and fluid dynamics to solve partial differential equations that represent the conservation laws of fluid dynamics. By utilizing computers for numerical solutions, CFD enables the analysis of complex fluid dynamics problems. For over five decades, CFD has been extensively employed in various industries, such as automotive and aerospace sectors and process industries. It has become a fundamental component of industrial research and development, providing cost-effective alternatives to performing costly experiments through rapid in silico prototyping. This approach reduces the number of required experiments, mitigating risks and costs. CFD serves multiple purposes, including (1) analyzing and gaining deeper insights into experimental results and observations; (2) supporting experiment design and planning; and (3) facilitating industrial process control. One significant advantage of CFD is its ability to provide detailed understanding of processes and phenomena that are impractical or impossible to directly observe in situ. CFD serves as a valuable tool for both forward (when the cause is known) and backward (when the effect is known) causality mapping. This versatility makes CFD an excellent diagnostics tool for applications in both industrial and medical domains. For example, CFD may be used to analyze the effects of virtual surgery on patient-specific computer models prior to actual surgery, in order to provide objective decision support for medical personnel.

All the steps in the creation of a high-quality CFD model are prone to errors and uncertainties, and an important part of the job as a CFD engineer is the reiteration and improvement of each step until the model performs adequately. While best practice guidelines exist and experience helps, model requirements may vary between different flow situations, and this can be a meticulous and time-consuming process. To utilize CFD as a clinical tool for decision support, there is a need for standardization of best practice guidelines. On one hand, due to relatively large variability between patients, there will be some degree of uncertainty regarding the accuracy of employed standard methods. On the other hand, this variability is an argument for employing the patient-specific diagnostics that only CFD can offer. Despite extensive work over the last decades, there is still controversy regarding best practice for CFD simulation of flow in the upper airways [83–85].

2.4.1. Virtual Surgery

The concept of virtual surgery envisions the use of digital, patient-specific models for the purpose of simulating the effect of surgical procedures or alternative treatment options on a computer. This may be carried out as part of clinical preoperative planning or theoretical research. This approach offers the potential to provide objective tools that can prove invaluable in optimizing individualized treatments while simultaneously reducing risks and costs. However, for clinical applications to be successful, it is imperative that these virtual surgery tools possess two key attributes: speed and accuracy. Furthermore, the software's user interface and automated workflow should be designed to eliminate the need for involvement from a CFD expert. It is in addressing these crucial requirements that significant challenges lie. In contrast, when it comes to scientific research, the demands for speed and user-friendliness may not be as stringent. This is because research activities often have access to cross-disciplinary expertise and the luxury of time.

Borojeni et al. [67] pointed out three main reasons that CFD-based virtual surgery is likely to have an important role in future clinical applications: (1) The subjective sense of nasal patency is primarily affected by local mucosal cooling, for which there are no available clinical measurement techniques. However, it can readily be estimated by CFD simulations. (2) Subjective assessment of nasal resistance correlates stronger with unilateral than bilateral airflow. (3) The inherent ability of CFD simulations to predict how anatomical changes will affect nasal flow distribution and other flow parameters. They proceeded to present normative ranges for selected airflow parameters to form targets for future nasal obstruction surgery planning.

Simulation-based virtual surgery software has already been demonstrated [86–88]. Vanhille et al. [89] created a virtual surgery planning software tool using CFD and tested it in a clinical setting by collecting feedback from nine surgeons. Moghaddam et al. [90] published a systematic virtual surgery method to select septoplasty candidates and predict surgical outcome using CFD. They foresee that their method can be used for fully automatic virtual septoplasty.

2.4.2. The Creation and Utilization of a CFD Model

Briefly, the process of creating and using a CFD model requires the following steps:

1. Acquisition and preparation of an adequately accurate digital model of the flow geometry (airway).
2. Spatial discretization of the geometry model to obtain a computational mesh on which the governing equations of the CFD model can be numerically solved.
3. Setting up the flow physics, e.g., which physical phenomena to include, boundary conditions, fluid and solid material properties, etc.
4. Determination of solution strategy, e.g., steady state or transient formulation, which numerical scheme to use, turbulence models, convergence criteria, etc.
5. Running the simulation until convergence.
6. Evaluation of the accuracy of the simulation. In case of unsatisfactory results, return to an earlier point, implement necessary improvements and modifications to the model, and repeat the process.

When the CFD model is finalized, it can be used to extract information about the flow, such as local pressures, temperatures, flow velocities, wall shear stresses and heat fluxes, etc.

It is outside the scope of the current paper to elaborate on the details of each step of the process, and the reader is referred to textbooks on CFD by, e.g., Patankar [91], Anderson [92], Versteeg and Malalasekera [93], Rodriguez [94], and Roychowdhury [95], as well as the user and theory guides of available CFD software. Selected topics are discussed briefly below.

2.4.3. Acquisition and Preparation of the Digital Airway Geometry Model

Realistic, digital airway geometry models can be acquired from medical imaging data (CT, MRI) through the process known as segmentation. The segmentation process typically produces a surface mesh consisting of triangles identified by the three-dimensional Cartesian coordinates of their vertices and normal vector (stereolithographic format). The surface mesh can be converted into volumetric models in CFD pre-processing software. A recent overview of the process was given by Cercos-Pita [96].

Airway geometry surface meshes are created from a set of two-dimensional bitmap images by tracking predefined contrast levels corresponding to the interface between air and tissue. The state of the art is to use semi-automatic segmentation software where only minor manual adjustments are needed after most of the segmentation is performed automatically, based on predefined default or user-provided parameters. The contrast level determining the air–tissue interface is typically given in terms of the Hounsfield unit [97,98]. HU = −1000 corresponds to air, while HU = 0 corresponds to water. Depending on its density, bone is represented by HU in the range 300 to 2500. There is no consensus about the appropriate level to describe the air–tissue interface (e.g., the mucous layer), and the literature reports HU levels used in the range −800 to −300 [99].

In general, an HU threshold closer to the value of air will result in narrower airway geometry, while an HU threshold closer to bone will provide more voluminous geometry. Due to the relatively coarse resolution of medical images compared with the width of the narrow passages in the nasal cavity, gross effects can be observed by inclusion or exclusion of a single layer of pixels around the edge of the airway. Aasgrav [4] used CFD to show that reducing the segmented airway cross-sectional area by removing one pixel around the perimeter in every CT slice used for segmentation corresponded to reducing the HU threshold from −300 to −600 and led to a twofold increase in flow resistance. This was later supported through observations reported by Cherobin et al. [100].

Quadrio et al. [101] indicated that CFD modelling results were robust with respect to the quality of the CT scan. Cherobin et al. [100] highlighted the uncertainties related to interpretation of CT images in the creation of 3D geometries for CFD modelling and evaluated the impact on various flow parameters by changing the Hounsfield unit threshold used in segmentation. They found that "CFD variables (pressure drop, flowrate, airflow resistance) are strongly dependent on the segmentation threshold".

Depending on the quality of the surface mesh resulting from the segmentation process, additional pre-processing might be required prior to subsequent steps towards a CFD model. e.g., it may be necessary to improve the quality of the surface mesh by eliminating geometrical artefacts, errors, and unnecessary/unphysical details and to convert the surface mesh into a volumetric format.

2.4.4. Computational Meshes

Computational mesh (or grid) refers to the spatial discretization required for the numerical solution of the governing equations of CFD. Rodriguez [94] presented best practice guidelines in establishing computational meshes for CFD simulations, and Lintermann [102] gave an introduction to the creation of computational meshes for the nasal cavity, in particular. There are three main aspects to consider when establishing a computational mesh for CFD simulations, namely, the mesh type, quality, and size.

Various CFD solvers may have different requirements and preferences regarding the mesh quality (e.g., length-to-width aspect ratios, skewness, and orthogonality) and permitted grid cell types (e.g., hexahedral, tetrahedral, polyhedral). The choice of mesh type may have implications for the efficacy and accuracy of the numerical solution. For complex three-dimensional geometries, such as the nasal cavity, tetra- or polyhedral cells are preferred due to the versatility of these grid generation algorithms. Near-wall boundary layers are commonly resolved using prismatic cells. Bass et al. [103] and Thomas and Longest [104] discussed the pros and cons of tetra- and polyhedral meshes in the CFD modelling of respiratory flows.

Flow structures smaller than the grid cells can, in general, not be captured by the numerical solution. Although there is no guarantee, the accuracy of the numerical solution can thus, to some extent, be expected to improve with the number of grid cells (mesh size), since this permits improved resolution of the flow fields. Caution is advised, however, when increasing mesh size, since there are many pitfalls associated with blindly increasing the number of grid cells. Grid convergence studies should be performed to assess the numerical solutions' dependency on the grid refinement. It is common practice to assume that the solution is accurate if grid refinement has little impact on key flow features.

The required computational power also generally increases with an increasing number of grid cells. Hence, numerical accuracy may be limited by available computational power. The literature review by Inthavong et al. [105,106] indicates that computational mesh sizes increased exponentially between 1993 and 2017. This corresponds well with Moore's law, which is based on the historical observation that available computational power has grown exponentially over time.

2.4.5. Flow Physics

After the geometry and mesh are established, which physical phenomena to include in the simulation must be determined. In Navier–Stokes-based CFD, the basic equations that need to be solved in a transient (time-dependent) respiratory flow problem are the transient continuity and momentum equations. It can be assumed that respiratory flow is incompressible, single-phase, and inert, so these can be formulated as

$$\nabla \cdot \mathbf{u} = 0, \tag{11}$$

and

$$\partial_t \mathbf{u} + (\mathbf{u} \cdot \nabla)\mathbf{u} = -(1/\rho)\nabla P + \nu\nabla^2\mathbf{u}, \tag{12}$$

respectively. Body forces (e.g., gravity) are neglected, ∇ and ∂_t denote the gradient and time derivative operators, $\mathbf{u}$ and P are the local instantaneous flow variables (velocity vector and pressure), and ρ and ν are the constant mass density and kinematic viscosity, respectively. If heating/cooling effects are included, an additional equation for the fluid temperature, T, must be considered,

$$\partial_t T + \nabla \cdot (\mathbf{u}T) = (k/\rho c_P)\nabla^2 T, \tag{13}$$

where it is assumed that air is thermally perfect and k and c_P are the constant thermal conductivity and specific heat capacity, respectively. If solid tissue, mucous, air humidity, airborne particles, non-constant material properties, or other complicating factors are considered, additional equations, variables, and terms are needed. In particular, if movement of the interface between air and soft tissue is considered, additional equations are needed to describe how the deformation of the airway and the soft tissue are interdependent and affected by local stresses on both sides of the air–tissue interface. This is known as fluid–structure interaction (FSI) modelling.

The Lattice–Boltzmann (LB) method is an alternative to the Navier–Stokes-based CFD and is mentioned here for completeness due to its suitability for fluid dynamics simulations in complex geometries. One of the method's main strengths is its scalability on high-performance computers [102].

In laminar flow, the flow variables are generally considered deterministic, and they can be predicted precisely from the governing flow equations given above. The flow is characterized by smooth, locally parallel streamlines. In the case that inertial terms are dominating over the viscous terms in the momentum equation, however, the flow may be turbulent. The ratio between inertial and viscous terms are typically expressed by the

Reynolds number (Equation (6)). Reynolds numbers above a certain threshold is commonly used as a criterium for considering turbulence or not, but care should be taken since the critical Reynolds number for laminar–turbulent transition can be sensitive to the flow configuration and fluid properties. In turbulent flow, the flow variables behave stochastically, and this is a notoriously difficult physical problem to describe. A wide range of modelling strategies and methods have been developed for the CFD modelling of turbulent flow, including popular approaches like Reynolds-averaged Navier–Stokes (RANS) methods, large eddy simulation (LES), and direct numerical simulation (DNS). It is beyond the scope of the current paper to discuss and compare the various turbulence modelling approaches in CFD, in detail. A systematic overview of advantages and limitations associated with the most popular turbulence models were presented by Ashraf et al. [107]. Details regarding the mathematical description of available turbulence models can be found in classical textbooks by, e.g., Tennekes and Lumley [108], Pope [109], or Wilcox [110], and in CFD simulation software user and theory guides. See, e.g., ANSYS Best Practice guidelines [111,112] and Theory guide [113] and the NASA turbulence modeling resource [114].

Boundary Conditions are required to describe the flow variables at all flow domain boundaries, i.e., inlets, outlets, and walls. Boundary conditions take the form of specifying variable values or gradients at the boundaries. Typical examples include specifying the mass flowrate at the inlet, the pressure at the outlet, and zero velocity at walls (no-slip condition), but other variants are also possible.

Initial conditions denote the initial flow variable fields used as a starting point for time evolution of transient solutions or for the search of a steady state solution.

Finally, if steady flow is considered, the time derivative terms are set to zero, $\partial_t \equiv 0$, and all variables are constant in time, everywhere. It is noted that steady boundary conditions do not generally guarantee steady flow alone if the flow is inherently unstable.

2.4.6. Correlations between CFD and Clinical Measures of Nasal Patency

The nature of CFD is to predict objective physical quantities, and it is natural to think that CFD must be able to reproduce objective clinical measures. Although correlations between subjective and objective clinical measures are disputed, some authors have reported that subjective measures and RMM are correlated [51]. This spurs optimism towards predicting subjective sensation of nasal patency through CFD simulations, adding clinical value to virtual surgery. However, it has been pointed out by several authors that the subjective sensing of nasal patency might not be a measure of the objective flow resistance, but rather the cooling effect of the nasal mucosa [68]. In this case, CFD must be correlated with subjective measures, directly, because no in vivo measurements exist to measure mucosal temperature or heat flux. Frank-Ito and Garcia [115] presented an in-depth review of the clinical implications of nasal airflow simulations, including their correlations with objective (RMM, AR) and subjective (NOSE, VAS) measures. They proposed that the complex nature of nasal diseases might prevent CFD-based nasal airway diagnostics using single CFD-derived variables alone, and that correlations should be based on combinations of CFD-derived variables.

Kimbell and co-authors presented the first comparisons between patient-reported subjective symptoms and CFD-based flow characteristics. They found moderate correlations between subjective measures (NOSE, VAS) and CFD-based unilateral nasal resistance [116] and heat flux [61] when considering data on the side affected by surgery. They only included a few patients in their investigation, however. Later, several studies demonstrated correlations between CFD simulation results (e.g., airflow patterns, mucosal cooling, and nasal resistance) and patients' subjective evaluation of nasal resistance [50,62,63,65–67]. Cherobin et al. [117] found good agreement between CFD and experimental results in a physical nasal replica. However, for their cohort consisting of 25 patients pre- and postoperatively, they found no correlation between subjective measures and RMM or CFD results.

The combined experience, that subjective and objective measures of nasal patency correlate with each other and also correlate with air flow variables obtained from CFD simulations, is a clear indication that CFD holds significant potential as a clinical decision support tool. However, there are obstacles that must be overcome. e.g., a major shortcoming in published CFD studies of nasal airflow is that CFD has not generally been able to reproduce in vivo RMM results.

Zachow et al. [118] and Hildebrandt [119] published CFD simulation data in excellent agreement with RMM data. However, it was later discovered that there were mistakes in the computations performed by an independent third party on which their conclusions were based [120]. In their later studies, CFD severely underpredicted the nasal resistance compared to RMM measurements [121]. More recently, Dong et al. [122] demonstrated perfect agreement between CFD and RMM.

Several authors have reported unexplained discrepancies between in vivo RMM measurements and in silico RMM based on CFD simulations [4,100,117,121,123–126]. Hemtiwakorn [123] reported RMM measurements to be one order of magnitude higher than CFD simulation data. Osman et al. [124] pointed out that the "bias between CFD and RMM seems to be a common problem" and "...it appears that the calculation of nasal resistance using CFD often leads to gross underestimation of nasal resistance compared to in-vivo measurements". Berger et al. [125] reported varying degrees of agreement between in vivo and in silico RMM when comparing pressure–flow curves from five patients. They found perfect matches in Subject 5, while in Subject 4, there was gross mismatch. In Subject 3, it was observed that good agreement was achieved on one side, but not on the other side, and in Subject 1, good agreement was achieved for inhalation but not for exhalation. Cherobin et al. [117] reported that CFD underpredicted nasal resistance compared with RMM.

2.5. A Review of Sources of Errors and Uncertainties Affecting Comparison of In Vivo and In Silico Rhinomanometry

Although a few studies report a good match between in vivo and in silico RMM, the general impression is that CFD-based models struggle to reproduce in vivo RMM pressure–flow curves [115]. It appears that in silico studies agree better with in vitro studies in rigid nasal cavity replicas, however [117,127].

It is useful to make a distinction between the terms uncertainty and error. An uncertainty is "a potential deficiency in any phase or activity in the modelling process that is due to lack of knowledge", whereas an error is "a recognizable deficiency in any phase or activity of modelling and simulation that is not due to lack of knowledge" [128]. It is noted that these definitions differ from typical definitions employed in experimental measurements.

Ideally, CFD simulation results can be validated against well-controlled experiments where uncertainties, e.g., regarding the flow geometry and mass and heat transfer, have been minimized, and measurement errors are under control. Under such conditions, the CFD model is subject to little uncertainty, and errors associated with poor choice of modelling strategies and submodels can readily be assessed, so that the model can be tuned to predict measured data with good accuracy. This might be the reason for the good agreement reported between in vitro and in silico RMM.

Due to difficulties and challenges associated with acquiring and assessing the quality and reproducibility of objective in vivo clinical data, including RMM and CT/MRI imaging data, there is substantial uncertainty related to the quantitative comparison of nasal resistance and pressure–flow curves obtained from in vivo and in silico RMM. e.g., an essential part of in silico RMM, not inherent in standard in vivo RMM, is the requirement of a detailed description of the nasal cavity geometry. In silico RMM typically utilizes medical imaging data to acquire the nasal cavity geometry, but unless specific actions and precautions are taken, it is unknown to what extent the medical images adequately describe the state of the nasal cavity during in vivo RMM.

Computed flow variables strongly depend on the flow geometry. Thus, the lack of knowledge about the instantaneous state of the nasal cavity during in vivo RMM causes major uncertainty regarding the quantitative comparison of nasal resistance and other flow parameters obtained from in vivo and in silico RMM. A discussion of uncertainties and errors in the comparison of in vivo and in silico RMM is therefore incomplete without a separate discussion of the relevance of medical imaging data with respect to describing the nasal cavity at the time of in vivo RMM.

Factors that may influence the comparison of in vivo and in silico RMM results are discussed briefly below, including physiological factors affecting the temporal variability of the nasal cavity geometry as well as uncertainties and errors associated with the procedures of acquisition of the digital nasal cavity geometry model and in vivo and in silico RMM.

2.5.1. Physiological Factors Affecting the Temporal Variability of the Nasal Cavity Geometry

To attain accurate predictions of in vivo RMM results through in silico RMM simulations, it is imperative to employ a digital nasal cavity model that faithfully represents the dynamic state of the nasal cavity during the in vivo RMM examination.

The alignment reported between in silico RMM and in vitro RMM in physical replicas of nasal cavities [127,129,130] starkly contrasts the observed disagreement with in vivo RMM [117,125]. This suggests that CFD models are correctly configured, but somehow fail to adequately represent the nasal cavity's actual state and function during in vivo RMM examination.

Two physiological mechanisms that can cause digital geometry models to misrepresent the nasal cavity geometry during in vivo RMM are (1) the nasal cycle, known to cause a periodic, temporal variation in nasal cavity volume, and (2) nasal cavity compliance, which may cause spontaneous expansion/contraction of the nasal cavity volume due to over-/under-pressure during respiration. In the following subsections, brief discussions are given about these two physiological phenomena. Other causes for errors and uncertainty associated with the acquisition of a digital nasal cavity geometry are discussed below.

Nasal Cycle

The nasal cycle causes spontaneous engorgement, hence cross-sectional variability, in the nasal cavities. Consequently, the nasal cycle affects the reproducibility of objective rhinometric measurements adversely and complicates the objective assessment of nasal patency when comparing pre- and postoperative measurements. Additionally, it poses challenges when comparing in vivo and in silico RMM results, as the nasal cavity's shape and volume may differ between in vivo RMM examination and the acquisition of medical imaging data used for in silico RMM.

In a study by Hasegawa and Kern [74], which involved 50 subjects, bilateral and unilateral nasal resistance measurements were conducted over a 6–7 h period. They observed that the bilateral nasal resistance remained relatively constant despite cyclic variations in unilateral resistances. The average ratio of highest to lowest unilateral resistances was 4.6 on the right side and 4.4 on the left side, with peak values reaching 16.3 and 13.7, respectively. Hence, accounting for the nasal cycle is crucial when assessing nasal resistance. It is noteworthy that the nasal cycle was found to be non-reproducible in all of the five subjects who underwent re-testing, as the durations and amplitudes of their nasal cycles varied. To quantify the nature of the nasal cycle, in numerical terms, Flanagan and Eccles [75] conducted hourly unilateral airflow measurements over 8 h periods in 52 subjects.

Patel et al. [131] employed CFD to investigate the impact of the nasal cycle on objective measures. They suggested that paradoxical postoperative worsening of NAO observed in simulations could be attributed to the nasal cycle. Gaberino et al. [65] created virtual mid-cycle models to correct for the nasal cycle, resulting in improved correlation between objective and subjective measures of nasal patency. Moghaddam et al. [90] pointed out that mucosal engorgement due to the nasal cycle can significantly affect CT images, thereby influencing the correlation between CFD and subjective and objective nasal patency scores. Susaman et al. [132] emphasized that rhinologists need to take the existence of the nasal cycle, which affects a large percentage of the population, into account when examining and measuring the nose.

Several authors [33,133,134] have pointed out that postural effects on the nasal resistance and the nasal cycle should be expected. The nasal resistance tends to be higher in the supine position. Consequently, differences in posture between medical imaging procedures and RMM examinations may lead to geometrical misrepresentation in in silico RMM simulations.

To mitigate the effects of the nasal cycle, it is common practice to perform RMM both before and after applying a decongestive nasal spray that shrinks the large veins in the nasal epithelium. This approach enables the evaluation of the anatomical nasal patency [45].

The presence of the nasal cycle suggests that if CT/MRI images and RMM measurements are not acquired within a short timeframe, in the same state of decongestion, and in the same posture, they may not reflect the same nasal geometries.

Nasal Cavity Compliance

In a study conducted by Fodil et al. [135], a simplified model of the nasal cavity was used to demonstrate that, depending on the pathological condition, the assumption of rigid nasal cavity walls is only valid for low flowrates. The rigid wall assumption generally failed for pressure drops above 20 Pa. In contradiction, Bailie et al. [136] claimed that during resting breathing, one can regard the nasal cavity as a rigid structure. More recent research by Akmenkalne et al. [137] corroborated the earlier work of Fodil et al. [135]. They investigated the mobility of the lateral nasal wall under the influence of breathing and emphasized that even during quiet breathing, we must take into account the deflection of the nasal walls. O'Neill and Tolley [72] used a simplified mathematical model based on Bernoulli's principle to compute the total pressure loss through the nasal cavity as a sum of minor losses. Their model allowed for the nasal gateway (valve) to dynamically adjust its cross-sectional area based on local pressure and a stiffness coefficient, providing a quantitative rationale for observed discrepancies between AR and RMM. Cherobin et al. [117] observed that while in silico RMM was in good agreement with in vitro RMM, it diverged significantly from in vivo RMM. This disparity was partly attributed to the rigid wall assumption in CFD, which matched the properties of the rigid nasal cavity replica used in in vitro experiments but might have failed to adequately represent physiological nasal cavity compliance. Schmidt et al. [121] reported systematic underprediction of nasal resistance in CFD simulations but found no significant difference between patients with or without nasal valve collapse or between inhalation and exhalation phases.

Considering that nasal cavity expansion/contraction in response to over-/underpressure during exhalation/inhalation, has an effect on the nasal resistance, it is anticipated that the pressure–flow curves will exhibit asymmetry between these respiratory phases. To assess this, one can compare mirrored exhalation curves with inhalation curves obtained from in vivo RMM. If the two sets of curves align well, it suggests that nasal cavity compliance is minimal and cannot account for the substantial differences between in silico and in vivo RMM results. However, it is important to note that this argument does not consider the Venturi effect. When the Venturi effect dominates over the hydrostatic effect, it can lead to contraction during both inhalation and exhalation, resulting in an overall increase in nasal resistance.

This line of reasoning is consistent with the observations of Akmenkalne et al. [137], who demonstrated contraction during both inhalation and exhalation in quiet breathing. For elevated breathing and forced sniffing, the hydrostatic pressure component appeared to dominate, causing contraction during inhalation and expansion during exhalation. An interesting observation in their Figure 4 was that after a period of forceful sniffing, the deflection of the lateral nasal wall reversed its direction, transitioning from negative to positive but trending downwards. However, this reversal and slow nasal wall relaxation time did not appear to correlate with flow or pressure curves, implying minimal impact on nasal resistance.

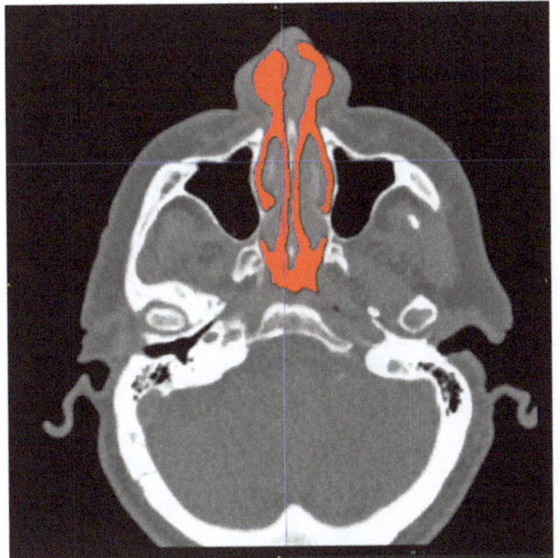

(**a**) Axial plane.

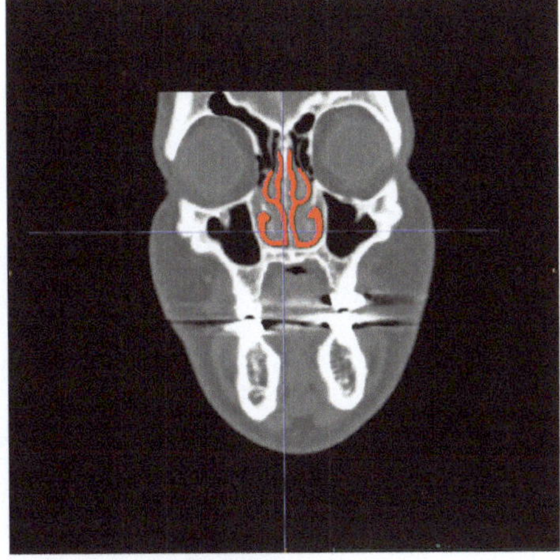

(**b**) Coronal Plane.

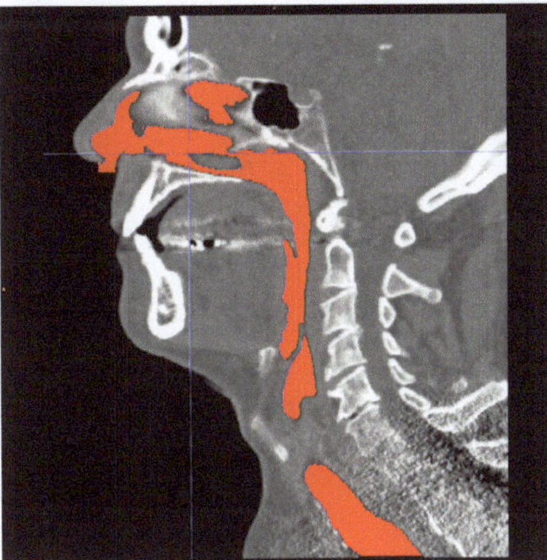

(**c**) Sagittal plane.

Figure 4. Computed tomography images of the current patient. The modelled airway is colored red, and the blue lines in each cross-section indicate the position of the other cross-sections.

The influence of nasal compliance on the hysteresis in RMM pressure–flow curves, illustrated in Section 2.3.3, has been discussed by Vogt and co-authors [77,138]. Wernecke et al. [77], Vogt and Zhang [138], Vogt et al. [139], Bozdemir et al. [76], and Frank-Ito and Garcia [115] presented pressure–flow curves featuring hysteresis where the portions of the curve corresponding to the accelerating and decelerating inspiratory phases were switched when compared to the simplified model showcased in Figure 2b of the present paper. Measurement results by Groß and Peters [140] support the time arrows in Figure 2b, but they attributed the observed pressure–flow curve hysteresis to the measurement technique, rather than nasal airflow dynamics. An adequate explanation for this disagreement is lacking.

It is anticipated that the influence of nasal compliance on RMM pressure–flow curves will manifest as asymmetry between inspiratory and expiratory pressure–flow curves as well as, referring to Figure 2b, a widening of the hysteresis loop. It may, however, be imagined a situation where the Venturi effect dominates over the static pressure such that local under-pressure is effectively independent of flow direction and causes (partial) collapse both during inhalation and exhalation. Owing to the phase disparity between volumetric flowrate and flow resistance, brief periods of counterintuitive over- and under-pressure may occur within the nasal cavity at the culmination of the inspiratory and expiratory phases, respectively (see Figure 2). Consequential local expansion/contraction may introduce complexity to the response of the pressure–flow curves.

2.5.2. Sources of Uncertainties and Errors in the Acquisition of Digital Nasal Cavity Geometry Models

The process of acquisition and preparation of the airway geometry was described in Section 2.4.3. Two of the main steps of the process are (1) the recording of medical imaging data by CT or MRI and (2) the establishment of a surface mesh through segmentation.

It is beyond the scope of the current paper to discuss the technology behind medical imaging, but the following aspects are highlighted:

- CT and MRI data have relatively low spatial resolution compared to the small-scale features of the nasal cavity. This may cause inaccurate description of the small features of the nasal cavity. Cone beam CT has been proposed as an alternative due to better resolution at lower radiation dosage [141].
- Low temporal resolution of CT and MRI data requires the patient to hold still while data are acquired to avoid blurred images. Effects of heartbeat, breathing, and swallowing may affect image quality adversely. This suggests that CT is preferred over MRI due to better temporal resolution.
- Good communication with the radiologist is required to ensure that the entirety of the nasal cavity is included in the data.
- For comparison of pre- and postoperative airways, the patient's posture and positioning in the CT/MRI scanner during postoperative examination should be identical to the preoperative situation. For instance, the apparent shape and volume of the pharyngeal tract may be affected by the relative tongue, jaw, head, and neck positions.
- The nasal cycle can be observed in medical imaging data. Medical imaging should thus be performed in the decongested state similar to decongested RMM, preferably in rapid succession after the clinical RMM procedure.

During segmentation of the imaging data, the following should be observed:

- The radiodensity threshold used to determine the interface between air and tissue can have a severe effect on the cross-sectional area, hence the nasal resistance. A low/high threshold will result in a narrower/more voluminous airway geometry, respectively, potentially affecting flow variables [100].
- Automatic segmentation methods may overlook important details or include secondary air spaces such as the paranasal sinuses, Eustachian tubes, or nasolacrimal ducts. It is recommended to confer with medical expertise such as radiology experts or surgeons to assess the resulting geometry model.

2.5.3. Sources of Uncertainties and Errors in In Vivo Rhinomanometry (Clinical)

RMM systems are typically proprietary systems where it is challenging to obtain access to raw measurement data and detailed information about the post processing of the measured data. Some studies have investigated the agreement between RMM measurements performed with devices by different manufacturers [79,121] or between RMM and inhouse benchmarks [142]. There is no evidence of systematic measurement errors in RMM measurement devices, but Hoffrichter et al. [77] pointed out that some rhinometers manipulate or average the measurements, suppressing hysteresis in the measured pressure–flow curves. Silkoff et al. [143] reported a high level of reproducibility in RMM.

Carney et al. [144], however, reported unacceptable variation and concluded that single RMM measurements are prone to large errors. Lack of reproducibility has also been reported by Thulesius et al. [145]. Bozdemir et al. [76] pointed out that the reproducibility of RMM measurements relies on ensuring identical conditions in subsequent measurements (e.g., air humidity and temperature, fit of the face mask, contralateral nostril closure, and avoiding oral breathing). This is best achieved by a skilled RMM operator.

Possible errors associated with improper conduction of the RMM procedure include false pressure and/or flowrate measurements due to, e.g.,:

- Air leakage along the edge of the face mask or contralateral nostril closure.
- Open mouth and oral breathing.
- Malfunction of the RMM equipment or post processing software.

In addition, there are uncertainties mainly associated with the geometrical/volumetric state of the nasal cavity during the RMM measurement, due to several factors:

- The nasal cycle may affect the unilateral nasal resistance.
- Posture has been shown to influence the nasal cycle [33]. Therefore, positioning of the patient may affect the RMM measurements.
- Compliance of the nasal walls can cause the nasal cavity to expand due to over-pressure during exhalation and contract due to under-pressure during inhalation or due to Venturi effect. This dynamic behavior may affect the nasal resistance and result in asymmetry and hysteresis in RMM pressure–flow curves. In situations where hysteresis is prominent, the inspiratory and expiratory segments of the pressure–flow curves may yield markedly distinct measurements of nasal resistance.
- Excessive temporal NAO due to inflammatory reactions or other causes may cause exaggerated nasal resistance that may affect RMM measurements and sometimes even prevent the patient from generating the required volumetric flowrate to conclude the RMM examination.

The uncertainties associated with acquisition of in vivo RMM data are mainly associated with the temporal variations in the geometrical state of the nasal cavity. It is therefore stressed that, for comparison between in vivo and in silico RMM, it should be ensured that medical imaging data correctly represent the nasal cavity during in vivo RMM. This can best be achieved by undertaking medical imaging examination in rapid succession of the RMM examination, in decongested state, and preferably in the same posture.

2.5.4. Sources of Uncertainties and Errors in In Silico Rhinomanometry (CFD)

When presented with experimental data from flow measurements, such as in vivo or in vitro RMM, it rests upon the CFD engineer to set up a CFD model that is able to reproduce the measured data, or to explain observed discrepancies between computed and measured data. A significant preparatory task in CFD modelling is to describe the flow system both qualitatively and quantitatively with respect to geometry (including flow restrictions/walls, inlets, and exits), material properties, and other factors that may affect the flow. Even if the geometry of the flow system is well known, there is a multitude of parameters and settings that must be chosen carefully when setting up the CFD model. Potential sources of error associated with setting up and running CFD simulations have been thoroughly covered by the European Research Community on Flow, Turbulence, and Combustion [146] and many others, e.g., Andersson et al. [147], Rodriguez [94], and Roychowdhury [95]. Inthavong et al. [105] reviewed in silico approaches to simulation of nasal airflow.

Besides fundamental errors and limitations in the program code of the CFD software, such as program bugs or truncation and rounding errors, errors in simulation results can be caused by, e.g.,:

- Poor computational mesh quality [94].
- Inadequate spatial or temporal refinement.
- Poorly selected solver settings and numerical schemes [148].

- Incorrect definition of flow physics, including, e.g., boundary conditions, material properties, and approximations.
- Inaccurate or incorrect solution due to poor convergence and/or failure to conserve mass, momentum, or energy.

The main uncertainties are related to the lack of knowledge about the flow problem to be modelled. These can be divided into four main categories related to (1) flow physics; (2) geometry; (3) required spatial and temporal numerical resolution; and (4) boundary conditions. Even if these are implemented correctly without errors, there may be uncertainty associated with their correct description. For simulation parameters associated with high uncertainty, sensitivity analysis may be required to assess the influence of variations in these parameters.

Flow Physics

Uncertainties surrounding the flow physics within the nasal cavity encompass aspects that, theoretically, could be elucidated through measurements or experiments. However, practical challenges arise in conducting in vivo measurements on patients, and a lack of in vitro experimental data complicates the matter. Consequently, an ongoing debate persists regarding fundamental aspects of nasal flow physics. This includes the deciding between quasi-steady and transient modeling, determining the optimal turbulence modeling strategy, and addressing other considerations such as the dependence of air's material and transport properties on pressure, temperature, and humidity.

(A) Modelling of temporal phenomena in respiratory flow

The physiological, respiratory flow in the nasal cavity is normally of pulsative nature. The literature review summarized in **Part II** (Section 3) suggests that steady flow modelling, by far, is the most popular approach in computational rhinology, however. It is appropriate to question the validity of the assumption of quasi-steady flow in respiratory flow modelling. e.g., how does the transient nature of the flow affect temporal effects such as hysteresis, developing flow boundary layers, and meandering of wakes or jets?

The simulation of transient flow adds complexity to CFD simulations compared to modelling steady state flow. Many authors have argued that nasal airflow can be approximated by quasi-steady flow [149–154].

In the current context, the concept of quasi-steady state implies that the time response of the overall flow phenomena within a system is much quicker than the variation in transient phenomena occurring in the system. The system's behavior can thus be assumed to be in instantaneous equilibrium with the transient phenomena, enabling its approximation by steady state simulations. In the case of nasal airflow, quasi-steady state suggests that the nasal flow parameters can be determined from the instantaneous respiratory pressure and velocity boundary conditions, at any given moment. This implies that pressure–flow curves in in silico RMM can be generated through a series of steady state simulations conducted at different volumetric flowrates, instead of relying on a transient simulation of the entire breathing cycle.

The Womersley number, named after J. R. Womersley [155], who studied pulsatile flow in arteries, is defined as the ratio between the transient inertial and viscous forces and is commonly expressed as follows:

$$\mathrm{Wo} = D_h \sqrt{\pi f / 2\nu}, \tag{14}$$

where D_h is the channel diameter, f is the pulsation frequency, and ν is the kinematic viscosity. The Womersley number can be used to characterize an unsteady flow as quasi-steady or not [156]. The flow may be considered quasi-steady if $\mathrm{Wo} < 1$. Inserting for $D_h = 5$ mm, $f = 0.2$ Hz, and $\nu = 1.5 \times 10^{-5}$ m^2/s, the expected Womersley number in unilateral nasal airflow is approximately $\mathrm{Wo} \approx 1$. This is in the intermediate range, where the oscillatory nature of the flow is not dominating but may have some influence.

Doorly et al. [157] discussed whether a series of quasi-steady simulations is sufficient to characterize tidal breathing. They referred to Shi et al. [153] and suggested that the quasi-steady assumption is valid for quiet breathing. Bosykh et al. [158] showed that a transient model produced almost identical results to steady state simulations produced by themselves as well as others. Furthermore, they observed that asymmetry in the respiratory cycle had little effect on the flow pattern in the nasal cavity compared to a sinusoidal inhalation/exhalation profile, which follows naturally from quasi-steady behavior. Bradshaw et al. [159] highlighted several phenomena observed in their transient simulations that cannot be seen in steady flow. In particular, their results indicate that transient simulations of the entire breathing cycle are essential in order to correctly capture air conditioning via heating/cooling and humidification.

A noteworthy characteristic of in vivo RMM pressure–flow curves is the presence of a hysteresis pattern [77]. This hysteresis has been attributed, among other factors, to unsteady/inertial pressure drop contribution stemming from varying flowrates during respiration, and it can naturally not be predicted by steady state flow simulations. See Section 2.3.3 for more details.

(B)   Modelling of turbulent, transitional, and laminar flow

The complex, dramatically varying flow channel cross-sections in the nasal cavity can have significant impact on the development of turbulent structures within the flow due to, e.g., flow separation, recirculation, varying pressure gradients, secondary flows, developing wall boundary layers, merging of separate flow streams, flow instabilities, etc. The understanding and prediction of turbulence in such scenarios typically requires very detailed CFD models and experiments. It can be expected that the behavior of such flow systems are highly non-linear and three-dimensional. e.g., Tretiakow et al. [160] found that the flow in the ostiomeatal complex (e.g., degree of turbulence) depended on the overall geometric features of the nasal cavity (e.g., nasal septum deviation).

Only DNS is an exact representation of the Navier–Stokes equations. All other turbulence models contain approximations with individual limitations and ranges of validity. e.g., while RANS models are ensemble averaged and unable to model individual turbulent eddies, LES models are able to track eddies larger than a given filter size (typically a function of the computational mesh size) and employ subgrid models to describe the effect of smaller eddies. In principle, LES should approach DNS in the limit of small filter sizes.

While RANS-based models are much cheaper than LES- or DNS-based models, in terms of computational power requirements, they are known to have many limitations. These models were typically created to solve specialized industrial problems with a good balance between accuracy and computational cost. Model parameters were thus tuned to predict standard, industrial flow scenarios. It is not given that these models are suitable for modelling of flow in complex geometries such as the nasal cavity. Moreover, these models are known to have severe limitations with respect to modelling transitional flow. Thus, if the nasal flow is transitioning between laminar and turbulent flow along the length of the nasal cavity and due to the respiratory variation in flow velocity, these models might not be able to predict the flow accurately.

Most authors discussing turbulence in the upper airways seem to consider the Reynolds number only as a criterium for the onset of turbulence. They fail to consider that it takes time to develop turbulence. In pipe flow, at Reynolds numbers above the critical Reynolds number, it generally takes more than 10 pipe diameters' flow length to fully develop the turbulent velocity profile. During restful tidal breathing, approximately 25 nasal volumes are inhaled/exhaled during one cycle [161] and considering that the length of the nasal passage is between five and fifteen times its hydraulic diameter, the flow field is thus unlikely to be fully developed. Even for steady flow, it seems unlikely that the flow can be fully developed due to the varying cross-sectional area along the nasal cavity, and the many anatomical features that affect the flow pattern. There may, however, be regions within the nasal cavity that experience periods of transitional/turbulent flow during a respiratory cycle. It can be expected that most ensemble-averaged turbulence models are unsuited for such complex spatially and temporally varying laminar/transitional/turbulent flow fields.

For flow channels with cross-sections that slightly deviate from a cylindrical shape, the hydraulic diameter has proven useful in predicting flow resistance using standard friction loss correlations, such as the Haaland correlation (Equation (7)). For cross-sectional shapes deviating from cylindrical shapes, inaccuracies have been reported [73]. This indicates that Reynolds numbers based on hydraulic diameter of complex cross-sections such as those in the nasal cavity may not be appropriate for predicting the transition from laminar to turbulent flow.

Transition between laminar and turbulent unsteady flow is still, despite its importance in many engineering applications, not fully understood [162]. Recently, Guerrero et al. reviewed the literature and performed DNS to investigate the transient behavior of accelerating [163] and decelerating [164] turbulent pipe flows. An early discussion of the transition between laminar and turbulent flow in pulsatile flow in the cardiovascular system was published by Yellin [165], who observed that large instantaneous Reynolds numbers did not result in transition to turbulence everywhere. Gündoğdu and Çarpinlioğlu [166,167] presented the theoretical background for pulsatile laminar, transitional, and turbulent flows, and reviewed theoretical and experimental investigations. Xu et al. [168] investigated the effect of pulsation on transition from laminar to turbulent flow for rigid, straight pipes. They observed that the delay in the transition to turbulence increases with decreasing Womersley number (Wo < 12) and increasing pulsation amplitude. The implication is that the turbulent transition threshold for Womersley numbers close to 1 is above Reynolds number 3000, in straight pipes, but turbulent puffs may exist due to the high Reynolds number time intervals of a respiratory cycle.

The combined effect of delayed transition and relatively short flow channel suggests that fully developed turbulent flow within the nasal cavity is improbable during resting respiratory flow. When utilizing RANS turbulence models that assume fully developed turbulent flow in cases where the flow is laminar, transitional, or developing, there is a risk of overestimating turbulent viscosity and, consequently, overall flow resistance. If the validation of in silico models relies solely on nasal resistance measured through in vivo RMM, there is a potential bias towards models that overestimate turbulent viscosity. This bias may help align in silico and in vivo RMM pressure–flow curves but could lead to an incomplete representation of other pertinent phenomena. This example illustrates the perils of overly simplistic analyses in the study of complex problems like nasal airflow and emphasizes the necessity for additional objective, measurable metrics in the assessment of nasal airflow.

Schillaci and Quadrio [148] compared laminar/RANS/LES simulations and concluded that the choice of numerical scheme is more important than the choice of turbulence model, although they emphasized that the chosen turbulence model should be able to handle three-dimensional, vortical, mostly laminar flow conditions. They suggested that LES or DNS is necessary to reliably simulate the full breathing cycle at intermediate intensity. Bradshaw et al. [159] performed hybrid RANS-LES simulations of the entire respiratory cycle and reported bilateral nasal airflow to be dominantly laminar. While LES is widely acknowledged as one of the most accurate turbulence modelling approaches, second only to DNS, it is worth noting that LES also encounters challenges in predicting transitional flow and the initiation of turbulence, as highlighted by Sayadi and Moin [169].

(C) Other aspects

Other aspects of minor importance are just mentioned briefly, for completeness.

- Temperature and humidity may affect the material properties of air. Some authors have suggested that these effects should be taken into account [76]. Other authors have dismissed these effects [151]. In the relevant temperature range, the mass density and viscosity of air varies by less than ten percent, and the effect of humidity is of the same order. For most situations, it is thus expected that this is of minor importance.

- Due to the small effect of pressure on air material properties within the relevant pressure range, and low flow velocities, it is safe to assume atmospheric ambient pressure and constant air material properties.

Geometry

The nasal cavity is a highly complex flow channel, with cross-sections that change dramatically in shape and area throughout the nose, and generally deviate significantly from cylindrical shape. Moreover, the nasal cavity is bounded by walls covered in mucosal lining, which may add a transient geometrical variation to the air–tissue interface, as well as constituting a non-rigid structure. The acquisition of a three-dimensional digital nasal cavity geometry model is prone to errors and uncertainties, as discussed in Section 2.5.2. In addition, there are uncertainties regarding the geometrical level of detail required to set up accurate CFD models. For the CFD model to be able to accurately predict the behavior of physical phenomena, it is essential that the geometry model does not misrepresent the actual flow geometry too much. When manufacturing the nasal cavity model, essential questions that should be considered include the following:

- How much of the surrounding volume outside the nose and in the oropharyngeal tract should be included? This consideration will affect to what extent the boundary conditions will affect the simulated flow fields. This question is closely intertwined with the discussion about boundary conditions, below.
- How much of the paranasal sinuses should be included? CFD simulation of the flow in the maxillary sinus was performed by Zang et al. [170]. Their conclusion was that the airflow inside the maxillary sinus was much lower (<5%) than the airflow in the nasal cavity. Due to the narrow passage connecting the paranasal sinuses with the nasal cavity, it is expected that negligible gas exchange takes place between the two [171]. This was supported by simulation results presented by Bradshaw et al. [159]. Kaneda et al. [126] reported that the inclusion of the paranasal sinuses did not improve the disagreement between computed and measured nasal resistances.
- What is the role of the oral cavity? Paz et al. [172] investigated the distribution between nasal and oral breathing under steady and unsteady flow. Chen et al. [173] concluded that the inclusion of the oral cavity in CFD simulations of steady and unsteady nasal cavity flow had very little impact. Open mouth and oral breathing may, however, affect the RMM pressure–flow curve.
- Should minor geometrical features such as nasal hair or the mucosal lining be considered?
  - ○ Hahn et al. [151] found that the inclusion of nasal hair increased turbulent intensity in the external nares during inspiratory flow but had little effect on downstream velocity profiles. Stoddard et al. [174] found that a reduction in nasal hair density had a positive impact on both subjective and objective measures of nasal obstruction, however.
  - ○ Lee et al. [175] illustrated how the mucous layer may affect local flow velocities in the nasal cavity.

Uncertainty associated with the geometrical (mis-)representation of the nasal cavity in CFD models may be due to unknown factors affecting the process of manufacturing three-dimensional geometries from medical imaging data, via segmentation, or uncertainty regarding how well the CT/MRI imaging data represent the actual nasal cavity geometry during the RMM procedure.

Required Spatial and Temporal Numerical Resolution

Spatial and temporal discretization is required in order to enable the numerical solution of the governing equations of CFD (Equations (9)–(11)). The rule of thumb is that the spatial and temporal resolution must be sufficient to resolve all spatial and temporal flow features of interest. In addition to determining the accuracy of CFD simulations, mesh and time step size may affect numerical stability and robustness of CFD solvers.

To assess the numerical solution's sensitivity to grid refinement, grid dependency tests should be performed. If the computed flow fields change negligibly by increasing the grid resolution, it is commonly assumed that grid independence is achieved. Frank-Ito et al. [176] reviewed the literature and investigated the requirements for grid independence in their own steady, inspiratory, laminar sinonasal cavity airflow with particle deposition. They emphasized the importance of mesh refinement analysis to obtain trustworthy computational solutions. Similarly, to assess transient solutions' sensitivity to time step size, comparison between simulations employing relatively short and long time steps should be performed.

Brief discussions of how the spatial and temporal resolution can introduce uncertainties in CFD simulations of nasal airflow are given below.

(A)   Spatial Resolution

The spatial resolution of the computational mesh determines the level of detail in the computed flow fields. e.g., in the presence of steep velocity gradients, refined meshes are needed to avoid numerical diffusion. Moreover, turbulent eddies smaller than the mesh size must be modelled by closure laws and subgrid models, which introduce approximations.

Particularly, to describe flow profiles accurately in the vicinity of walls, the near-wall mesh must honor requirements by the turbulence model employed. The theory behind this is described in classical text books on turbulence by, e.g., Tennekes and Lumley [108], Pope [109], or Wilcox [110], and in CFD simulation software user and theory guides.

Distance to the wall is commonly expressed in wall units, where the dimensionless wall distance is expressed as

$$y^+ = u_\tau y/\nu , \tag{15}$$

where $u_\tau = \sqrt{\tau_w/\rho}$ is the shear velocity, τ_w is the wall shear stress, and y is the distance to the wall. The Law of the wall states that the dimensionless velocity parallel to the wall is given by

$$u^+ = u/u_\tau = \begin{cases} y^+ & \text{for} \quad y^+ \lesssim 5 \text{ (viscous sublayer)} \\ \frac{1}{0.41} \ln y^+ + 5 & \text{for} \quad y^+ \gtrsim 35 \text{ (log layer)} \end{cases}, \tag{16}$$

where u denotes the flow velocity parallel to the wall. The intermediate range between the viscous sublayer and the log layer is known as the buffer layer. The original idea by Launder and Spalding [177] was to use Equation (14) as a wall function for the velocity boundary conditions at the wall. This approach, which is used in some classic RANS turbulence models, such as the $k\varepsilon$ type turbulence models, requires that the centroids of computational grid cells residing at the wall are in the log layer ($y^+ \gtrsim 50$). Other RANS turbulence model types, such as the $k\omega$ and $k\omega$ SST models and $k\varepsilon$ with enhanced wall treatment, require (or permit) that the near-wall grid cells are within the viscous sublayer ($y^+ \lesssim 5$). While near wall grid cells in the buffer layer may be handled with blending functions, they are a major source of misrepresentation of wall shear stresses and should be avoided. LES and DNS approaches generally require $y^+ < 1$. Near-wall grid cells outside the appropriate y^+ range may result in incorrect turbulence production and turbulent viscosity, hence the flow resistance.

Due to the complex geometrical nature of the nasal cavities, it is expected that boundary layer thicknesses will experience significant spatial and temporal variations due to the breathing cycle. The best option might therefore be to ensure that the computational mesh is fine enough to maintain near wall cells in the viscous sublayer, everywhere, for all flowrates, and to utilize a suitable turbulence model. Inthavong et al. [106] discussed mesh resolution requirements for laminar nasal air flow and suggested that $y^+ < 0.27$ in the near-wall grid cells.

Outside the near-wall region, the spatial resolution must be sufficient to resolve all relevant flow structures. It has been suggested that 4–6 million grid cells is generally sufficient to achieve mesh independence [85,106,176], but this is a generalization that should be accepted with caution, since it might not be appropriate for all nasal geometries and volumetric flowrates.

Adaptive meshing, which is a technique that regenerates and/or adapts the mesh based on predefined flow field criteria, is an approach that may be well suited to respiratory breathing, where a wide range of flow characteristics and features can be expected, and a fixed mesh might not be the best choice for the entire range of volumetric flowrates. This technique allows for the refinement of the mesh in regions of steep gradients or small flow features as well as coarsening of the mesh where spatial variations are modest. This further allows for a non-constant number of grid cells, reducing computational cost and giving shorter computation times when the flowrates are lower. It does come at the computational cost of remeshing/adjusting the computational mesh, however. Adaptive meshing is not exclusive to transient simulations but may also be used to improve accuracy in steady state simulations. The present author is not aware of any studies investigating this for nasal airflow.

(B)   Temporal Resolution

The temporal resolution determines simulations' ability to correctly describe transient variations in the flow fields. e.g., long time steps might not be able to capture quickly fluctuating phenomena. The time step size is commonly characterized by the dimensionless Courant–Friedrich–Lewy (CFL) number [178,179], which expresses the ratio of the advected distance during one time step, $u\Delta t$, to the characteristic grid size, Δx,

$$CFL = u\Delta t/\Delta x. \tag{17}$$

Here, u denotes the flow velocity through the grid cell, and Δt is the time step size.

The *CFL* number plays a critical role in ensuring the numerical stability and accuracy of CFD solvers. Explicit CFD solvers restrict information propagation to the maximum of one grid cell per time step ($CFL \leq 1$). Consequently, this limitation forces the use of exceedingly short time steps when dealing with small geometry features resolved by small grid cells, resulting in prolonged and costly simulations. Implicit solvers, on the other hand, permit longer time steps, but incorrect simulation results can be the result for too large *CFL* numbers.

On a fixed computational mesh, with a fixed time step the CFL number tends to zero at the culmination of the inspiratory and expiratory phases of the respiratory cycle. Depending on the numerical scheme employed, short time steps can introduce numerical diffusion, blurring the details of the flow, but the main downside is an unnecessarily high number of time steps. To mitigate this issue, adaptive time stepping strategies can be employed, where the time step size increases as the volumetric flowrate decreases. This approach helps maintain favorable *CFL* numbers and reduces computational cost. However, it is important to note that since the volumetric flowrate eventually dwindles to zero, a numerical scheme capable of handling low *CFL* numbers remains essential.

Boundary Conditions

Setting appropriate boundary conditions is a crucial step in configuring CFD models. Accurately describing flow parameters such as velocity, turbulence intensity, pressure, and temperature at the boundaries often presents a challenging task, leaving CFD engineers to rely on best available estimates or educated guesses. Consequently, it is important to position the boundaries at sufficient distance from the region of interest to prevent undue interference with the essential details of the flow. However, expanding the simulation domain to achieve this boundary distance unavoidably incurs higher computational costs. Thus, the determination of boundaries' locations necessitates careful balance between accuracy and cost efficiency.

In the specific context of nasal airflow analysis, the boundaries include the walls of the nasal cavity and the flow in- and outlets.

- The walls are typically treated as smooth non-slip boundaries. Nevertheless, the presence of the mucosal lining introduces the possibility that surface roughness and slip conditions might need consideration. While the nasal wall temperature is commonly assumed to fall within the range of normal body core temperature, this assumption may require more careful consideration if the inhaled air is significantly colder.

- The nostrils serve as inlets to the nasal cavity during inhalation and outlets during exhalation. However, it is reasonable to suspect that truncating the computational domain at the nostrils may compromise the accurate description of airflow entering or exiting the nasal cavity. An alternative approach is to extend the computational domain to encompass the external airspace around the nose to achieve a more realistic airflow distribution at the nostrils. A study by Taylor et al. [180] suggested that the qualitative description of the inflow conditions at the nares may not be critical when computing general flow patterns and overall measures, but for detailed regional flow patterns, carefully chosen inflow conditions may be necessary.
- Modeling the entire airway, including the lungs and alveoli, is impractical in nasal airflow studies. Therefore, the airway is typically truncated somewhere in the laryngopharyngeal tract. The location of this truncation has traditionally been based on available computational resources and the specific phenomena of interest. Although the location of truncation may be less critical during inhalation, more attention may be warranted during exhalation. Wu et al. [181] demonstrated, in a physical experiment, that the flow in the pharynx is laminar during normal breathing, but Bradshaw et al. [159] highlighted the importance of including a realistic pharyngeal tract to achieve accurate flow conditions in the nasopharynx during exhalation. The pharyngeal tract is a complex, soft-tissue-enclosed flow channel susceptible to head and neck movements, swallowing, tongue movement, and compliance with over-/underpressure due to breathing. The exhalatory flow pattern entering the nasopharynx is likely to be affected by this. The level of realism required in the pharyngeal tract to attain acceptable inflow to the nasopharynx is still unresolved.

Once the boundary locations are determined, careful selection of boundary types (e.g., Dirichlet or Neumann) is required to ensure uniqueness of solution before defining boundary values. These boundary values can be constant or vary with time and/or position along the boundaries. There is generally significant uncertainty associated with the determination of the local boundary values, necessitating sensitivity analysis to evaluate the impact of boundary conditions.

2.6. Summary of Part I

- Rhinology, a specialized branch of otorhinolaryngology, is dedicated to advancing diagnostics and treatment methods for nasal and sinonasal disorders, including conditions like nasal airway obstruction (NAO).
- The discord between objective and subjective clinical assessments of NAO severity has created an opening for mathematical modeling tools, such as computational fluid dynamics (CFD), to enhance our understanding of nasal function.
- Computational rhinology, a subfield of biomechanics, employs numerical simulations, like CFD, to gain deeper insights into nasal and sinus function and pathology.
- Computational rhinology is poised to exert a substantial influence on clinical medicine by offering objective, simulation-based decision support for tailoring patient-specific treatment options within the realm of otorhinolaryngology. Furthermore, it may facilitate research and development of novel or improved treatment methods, as well as comparisons between patient-specific and cohort studies.
- While substantial progress has been made over the past three decades toward the clinical application of CFD, there is a lack of robust evidence supporting its applicability and value, and it is yet to attain widespread acceptance as a viable clinical decision support tool.
- Despite significant collaborative efforts from experts in both rhinology and CFD over several decades, CFD is not able to reproduce results from objective clinical measurements, such as rhinomanometry (RMM). In particular, in silico RMM consistently underpredicts nasal resistance compared to in vivo RMM.

- A comprehensive overview of sources of error and uncertainty affecting the comparison of in vivo and in silico RMM has been presented. The observed discrepancies may be the result of a combination of multiple independent factors, rather than a single, isolated cause. Major sources of uncertainty include the following:
  - ○ Comparability of nasal cavity geometry during RMM and medical imaging examinations.
  - ○ The impact of nasal compliance.
  - ○ CFD modelling strategies (e.g., turbulence modelling, unsteady/steady flow).
- Regardless of RMM's capability to predict a patient's subjective sensation of nasal patency, it serves as one of few opportunities for validating in silico nasal airflow models. Consequently, RMM plays an indispensable role in the field of computational rhinology.
- The lacking agreement between in vivo and in silico RMM results is a fundamental problem that must be addressed for CFD to gain recognition as a reliable, objective clinical decision support tool.

3. Part II—Overview of Published Literature on In Vitro and In Silico Nasal Airflow Studies

In the field of rhinology, CFD is considered an emerging technology with significant potential. Three decades ago, Keyhani et al. [152] published the first anatomically accurate CFD model of nasal airflow, marking the beginning of advancements in the workflow from medical imaging (such as CT or MRI) to CFD analysis. Since then, several reports have highlighted successful applications of CFD in areas such as surgical intervention planning and improved understanding of nasal airflow. However, despite these advancements, CFD has not yet gained widespread use in clinical practice. One of the main obstacles is the lack of consensus among otolaryngologists regarding objective evaluation criteria for assessing nasal function and guiding surgical decisions [90,182]. This poses a hurdle for utilizing CFD as a clinical decision support tool, because CFD relies on clearly defined questions and produces mainly quantitative results. Despite groundbreaking scientific progress and increasing interest within the otorhinolaryngology community [183], CFD technology is still considered immature. This is evident from the fact that recent reviews of diagnostic tools in rhinology [46] and of recent advances in surgical treatments for obstructive sleep apnea [184] did not even mention CFD. A specific concern raised by Vicory et al. [185] is the labor-intensive nature of manual segmentation in CFD-based virtual surgery, which adds complexity and time requirements to the process. However, there is optimism for future improvements in this area. It is anticipated that segmentation methods will undergo advancements, leveraging novel techniques in machine learning. These advancements have the potential to optimize and automate the workflow from medical images to CFD analysis, streamlining the process for greater efficiency. Wong et al. [20] argue that the maturity of CFD in rhinology is approaching the level where its successful implementation in clinical practice is feasible. They suggested that CFD should be recognized as a valuable diagnostic tool within rhinology. However, despite tremendous efforts in realizing clinical relevance of CFD, there is still no proof that CFD-based clinical decision support will actually improve patient outcomes [115].

Bailie et al.'s overview of numerical modelling of nasal airflow [136], aimed at the otolaryngology community, continues to hold relevance today. However, it is important to keep in mind the significant scientific and engineering advances that have taken place in every stage of the process from generating realistic 3D computer models from medical imaging data to conducting multiphysics simulations and analysis involving, e.g., turbulence and soft tissue movement. These breakthroughs have been accompanied by notable improvements in computational power, facilitating the use of high-fidelity computational meshes with tens of millions of grid cells, while maintaining relatively short computation times. In a more recent publication, Lintermann [186] addresses the general application of CFD in rhinology. While it did not cite the most recent scientific publications in the field, it serves as an accessible introduction to researchers who are new to CFD, providing a valuable starting point for further exploration. Tu et al. [187] published the first textbook presenting a comprehensive overview of the possibilities of utilizing engineering CFD to enhance the medical understanding of respiratory airflow and particle transport. A more recent text book edited by Inthavong et al. [188] "explores computational fluid dynamics in the context of the human nose, ..." and "focuses on advanced research topics, such as virtual surgery, AI-assisted clinical applications and therapy, as well as the latest computational modeling techniques, controversies, challenges and future directions in simulation using CFD software".

In the past fifteen years, the accessibility of generating patient-specific 3D geometry models using CT imaging techniques or similar methods has significantly increased [18,96]. This progress, driven by advancements in automatic segmentation routines, has greatly facilitated the implementation of CFD simulations for studying airflow in the human upper airways by fluid dynamics researchers worldwide. Additionally, simplified CFD simulation setups have made it possible for non-specialists to configure and execute reasonable simulations, even without extensive training in the field. These trends are reflected in the rapidly growing number of publications discussing various topics relevant to otolaryngology, including targeted drug delivery via nasal spray, OSA, NAO, virtual surgery, and more. However, a notable limitation observed in many of these publications is the lack of clinical grounding. For example, numerous publications present simulation results without adequately interpreting their clinical value or applicability. Additionally, some publications overlook the inclusion of all relevant parts of the airway or fail to exclude irrelevant anatomical features. These shortcomings could have been addressed through closer collaboration with clinical personnel to ensure a more comprehensive approach.

In a review paper by Inthavong et al. [105], it was noted that the annual publication count of in silico studies on nasal flows exhibited nearly exponential growth from 1993 to 2018. In 2018 alone, close to eighty articles were published on this subject. The rapidly increasing number of publications has led to a host of review papers being published over the last two decades. Baile et al. [136], Zhao and Dalton [18], and Leong et al. [189] reviewed the earliest literature on the topic and discussed the implications and prospects of mathematical modelling of nasal airflow. Zhao and Dalton pointed out that CFD modelling can have important implications for (1) predicting how inflammation or anatomy can affect airflow patterns and olfaction; (2) optimization of treatment; and (3) prediction of particle transport (hereunder both pollutants and medical drugs). Kim et al. [171] reviewed studies that employed physical models as well as studies focusing on numerical modelling. They gave an overview of available methods and challenges associated with the employment of CFD in patient-specific diagnostics and analysis in rhinology and underlined that experienced otolaryngologists should be involved in quality assurance of 3D airway models. They proceeded to discuss how the nasal cycle may affect the evaluation of nasal patency and reviewed how inaccuracies stemming from various segmentation methods and 3D modelling methods may affect the CFD geometry. A recent review aimed at rhinologists [19] highlighted the potential applicability of CFD in clinical applications as a diagnostic tool. Radulesco et al. [190] gave a literature review focusing on the usefulness of CFD in the assessment of NAO. Other recent literature reviews were given by, e.g., Faizal et al. [191] and Ayodele et al. [99]. Recent publications considering patient-specific mathematical modelling and simulation for clinical decision support include, e.g., [89,90,192–195]. Interestingly, many research questions raised by Quadrio et al. [196] a decade ago regarding nasal airflow, such as the selection of turbulence models and boundary conditions, as well as the validity of assuming rigid walls, remain unresolved.

The bibliography in the current paper was compiled through a combination of ancestry and descendancy literature review approaches, utilizing internet-based publication databases such as Google Scholar, Researchgate, PubMed, and journal web pages. In particular, recent review papers and textbooks were used as basis for discovering relevant publications. Because this paper focuses on the nasal cavity, publications that did not include the nasal cavity were disregarded. Also, the current paper does not review the extensive scientific literature concerning FSI, acoustic phenomena, or particle transport and deposition, in human airways. Recent literature overviews on these topics were provided by Ashraf et al. [107] and Le et al. [197] for FSI, Xi et al. [198] for acoustics, and Larimi et al. [199] for particle deposition.

Below, a tabulated overview of literature concerned with various topics relevant to CFD modelling in rhinology is given, based on a combination of ancestry and descendancy literature review approaches. In Table 1, cited papers are classified and grouped depending on their approach to physical or numerical modelling of nasal airflow, e.g., steady vs. transient volumetric flowrate and type of modelling approach.

3.1. In Vitro Studies in Physical Nasal Replicas

Prior to the advent of CFD capabilities enabling accurate nasal airflow simulations, researchers relied on flow experiments conducted in physical replicas to improve their understanding of nasal airflow distribution during respiration. Pioneering experimental investigations were carried out by Proetz [149], Stuiver [200], Masing [201], Hornung et al. [202], and Hahn et al. [151]. More recently, Kelly et al. [203] utilized particle image velocimetry (PIV) to examine airflow patterns in a physical replica of a nasal cavity and provided an overview of previous experimental work in this domain. Several studies have reported favorable agreement between flow fields obtained from CFD modeling and measurements in physical models [157,204]. Notable recent experimental studies have been carried out by Le et al. [197], Ormiskangas et al. [205], Berger et al. [42], van Strien et al. [206], and Reid et al. [130]. Experimental work continues to play a crucial role in the validation of CFD models.

3.2. In Silico Cohort Studies

A few studies have made an effort to compare and evaluate CFD results for a cohort of multiple individuals. Zhao and co-authors [207] made an attempt to classify what constitutes normal nasal airflow in a cohort of 22 healthy subjects. Ramprasad and Frank-Ito [208] used CFD to study the role of three nasal vestibule phenotypes found in their cohort consisting of 16 subjects, on nasal physiology. Gaberino et al. [65] used virtually created nasal cycle mid-point CFD models to study 12 patients. Sanmiguel-Rojas et al. [209] proposed two non-dimensional estimators representing geometrical features and nasal resistance and used CFD simulations to investigate how these estimators varied in 24 healthy and 25 deceased subjects. Radulesco et al. [66] used CFD to evaluate NAO due to septal deviation in 22 patients. Borojeni et al. [67] used CFD to establish normative ranges for nasal air flow variables in a cohort consisting of 47 healthy adults. Li et al. [210] used CFD to study 30 patients with obstructive sleep apnea and NAO pre- and postoperatively to find the effect of nasal surgery on the airflow in the nasal and palate pharyngeal cavities. Ormiskangas et al. [211] compared CFD-based wall shear stress and heat flux results to subjective evaluation of NAO pre- and post-inferior turbinate surgery for 25 patients.

3.3. Modelling Approaches in In Silico Studies

It is expected that CFD can provide objective decision support in treatment of breathing-related disorders. However, CFD requires experimental validation to gain full confidence due to the many tuning parameters in the various submodels and solvers involved. Several authors have experienced that CFD results generally deviate from in vivo measurements but display better agreement with in vitro studies (see **Part I**, Section 2.4.6). So far, the reasons for this are unclear. A review of sources of errors and uncertainties affecting comparison between in vivo and in silico RMM was presented in **Part I** (Section 2.5).

There is an ongoing debate about the nature of the flow in the human upper airways in the scientific literature [83–85]. One of the main concerns has been whether the flow is laminar, transitional, or turbulent. A wide variety of turbulence models exist, and many of them have been employed in in silico nasal airflow studies, including laminar, RANS, LES, and DNS modelling concepts.

The selection of a turbulence model should generally be based on a good understanding of the flow phenomena that is to be modelled. e.g., the many different RANS models may only differ by small nuances, but they are all tailored to specific calibration cases. It is somewhat naïve to expect a RANS model developed for fully developed pipe flow to perform accurately in a very different flow situation such as unsteady non-equilibrium flow in a complex flow channel such as the nasal cavity. CFD modelling of transitional flow is considered one of the most difficult tasks, and only a few turbulence models are able to handle this with reasonable accuracy.

There is no consensus regarding the most appropriate modelling technique, but the current review indicates that steady state laminar modelling has been the predominant modelling strategy. A rule of thumb adopted by many authors over the last three decades, when faced with the difficult choice of which turbulence model to select in CFD modelling of nasal airflow, is that unilateral nasal airflow is laminar below 15–20 L/min and turbulent above 20 L/min. This generalization has become so common that the original sources are not evident, and some authors do not even reference a source [105,106,212,213]. It appears that the original references of this assumption are the early in vitro unilateral nasal airflow studies by Swift and Proctor [214], Schreck et al. [215], and Hahn et al. [151]. Swift and Proctor [214] observed laminar flow at 125 mL/s and turbulence at 208 mL/s; Schreck et al. [215] observed turbulence at flowrates corresponding to 200 mL/s; and Hahn et al. [151] observed laminar flow at 180 mL/s and turbulence at 560 and 1100 mL/s. Numerical simulations by Li et al. [83] demonstrated that a laminar model achieved good agreement with LES and DNS simulations up to a unilateral flowrate of 180 mL/s. Simmen et al. [216] observed partial turbulence generated in the nasal gateway (valve) region even at low velocities, however, and it appears natural, from an evolutionary perspective, to promote turbulence in the nasal cavities, since this will aid in fulfilling the main functionalities of the nose (heat exchange, humidification, and dust particle deposition).

Aasgrav et al. [3] showed that laminar and RANS CFD models produced similar results. The conclusion has been supported by many researchers including, e.g., Larimi et al. [199], who reported negligible turbulence intensity in the nasal cavity for normal, inspiratory breathing rates, and Schillaci and Quadrio [148] and Bradshaw et al. [159], who reported mostly laminar flow for normal resting breathing conditions.

Turbulence intensity has been reported to be low in the nasal cavity, but the ability to model vortexes has been promoted as an important feature of nasal air flow models, e.g., to predict particle deposition, heat exchange, and humidification correctly [217]. Li et al. [83] reported from a validation study including laminar, RANS, LES, and DNS models, where they concluded that DNS generally reproduced experimental data best, but LES and laminar models were also accurate, depending on the flowrate. They emphasize the need to select the turbulence model carefully, to be able to compute flow quantities of interest accurately. Berger, Pillei et al. [42] compared Lattice–Boltzmann (LB)-, LES-, and Navier–Stokes-based RANS models to experiments in a physical replica of a nasal cavity and reported that the different techniques "agree with some caveats". Moreover, they demonstrated that LB LES was computationally very cheap and more accurate than Navier–Stokes based RANS. In **Part III** (Section 4) of the present paper, laminar and RANS steady state models utilizing relatively coarse computational meshes are compared to fully resolved transient LES simulations. The results indicate that the flow is near-laminar at all flowrates for the current patient.

Another dispute is related to whether respiratory flow can be approximated by quasi-steady models or not [158,159,218]. The quasi-steady approach assumes that transient, tidal breathing can be approximated by a series of steady state simulations and is typically much cheaper, in terms of computational cost, than transient simulations. Steady state simulations may, however, overlook important transient physical phenomena such as flow instabilities due to unsteady vortices or jets, development of flow structures, and heating and humidification [159]. Many authors have argued that normal, tidal breathing is quasi-steady [149–154] (see **Part I**, Section 2.5.4). **Part III** (Section 4) of the present paper concludes that transient effects are negligible with respect to in silico RMM modelling, based on the observation that steady state models performed almost identical to a transient model.

In Table 1, a non-exhaustive overview of in vitro and in silico nasal airflow studies is presented. Each cited paper is classified by its modelling strategy. The left and right columns refer to papers employing steady or transient volumetric flowrates, respectively. Row-wise, a distinction is made between in vitro flow experiments, in physical replicas of the nasal cavity, and in silico CFD studies of nasal flow. In silico studies are further grouped according to their modelling approach, Navier–Stokes- or Lattice–Boltzmann-based, as well as the turbulence modelling approach employed (laminar, RANS, LES, DNS).

Table 1. A non-exhaustive overview of experimental and numerical studies of human nasal airflow employing steady and transient flow boundary conditions.

		Volumetric Flowrate	
		Steady	**Transient**
Experiments in physical replicas (in vitro)		[42,83,117,151,173,203–205,214–216,219–221]	[42,127,129,130,149,151,157,197,200–204,206,222–224]
Navier–Stokes-based models (in silico) — Laminar		[5,50,57–59,61–63,65,68,83,86,88,90,100,101,106,116,117,123,126,129,131,148,152,158,172,175,176,180,192,199,205,207–209,211,219,225–236] *The present study*	[126,129,153,158,172]
RANS	$k\varepsilon$	[3,4,42,83,170,204,210,220,226,230,237–239] *The present study*	[224]
RANS	$k\omega$	[83,117,121,122,173,204,205,207,226,229,237,240–242]	[173,197]
RANS	$k\omega$ SST	[83,101,124,148,172,175,199,204,205,212,213,220,243–245] *The present study*	[118,160,172,194,223]
RANS	Spalart–Allmaras	[204,226,246]	
RANS	Reynolds stress model	[83]	
LES and RANS-LES hybrid models		[83,148,204,206,217,237,247]	[159,248–250] *The present study*
DNS		[83]	
Lattice–Boltzmann-based methods (in silico)		[42,125,195,251–253]	[127]

3.4. Summary of Part II

- An increasing number of scientific publications are being published in the cross-disciplinary field of computational rhinology. As of 2018, the publication rate was approximately 80 papers per year after several years of near-exponential increase.
- The rapid increase in the publication rate is mainly attributed to the advancements made in automatic segmentation of medical imaging data, streamlining the process of 3D geometry generation, and automatic, unstructured meshing of complex geometries.
- There is no consensus regarding the choice of turbulence model in nasal airflow simulations. Laminar flow modelling appears to be the most popular approach, by far, followed by RANS models. Few publications have reported from LES or DNS modelling.
- Most studies have investigated steady, inspiratory flow. Relatively few publications have reported from expiratory or transient/respiratory flow.

4. Part III—In Silico RMM Simulation Results

Computational fluid dynamics (CFD) is a thoroughly proven methodology/technology widely used in engineering and scientific research to study, design, and optimize complex fluid systems. Over the last fifty years, this methodology has warranted technological leaps, which we take for granted in the modern society, within a wide range of application such as weather forecasting and climate research, automotive and aerospace industries, the energy sector, industrial design, and many others.

CFD simulations have contributed to improving the general understanding of nasal function and have been proposed as a promising tool for enhancing diagnostics and treatment planning in otorhinolaryngology. A severe limitation to the employment of CFD as an objective clinical tool is, however, the reported lack of agreement between in vivo measurements such as rhinomanometry (RMM) and CFD (see **Part I**, Section 2.4.6).

In **Part I** (Section 2) of the current paper, the foundations for and challenges within the emerging scientific field of computational rhinology were elaborated. Sources of errors and uncertainties that may affect the comparison of in vivo and in silico RMM were discussed. In **Part II** (Section 3), a literature review of CFD studies involving nasal airflow was provided. This part of the paper showcases previously unpublished high-fidelity patient-specific CFD simulations of nasal airflow in Active Anterior Rhinomanometry (AAR).

The aim of this study was to scrutinize selected possible reasons for the observed discrepancies between in vivo and in silico RMM (refer also to **Part I**, Section 2.5.4). Specifically, this study investigated the threefold hypothesis that these discrepancies can be explained by (1) *poor choice of turbulence model*, (2) *insufficient spatial or temporal resolution*, or (3) *neglecting transient effects*, in CFD models. To do this, finely resolved LES-based transient CFD simulations were performed to obtain benchmark in silico RMM pressure–flow curves on both sides of a patient-specific nasal cavity geometry obtained from CT images. These simulations, which are among the most detailed simulations of nasal airflow reported in the scientific literature, to date, were used to assess more simplistic steady state laminar- and RANS-based CFD simulations utilizing a relatively coarse mesh.

Notably, almost perfect agreement between the various CFD models was observed with respect to overall flow parameters. However, the CFD results deviated significantly from in vivo RMM data. Analysis of the transient LES model indicates that errors attributable to spatial or temporal resolution were insignificant. The main findings from the simulations were thus that the flow was predominantly near-laminar and quasi-steady. Hence, the initial hypotheses were effectively disproved, directing the attention towards other main sources of errors such as the digital geometry model. The most plausible reason for gross disagreement between in silico and in vivo RMM appears to be misrepresentation of the actual nasal cavity by the digital geometry model. It is proposed that nasal cavity compliance and the drag effect of nasal hair are key components in this regard.

4.1. Methods

The present study used patient-specific clinical data for creation of and comparison with CFD simulations. Computed tomography (CT) images were utilized to create a digital 3D geometry model of the patient's nasal cavity, which was used as basis for CFD simulations of active anterior rhinomanometry (AAR). In silico pressure–flow curves were obtained from the CFD simulations and compared with the patient's AAR data.

4.1.1. Clinical Data

The clinical data were obtained from a 67-year-old male patient with a body mass index (BMI) of 28. Polysomnography resulted in a measured apnea–hypopnea index (AHI) of 23. He was diagnosed with obstructive sleep apnea and scheduled for nasal surgery to correct septum deviation and increase the volume of the left vestibule. The surgical intervention resulted in a significant improvement in AHI. The current paper presents preoperative patient data and patient-specific CFD simulation results. Preoperative clinical examination was performed at St. Olav's hospital, the university hospital in Trondheim, and included AAR and CT imaging.

AAR was performed with the patient in a seated position using Otopront® RHINO-SYS rhinometry system [254] before and after decongestion to establish in vivo pressure-velocity curves. Decongestion was achieved through the application of xylometazoline via nasal spray, which was allowed to work for 15 min. Figure 1 shows the resulting RMM measurement output, where blue and red curves correspond to left and right sides of the nose, respectively, and light/dark colors correspond to before/after decongestion. Representative pressure–flow curves corresponding to the decongested state of the nose, used for comparison with CFD simulation data, are shown as dashed black curves.

CT images were obtained with a Siemens Sensation 64, with the patient in the supine position (Figure 4). The scan provided 342 slices of 1.0 mm thickness. The 2D CT images consisted of 512×512 pixels. The volume of each voxel was 0.167 mm^3.

The patient was part of a prospective study approved by the Norwegian Ethical Committee and registered in clinicaltrials.gov (NCT01282125). Written informed consent was obtained from the patient.

4.1.2. Geometry Retrieval

A 3D geometry suitable for CFD simulations was obtained through segmentation of the patient's CT images. Segmentation was performed using the software ITK-SNAP 3.4.0 [255]. Automatic segmentation using a Hounsfield unit (HU) threshold of -300 was supplemented with manual segmentation in cooperation with medical experts to obtain high-quality geometry. For simplicity, the paranasal sinuses were manually excluded. Due to the narrow passage connecting the paranasal sinuses with the nasal cavity, it is expected that negligible gas exchange takes place between the two [171]. Furthermore, the inclusion of the paranasal sinuses will increase the complexity and size of the computational mesh, hence require greater computational effort. The entire procedure of segmentation and geometry preparation was described by Jordal et al. [5]. The 3D geometry was truncated at the nostrils and just below the nasopharynx. The nasopharyngeal boundary was extended by a length corresponding to approximately six hydraulic diameters to distance the flow boundary from the region of interest and to dampen possible effects of the boundary condition. The geometry was considered rigid and is shown in Figure 5.

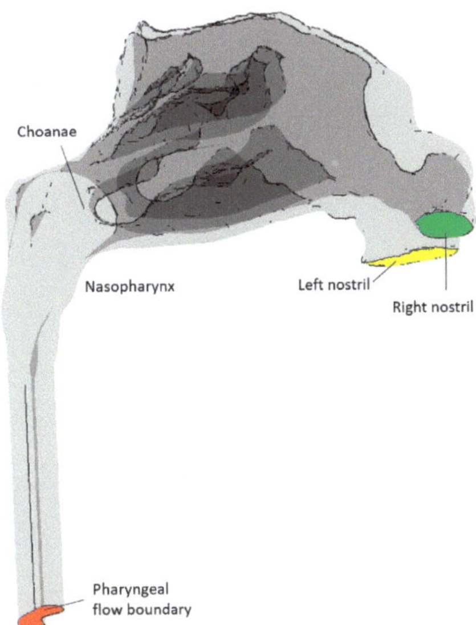

Figure 5. Flow geometry obtained from segmentation of CT images. Paranasal sinuses and the oral

cavity are omitted from the model. Computational boundaries are indicated by color: Air–tissue wall boundary (gray), pharyngeal flow boundary (red); right nostril (green); left nostril (yellow).

4.1.3. In Silico Rhinomanometry (CFD Modelling)

In silico RMM was performed by CFD simulation of unilateral nasal airflow in the patient-specific 3D geometry. During the simulations, both nasal cavities were included, but one nostril was closed such that there was only airflow through one nasal cavity at the time. The transnasal pressure drop was thus approximated by the pressure difference between the open and the closed nostril, similar to in vivo AAR. See **Part I** (Section 2.3.2) for additional discussion of AAR.

CFD modelling was performed in ANSYS Fluent 2019 R2 [256]. Steady state simulations were performed to produce quasi-steady RMM curves. These simulations utilized laminar, $k\varepsilon$ realizable, and $k\omega$ SST models. Transient simulations were performed with the large eddy simulation (LES) to produce the entire RMM curves including potential effects of unsteady, respiratory flow. Three breathing cycles were simulated on each side.

All simulations were isothermal with constant air properties ($\rho = 1.225$ kg/m^3, $\mu = 1.7894 \times 10^{-5}$ Pas). Additional details regarding the simulations are summarized in Table 2. Most simulation settings were kept at default values as proposed by ANSYS Fluent. Refer to ANSYS Fluent user and theory guides for more details regarding the model settings and theory background [113,257]. In all simulations, convergence was assumed when residuals dropped below 0.001. In the transient simulations, this required between 1 and 7 iterations per time step, depending on the flowrate.

The simulation case-files and simulation results from the LES simulations have been made available to the public under the Creative Commons Attribution-Noncommercial (CC BY-NC) license [14].

Table 2. Overview of simulations, including simulation type, turbulence model, options activated, and computational mesh utilized.

Turbulence Model	Turbulence Model Options Activated	Computational Mesh	Volumetric Flowrate	Solver Settings	Discretization
Laminar		Coarse	Steady	- SIMPLE pressure-velocity coupling - Rhie–Chow distance-based flux-type	- Pressure: standard - Advected variables: first order upwind - Gradients: cell-based least squares
$k\varepsilon$ realizable	- Enhanced wall treatment - Curvature correction - Production limiter	Coarse	Steady		
$k\omega$ SST	- Curvature correction - Production Kato–Launder - Production limited - Intermittency transition model	- Coarse - Fine (peak insp. flow)	Steady		
LES	WALE	Fine	Transient	SIMPLE pressure-velocity coupling	- Pressure: second order discretization - Momentum: bounded central differencing - Gradients: node-based Green–Gauss method - Time stepping: bounded second order implicit scheme

Turbulence Modelling

For laminar flow, Equations (11) and (12) are solved directly. For turbulent flow employing Reynolds-averaged Navier–Stokes (RANS) turbulence models, such as the $k\varepsilon$ and $k\omega$ model families, the velocity is considered a sum of a mean velocity and a fluctuating component whose ensemble average is zero, $\mathbf{u} \rightarrow \overline{\mathbf{u}} + \mathbf{u}'$ with $\langle \mathbf{u}' \rangle = 0$, such that $\langle \mathbf{u} \rangle = \overline{\mathbf{u}}$, where the brackets indicate ensemble averaging. Inserting for the fluctuating velocity in Equations (11) and (12) and performing ensemble averaging results in the RANS equations, in which the Reynolds stress term, $\langle (\mathbf{u}' \cdot \nabla)\mathbf{u}' \rangle$, represents the effect of turbulence. In order to close Equation (12), these stress terms are modelled by the Boussinesq hypothesis, where the Reynolds stress terms are assumed to be related to the mean velocity gradients such that the effect of turbulence can be modelled through an increased effective viscosity (eddy viscosity model). Equations (11) and (12) are thus written as the RANS equations,

$$\nabla \cdot \overline{\mathbf{u}} = 0 , \tag{18}$$

and

$$\partial_t \overline{\mathbf{u}} + (\overline{\mathbf{u}} \cdot \nabla)\overline{\mathbf{u}} = -(1/\rho)\nabla \overline{P} + (\nu + \nu_t)\nabla^2 \overline{\mathbf{u}} , \tag{19}$$

where the ν_t is the isotropic turbulent kinematic viscosity. In the $k\varepsilon$ and $k\omega$ model families, additional equations are solved for the turbulent kinetic energy, k, dissipation rate, ε, and specific dissipation rate, ω, which are used to compute the turbulent kinematic viscosity. $k\varepsilon$ realizable [258] is an improved variant of the standard $k\varepsilon$ model that has been shown to have superior performance in modelling complex flows with, e.g., separation and secondary flow features [113]. $k\omega$ SST is a hybrid model combining the strengths of the $k\omega$ model, which is considered more accurate in the near-wall region, and the $k\varepsilon$ model, which is considered better in the outer parts of the turbulent boundary layer. The enhanced wall treatment (EWT) option in ANSYS Fluent also enables near-wall modelling capabilities in the $k\varepsilon$ models. Thus, both $k\varepsilon$ with EWT and $k\omega$ SST are y^+-insensitive models [112].

For LES turbulence models, the governing equations (Equations (11) and (12)) are filtered through the finite-volume discretization. The LES continuity equation becomes identical to Equation (18), but the Reynolds stress term in the momentum equation is replaced by the subgrid-scale stress tensor, $\boldsymbol{\tau}_{SGS}$,

$$\partial_t \overline{\mathbf{u}} + (\overline{\mathbf{u}} \cdot \nabla)\overline{\mathbf{u}} = -(1/\rho)\left[\nabla \overline{P} + \nabla \cdot \overline{\boldsymbol{\tau}_{SGS}}\right] + \nu\nabla^2 \overline{\mathbf{u}}. \tag{20}$$

The RANS and LES momentum equations are, however, formally identical. In ANSYS Fluent, the subgrid-scale stress models employ the Boussinesq hypothesis, as the RANS models, and the main difference between the available LES variants is in the calculation of the subgrid-scale turbulent viscosity. Here, the Wall-Adapting Local Eddy-Viscosity (WALE) subgrid-scale stress model [259] was employed. This model has the trait that it predicts zero turbulent viscosity in laminar flow, and it is therefore considered suitable for modelling wall bounded flows with laminar turbulent transition [111,260].

The core idea of the LES modelling strategy is that the majority of the energy carrying turbulent eddies are modelled directly on a sufficiently fine grid, while the remaining (small) fraction of eddies is modelled via the subgrid-scale model. LES is widely accepted as the most accurate turbulence model, second to direct numerical simulation (DNS), which resolves even the smallest turbulent eddies. It is assumed that LES approaches DNS in the limit of high spatial resolution (small grid cells). Additional discussion of turbulence models was provided in **Part I** (Section 2.4.5), and the turbulence models available in ANSYS Fluent are well documented in the ANSYS Fluent User and Theory guides [113,257].

Boundary and Initial Conditions

The boundaries of the flow geometry consist of the nasal cavity walls, the left and right nostrils, and the pharyngeal flow boundary (see Figure 5).

CFD simulations were set up to replicate unilateral RMM, where one nostril is occluded while measuring the volumetric flowrate and pressure drop in the contralateral nasal cavity. Hence, all simulations were performed unilaterally, with one nostril closed and the other open, leading to two specific scenarios: left open–right closed (LO-RC) and left closed–right open (LC-RO).

The nasal cavity walls and closed nostril were treated as standard smooth non-slip wall boundaries. The open nostril was specified as a pressure inlet with a specified, constant pressure of 0 Pa(g). In ANSYS Fluent, the pressure formulation at a pressure inlet depends on the flow direction across the boundary. Thus, the total and static pressures at the nostril boundary were alternatingly equal to the specified pressure, during inhalation and exhalation, respectively (see the ANSYS Fluent user guide for details [257]). This choice of boundary condition was aimed at mimicking the difference between inhalation from and exhalation into a stagnant reservoir of air. During inhalation, it is expected that air will experience loss-free acceleration from the far field to the nostril, conserving the total pressure. This implies a negative static gauge pressure of approximately $\rho u^2/2$ during inhalation, where u is the mean velocity at the nostril [81]. During exhalation, however, it is expected that the jet of air exiting the nostril will dissipate and lose all its kinetic energy to the surrounding air, and the pressure at the nostril is approximately equal to the far field static pressure.

The pharyngeal flow boundary was specified as an inlet velocity boundary condition, with a uniform velocity corresponding to the specified volumetric flowrate. Negative and positive flowrates indicated inhalation and exhalation, respectively. For steady state simulations, the constant, volumetric flowrate was specified for each simulation, but for the transient simulations, tidal respiration was approximated by

$$Q(t) = -Q_{\max}\cos\left(\frac{2\pi(t-t_0)}{\tau}\right), \tag{21}$$

where t was the flow time, $t_0 = 1$ s was the duration of the pseudo-steady initialization described below, $\tau = 5$ s was the duration of one breathing cycle, and the peak volumetric flowrate was $Q_{\max} = 600$ mL/s. This corresponds to a tidal volume of 955 mL, which is above what would be expected in normal breathing [161,261].

The same turbulence boundary conditions were employed for the open nostril and the pharyngeal flow boundary. Default values were used—5% turbulent intensity, turbulent viscosity ratio of 10, and intermittency of 1 ($k\omega$ SST, only). For the LES simulations, the synthetic turbulence generator option was selected. See the ANSYS Fluent user and theory guides for details [113,257].

Initial conditions for the transient LES-based simulations were obtained by first performing a steady state fine-mesh $k\omega$ SST simulation with peak inspiratory flow, $Q = -Q_{\max}$. Next, the resulting flow fields were converted by employing the text user interface command *solve/initialize/init-instantaneous-vel* before pseudo-steady LES simulation with constant peak inspiratory flow, was run for 1 s flow time ($t_0 = 1$ s). This initialization procedure allowed for the development of the flow fields, and no pronounced start-up effects were observed, in contrast to the report by Bradshaw et al. [159].

Spatial and Temporal Resolution

Ansys Meshing [256] was employed to create coarse and fine computational meshes. Default software settings were used, but the capture proximity and capture curvature options were enabled. Inflation layers consisting of five layers of pentahedral prismatic cells and a growth ratio of 1.2 were established along all wall surfaces and the closed nostril, and tetrahedral cells were used for the remaining geometry. The coarse and fine meshes used element sizes of 1 and 0.2 mm, respectively, which resulted in 876 thousand

and 44.7 million grid cells. Figure 6 showcases coarse and fine surface meshes. Figure 7 illustrates the computational meshes at selected cross-sections throughout the nasal cavity. For the coarse mesh, the gap between opposing inflation layers was approximately the width of 2–3 grid cells, in the narrowest passages. On the fine mesh, however, even the narrowest passages were finely resolved by many grid cells.

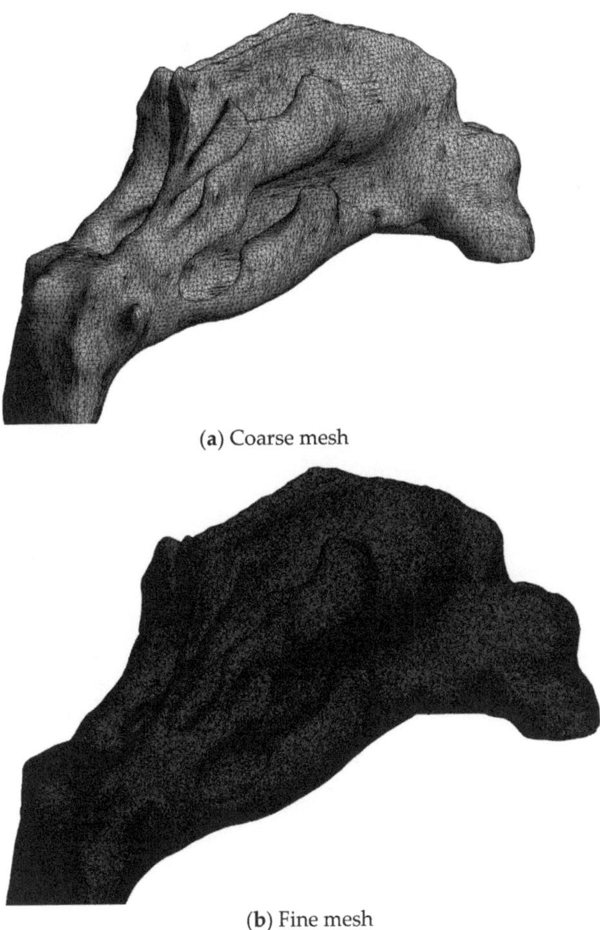

(**a**) Coarse mesh

(**b**) Fine mesh

Figure 6. Surface mesh.

(**a**) Positions of cross-section on which mesh resolution is illustrated.

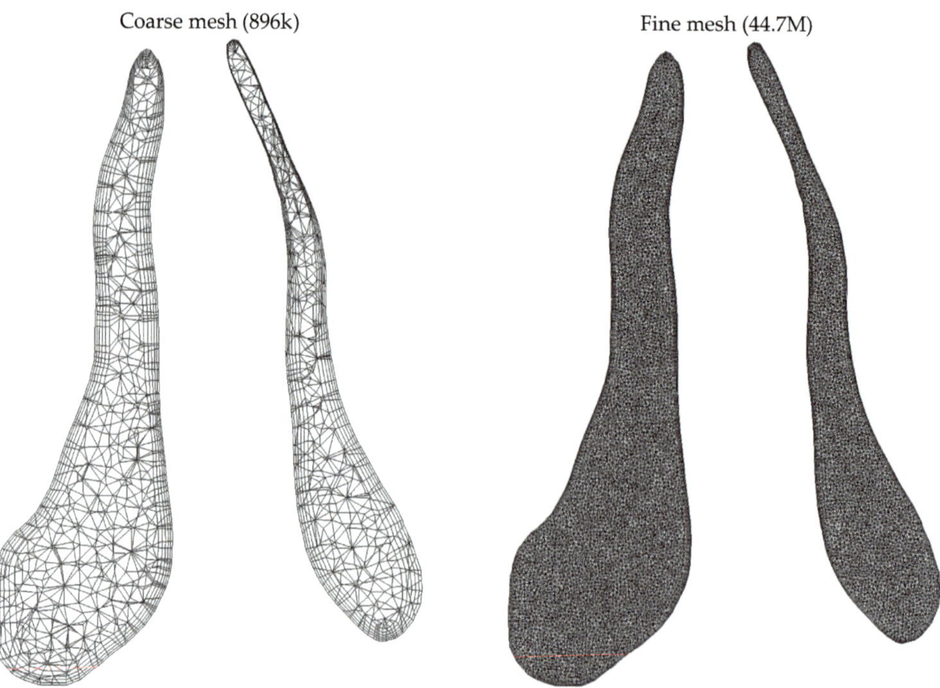

Figure 7. *Cont.*

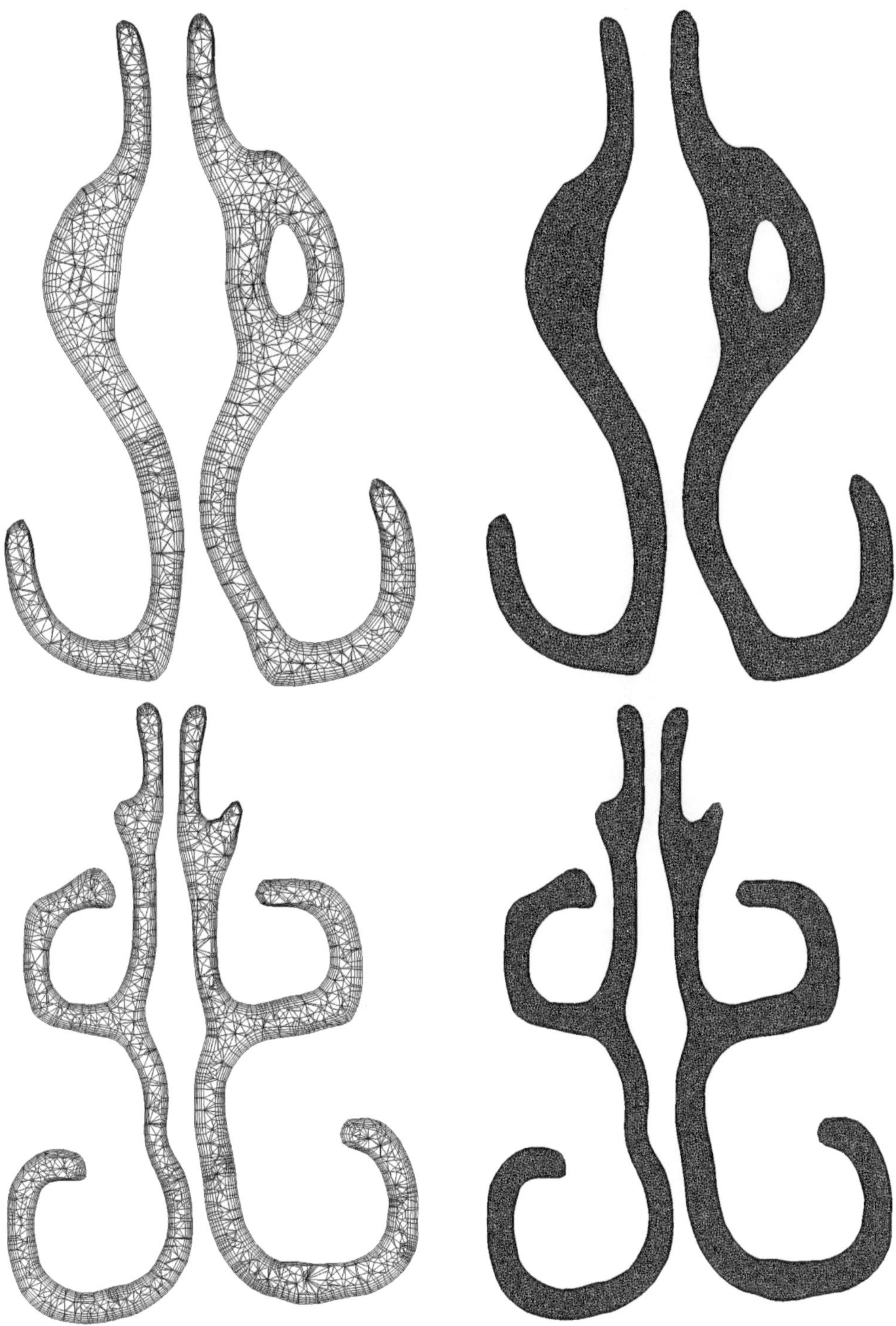

Figure 7. *Cont.*

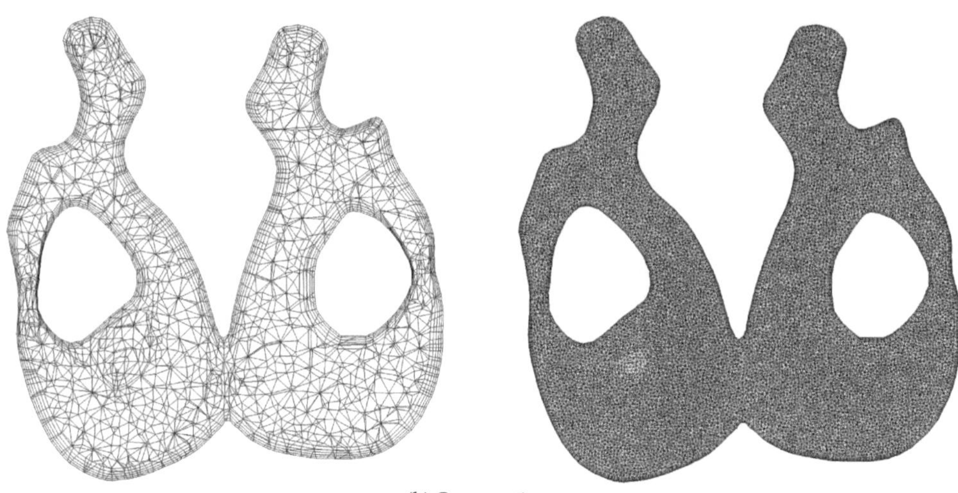

(**b**) Cross-sections

Figure 7. Visualization of the coarse and fine mesh at selected cross-sections through the nasal cavity.

Time steps of 10 microseconds were used for the transient LES simulations.

4.2. Results

The main content of the CFD study presented in this part of the paper was the generation of in silico RMM pressure–flow curves from high-fidelity CFD modelling and their comparison to in silico curves obtained from simpler CFD models as well as patient-specific, clinical in vivo measurements.

Most of the simulation results presented here were obtained from the fine-mesh CFD models described earlier. Unless explicitly stated otherwise, the presented data were obtained from the fine-mesh $k\omega$ SST simulations at peak inspiratory flow. Results from the coarse-mesh steady state $k\varepsilon$ realizable and $k\omega$ SST models are limited to the in silico RMM pressure–flow curves. The coarse-mesh laminar simulations were, in addition, used to compare pressure profiles to the LES simulations. Animations showcasing the transient behavior of selected flow field variables, obtained from the LES simulations, are available in the Supplementary Material.

During the transient simulations, data were systematically saved at regular intervals. For instance, the differential pressure difference between the right and left nostrils, along with pressure and velocity data at selected cross-sections throughout the geometry, was saved at every time step. Additionally, simulation data files containing instantaneous values of all flow variables as well as recorded turbulence statistics, were saved every ten thousandth time step (every tenth second). The entire dataset is accessible in the public domain for non-commercial use [14]. A separate document describing the contents of the dataset is available in the data repository.

Despite capturing turbulence statistics during the transient LES simulations, they are not incorporated or discussed in the current report. This decision was rooted in the continuous variation of mean flowrate in accordance with its sinusoidal boundary condition, making it unclear how turbulence statistics could contribute to the present analysis. Nevertheless, these data are provided in the published dataset, inviting other researchers to scrutinize and engage in discussion regarding its potential applications.

4.2.1. In Silico RMM-Results

In the transient LES-based simulations, three breathing cycles were simulated on each side of the nose. Throughout these simulations, the differential pressure between the right and left nostrils were monitored throughout the simulations. Figure 8 shows the transient behavior of the differential pressure. There were no evident differences between the three breathing cycles on either side. For steady state simulations, repeated simulations with different pharyngeal flowrate conditions were performed.

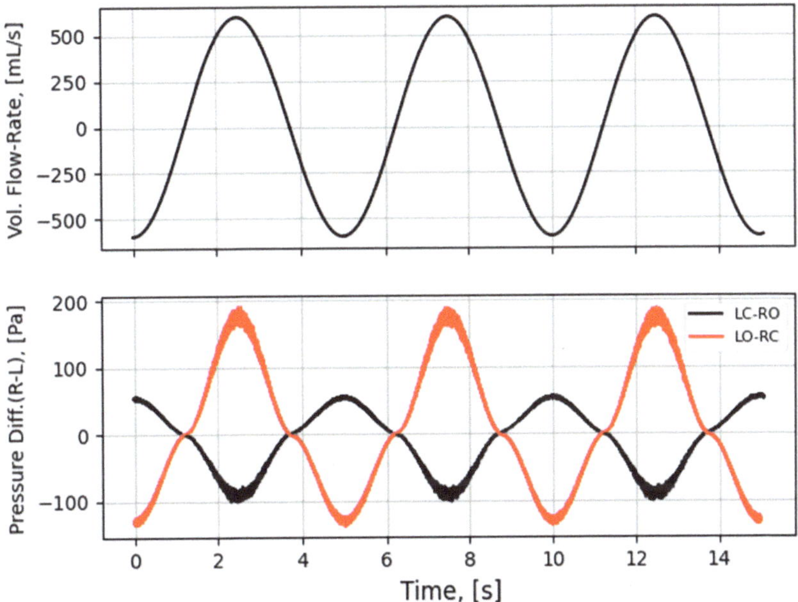

Figure 8. Volumetric flowrate (**top**) and nasal pressure drop (**bottom**) vs. time, in LES-based in silico unilateral AAR. Negative and positive flowrates indicate inhalation and exhalation, respectively. LC/O = left closed/open. RC/O = right closed/open.

In generating in silico RMM pressure–flow curves, the volumetric flowrate was plotted against the differential pressure data. During inhalation, the pressure at the closed nostril is lower than at the open nostril, while during exhalation, the opposite situation occurs. Consequently, the differential pressure between the right and left nostril will be positive during inhalation and negative during exhalation on the right side, and vice versa on the left side. It is noted that in the clinical in vivo RMM pressure–flow curves depicted in Figure 1, both differential pressure and volumetric flowrate are non-negative. In the in silico RMM pressure–flow curves presenting the current simulation data, the volumetric flowrate remains non-negative throughout the respiratory cycle. However, the sign of the differential pressure was adjusted to align with the convention where inhalation corresponds to a positive pressure difference between the open nostril and the nasopharynx, and exhalation corresponds to a negative pressure difference. The resulting in silico RMM pressure–flow curves thus emulate clinical RMM curves, ensuring that pressure–flow curves for inhalatory and exhalatory breathing phases on the left and right sides are appropriately plotted in the corresponding quadrants of the pressure–flow chart. Specifically, pressure–flow curves for the inhalation phase on the right and left sides are represented in the upper and lower right quadrants, respectively. Conversely, pressure–flow curves for the exhalation phase on the right and left sides are plotted in the lower and upper left quadrants, respectively.

Figure 9 illustrates how the various steady state coarse-mesh simulations (laminar, $k\varepsilon$ realizable, and $k\omega$ SST models) produced almost identical RMM pressure–flow curves, on both sides of the nose. These curves closely agree with the fine-mesh $k\omega$ SST model, at peak inspiratory flow. During exhalation, the agreement among the models is somewhat diminished, with the laminar model predicting the highest nasal resistance and the $k\varepsilon$

realizable model predicting the lowest. Figure 10 reveals that the transient fine-mesh LES yields RMM pressure–flow curves nearly identical to those generated by the steady state coarse-mesh laminar simulations. Figure 11 provides a comparison between in silico RMM pressure–flow curves and representative, decongested in vivo RMM pressure–flow curves for the specific patient in this study. It is evident that the CFD simulations severely underpredicted the pressure drop, hence nasal resistance, displaying a discrepancy exceeding one order of magnitude.

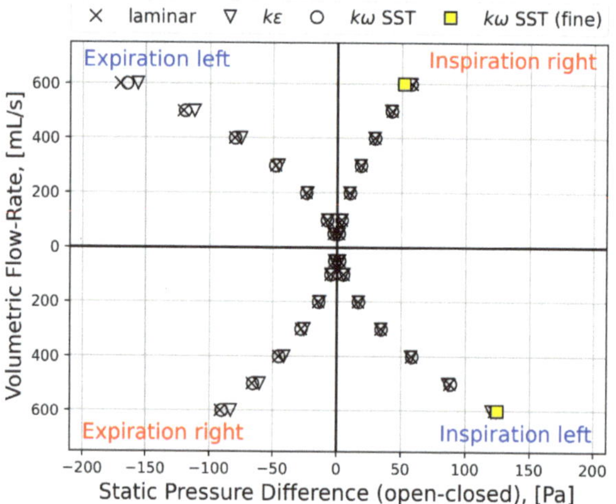

Figure 9. Quasi-steady in silico RMM pressure–flow curves obtained from CFD simulations utilizing steady state laminar (×), *kε* realizable with enhanced wall treatment (∇), and *kω* SST (○) on the coarse mesh, and *kω* SST (▢) on the fine mesh.

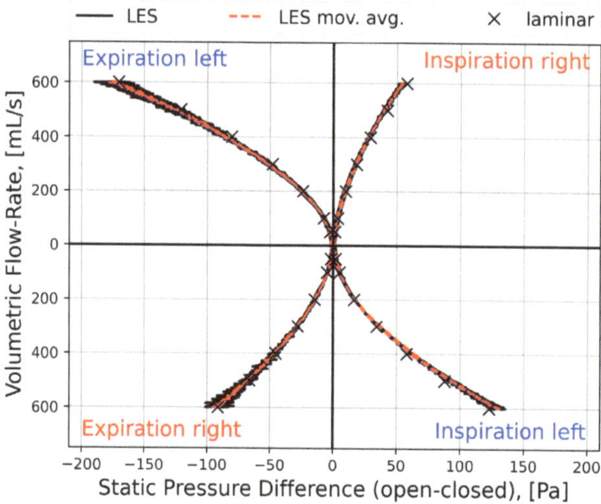

Figure 10. In silico RMM pressure–flow curves obtained from transient LES-based (solid line) and quasi-steady laminar (×) CFD simulations.

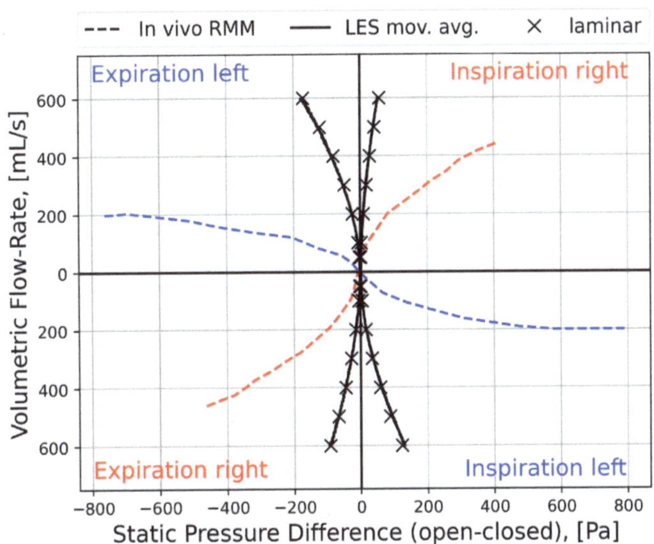

Figure 11. Comparison of in vivo and in silico RMM pressure–flow curves obtained from clinical measurements (dashed line) and moving average of the transient LES-based (solid line) and quasi-steady laminar (×) CFD simulations.

Assuming a defined channel length and curve-fitting Equation (4) to RMM data, representative hydraulic diameters for the nasal cavities can be determined. In Figure 12, best-fit curves and best-fit hydraulic diameters are presented for current in vivo and in silico RMM data utilizing a channel length of $L = 0.1$ m. The nasal resistance on the left side was higher than on the right, resulting in a smaller hydraulic diameter on the left side. The hydraulic diameters that best capture the in silico data are notably higher than those representing the in vivo data. Additionally, a discernible difference between inhalation and exhalation, particularly on the right side, is observed in the in silico data. Examination of the in vivo data reveals that on the left side, the curves flatten out for high pressure drops. The inability of the volumetric flowrate to increase with rising pressure drop indicates a proportional relationship between hydraulic resistance and pressure drop, leading to a constant volumetric flowrate (see Equation (1)). This phenomenon may be attributed to a shrinking cross-sectional area caused by the collapse of the nasal cavity. It is interesting to note that despite the observed effect on the left side, the in vivo RMM curves exhibit symmetry with respect to inhalation and exhalation. This symmetry suggests that nasal compliance occurs due to strong Venturi effect, which remains insensitive to the flow direction.

Figure 13 illustrates hysteresis loop resulting from the by the unsteady flow in the transient LES simulation, as discussed in **Part I** (Section 2.3.3). The hysteresis width introduces a fitting restraint on the channel length and hydraulic diameter, and the channel length L is eliminated. The combination of Equations (4) and (9) results in

$$\Delta P = \frac{W_{\Delta P}}{\omega Q_{\max}} \left[\frac{f_D}{\pi D_h^3} Q^2 + \frac{1}{2} \frac{\partial Q}{\partial t} \right] , \qquad (22)$$

where $W_{\Delta P}$ is the hysteresis width. By applying Equation (21) to describe the unsteady volumetric flowrate and employing the hysteresis widths observed in the LES-based RMM curves, the hydraulic diameter was adjusted to achieve the best-fit curves shown in Figure 13. The curve fitting process was conducted separately for the inspiratory and expiratory phases on both sides. To optimize the fit, the Haaland equation (Equation (7)) for the (turbulent) friction factor was employed across all flowrates. While the best-fit curve accurately represented the hysteresis loop on the right side, the fit on the left side was less precise. The resulting channel lengths derived by inserting the best-fit hydraulic diameter, measured hysteresis width, and air properties into Equation (9) were $L = 3.3$ cm and 2.6 cm for inhalation and exhalation on the left side, respectively. On the right side, the corresponding lengths were $L = 4.2$ cm and 2.9 cm. These numbers are significantly lower than the physical length of the nasal cavity, underscoring a limitation in the comprehensiveness of Equation (22) in capturing essential physical aspects of the intricacies of nasal air flow. Hydraulic diameters derived from curve-fits of the laminar simulation data demonstrated good agreement with those representing the LES hysteresis when similar channel lengths were employed. This consistency suggests that the laminar simulation effectively captures pertinent physics.

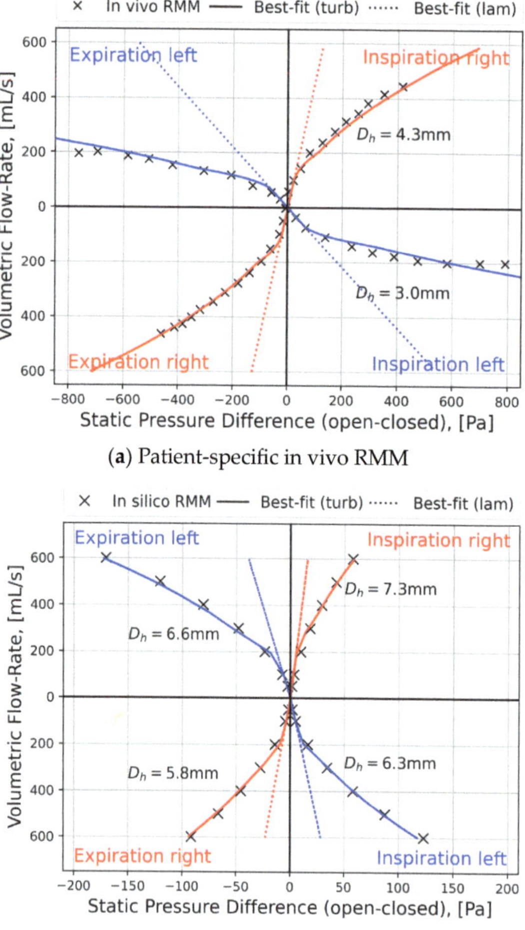

(**a**) Patient-specific in vivo RMM

(**b**) In silico RMM based on the laminar CFD model on the coarse mesh

Figure 12. RMM curves ($\times$) with best-fit (solid) curves obtained by assuming a channel length of $L = 0.1$ m, neglecting the unsteady term, and adjusting the hydraulic diameter, D_h, in Equation (4). Best-fit hydraulic diameters are printed in the figures. For the in vivo data, a single hydraulic diameter was used for in- and exhalation, but for the in silico data, in- and exhalation data were curve-fitted separately. The in vivo RMM curves are representative of patient-specific RMM pressure–flow curves obtained in the decongested state (see Figure 1).

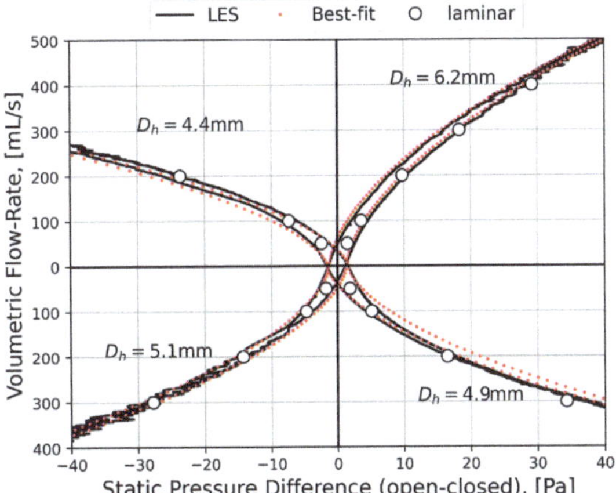

Figure 13. Comparison between LES (solid black), LES best-fit curve (dotted red), and laminar (O) in silico RMM simulation results.

4.2.2. Spatial and Temporal Resolution

Based on peak inspiratory and expiratory flowrates, the coarse-mesh simulations yielded wall y^+ values below 5, almost everywhere. The fine-mesh simulations yielded wall y^+ values below 1. Hence, the near wall grid cells were situated in the viscous sublayer almost everywhere, for all flowrates (see Equation (16)), in all the simulations. Wall y^+ contour plots obtained from the fine-mesh $k\omega$ SST model are shown in Figures 14 and 15 for the LO-RC and LC-RO peak inspiratory flow simulations, respectively.

Due to the heightened velocities resulting from the narrower nasal cavity, y^+ values on the left side were slightly higher than on the right side. The narrowest sections exhibited the highest wall shear stress, hence y^+ values, due to the relatively elevated flow velocities in these regions.

The subgrid-scale turbulent viscosity in the WALE LES model, as implemented in ANSYS Fluent and used in the current simulations, is proportional to the square of the mixing length for subgrid scales [113],

$$l_{LES} = \min(\kappa y, 0.325\Delta) , \tag{23}$$

where $\kappa = 0.4187$ is the von Kármán constant, y represents the distance to the nearest wall, κy is thus the length scale of the largest eddies, and Δ is the cube root of the grid cell volume. On the fine mesh, it was only in the two first layers of prismatic grid cells along the walls that $\Delta > \kappa y/0.325$. Thus, the subgrid-scale turbulent viscosity was predominantly determined by the computational grid cell size. The two first layers of near-wall grid cells were inside the viscous sublayer for all flowrates, which aligns with the capabilities of the WALE LES model to correctly represent the behavior of wall-bounded flows [113]. The Kolmogorov length and time scales, representing the smallest scales of turbulence, can be expressed as

$$\eta = \left(v^3/\varepsilon\right)^{1/4} , \tag{24}$$

and

$$\tau_\eta = (v/\varepsilon)^{1/2} , \tag{25}$$

respectively, where v is the kinematic viscosity, $\varepsilon = -dk/dt$ is the turbulent dissipation rate, and k is the turbulent kinetic energy [109]. Utilizing results from the fine-mesh $k\omega$ SST simulations at peak inspiratory flow, the Kolmogorov microscales were computed from

Equations (24) and (25). Contour plots at selected cross-sections are depicted in Figure 16 for LO-RC and LC-RO simulations.

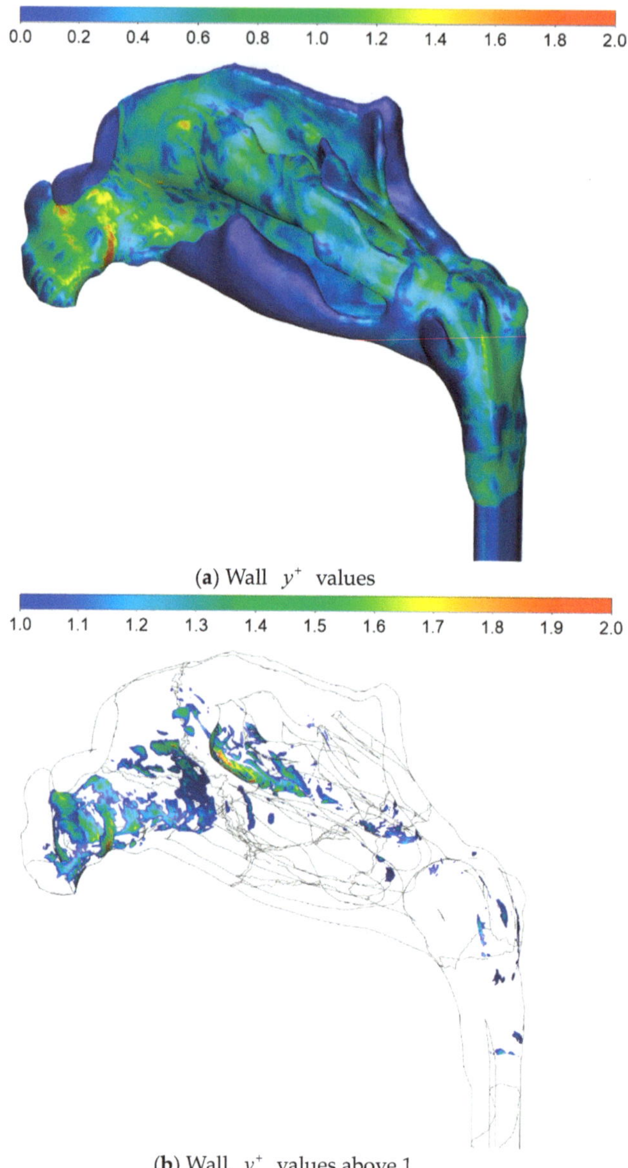

(**a**) Wall y^+ values

(**b**) Wall y^+ values above 1

Figure 14. Estimated wall y^+ values based on the LO-RC steady state fine-mesh $k\omega$ SST simulation of peak inspiratory flow (see Equation (15)).

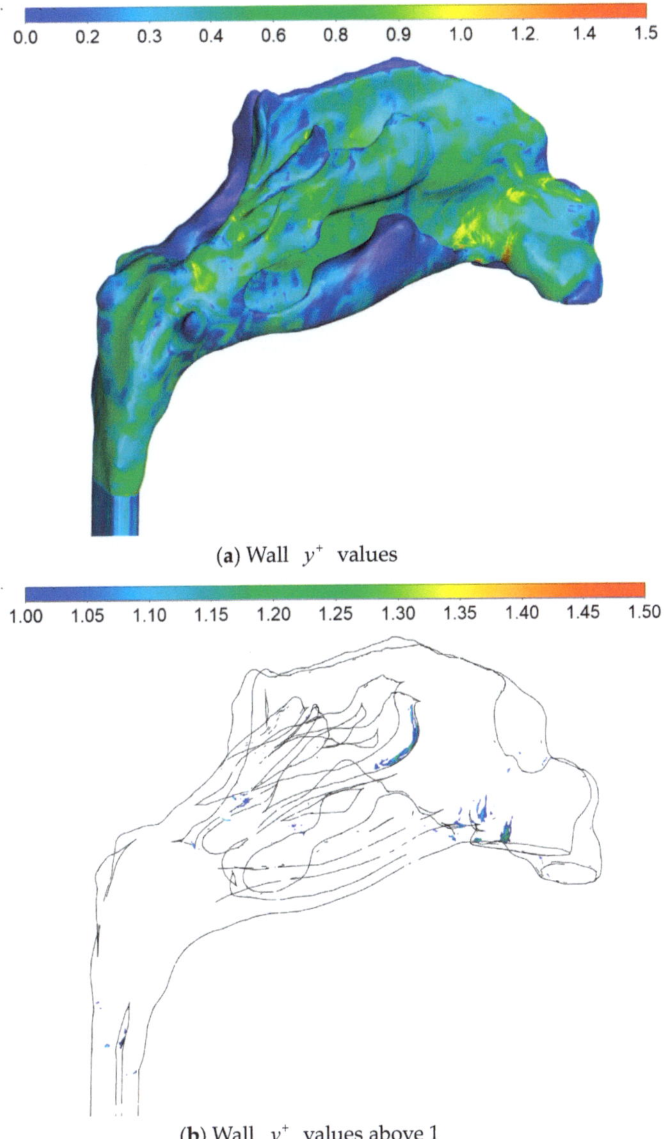

(**a**) Wall y^+ values

(**b**) Wall y^+ values above 1

Figure 15. Estimated wall y^+ values based on the LC-RO steady state fine-mesh $k\omega$ SST simulation of peak inspiratory flow (see Equation (15)).

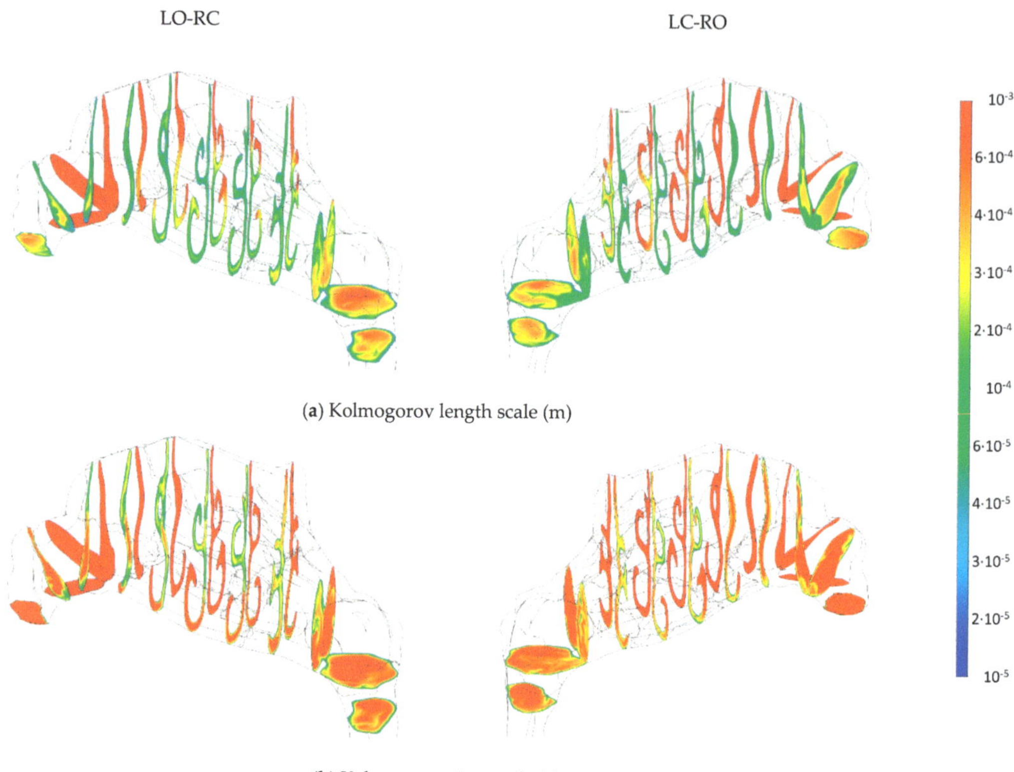

Figure 16. Kolmogorov micro scales: (**a**) length scale; and (**b**) time scale computed from the LO-RC and LC-RO fine-mesh $k\omega$ SST peak inspiratory flow simulations.

To assess the spatial and temporal resolution of the fine-mesh simulations, local subgrid scale mixing lengths determined from Equation (23) and time step size were compared to the Kolmogorov length and time scales determined by Equations (24) and (25), respectively. The turbulent dissipation rates needed to compute the Kolmogorov microscales were obtained from the fine-mesh $k\omega$ SST simulations of peak inspiratory flow. As the turbulent dissipation rate varies with location and flow velocity, the Kolmogorov microscales also exhibit corresponding variations. The contour plots in Figure 17 depict the ratio of (a) the LES and Kolmogorov length scales, l_{LES}/η; and (b) the time step size and the Kolmogorov time scale, $\Delta t/\tau_\eta$, at selected cross sections for LO-RC and LC-RO simulations. Figure 18 presents the distributions of length and time scale ratios through histogram plots with a bin size of 0.1. It is evident that the vast majority of grid cells exhibited subgrid scale mixing lengths smaller than the local Kolmogorov length scale ($l_{LES}/\eta < 1$) and had time steps shorter than the Kolmogorov time scale ($\Delta t/\tau_\eta < 1$). However, the LO-RC simulation generally demonstrated higher ratios compared to the LC-RO simulation, attributed to the overall higher flow velocities in the former. In the LO-RC and LC-RO simulations, 1.2% and 0.2% of the grid cells, respectively, featured subgrid scale mixing lengths exceeding the Kolmogorov length scale. Only the LO-RC simulation exhibited a minor fraction of grid cells (0.01%) where the Kolmogorov time scale was smaller than the time step size. Grid cells with elevated length scale ratios were generally situated within the wall boundary layers, albeit not directly at the wall, in regions characterized by relatively high wall shear stress. The grid cells with the highest time scale ratios were found at the anterior of the middle meatus.

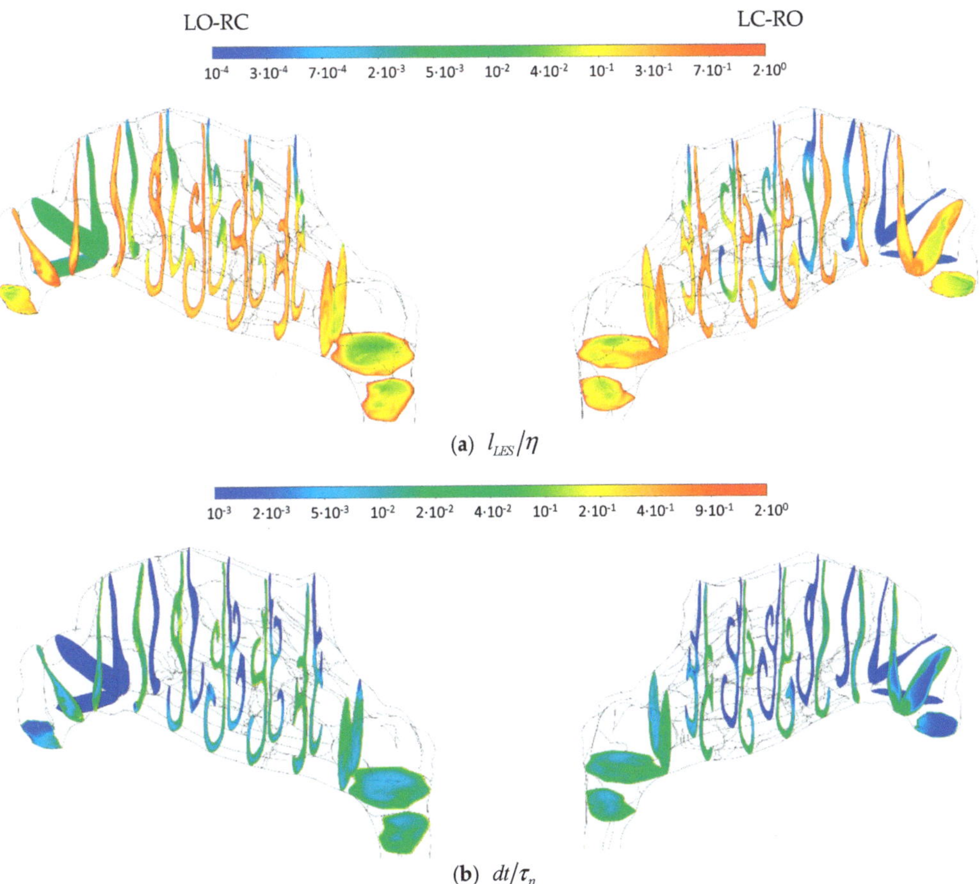

Figure 17. Contour plots showing the ratios of (**a**) LES simulations mixing lengths (Equation (23)) to the Kolmogorov length scales (Equation (24)); (**b**) LES simulations time step size (10 μs) to Kolmogorov time scales (Equation (25)), based on the LO-RC and LC-RO fine-mesh $k\omega$ SST peak inspiratory flow simulations.

The subgrid-scale turbulent viscosity ratio was close to zero almost everywhere in the nasal cavity, as shown in Figure 19. In the nasopharynx, however, the turbulent viscosity ratio is substantially higher. The grid cells displaying the poorest spatial and temporal resolution were in concordance with those featuring the highest subgrid scale turbulent viscosity.

CFL numbers (see Equation (17)), as calculated by ANSYS Fluent, were obtained from the fine-mesh $k\omega$ SST simulations of peak inspiratory flow. Contour plots presented in Figure 20 reveal values below 2 on the right side and below 4 on the left side. Hence, the chosen time steps exceeded the recommended range for LES (*CFL* ~ 1). It has, however, been shown that *CFL* ~ 5 can work in some cases [111]. Given the time-varying flowrates during the respiratory cycle, the *CFL* number varied periodically between 0 and these maximum values. The implicit formulation of the CFD simulation is robust with respect to high *CFL* numbers, and no issues regarding numerical stability were experienced.

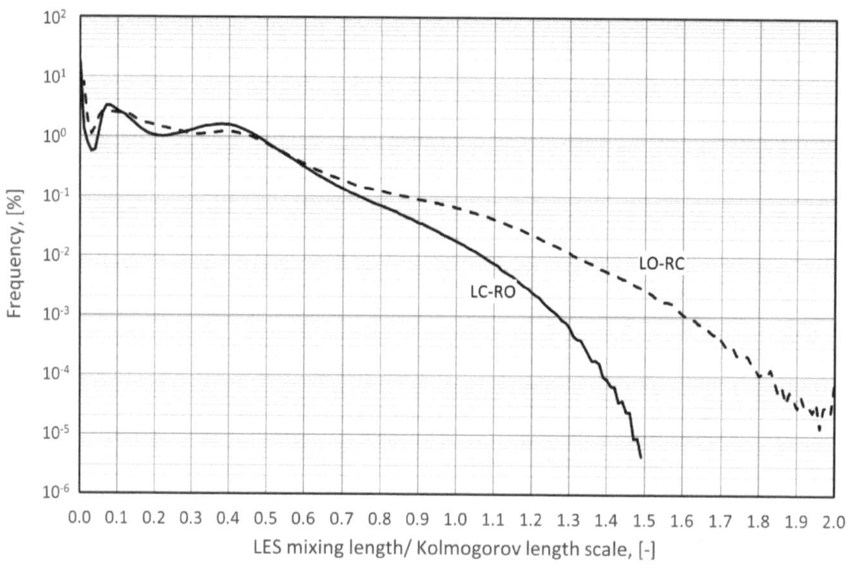

(**a**) l_{LES}/η-histogram

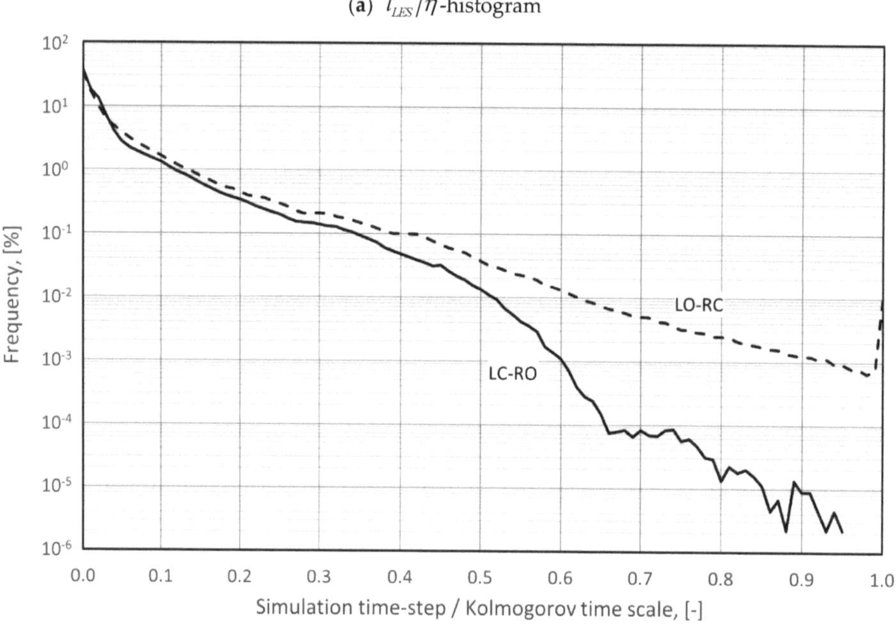

(**b**) dt/τ_η-histogram

Figure 18. Histograms illustrating the distribution of ratios of (**a**) LES simulations mixing lengths (Equation (23)) to the Kolmogorov length scales (Equation (24)); (**b**) LES simulations time step size (10 µs) to Kolmogorov time scales (Equation (25)), based on the LO-RC and LC-RO fine-mesh $k\omega$ SST peak inspiratory flow simulations. Bin sizes are 0.01.

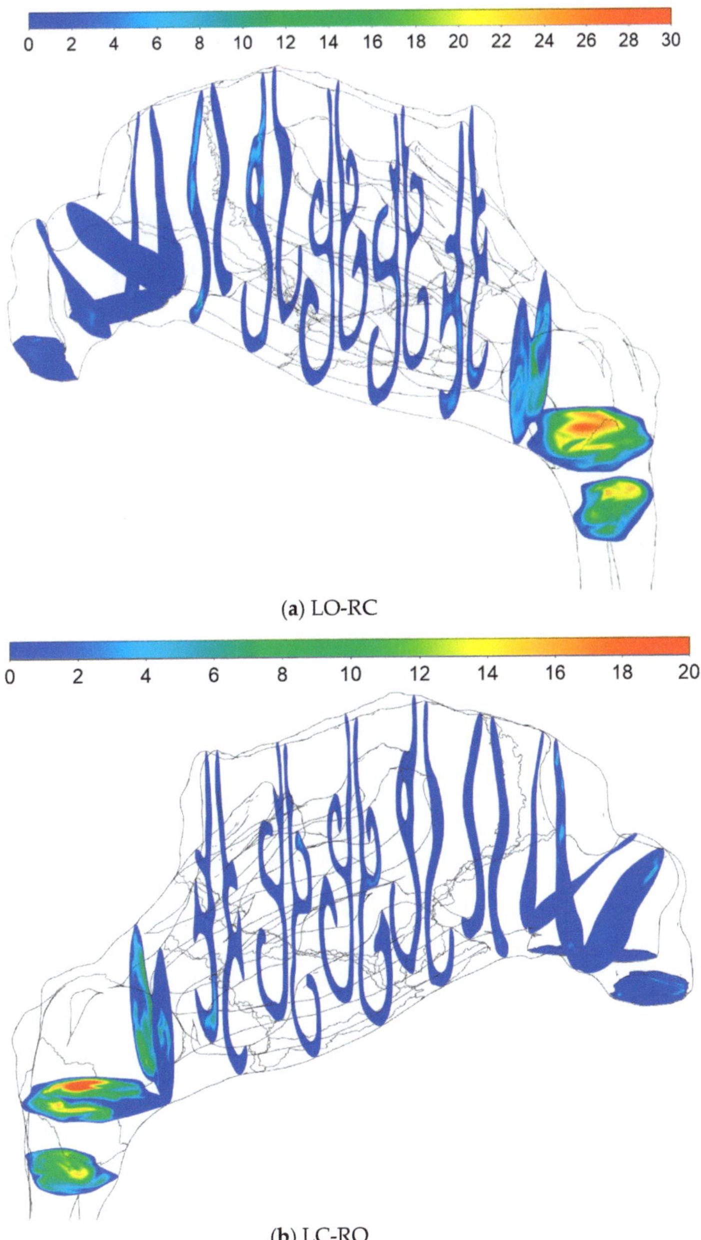

(**a**) LO-RC

(**b**) LC-RO

Figure 19. Estimated turbulent viscosity ratio values based on the fine-mesh $k\omega$ SST peak inspiratory flow simulations.

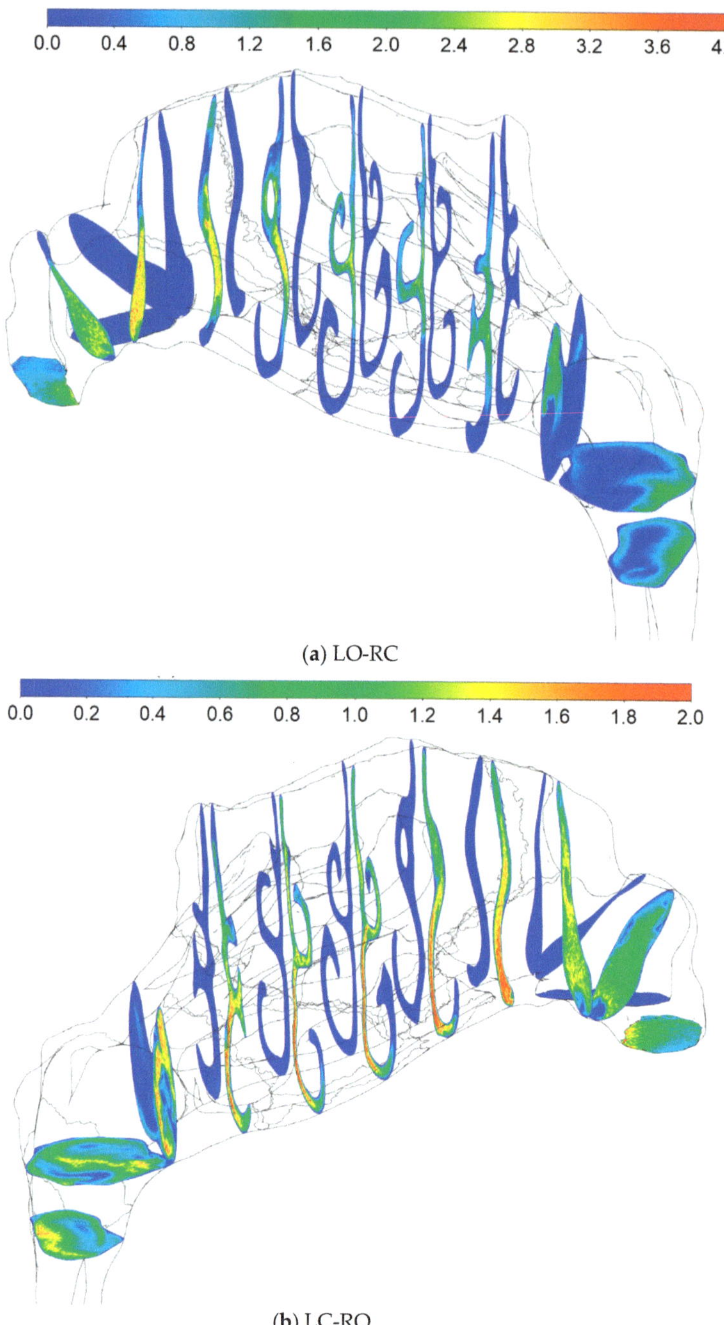

(**a**) LO-RC

(**b**) LC-RO

Figure 20. *CFL* number contour plots at selected cross-sections based on the fine-mesh $k\omega$ SST peak inspiratory flow simulations and time step size of $\Delta t = 10\ \mu s$ (see Equation (17)).

4.2.3. Flow Velocity Fields

In Figure 21, flow velocity contours on selected cross sections for the LO-RC and LC-RO fine-mesh $k\omega$ SST simulations of peak inspiratory flow are presented. Notable variability is evident within each cross-section, reflecting a complex flow pattern. The lower part of the nasal cavity, adjacent to the inferior turbinate, emerges as the preferred flow path. Qualitatively, it was noted that the favored flow path on the left side was slightly higher than on the right side, closer to the middle turbinate.

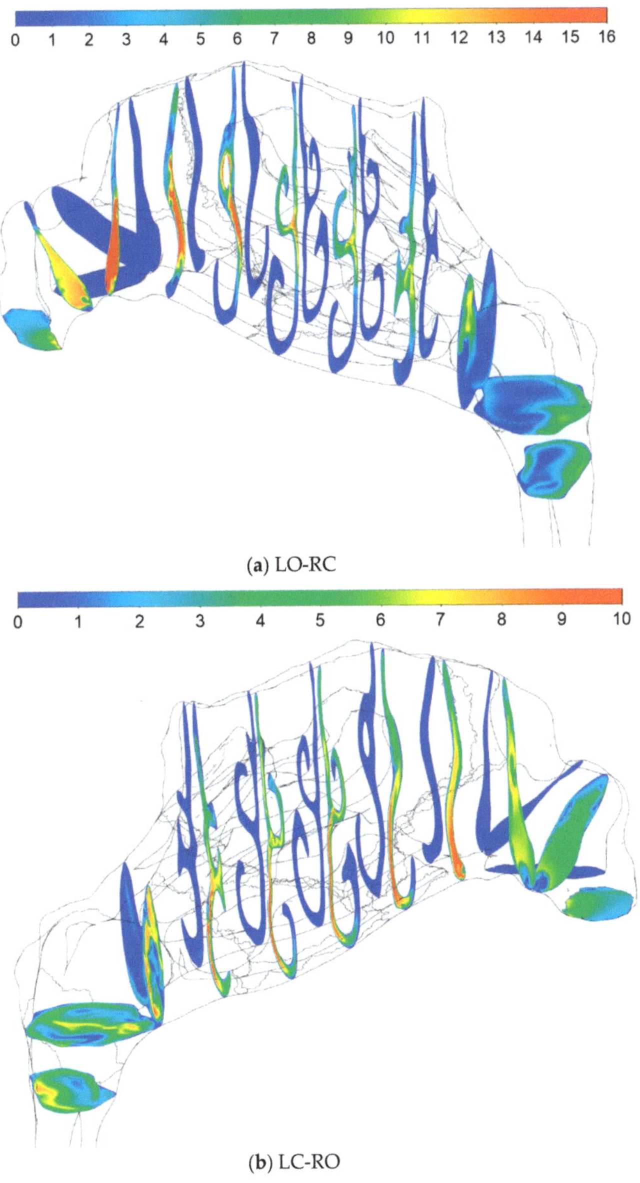

(**a**) LO-RC

(**b**) LC-RO

Figure 21. Flow velocity magnitude (m/s) contour plots at selected cross-sections for (**a**) LO-RC and (**b**) LC-RO, fine-mesh $k\omega$ SST peak inspiratory flow simulations.

Figure 22 displays velocity contours and vectors on the sagittal symmetry plane, illustrating distinct flow patterns in the nasopharynx at peak inspiratory and expiratory flow. During inspiration, significant mixing occurs, dominated by relatively large unsteady vortical structures. During expiration, however, the nasopharyngeal flow is dominated by the nasopharyngeal jet, as highlighted by Bradshaw et al. [159]. Airflow velocities were generally low enough that the assumption of incompressible flow was valid.

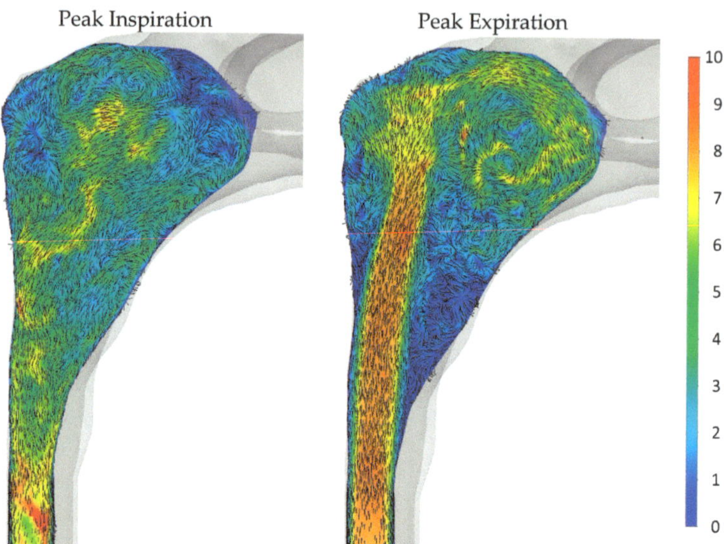

Figure 22. Nasopharyngeal velocity contours (m/s) and vectors drawn at the sagittal symmetry plane for peak inspiratory and expiratory flow for the transient LES-based LC-RO simulation on the fine mesh.

Using unilateral hydraulic diameters from the open side of the cross-sections in Figure 23, Reynolds numbers based on the peak flowrate (600 mL/s) were computed (see Equation (6)). Figure 24 depicts these results for unilateral flow on the left and right sides, respectively. In the anterior part of the nose (nasal gateway), the Reynolds number is notably higher on the left side, whereas in the posterior nasal cavity, it is slightly higher on the right side. Throughout the nasal cavity, the estimated peak Reynolds numbers are in the range that would be expected to be turbulent for fully developed flow in straight channels (>4000).

In Figures 25 and 26, Q-criterion isosurfaces colored by velocity provide a visual representation of vortical structure generation in the transient LES simulations. Notably, during exhalation, there is minimal production of vortexes in the straight pharyngeal inlet section. Conversely, the nasal cavity exhibits substantial vortex production and mixing. During inhalation, the vortexes produced in the nasal cavity are transported downstream through the pharyngeal section. Intriguingly, some vortexes penetrate deeply into the closed nasal cavity. Although their velocities are low, vortexes facilitate the movement of air toward the olfactory region at the top of the nasal cavity, where the sense of smell is located.

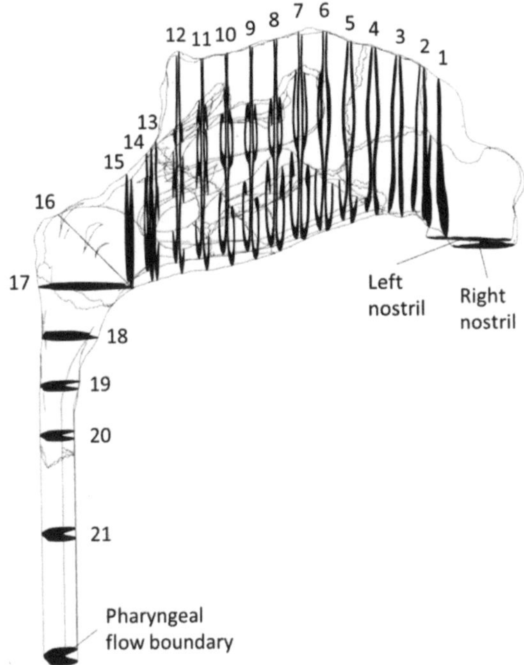

Figure 23. Illustration of cross section positions used in subsequent figures. The cross-sections 1–13 were positioned at regular intervals of 0.5 mm. The two nasal cavities merge into the nasopharynx between cross-sections 13 and 14. Nasopharynx is thus represented by cross-sections 14–19. The cross-sections 20 and 21 are identical in shape and size to the pharyngeal outflow boundary.

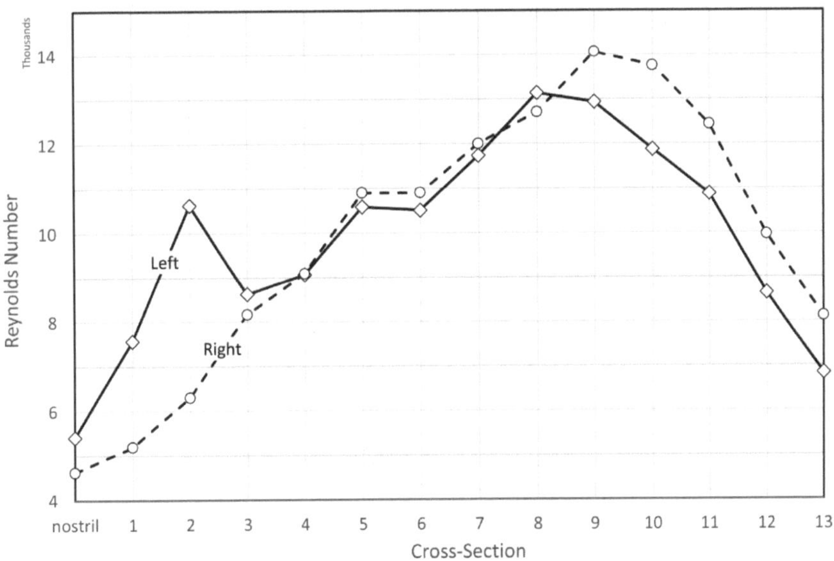

Figure 24. Estimated Reynolds number based on unilateral peak flow (600 mL/s) and the unilateral hydraulic diameter of the open side (Equation (6)). Cross-section numbers refer to positions given in Figure 23.

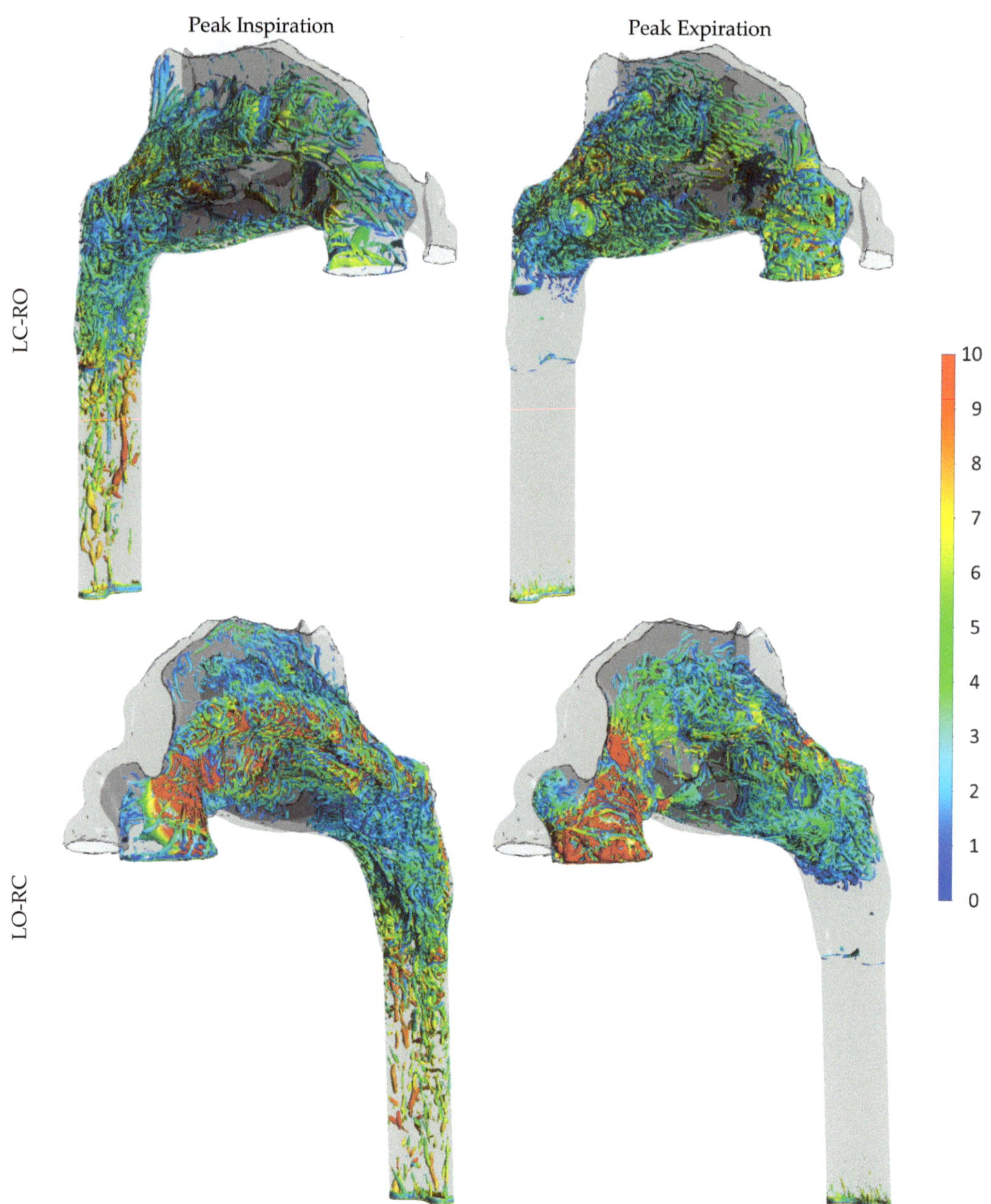

Figure 25. Q-criterion iso surfaces (value of 3×10^6 L/s^2) viewed from the side (sagittal-coronal view) and colored by velocity magnitude (m/s), at peak inspiration (**left**) and expiration (**right**) during the third respiratory cycle in in silico unilateral rhinomanometry, based on transient LES simulations on the fine mesh.

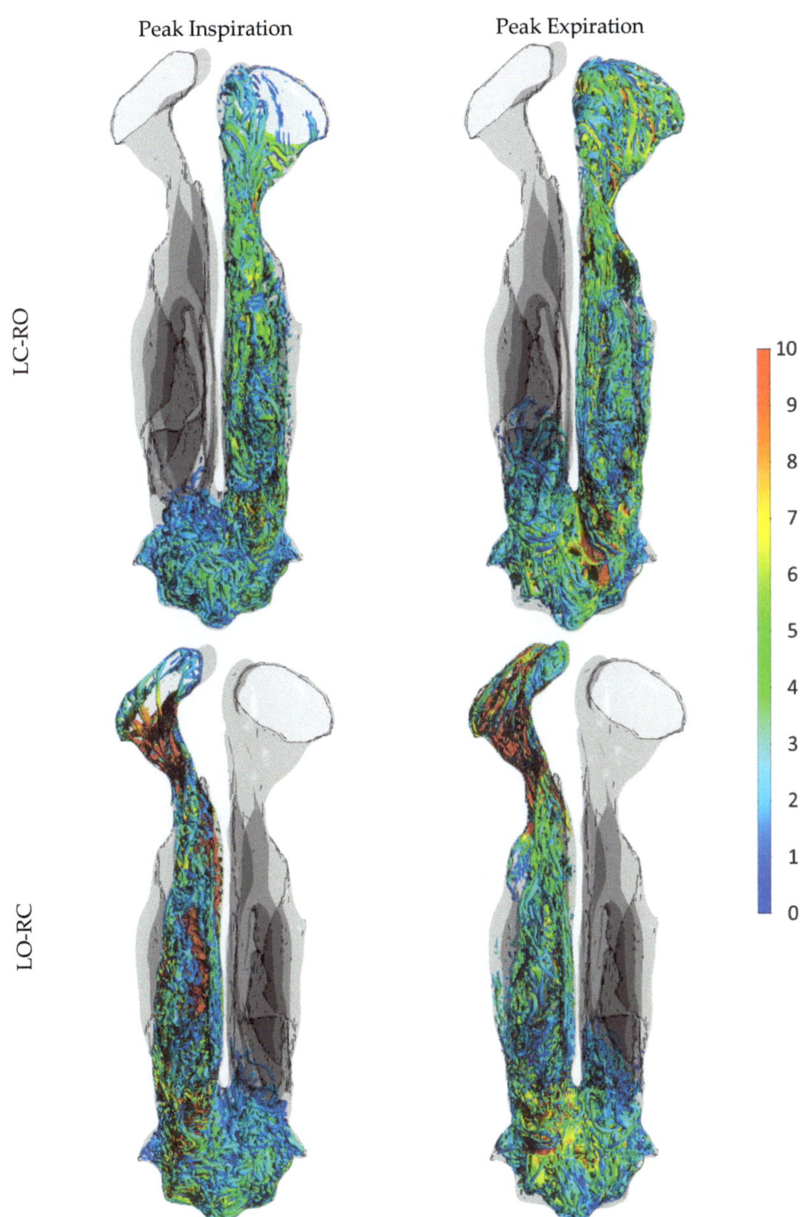

Figure 26. Q-criterion iso surfaces (value of $3 \cdot 10^6$ L/s^2) viewed from above (axial view) and colored by velocity magnitude, at peak inspiration (**left**) and expiration (**right**) during the third respiratory cycle in in silico unilateral rhinomanometry, based on transient LES simulations on the fine mesh.

4.2.4. Stress Fields

Figure 27 displays static pressure contours on selected cross sections, for the LO-RC and LC-RO fine-mesh $k\omega$ SST simulations of peak inspiratory flow. In Figures 28 and 29, the distributions of normal stress (static pressure) and wall shear stress on the nasal cavity wall are presented. The most notable pressure drops occur at the nasal gateways, coinciding with the smallest cross-sectional areas. These regions also exhibit the highest wall shear stresses, consistent with the identified y^+ hot spots in Figures 14 and 15, due to the elevated flow velocity. The significantly higher pressure drop observed in the left nasal gateway, compared to the right, is attributed to the narrower passage on the left due to the deviating septum (see Figure 7).

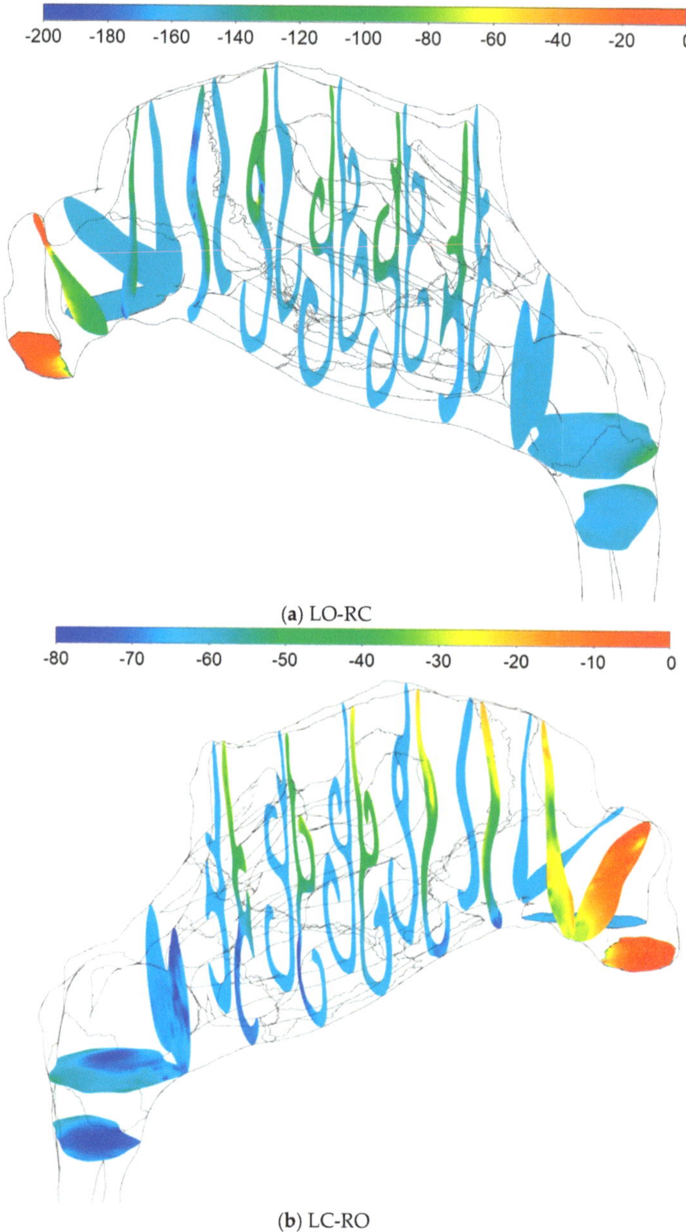

(a) LO-RC

(b) LC-RO

Figure 27. Static pressure contour plots (Pa) at selected cross-sections for (**a**) LO-RC and (**b**) LC-RO, steady state fine-mesh $k\omega$ SST simulation of peak inspiratory flow.

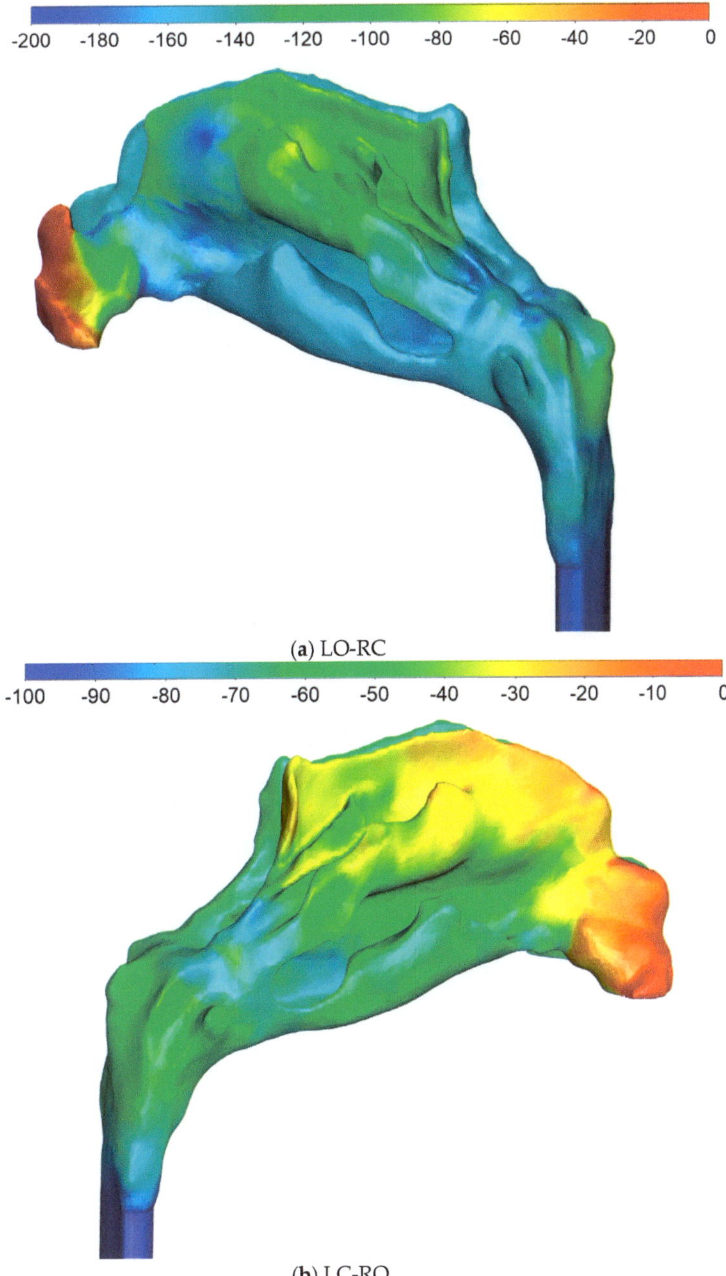

Figure 28. Static pressure contour plots (Pa) at the nasal cavity wall for (**a**) LO-RC and (**b**) LC-RO, steady state fine-mesh $k\omega$ SST simulation of peak inspiratory flow.

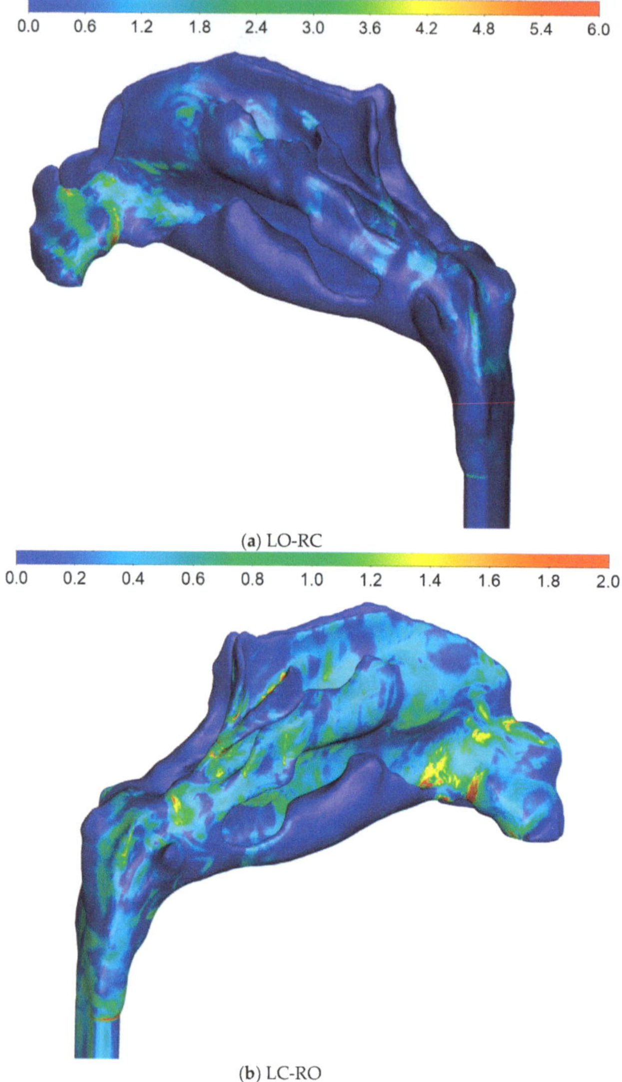

Figure 29. Wall shear stress contour plots (Pa) at the nasal cavity wall for (**a**) LO-RC and (**b**) LC-RO, steady state fine-mesh $k\omega$ SST simulation of peak inspiratory flow.

Figures 30 and 31 show the area-averaged static and total pressures at the cross-sections indicated in Figure 23 for peak inspiratory and expiratory flow in the fine-mesh LES and coarse-mesh laminar simulations, respectively. The averaging was exclusively based on the open nasal cavity. The figures reveal a generally good agreement between the laminar and LES-based models. Additionally, two notable observations emerge from these curves:

1. Comparison of the total pressure development along the length of the nasal cavity indicates varying pressure losses and local flow resistance in different parts of the nose, depending on the flow direction. This suggests that the total unilateral nasal resistance, derived from the integral of the local resistances along the nasal cavity passage, may exhibit dependence on the flow direction.

2. During inhalation, the stagnation pressure at the occluded nostril corresponds well to the static pressure in the nasopharynx for both sides of the nose. Conversely, during exhalation, a closer correspondence is observed between the stagnation pressure at the occluded nostril and the total pressure in the nasopharynx.

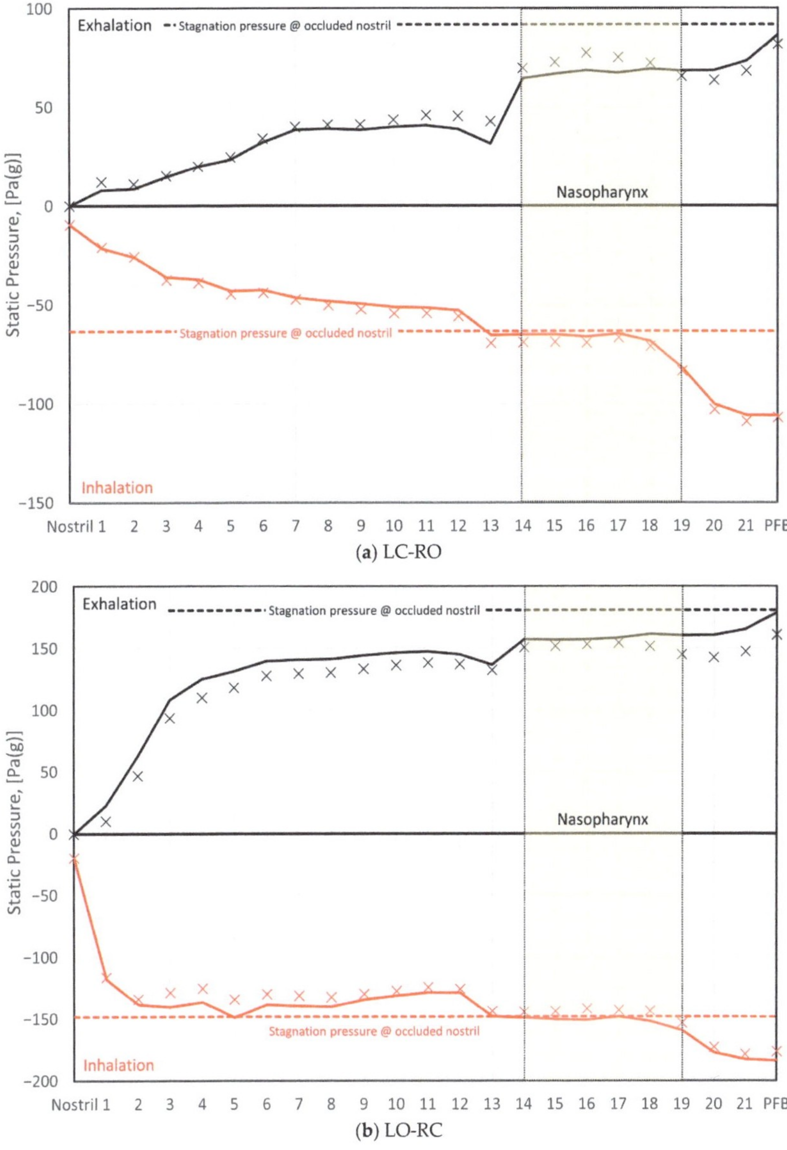

Figure 30. Area-averaged static pressures computed at cross-sections indicated in Figure 23. The data were obtained from fine-mesh LES (solid line) and coarse-mesh laminar (X) simulations, at peak expiratory (black) and inspiratory (red) unilateral flow. The dashed, horizontal lines indicate the area-averaged static pressure at the occluded nostril, obtained from the LES simulation. The hatched area highlights the nasopharyngeal cross-sections.

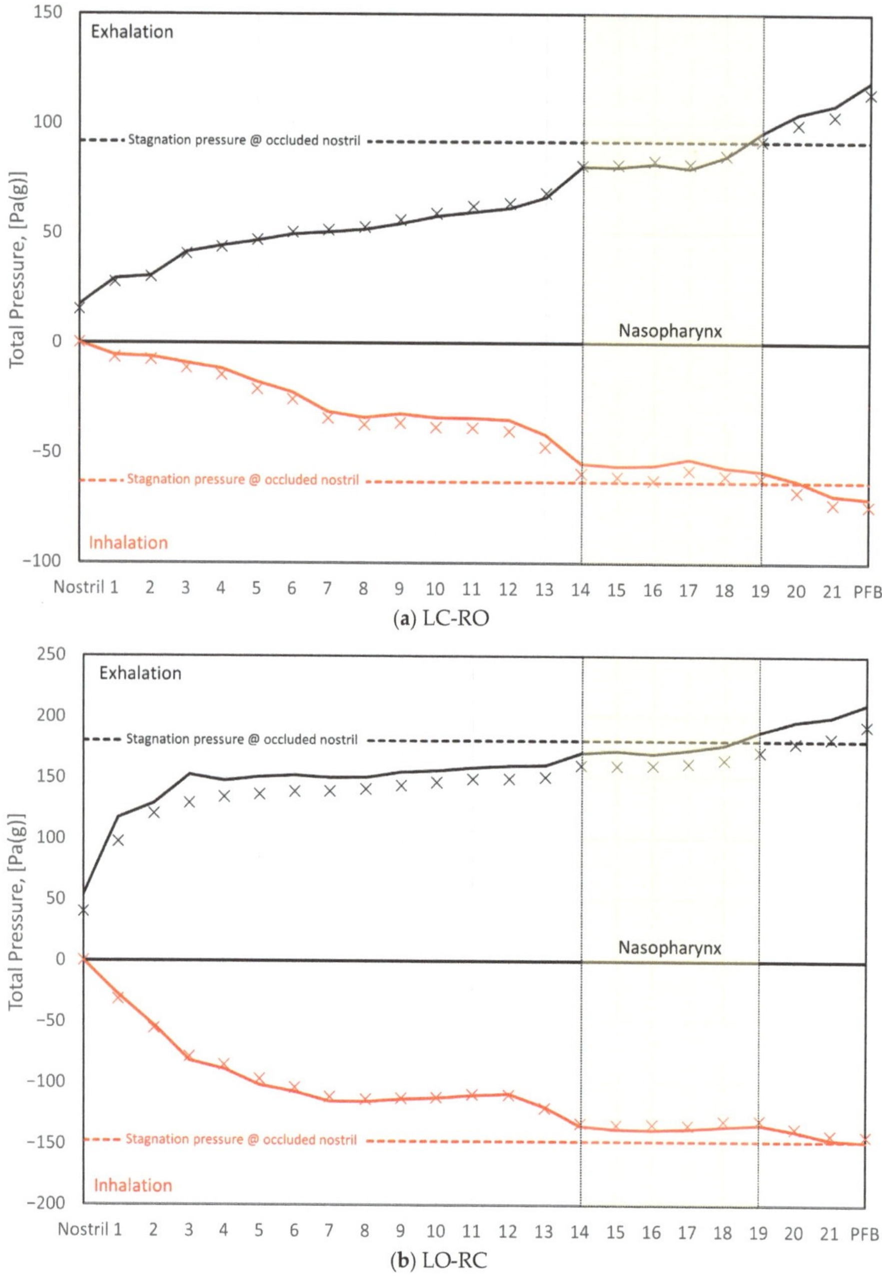

Figure 31. Area-averaged total pressures computed at cross-sections indicated in Figure 23. The data were obtained from fine-mesh LES (solid line) and coarse-mesh laminar (X) simulations, at peak expiratory (black) and inspiratory (red) unilateral flow. The dashed horizontal lines indicate the area-averaged static pressure at the occluded nostril, obtained from the LES simulation. The hatched area highlights the nasopharyngeal cross-sections.

4.3. Discussion

The simulation of nasal airflow using CFD poses a challenging task owing to the intricate geometry of the nasal cavity, the dynamic nature of respiratory flow, and limited availability of validation measurements. Despite the growing popularity of CFD in computational rhinology, there remains a shortage of high-fidelity studies assessing the validity of prevalent modeling approaches, such as steady state laminar flow or RANS turbulence modeling, in nasal airflow.

This paper presents highly detailed nasal airflow simulations, representing some of the most comprehensive work published in this field to date. The aim is to contribute significantly to ongoing debates on optimal simulation set-up. Specifically, focus has been on establishing a benchmark for evaluating laminar and RANS-based models, addressing observed gross discrepancies between in vivo and in silico RMM. A key finding is the minimal variation between in silico RMM pressure–flow curves resulting from different turbulence models, including coarse-mesh steady state laminar, $k\varepsilon$ realizable, and $k\omega$ SST and fine-mesh steady state $k\omega$ SST and transient LES. This challenges this study's initial hypotheses that the disagreement between in vivo and in silico RMM may be explained by poor choice of turbulence model, inadequate spatial/temporal resolution, or unsteady effects.

The following discussion delves into detailed considerations for each element of the initial hypothesis, followed by exploration of factors related to the nasal cavity geometry model. The section is concluded with some final remarks regarding the patient-specific RMM pressure–flow curves and the curve-fitting of the Bernoulli equation (Equation (4)). The discussion should be understood in the context of the broader discussion provided in **Part I** (Section 2.5).

4.3.1. Turbulence Modelling

While the assumption of laminar flow has traditionally dominated CFD modelling of nasal airflow (refer to **Part II**, Section 3), a consensus on whether its characteristic behavior is laminar, transitional, or turbulent remains elusive [83–85]. Published comparisons between laminar and turbulent modelling approaches have typically employed steady state RANS-based turbulence models relying on the Boussinesq hypothesis. A big disadvantage with this approach is its assumption that turbulent viscosity is an isotropic scalar quantity, which is not generally true. More sophisticated modelling approaches that account for transient phenomena or employ turbulence models such as Reynolds stress models (RSMs), large eddy simulation (LES), or direct numerical simulation (DNS) tend to be computationally expensive. Few publications have investigated the effects of imposing unsteady flow boundary conditions, such as respiratory tidal flow, or applied these turbulence models to simulate nasal airflow. In addition to the present study, recent investigations utilizing finely resolved LES models with time-varying flow have been conducted by Lu et al. [248], Calmet et al. [249,250], Bradshaw et al. [159], and Hebbink et al. [127]. Other publications have also utilized high-fidelity CFD models under steady flow conditions; refer to Table 1 for an overview.

The present paper's findings, asserting the similarity between laminar CFD models and more complex turbulence modeling approaches in predicting gross flow features, find support in several other studies. Li et al. [83] demonstrated that their laminar model exhibited good agreement with LES and DNS results at a volumetric flowrate of 180 mL/s. Calmet et al. [250] similarly concluded that a laminar flow model accurately predicted gross flow features, albeit with some underestimation of local flow fluctuations and turbulence intensity. Calmet et al. [250] and Bradshaw et al. [159] observed transitional and dominantly laminar nasal airflow, respectively. In the current study, the unilateral pressure drop exhibited low sensitivity to the choice of computational mesh or turbulence modelling approach on both sides of the nose across a wide range of volumetric flowrates.

Considering the calculated Reynolds numbers for the current patient-specific model alone, turbulent flow would be anticipated in both nasal cavities at peak flowrates (refer to Figure 24). However, when factoring in the respiratory frequency of $f = 0.2$ Hz, the kinematic viscosity of air, and a hydraulic diameter of less than 1 cm into Equation (14), it becomes evident that the Womersley number for the current simulations was of the order of 1. This falls within the low range, as discussed by Xu et al. [168], suggesting a delayed transition to turbulence. These findings, coupled with the simulation results indicating near-laminar flow, illustrate that the critical Reynolds number for turbulent pipe flow is not a reliable indicator for selecting an appropriate turbulence model for respiratory nasal airflow.

In wall-bounded flows, there is a risk of underestimating turbulence intensity in RANS- and LES-based CFD models if the wall boundary layer is not treated appropriately. The nasal cavity is characterized by intricate flow cross-sections featuring high aspect ratios between the axial, spanwise, and wall-normal length scales. Hence, geometrical complexity alone makes it challenging to achieve a computational mesh that satisfies the wall requirements of conventional wall-function-based turbulence models. For varying volumetric flowrates, the challenge amplifies. It is, therefore, recommended to employ a computational grid that resolves the viscous sublayer and utilize y^+-insensitive turbulence models capable of accurate representation of the wall shear stresses. In the present study, these criteria were met for all turbulence models employed, indicating the attainment of accurate and reliable simulation results. A discussion of spatial and temporal resolution is provided below.

Lastly, it is acknowledged that turbulence requires initiation. This can occur through boundary effects such as wall roughness, turbulent fluctuations, e.g., at inlets, geometric irregularities such as bends, edges, or steps, or flow obstacles such as nasal hair. If crucial turbulence triggers are missing from the model, predicted laminar-like flow might not accurately reflect the physiological situation. As shown in Figure 25, during exhalation, little vortical flow existed in the pharyngeal inlet section. The vortices generated at the inlet dissipated before reaching the nasopharynx. Nevertheless, the significant vortex production in the nasopharynx and nasal cavities implies that there were sufficient triggers present to generate turbulence in the nasal cavity.

4.3.2. Spatial and Temporal Resolution

The wall y^+-insensitive RANS models provided in ANSYS Fluent are expected to perform well in the wall y^+ ranges seen in the current simulations, both on the coarse and fine meshes. Using these RANS models avoids the problems associated with employing wall functions [112]. Wall $y^+ < 1$, which was seen almost everywhere at peak flow in the fine-mesh simulations, generally satisfies the requirement of fully resolved LES models [111], where the wall shear stress is computed directly from the laminar stress–strain relationship [113].

The Kolmogorov length and time scales (Equations (24) and (25)) offer a valuable metric for evaluating the spatial and temporal resolution in numerical models of turbulent flow [108–110]. At the Kolmogorov scale, viscous forces dominate, leading to dissipation of turbulent energy into thermal energy. These scales characterize the smallest turbulent eddies, and detailed models such as DNS must therefore resolve the Kolmogorov microscales [262]. Kolmogorov's hypotheses [263] posit that turbulence is isotropic at the smallest scales, implying that the Boussinesq hypothesis can be reasonably accurate. Consequently, y^+-insensitive RANS- and LES-based models exhibit high accuracy, approaching DNS, when the spatial and temporal resolution is of the order of the Kolmogorov microscales.

Figure 18 indicates that the majority of fine-mesh grid cells featured LES mixing lengths and timesteps smaller than the Kolmogorov length and time scales, respectively. The spatial and temporal resolution, comparable to the Kolmogorov microscales, along with a finely resolved wall boundary layer, suggests that the fine-mesh LES simulations were finely resolved, albeit not fully, during peak flow. Because the Kolmogorov microscales increase with decreasing velocities, the LES was most likely fully resolved at lower volumetric flowrates, however. As the computational mesh exhibited sufficient resolution to resolve the smallest turbulent eddies, and the temporal resolution effectively resolved the turbulent time scale, the LES can be considered near-DNS, within the nasal cavity. This view is further supported by the relatively small turbulent viscosity ratios depicted in Figure 19. Consequently, errors attributable to spatial or temporal resolution are deemed negligible in the fine-mesh LES.

It is noted, however, that the turbulent dissipation rate is not readily available from the present LES simulations, since it relies on statistical analysis of fully developed turbulent flow. In the current simulations, the volumetric flowrate was continuously varying due to the oscillatory nature of respiratory flow, and pseudo-steady statistics were thus not available. Hence, the formulae for calculating the Kolmogorov microscales (Equations (24) and (25)) are not directly applicable for assessing the LES simulations. To address this limitation, the Kolmogorov microscales used in Figures 16–18 were derived from the steady state fine-mesh $k\omega$ SST simulations at peak inspiratory flow. These simulations assumed constant volumetric flowrate, hence pseudo-steady turbulence statistics. This approach, while a necessary adaptation, provides valuable insights despite the dynamic nature of the respiratory flow in the current simulations.

Rigorous statistical analysis of the LES data was not performed in the current study, but the simulation data have been made freely available to the public [14]. Other investigators are encouraged to examine and analyze the dataset. It is worth noting that in the continuously developing flow fields due to the oscillatory nature of respiratory flow, the steady state statistics of classical turbulence theory might not hold. The temporal (and spatial) development of the out-of-equilibrium turbulent energy spectrum is a topic for further study [264].

4.3.3. Transient Effects

The physiological respiratory cycle deviates from a perfectly sinusoidal function of time. Notably, there exists an inherent asymmetry between inhalation and exhalation, leading to a marginally higher peak inspiratory flowrate compared to the peak expiratory rate. Consequently, during exhalation, a plateau emerges, characterized by an almost constant flowrate. Realistic breathing cycles have been presented by, e.g., Van Hove et al. [223], Calmet et al. [250], and Pawade [129]. In transient simulations, obtaining turbulence statistics benefits from prolonged intervals of near-constant flowrate, enabling a broader time-averaging window. With a continuously varying mean flow velocity, as in the current LES, turbulence statistics sampling becomes challenging with respect to computational cost, since it must rely on a high number of breathing cycles.

Turbulence evolves distinctively in continuously accelerating/decelerating flow compared to steady flow [167]. A period of near-steady flow, as seen in physiological expiration curves, could potentially promote turbulence development if the Reynolds number is sufficiently high. Nevertheless, in the current simulations, the RANS-based models, representing fully developed steady flow, contradict this notion concerning peak flow in the specific patient-specific model.

Inthavong et al. [105] proposed that eliminating start-up effects in transient simulations, starting from a quiescent state, required two to three breathing cycles, whereas Bradshaw et al. [159] dismissed the first cycle. In the current simulations, a fine-mesh steady state $k\omega$ SST model was employed to establish initial peak inspiratory flow conditions for the transient LES. Notably, no start-up effects were observed, and there were no discernible differences between successive breathing cycles. Consequently, it can be inferred that the start-up effects were either enduring or entirely absent.

The nasal airflow simulation results presented in this paper show that a series of steady state simulations closely replicated the transient simulation of the entire (sinusoidal) breathing cycle. Thus, for the specific patient considered, the respiratory nasal airflow could be characterized as quasi-steady. This observation aligns with expectations, given the low Womersley number (Equation (14)). Quasi-steadiness implies that the shape of the respiratory cycle is non-substantial. However, it is important to note that the width and shape of the unsteady hysteresis affecting RMM pressure–flow curves are influenced by the time-derivative of the volumetric flowrate, hence the slope of the breathing profile (refer to **Part I**, Section 2.3.3 and Figure 13). It is underlined that the hysteresis loops affecting the pressure–flow curves resulting from the transient LES were solely due to the non-zero time derivative of the flowrate, not start-up effects or nasal compliance.

To sum up, while the present simulation data revealed some transient effects, their influence on the in silico RMM results was found to be minimal. Therefore, transient effects are deemed highly improbable as an explanation for the significant disparity observed between in silico and in vivo RMM results.

4.3.4. Geometry

One potential source of error in the comparison of in silico and in vivo RMM data, not investigated in this study, pertains to the consistency between the digital geometry model used in CFD simulations and the actual nasal cavity geometry during RMM. For instance, Aasgrav [4] highlighted the substantial impact of segmentation settings on nasal resistance by contrasting CFD models based on different HU threshold settings. Beyond this, there are additional points that warrant discussion in relation to patient-specific geometry models. This section aims to underscore key aspects to be mindful of when evaluating the congruence between the digital representation and the real anatomical features. Refer also to the broader discussions in **Part I** (see Sections 2.4.3, 2.5.2 and 2.5.4).

A notable deviation from realistic airway geometry in the current digital patient-specific model is the truncation of the geometry at the nostrils and the subsequent truncation and extrusion of the pharyngeal tract. The former restricts the faithful representation of airflow distribution across the open nostril during inhalation and exhalation, as well as the turbulent intensity during inhalation. However, Taylor et al. [180] observed that overall flow patterns and measures were insensitive to the detailed prescription of inflow conditions at the nares. The latter predominantly affects the turbulent intensity during exhalation, as illustrated in Figure 25, where the flow featured minimal vorticity when it entered the nasopharynx, during exhalation. This indicates that production of vorticity and turbulent structures was delayed until the nasopharynx and nasal cavity. While the precise influence on simulation results remains unclear, this observation aligns with Bradshaw et al.'s findings [159], underscoring the importance of accurately describing and including the pharyngeal tract to model vortex generation and transport during both inhalation and exhalation. Neglecting the paranasal sinuses is consistent with the perspective of several other researchers [126,159,170,171].

When exploring additional factors that could influence the comparability of the patient-specific geometry employed in in silico RMM and the actual airway geometry during in vivo RMM, two primary considerations emerge: (1) the condition of the nasal cavity during CT image acquisition and (2) the impact of nasal compliance.

1.   During in vivo RMM, nasal cavity decongestion was achieved through the application of xylometazoline. This is expected to maximize nasal cavity volume, through a reduction in turbinate swelling, and minimize nasal resistance. However, the CT image acquisition, the basis for the digital patient-specific geometry model, took place in the natural, non-decongested state. As a result, it was subject to the nasal cycle and various factors affecting spontaneous turbinate and nasal mucosa swelling. For instance, in the natural non-decongested state, postural effect on the nasal resistance and nasal cycle may be anticipated [33,133,134]. The nasal resistance is expected to be higher in the supine position, in which CT images were obtained, than in the sitting position, in which the in vivo RMM data were obtained. The lack of decongestion and the postural effect are both expected to increase the nasal resistance. Hence, the correction of these sources of error would presumably reduce the nasal resistance predicted by the in silico RMM, further increasing the disagreement with the in vivo RMM data.

2. It has been proposed that nasal compliance may affect RMM curves (see **Part I**, Section 2.5.1), and that rigid CFD geometries will fail to reproduce in vivo RMM curves due to this. However, for the current patient, it is observed that the patient-specific in vivo RMM curves are almost symmetrical with respect to in/exhalation (see Figure 12a). At the same time, particularly on the left side, the in vivo RMM pressure–flow curve plateaus, suggesting a significant increase in nasal resistance beyond a critical flowrate. To explain these observations by nasal compliance, a collapsible constriction is required, where the Venturi effect dominates over the static pressure in such a way that the collapse is independent of the flow direction. Without such a constriction, asymmetrical collapse would be expected due to the under-/over-pressures in the nasal cavity during the inspiratory and expiratory phases, respectively. For instance, the phenomenon of nasal gateway collapse exemplifies this, where the collapse occurs exclusively during inhalation [265,266].

The first point suggests that the discordance between in silico and in vivo RMM may be more prominent than indicated by the current findings. On the other hand, the second point provides a plausible rationale for the significant misalignment between simulations and measurements. This contradicts the conventional notion that a decongested nasal cavity retains rigidity, calling for robust evidence to demonstrate the (partial) symmetrical collapse of certain nasal cavity parts during both inhalation and exhalation.

For future studies comparing in silico and in vivo RMM, it is recommended to exercise caution in ensuring that the patient-specific geometry model accurately represents the airway geometry during in vivo RMM. This can be achieved by minimizing the time gap between clinical RMM and CT data acquisition and ensuring that both are obtained in a decongested state, preferably in the same position (either supine or sitting). This approach will contribute to reducing a major source of error in the analysis of in silico nasal airflow data.

As a final point, the role of nasal hair should be considered. Hahn et al. [151] and Stoddard et al. [174] presented differing conclusions regarding the significance of nasal hair. Hahn et al. found little impact on overall airflow, while Stoddard et al. observed a positive effect on both subjective and objective measures of nasal obstruction following the removal of nasal hair, suggesting its noteworthy contribution to nasal resistance. A recent CFD study by Haghnegahdar et al. [267] indicated that nasal hair significantly influenced nasal airflow patterns, although they did not report nasal resistance. The potential impact of nasal hair on nasal resistance appears to warrant increased attention.

4.3.5. Final Remarks and Observations Regarding the Analysis of Rhinomanometry Pressure Measurements

In the course of anterior rhinomanometry (RMM), the assessment of nasopharyngeal pressure involves measuring the stagnation pressure at the occluded nostril. This pressure is then used as an approximation of the nasopharyngeal pressure. The differential pressure between the open nostril and the occluded nostril thus serves as an indicator of the pressure drop across the nasal cavity, facilitating the calculation of nasal resistance (refer to **Part I**, Section 2.3 for more details). However, the simulation findings discussed in the preceding section (see Figures 30 and 31) indicate that interpreting these measurements may not be straightforward. Firstly, the results indicate that nasal resistance could be influenced by the direction of airflow (inhalation versus exhalation). Additionally, it is noted that the measured stagnation pressure corresponds primarily to the static nasopharyngeal pressure during inhalation, while during exhalation, it aligns more closely with the total pressure.

Anticipation of the former aligns with classical potential theory, where symmetry in flow direction reversal only occurs in ideal situations. As illustrated in Figures 25 and 26, the vorticity of nasal airflow is substantial, and non-recoverable pressure losses, such as those caused by vortex shedding, are expected to vary based on the flow direction, particularly when there are abrupt changes in cross-sectional areas or flow direction [81].

The latter phenomenon can be elucidated using principles akin to those governing airspeed measurements with Pitot tubes in, for example, aviation (see Figure 32). In this context, when the flow of air is directed towards the cavity with its opening facing the flow, it is deflected due to a force exerted by the fluid inside the cavity. This force is counteracted by the stagnation pressure at the cavity's bottom. Conversely, for the cavity with its opening facing the opposite direction, the wake behind the cavity may induce a suction force, reducing the pressure at the cavity's bottom. The observed disparity in simulation results between the static pressure at the occluded nostril and the total pressure in the nasopharynx during exhalation can be attributed to flow curvature. As the airflow enters the nasopharynx, it bends towards the open nasal cavity, and vortical structures penetrate into the occluded nasal cavity (see Figures 25 and 26). These factors necessitate the application of a force that influences the stagnation pressure. In the context of RMM, this implies a potential overestimation of nasopharyngeal pressure during exhalation, leading to an underestimation of pressure drop and nasal resistance.

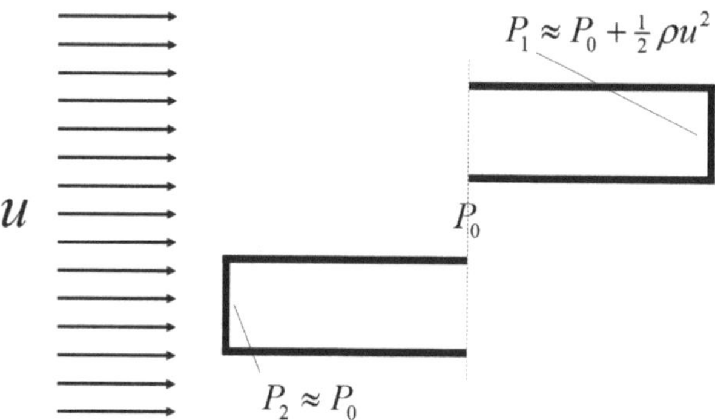

Figure 32. Two identical containers are placed in a flowing fluid of mass density, ρ, and uniform velocity, u, such that one (1) has its opening facing upwind and the other (2) faces downwind. The containers are aligned such that the static pressures at the openings of the containers are identical. The static pressures at the bottom of the two containers will differ by the dynamic pressure of the flowing fluid, $\tfrac{1}{2}\rho u^2$, since this is the pressure necessary to be exerted on the fluid to bring it to stagnation at the opening of the container. Thus, pressures at the bottom of the upwind and downwind facing container are approximately equal to the total and static pressures of the fluid, respectively, at the container openings. The described scenario is ideal, and "approximate" indicates that entrance effects, such as vortices in the openings of the containers, may affect the actual pressures.

In standard RMM, there is unfortunately insufficient information to rectify the inherent overprediction of nasopharyngeal pressure during exhalation. However, if the cross-sectional area of the open nostril were measured, it would allow for the estimation of total pressure at the nostril, enabling the calculation of pressure drop as a change in total pressure. In channels with varying cross-sectional areas, recoverable changes in static pressure may result from the acceleration or deceleration of the flow. Conversely, changes in total pressure are solely due to non-recoverable losses, suggesting that total pressure measurements may provide a more reliable basis for estimating nasal resistance compared to static pressure measurements.

From this discussion, it becomes evident that interpretation and application of standard clinical anterior RMM examinations should consider two key factors: (1) the physical nasal resistance may be influenced by the direction of airflow, and (2) exhalatory pressure measurements are likely to underpredict the pressure drop and unilateral nasal resistance.

The Use of Bernoulli's Equation

The patient-specific clinical in vivo RMM pressure–flow data presented in Figure 1 shows that there was significant effect on the nasal resistance by decongesting the nose, on both sides. However, the data exhibits significant variability, with evident large hysteresis loops affecting each breathing cycle. Particularly on the left side, where the nasal resistance was highest, the scatter in differential pressure values span a wide range, for given volumetric flowrates. The characteristic loops observed in the in vivo RMM pressure–flow curves have been previously discussed by Vogt et al. [77,139], who proposed that nasal compliance contributes to the hysteresis, and Groß and Peters [140], who suggested that measurement techniques in RMM may be a contributing factor. In **Part I** (Section 2.3.3) it was illustrated how hysteresis might result from unsteady flow, alone.

The representative in vivo RMM curves presented in Figure 1 were created by manual placement of pressure–flow points, to represent the minimal nasal resistance observed in the data. It is evident that this representative nasal resistance significantly underestimates the actual nasal resistance during certain recorded breathing cycles.

The discrepancy between in silico and in vivo RMM is evident not only in terms of nasal resistance but also in other aspects. First, there is much less scatter in the in silico pressure data. Although pressure fluctuations are present in the LES-based pressure–flow curves (see Figures 8 and 10), they manifest as a noisy signal typical of turbulent fluctuations. Additionally, the tight hysteresis envelopes inherent in the LES data appear quite different from the large hysteresis loops observed in the in vivo data (compare Figures 1 and 13).

Excellent curve-fits of Bernoulli's equation (Equation (4)) were obtained for both in silico and in vivo RMM pressure–flow curves (refer to Figures 12 and 13). Deviation from the straight lines associated with laminar flow in a straight tube, have been attributed to the transition to turbulent flow [78]. Such behavior was observed both in in vivo and in silico RMM curves presented here. The present CFD simulations have, however, provided compelling evidence that the flow was laminar. Conversely, the optimal curve-fit was obtained using the (turbulent) friction factor obtained from the Haaland equation (Equation (7)) regardless of Reynolds number. These findings suggest that considerations of turbulence alone may not be sufficient for assessing RMM pressure–flow curves. Without delving into an exhaustive analysis, it is suggested that the apparent turbulent behavior, when assessing the Bernoulli equation curve-fit in isolation, may be influenced by other factors related to the deviation between the nasal cavity and a straight pipe of constant diameter. For instance, the non-circular shape of the nasal cavity, numerous abrupt changes in cross-sectional area, bends, and curves, along with the presence of nasal hair and mucosal lining, may introduce minor friction losses that should be accounted for [81]. Additionally, the impact of unsteady flow, such as significant vortex shedding, as illustrated in Figures 25 and 26, and the possibility of nasal compliance should be considered.

4.4. Summary and Conclusions

Some of the most detailed CFD simulations of nasal airflow, to date, were performed to investigate commonly reported gross discrepancies between in vivo unilateral anterior rhinomanometry and in silico CFD simulation results. Specifically, the aim of this study was to investigate the threefold hypothesis that these discrepancies can be explained by (1) poor choice of turbulence model, (2) insufficient spatial or temporal resolution, or (3) neglecting transient effects, in CFD models.

A patient-specific digital nasal airway geometry model was obtained by segmentation of preoperative CT images from an OSA patient. Finely resolved transient LES was employed to assess steady state laminar and y^+ insensitive RANS simulations on a relatively coarse mesh. A remarkable agreement was observed between pressure–flow curves obtained with the various CFD models. Nasal pressure drop, hence nasal resistance, was severely underpredicted compared to in vivo RMM, however.

Although local Reynolds numbers ranged from 4 to 14 thousand throughout the nasal cavities at peak flow, the simulation results imply that the modelled flow was predominantly near-laminar/transitional. The delayed development of turbulence may be explained by the oscillatory nature of respiratory flow as well as the complex shape of the nasal airway, preventing the development of turbulent boundary layers.

Spatial and temporal resolution of the LES model was of the order of the Kolmogorov microscales at peak flow, suggesting that errors attributable to spatial or temporal resolution were insignificant.

Although transient phenomena such as pressure–flow curve hysteresis and unsteady vortex shedding were observed in the transient LES results, comparison of pressure–flow curves from steady state and transient simulations revealed that the flow could be considered quasi-steady.

In summary, the present simulation results effectively disproved the paper's initial hypothesis, and the cause for disagreement between in silico and in vivo RMM must be explained by other means. The most plausible factor appears to be that the digital geometry model was a poor representation of the actual nasal cavity. Key aspects in this regard may be nasal cavity compliance or drag caused by nasal hair.

Supplementary Materials: The following supporting information can be downloaded at: https://www.mdpi.com/article/10.3390/bioengineering11030239/s1, Animation A1: A1_pressure_at_wall_LCRO.mpg; Animation A2: A2_pressure_at_wall_LORC.mpg; Animation A3: A3_Q-criterion_LCRO.mpg; Animation A4: A4_Q-criterion_LORC.mpg; Animation A5: A5_nasopharynx_velocity_contour_vectors_LCRO.mpg. Transient LES case files and data presented in this study are publicly available in the NIRD Research Data Archive (https://archive.sigma2.no, accessed on 1 January 2024) at doi:10.11582/2023.00126 [14].

Funding: This study was part of the research projects Modeling of obstructive sleep apnea by fluid–structure interaction in the upper airways [268] and Virtual Surgery in the Upper Airways—New Solutions to Obstructive Sleep Apnea Treatment [269] financed by the Research Council of Norway, under the FRINATEK program (grants 231741 and 303218). Simulation resources were financed by SINTEF [270] and Sigma2—the National Infrastructure for High Performance Computing and Data Storage in Norway [271]. Simulation data sharing and storage is made possible by Norwegian research infrastructure services (NRIS) [272] via NIRD—the National Infrastructure for Research Data.

Institutional Review Board Statement: The patient included in this paper was part of a prospective study approved by the Norwegian Regional Ethics Committee, REK sør-øst (9 October 2014, 2014/1450) and registered in clinicaltrials.gov (NCT01282125). Written, informed consent was obtained from all patients involved in this study.

Informed Consent Statement: Not applicable.

Data Availability Statement: Transient LES case files and data presented in this study are publicly available in the NIRD Research Data Archive [14] (accessed on 1 January 2024). Additional simulation files and data are available upon request.

Acknowledgments: The author is grateful for fruitful discussions and a vivid, curiosity-driven research environment within the project consortium and at dept. Flow technology at SINTEF. In particular, the input from the following colleagues is acknowledged: prof. Bjørn H. Skallerud (NTNU), prof. Bernhard Müller (NTNU), prof. Ståle Nordgård (NTNU, St. Olavs hospital the university hospital in Trondheim), Mads Moxness (Aleris), Sigrid K. Dahl (SINTEF), Stein Tore Johansen (SINTEF), Are J. Simonsen (SINTEF), Paal Skjetne (SINTEF), Balram Panjwani (SINTEF), and John C. Morud (SINTEF).

Conflicts of Interest: The author declares that there are no conflicts of interest regarding the publication of this article.

Nomenclature

∂_t Time derivative operator, $[1/s]$
∇ Gradient operator, $[1/m]$
A Channel cross-sectional area, $[m^2]$

c_P	Specific heat capacity, $[\text{J/kgK}]$
CFL	Courant–Friedrich–Lewy number, *dimensionless*
D_h	Channel hydraulic diameter, $[\text{m}]$
ε_r	Relative wall roughness, *dimensionless*
ε	Turbulent dissipation rate, $[\text{m}^2/\text{s}^3]$
f	Frequency, $[\text{Hz}]$
f_D	Darcy friction factor, *dimensionless*
η	Kolmogorov length scale, $[\text{m}]$
k	Thermal conductivity, $[\text{W/mK}]$
k	Turbulent kinetic energy, $[\text{m}^2/\text{s}^2]$
λ	Taylor length scale, $[\text{m}]$
L	Channel length, $[\text{m}]$
μ	Dynamic viscosity, $[\text{Pas}]$
ν	Kinematic viscosity, $[\text{m}^2/\text{s}]$
ν_t	Turbulent kinematic viscosity, $[\text{m}^2/\text{s}]$
ω	Angular frequency, $[\text{Hz}]$
ω	Specific turbulent dissipation rate, $[\text{Hz}]$
O	Channel perimeter, $[\text{m}]$
P	Pressure, $[\text{Pa}]$
ΔP	Pressure difference, $[\text{Pa}]$
Q	Volumetric flowrate, $[\text{m}^3/\text{s}]$
ρ	Mass density, $[\text{kg/m}^3]$
R	Flow resistance, $[\text{Pas/m}^3]$
Re	Reynolds number, *dimensionless*
Δt	Time step size, $[\text{s}]$
τ	Period of the breathing cycle, $[\text{s}]$
τ_η	Kolmogorov time scale, $[\text{s}]$
$\boldsymbol{\tau}_{SGS}$	Subgrid-scale stress tensor, $[\text{Pa}]$
τ_w	Wall shear stress, $[\text{Pa}]$
T	Temperature, $[\text{K}]$
$\mathbf{u}$	Flow velocity vector, $[\text{m/s}]$
u	Flow velocity, $[\text{m/s}]$
u_τ	Shear velocity, $[\text{m/s}]$
Δx	Grid size, $[\text{m}]$
y	Distance to the wall, $[\text{m}]$
Wo	Womersley number, *dimensionless*
In silico	In a digital computer model, e.g., CFD
In vitro	In a physical replica
In vivo	In a living patient
AR	Acoustic rhinometry
AAR	Active anterior rhinomanometry
AHI	Apnea–hypopnea Index
CFD	Computational fluid dynamics
CT	Computed tomography
DNS	Direct numerical simulation
LB	Lattice–Boltzmann
LES	Large eddy simulation
MRI	Magnetic resonance imaging
NAO	Nasal airway obstruction
NR	Nasal resistance
NS	Navier–Stokes
NOSE	Nasal obstruction symptom evaluation
OSA	Obstructive sleep apnea
PNIF	Peak nasal inspiratory flow
RANS	Reynolds-averaged Navier–Stokes
RMM	Rhinomanometry
VAS	Visual analogue scale

References

1. Broschek, B.; Sonora, C.M.G. Experimental Study of Fluid-Structure Interaction in a Simplified Geometry of the Human Upper Airways. Master's Thesis, NTNU, Trondheim, Norway, 2015. Available online: https://bibsys-almaprimo.hosted.exlibrisgroup.com/permalink/f/13q4kuj/BRAGE11250/2385338 (accessed on 1 January 2024).
2. Jordal, M.R. Patient Specific Numerical Simulation of Flow in the Human Upper Airways. Master's Thesis, NTNU, Trondheim, Norway, 2016. Available online: https://bibsys-almaprimo.hosted.exlibrisgroup.com/permalink/f/13q4kuj/BRAGE11250/2405975 (accessed on 1 January 2024).
3. Aasgrav, E. CFD Simulations of Turbulent Flow in the Human Upper Airways. In Proceedings of the 12th International Conference on CFD in Oil & Gas, Metallurgical and Process Industries, Trondheim, Norway, 30 May 2017.
4. Aasgrav, E. Investigation of CFD Simulations of Flow in the Upper Airways. Master's Thesis, NTNU, Trondheim, Norway, 2017. Available online: https://bibsys-almaprimo.hosted.exlibrisgroup.com/permalink/f/13q4kuj/BRAGE11250/2454893 (accessed on 1 January 2024).
5. Jordal, M.R.; Johnsen, S.G.; Dahl, S.K.; Müller, B. Patient Specific Numerical Simulation of Flow in the Human Upper Airways for Assessing the Effect of Nasal Surgery. In Proceedings of the 12th International Conference on CFD in Oil & Gas, Metallurgical and Process Industries, Trondheim, Norway, 30 May 2017; p. 10.
6. Liu, H.; Moxness, M.H.S.; Prot, V.E.; Skallerud, B.H. Palatal Implant Surgery Effectiveness in Treatment of Obstructive Sleep Apnea: A Numerical Method with 3D Patient-Specific Geometries. *J. Biomech.* **2018**, *66*, 86–94. [CrossRef] [PubMed]
7. Moxness, M.H.S. The Influence of the Nasal Airway in Obstructive Sleep Apnea. Ph.D. Thesis, NTNU, Trondheim, Norway, 2018.
8. Moxness, M.H.S.; Wülker, F.; Helge Skallerud, B.; Nordgård, S. Simulation of the Upper Airways in Patients with Obstructive Sleep Apnea and Nasal Obstruction: A Novel Finite Element Method: Novel FE Method for OSA and Nasal Obstruction. *Laryngoscope Investig. Otolaryngol.* **2018**, *3*, 82–93. [CrossRef] [PubMed]
9. Khalili, M.E.; Larsson, M.; Müller, B. High-Order Ghost-Point Immersed Boundary Method for Viscous Compressible Flows Based on Summation-by-Parts Operators. *Int. J. Numer. Methods Fluids* **2019**, *89*, 256–282. [CrossRef]
10. Akbar, B. Mathematical Modelling and Simulation of Flow in Collapsible Tubes. Master's Thesis, NTNU, Trondheim, Norway, 2022. Available online: https://bibsys-almaprimo.hosted.exlibrisgroup.com/permalink/f/13q4kuj/BRAGE11250/3023112 (accessed on 1 January 2024).
11. Ayyalasomayajula, V.; Moxness, M.; Skallerud, B. Potential of Computational Models in Personalized Treatment of Obstructive Sleep Apnea: A Patient-Specific Partial 3D Finite Element Study. *Biomech. Model. Mechanobiol.* **2023**. [CrossRef] [PubMed]
12. Wu, J.; Zhao, G.; Li, Y.; Zang, H.; Wang, T.; Wang, D.; Han, D. Apnea–Hypopnea Index Decreased Significantly after Nasal Surgery for Obstructive Sleep Apnea: A Meta-Analysis. *Medicine* **2017**, *96*, e6008. [CrossRef] [PubMed]
13. Schoustra, E.; Van Maanen, P.; Den Haan, C.; Ravesloot, M.J.L.; De Vries, N. The Role of Isolated Nasal Surgery in Obstructive Sleep Apnea Therapy—A Systematic Review. *Brain Sci.* **2022**, *12*, 1446. [CrossRef]
14. Johnsen, S.G. *In Silico Rhinomanometry—A High Fidelity LES CFD Simulation Study [Data Set]*; Norstore, Sigma2: Trondheim, Norway, 2023. [CrossRef]
15. Zwicker, D.; Ostilla-Mónico, R.; Lieberman, D.E.; Brenner, M.P. Physical and Geometric Constraints Shape the Labyrinth-like Nasal Cavity. *Proc. Natl. Acad. Sci. USA* **2018**, *115*, 2936–2941. [CrossRef] [PubMed]
16. Cingi, C.; Bayar Muluk, N. (Eds.) *All Around the Nose: Basic Science, Diseases and Surgical Management*; Springer International Publishing: Cham, Switzerland, 2020; ISBN 978-3-030-21216-2.
17. Singh, S.; Awasthi, N.; Gupta, T. Mouth Breathing-Its Consequences, Diagnosis & Treatment. *Acta Sci. Dent. Sci.* **2020**, *4*, 32–41. [CrossRef]
18. Zhao, K.; Dalton, P. The Way the Wind Blows: Implications of Modeling Nasal Airflow. *Curr. Allergy Asthma Rep.* **2007**, *7*, 117–125. [CrossRef]
19. Leite, S.H.P.; Jain, R.; Douglas, R.G. The Clinical Implications of Computerised Fluid Dynamic Modelling in Rhinology. *Rhinol. J.* **2018**, *57*, 2–9. [CrossRef]
20. Wong, E.; Inthavong, K.; Singh, N. Comment on the European Position Paper on Diagnostic Tools in Rhinology—Computational Fluid Dynamics. *Rhinol. J.* **2019**, *57*, 477–478. [CrossRef] [PubMed]
21. Spicuzza, L.; Caruso, D.; Di Maria, G. Obstructive Sleep Apnoea Syndrome and Its Management. *Ther. Adv. Chronic Dis.* **2015**, *6*, 273–285. [CrossRef] [PubMed]
22. Trosman, I.; Trosman, S.J. Cognitive and Behavioral Consequences of Sleep Disordered Breathing in Children. *Med. Sci.* **2017**, *5*, 30. [CrossRef] [PubMed]
23. Abbasi, A.; Gupta, S.S.; Sabharwal, N.; Meghrajani, V.; Sharma, S.; Kamholz, S.; Kupfer, Y. A Comprehensive Review of Obstructive Sleep Apnea. *Sleep Sci. Sao Paulo Braz.* **2021**, *14*, 142–154.
24. Reutrakul, S.; Mokhlesi, B. Obstructive Sleep Apnea and Diabetes. *Chest* **2017**, *152*, 1070–1086. [CrossRef] [PubMed]
25. Yeghiazarians, Y.; Jneid, H.; Tietjens, J.R.; Redline, S.; Brown, D.L.; El-Sherif, N.; Mehra, R.; Bozkurt, B.; Ndumele, C.E.; Somers, V.K.; et al. Obstructive Sleep Apnea and Cardiovascular Disease: A Scientific Statement from the American Heart Association. *Circulation* **2021**, *144*, e56–e67. [CrossRef] [PubMed]
26. Castaneda, A.; Jauregui-Maldonado, E.; Ratnani, I.; Varon, J.; Surani, S. Correlation between Metabolic Syndrome and Sleep Apnea. *World J. Diabetes* **2018**, *9*, 66–71. [CrossRef]

27. Bonsignore, M.R.; Randerath, W.; Riha, R.; Smyth, D.; Gratziou, C.; Goncalves, M.; McNicholas, W.T. New Rules on Driver Licensing for Patients with Obstructive Sleep Apnoea: EU Directive 2014/85/EU. *Eur. Respir. J.* **2016**, *47*, 39–41. [CrossRef]
28. Sawada, D.; Tomooka, K.; Tanigawa, T. Changes in Attitudes of Life Insurance Companies Towards Patients with Sleep Apnea Syndrome Undergoing Continuous Positive Airway Pressure in Japan. *Juntendo Med. J.* **2022**, *68*, 606–612. [CrossRef]
29. Young, T.; Finn, L.; Kim, H. Nasal Obstruction as a Risk Factor for Sleep-Disordered Breathing. *J. Allergy Clin. Immunol.* **1997**, *99*, S757–S762. [CrossRef] [PubMed]
30. Li, H.-Y.; Wang, P.-C.; Hsu, C.-Y.; Cheng, M.; Liou, C.-C.; Chen, N.-H. Nasal Resistance in Patients with Obstructive Sleep Apnea. *ORL* **2005**, *67*, 70–74. [CrossRef]
31. Tagaya, M.; Nakata, S.; Yasuma, F.; Noda, A.; Morinaga, M.; Yagi, H.; Sugiura, M.; Teranishi, M.; Nakashima, T. Pathogenetic Role of Increased Nasal Resistance in Obese Patients with Obstructive Sleep Apnea Syndrome. *Am. J. Rhinol. Allergy* **2010**, *24*, 51–54. [CrossRef]
32. Blomster, H.; Kemppainen, T.; Numminen, J.; Ruoppi, P.; Sahlman, J.; Peltonen, M.; Seppa, J.; Tuomilehto, H. Impaired Nasal Breathing May Prevent the Beneficial Effect of Weight Loss in the Treatment of OSA. *Rhinol. J.* **2011**, *49*, 587–592. [CrossRef]
33. Hueto, J.; Santaolalla, F.; Sanchez-del-Rey, A.; Martinez-Ibargüen, A. Usefulness of Rhinomanometry in the Identification and Treatment of Patients with Obstructive Sleep Apnoea: An Algorithm for Predicting the Relationship between Nasal Resistance and Continuous Positive Airway Pressure. a Retrospective Study. *Clin. Otolaryngol.* **2016**, *41*, 750–757. [CrossRef]
34. Hoven, K.M.; Aarstad, H.-J.; Steinsvag, S.K. Associations between Nasal Characteristics and Sleep Polygraphic Data in Patients Suspected Obstructive Sleep Apnea. *Rhinol. Online* **2020**, *3*, 79–86. [CrossRef]
35. Hoel, H.C.; Kvinnesland, K.; Berg, S. Impact of Nasal Resistance on the Distribution of Apneas and Hypopneas in Obstructive Sleep Apnea. *Sleep Med.* **2020**, *71*, 83–88. [CrossRef] [PubMed]
36. Hoel, H.C.; Kvinnesland, K.; Berg, S. Outcome of Nasal Measurements in Patients with OSA—Mounting Evidence of a Nasal Endotype. *Sleep Med.* **2023**, *103*, 131–137. [CrossRef] [PubMed]
37. Scott, W.C.; Kent, D.T. Nasal Obstruction and Sleep-Disordered Breathing. In *Management of Obstructive Sleep Apnea*; Kim, K.B., Movahed, R., Malhotra, R.K., Stanley, J.J., Eds.; Springer International Publishing: Cham, Switzerland, 2021; pp. 243–257. ISBN 978-3-030-54145-3. [CrossRef]
38. Chambers, K.J.; Horstkotte, K.A.; Shanley, K.; Lindsay, R.W. Evaluation of Improvement in Nasal Obstruction Following Nasal Valve Correction in Patients With a History of Failed Septoplasty. *JAMA Facial Plast. Surg.* **2015**, *17*, 347–350. [CrossRef] [PubMed]
39. Nakata, S.; Noda, A.; Yagi, H.; Yanagi, E.; Mimura, T.; Okada, T.; Misawa, H.; Nakashima, T. Nasal Resistance for Determinant Factor of Nasal Surgery in CPAP Failure Patients with Obstructive Sleep Apnea Syndrome. *Rhinology* **2005**, *43*, 296–299. [PubMed]
40. Moxness, M.H.S.; Nordgård, S. An Observational Cohort Study of the Effects of Septoplasty with or without Inferior Turbinate Reduction in Patients with Obstructive Sleep Apnea. *BMC Ear Nose Throat Disord.* **2014**, *14*, 11. [CrossRef] [PubMed]
41. Pang, K.P.; Montevecchi, F.; Vicini, C.; Carrasco-Llatas, M.; Baptista, P.M.; Olszewska, E.; Braverman, I.; Kishore, S.; Chandra, S.; Yang, H.C.; et al. Does Nasal Surgery Improve Multilevel Surgical Outcome in Obstructive Sleep Apnea: A Multicenter Study on 735 Patients. *Laryngoscope Investig. Otolaryngol.* **2020**, *5*, 1233–1239. [CrossRef]
42. Berger, M.; Pillei, M.; Mehrle, A.; Recheis, W.; Kral, F.; Kraxner, M.; Bardosi, Z.; Freysinger, W. Nasal Cavity Airflow: Comparing Laser Doppler Anemometry and Computational Fluid Dynamic Simulations. *Respir. Physiol. Neurobiol.* **2021**, *283*, 103533. [CrossRef]
43. Malm, L. Measurement of Nasal Patency. *Allergy* **1997**, *52*, 19–23. [CrossRef]
44. Ottaviano, G.; Fokkens, W.J. Measurements of Nasal Airflow and Patency: A Critical Review with Emphasis on the Use of Peak Nasal Inspiratory Flow in Daily Practice. *Allergy* **2016**, *71*, 162–174. [CrossRef]
45. Eccles, R. Measurement of the Nasal Airway. In *Scott-Brown's Otorhinolaryngology and Head and Neck Surgery*; CRC Press: Boca Raton, FL, USA, 2018; Volume 1, p. 1402. ISBN 978-0-203-73103-1.
46. Rimmer, J.; Hellings, P.; Lund, V.J.; Alobid, I.; Beale, T.; Dassi, C.; Douglas, R.; Hopkins, C.; Klimek, L.; Landis, B.; et al. European Position Paper on Diagnostic Tools in Rhinology. *Rhinol. J.* **2019**, *57*, 1–41. [CrossRef]
47. Rhee, J.S.; Sullivan, C.D.; Frank, D.O.; Kimbell, J.S.; Garcia, G.J.M. A Systematic Review of Patient-Reported Nasal Obstruction Scores: Defining Normative and Symptomatic Ranges in Surgical Patients. *JAMA Facial Plast. Surg.* **2014**, *16*, 219–225. [CrossRef] [PubMed]
48. André, R.F.; Vuyk, H.D.; Ahmed, A.; Graamans, K.; Nolst Trenité, G.J. Correlation between Subjective and Objective Evaluation of the Nasal Airway. A Systematic Review of the Highest Level of Evidence: Subjective and Objective Evaluation of Nasal Patency. *Clin. Otolaryngol.* **2009**, *34*, 518–525. [CrossRef] [PubMed]
49. Zhao, K.; Blacker, K.; Luo, Y.; Bryant, B.; Jiang, J. Perceiving Nasal Patency through Mucosal Cooling Rather than Air Temperature or Nasal Resistance. *PLoS ONE* **2011**, *6*, e24618. [CrossRef] [PubMed]
50. Casey, K.P.; Borojeni, A.A.T.; Koenig, L.J.; Rhee, J.S.; Garcia, G.J.M. Correlation between Subjective Nasal Patency and Intranasal Airflow Distribution. *Otolaryngol. Neck Surg.* **2017**, *156*, 741–750. [CrossRef] [PubMed]
51. Mozzanica, F.; Gera, R.; Bulgheroni, C.; Ambrogi, F.; Schindler, A.; Ottaviani, F. Correlation between Objective and Subjective Assessment of Nasal Patency. *Iran. J. Otorhinolaryngol.* **2016**, *28*, 313–319. [PubMed]
52. Hellgren, J.; Lundberg, M.; Rubek, N.; von Buchwald, C.; Steinsvåg, S.; Mäkitie, A. Unmet Challenges in Septoplasty—Nordic Studies from a Uniform Healthcare and Geographical Area. *Front. Surg.* **2022**, *9*, 1061440. [CrossRef] [PubMed]

53. Quine, S.M.; Eccles, R. Nasal Resistance from the Laboratory to the Clinic. *Curr. Opin. Otolaryngol. Head Neck Surg.* **1999**, *7*, 20–25. [CrossRef]

54. Eccles, R.; Jones, A.S. The Effect of Menthol in Nasal Resistance to Air Flow. *J. Laryngol. Otol.* **1983**, *97*, 705–709. [CrossRef] [PubMed]

55. Eccles, R.; Jawad, M.S.; Morris, S. The Effects of Oral Administration of (—)-Menthol on Nasal Resistance to Airflow and Nasal Sensation of Airflow in Subjects Suffering from Nasal Congestion Associated with the Common Cold. *J. Pharm. Pharmacol.* **1990**, *42*, 652–654. [CrossRef] [PubMed]

56. Gill, A.S.; Said, M.; Tollefson, T.T.; Steele, T.O. Update on Empty Nose Syndrome: Disease Mechanisms, Diagnostic Tools, and Treatment Strategies. *Curr. Opin. Otolaryngol. Head Neck Surg.* **2019**, *27*, 237–242. [CrossRef] [PubMed]

57. Malik, J.; Otto, B.A.; Zhao, K. Computational Fluid Dynamics (CFD) Modeling as an Objective Analytical Tool for Nasal/Upper Airway Breathing. *Curr. Otorhinolaryngol. Rep.* **2022**, *10*, 116–120. [CrossRef]

58. Di, M.-Y.; Jiang, Z.; Gao, Z.-Q.; Li, Z.; An, Y.-R.; Lv, W. Numerical Simulation of Airflow Fields in Two Typical Nasal Structures of Empty Nose Syndrome: A Computational Fluid Dynamics Study. *PLoS ONE* **2013**, *8*, e84243. [CrossRef]

59. Li, C.; Farag, A.A.; Leach, J.; Deshpande, B.; Jacobowitz, A.; Kim, K.; Otto, B.A.; Zhao, K. Computational Fluid Dynamics and Trigeminal Sensory Examinations of Empty Nose Syndrome Patients: Computational and Trigeminal Studies of ENS. *Laryngoscope* **2017**, *127*, E176–E184. [CrossRef]

60. Li, C.; Farag, A.A.; Maza, G.; McGhee, S.; Ciccone, M.A.; Deshpande, B.; Pribitkin, E.A.; Otto, B.A.; Zhao, K. Investigation of the Abnormal Nasal Aerodynamics and Trigeminal Functions among Empty Nose Syndrome Patients: Abnormal Nasal Aerodynamics in ENS Patients. *Int. Forum Allergy Rhinol.* **2018**, *8*, 444–452. [CrossRef]

61. Kimbell, J.S.; Frank, D.O.; Laud, P.; Garcia, G.J.M.; Rhee, J.S. Changes in Nasal Airflow and Heat Transfer Correlate with Symptom Improvement after Surgery for Nasal Obstruction. *J. Biomech.* **2013**, *46*, 2634–2643. [CrossRef]

62. Sullivan, C.D.; Garcia, G.J.M.; Frank-Ito, D.O.; Kimbell, J.S.; Rhee, J.S. Perception of Better Nasal Patency Correlates with Increased Mucosal Cooling after Surgery for Nasal Obstruction. *Otolaryngol. Neck Surg.* **2014**, *150*, 139–147. [CrossRef]

63. Zhao, K.; Jiang, J.; Blacker, K.; Lyman, B.; Dalton, P.; Cowart, B.J.; Pribitkin, E.A. Regional Peak Mucosal Cooling Predicts the Perception of Nasal Patency. *Laryngoscope* **2014**, *124*, 589–595. [CrossRef] [PubMed]

64. Bailey, R.S.; Casey, K.P.; Pawar, S.S.; Garcia, G.J.M. Correlation of Nasal Mucosal Temperature With Subjective Nasal Patency in Healthy Individuals. *JAMA Facial Plast. Surg.* **2017**, *19*, 46–52. [CrossRef]

65. Gaberino, C.; Rhee, J.S.; Garcia, G.J.M. Estimates of Nasal Airflow at the Nasal Cycle Mid-Point Improve the Correlation between Objective and Subjective Measures of Nasal Patency. *Respir. Physiol. Neurobiol.* **2017**, *238*, 23–32. [CrossRef] [PubMed]

66. Radulesco, T.; Meister, L.; Bouchet, G.; Varoquaux, A.; Giordano, J.; Mancini, J.; Dessi, P.; Perrier, P.; Michel, J. Correlations between Computational Fluid Dynamics and Clinical Evaluation of Nasal Airway Obstruction Due to Septal Deviation: An Observational Study. *Clin. Otolaryngol.* **2019**, *44*, 603–611. [CrossRef]

67. Borojeni, A.A.T.; Garcia, G.J.M.; Moghaddam, M.G.; Frank-Ito, D.O.; Kimbell, J.S.; Laud, P.W.; Koenig, L.J.; Rhee, J.S. Normative Ranges of Nasal Airflow Variables in Healthy Adults. *Int. J. Comput. Assist. Radiol. Surg.* **2020**, *15*, 87–98. [CrossRef] [PubMed]

68. Tjahjono, R.; Salati, H.; Inthavong, K.; Singh, N. Correlation of Nasal Mucosal Temperature and Nasal Patency—A Computational Fluid Dynamics Study. *Laryngoscope* **2023**, *133*, 1328–1335. [CrossRef]

69. Frank, D.O.; Kimbell, J.S.; Cannon, D.; Pawar, S.S.; Rhee, J.S. Deviated Nasal Septum Hinders Intranasal Sprays: A Computer Simulation Study. *Rhinol. J.* **2012**, *50*, 311–318. [CrossRef]

70. Al_Omari, A.K.; Saied, H.F.I.; Avrunin, O.G. Analysis of Changes of the Hydraulic Diameter and Determination of the Air Flow Modes in the Nasal Cavity. In *Image Processing and Communications Challenges 3*; Advances in Intelligent and Soft, Computing; Choraś, R.S., Ed.; Springer: Berlin/Heidelberg, Germany, 2011; Volume 102, pp. 303–310. ISBN 978-3-642-23153-7.

71. Pirozzoli, S. On Turbulent Friction in Straight Ducts with Complex Cross-Section: The Wall Law and the Hydraulic Diameter. *J. Fluid Mech.* **2018**, *846*, R1. [CrossRef]

72. O'Neill, G.; Tolley, N.S. The Complexities of Nasal Airflow: Theory and Practice. *J. Appl. Physiol.* **2019**, *127*, 1215–1223. [CrossRef]

73. Benim, A.C.; Maddala, S.B. Numerical Investigation of Friction Laws for Laminar and Turbulent Flow in Undulated Channels. *Int. J. Numer. Methods Heat Fluid Flow* **2021**, *31*, 1837–1856. [CrossRef]

74. Hasegawa, M.; Kern, E.B. Variations in Nasal Resistance in Man: A Rhinomanometric Study of the Nasal Cycle in 50 Human Subjects. *Rhinology* **1978**, *16*, 19–29. [PubMed]

75. Flanagan, P.; Eccles, R. Spontaneous Changes of Unilateral Nasal Airflow in Man. A Re-Examination of the 'Nasal Cycle'. *Acta Otolaryngol. (Stockh.)* **1997**, *117*, 590–595. [CrossRef]

76. Bozdemir, K.; Korkmaz, H.; Franzese, C.B. The Evaluation of the Nose, Nasal Cavity and Airway. In *All Around the Nose*; Cingi, C., Bayar Muluk, N., Eds.; Springer International Publishing: Cham, Switzerland, 2020; pp. 85–91. ISBN 978-3-030-21216-2. [CrossRef]

77. Vogt, K.; Jalowayski, A.A.; Althaus, W.; Cao, C.; Han, D.; Hasse, W.; Hoffrichter, H.; Mösges, R.; Pallanch, J.; Shah-Hosseini, K.; et al. 4-Phase-Rhinomanometry (4PR)—Basics and Practice 2010. *Rhinol. Suppl.* **2010**, *21*, 1–50. [PubMed]

78. Mlynski, G.; Löw, J. Die Rhinoresistometrie—Eine Weiterentwicklung der Rhinomanometrie. *Laryngo-Rhino-Otol.* **1993**, *72*, 608–610. [CrossRef] [PubMed]

79. Naito, K.; Iwata, S.; Cole, P.; Fraschetti, J.; Humphrey, D. An International Comparison of Rhinomanometry between Canada and Japan. *Rhinology* **1991**, *29*, 287–294.

80. Moore, M.; Eccles, R. Objective Evidence for the Efficacy of Surgical Management of the Deviated Septum as a Treatment for Chronic Nasal Obstruction: A Systematic Review: Objective Evidence for the Efficacy of Surgical Management of the Deviated Septum: A Systematic Review. *Clin. Otolaryngol.* **2011**, *36*, 106–113. [CrossRef]

81. Blevins, R.D. *Applied Fluid Dynamics Handbook*; Van Nostrand Reinhold Co.: New York, NY, USA, 1984; ISBN 978-0442212964.

82. Haaland, S.E. Simple and Explicit Formulas for the Friction Factor in Turbulent Pipe Flow. *J. Fluids Eng.* **1983**, *105*, 89–90. [CrossRef]

83. Li, C.; Jiang, J.; Dong, H.; Zhao, K. Computational Modeling and Validation of Human Nasal Airflow under Various Breathing Conditions. *J. Biomech.* **2017**, *64*, 59–68. [CrossRef]

84. Fletcher, D.F. Use of CFD to Stimulate Flow in the Nose and Airway: Best Practices, Quality and Future Perspectives. In Proceedings of the SCONA 2023 Conference, Brisbane, Australia, 8 March 2023.

85. Inthavong, K. Current State of the Art, Controversies, and the Future of Nose and Airway Simulations. In Proceedings of the SCONA 2023 Conference, Brisbane, Australia, 8 March 2023.

86. Burgos, M.A.; Sanmiguel-Rojas, E.; Singh, N.; Esteban-Ortega, F. DigBody®: A New 3D Modeling Tool for Nasal Virtual Surgery. *Comput. Biol. Med.* **2018**, *98*, 118–125. [CrossRef] [PubMed]

87. Quammen, C.W.; Taylor, R.M.; Krajcevski, P.; Mitran, S.; Enquobahrie, A.; Superfine, R.; Davis, B.; Davis, S.; Zdanski, C. The Virtual Pediatric Airways Workbench. *Stud. Health Technol. Inform.* **2016**, *220*, 295–300. [PubMed]

88. Sanmiguel-Rojas, E.; Burgos, M.A.; Esteban-Ortega, F. Nasal Surgery Handled by CFD Tools. *Int. J. Numer. Methods Biomed. Eng.* **2018**, *34*, e3126. [CrossRef] [PubMed]

89. Vanhille, D.L.; Garcia, G.J.M.; Asan, O.; Borojeni, A.A.T.; Frank-Ito, D.O.; Kimbell, J.S.; Pawar, S.S.; Rhee, J.S. Virtual Surgery for the Nasal Airway: A Preliminary Report on Decision Support and Technology Acceptance. *JAMA Facial Plast. Surg.* **2018**, *20*, 63–69. [CrossRef] [PubMed]

90. Moghaddam, M.G.; Garcia, G.J.M.; Frank-Ito, D.O.; Kimbell, J.S.; Rhee, J.S. Virtual Septoplasty: A Method to Predict Surgical Outcomes for Patients with Nasal Airway Obstruction. *Int. J. Comput. Assist. Radiol. Surg.* **2020**, *15*, 725–735. [CrossRef] [PubMed]

91. Patankar, S.V. *Numerical Heat Transfer and Fluid Flow*; Series in Computational Methods in Mechanics and Thermal Sciences; Hemisphere Publishing Corporation/Taylor & Francis: New York, NY, USA, 1980; ISBN 0-89116-522-3.

92. Anderson, J.D. *Computational Fluid Dynamics: The Basics with Applications*; McGraw-Hill Series in Mechanical Engineering; McGraw-Hill: New York, NY, USA, 1995; ISBN 978-0-07-001685-9.

93. Versteeg, H.K.; Malalasekera, W. *An Introduction to Computational Fluid Dynamics: The Finite Volume Method*; Pearson Education Ltd.: Harlow, UK, 2007; ISBN 978-0-13-127498-3.

94. Rodriguez, S. *Applied Computational Fluid Dynamics and Turbulence Modeling: Practical Tools, Tips and Techniques*; Springer International Publishing: Cham, Switzerland, 2019; ISBN 978-3-030-28690-3.

95. Roychowdhury, D.G. Best Practice Guidelines in CFD. In *Computational Fluid Dynamics for Incompressible Flow*; CRC Press: Boca Raton, FL, USA, 2020; pp. 323–334. ISBN 978-0-367-80917-1.

96. Cercos-Pita, J.L. Computational Reconstruction of the Human Nasal Airway. In *Clinical and Biomedical Engineering in the Human Nose*; Inthavong, K., Singh, N., Wong, E., Tu, J., Eds.; Biological and Medical Physics, Biomedical Engineering; Springer: Singapore, 2021; pp. 63–84. ISBN 9789811567155.

97. DenOtter, T.D.; Schubert, J. Hounsfield Unit. In *StatPearls [Internet]*; StatPearls Publishing: Treasure Island, FL, USA, 2023. Available online: https://www.ncbi.nlm.nih.gov/books/NBK547721/ (accessed on 1 January 2024).

98. Dance, D.R. (Ed.) *Diagnostic Radiology Physics: A Handbook for Teachers and Students*; Non-serial Publications; International Atomic Energy Agency: Vienna, Austria, 2014; ISBN 978-92-0-131010-1.

99. Ayodele, O.J.; Oluwatosin, A.E.; Taiwo, O.C.; Dare, A.A. Computational Fluid Dynamics Modeling in Respiratory Airways Obstruction: Current Applications and Prospects. *Int. J. Biomed. Sci. Eng.* **2021**, *9*, 16–26. [CrossRef]

100. Cherobin, G.B.; Voegels, R.L.; Gebrim, E.M.M.S.; Garcia, G.J.M. Sensitivity of Nasal Airflow Variables Computed via Computational Fluid Dynamics to the Computed Tomography Segmentation Threshold. *PLoS ONE* **2018**, *13*, e0207178. [CrossRef]

101. Quadrio, M.; Pipolo, C.; Corti, S.; Messina, F.; Pesci, C.; Saibene, A.M.; Zampini, S.; Felisati, G. Effects of CT Resolution and Radiodensity Threshold on the CFD Evaluation of Nasal Airflow. *Med. Biol. Eng. Comput.* **2016**, *54*, 411–419. [CrossRef]

102. Lintermann, A. Computational Meshing for CFD Simulations. In *Clinical and Biomedical Engineering in the Human Nose*; Inthavong, K., Singh, N., Wong, E., Tu, J., Eds.; Biological and Medical Physics, Biomedical Engineering; Springer: Singapore, 2021; pp. 85–115. ISBN 9789811567155.

103. Bass, K.; Boc, S.; Hindle, M.; Dodson, K.; Longest, W. High-Efficiency Nose-to-Lung Aerosol Delivery in an Infant: Development of a Validated Computational Fluid Dynamics Method. *J. Aerosol Med. Pulm. Drug Deliv.* **2019**, *32*, 132–148. [CrossRef]

104. Thomas, M.L.; Longest, P.W. Evaluation of the Polyhedral Mesh Style for Predicting Aerosol Deposition in Representative Models of the Conducting Airways. *J. Aerosol Sci.* **2022**, *159*, 105851. [CrossRef]

105. Inthavong, K.; Das, P.; Singh, N.; Sznitman, J. In Silico Approaches to Respiratory Nasal Flows: A Review. *J. Biomech.* **2019**, *97*, 109434. [CrossRef] [PubMed]

106. Inthavong, K.; Chetty, A.; Shang, Y.; Tu, J. Examining Mesh Independence for Flow Dynamics in the Human Nasal Cavity. *Comput. Biol. Med.* **2018**, *102*, 40–50. [CrossRef]

107. Ashraf, W.; Jacobson, N.; Popplewell, N.; Moussavi, Z. Fluid–Structure Interaction Modelling of the Upper Airway with and without Obstructive Sleep Apnea: A Review. *Med. Biol. Eng. Comput.* **2022**, *60*, 1827–1849. [CrossRef]

108. Tennekes, H.; Lumley, J.L. *A First Course in Turbulence*; Nachdruck des Originals von 1972; The MIT Press: Cambridge, MA, USA, 1972; ISBN 978-0-262-53630-1.

109. Pope, S.B. *Turbulent Flows*, 1st ed.; Cambridge University Press: Cambridge, UK, 2000; ISBN 978-0-521-59125-6.

110. Wilcox, D.C. *Turbulence Modeling for CFD*, 3rd ed.; 2 print; DCW Industries: La Cañada, CA, USA, 2010; ISBN 978-1-928729-08-2.

111. Menter, F.R. *Best Practice: Scale-Resolving Simulations in Ansys CFD*; ANSYS Rep.; ANSYS, Inc.: Canonsburg, PA, USA, 2015.

112. Menter, F.R.; Lechner, R.; Matyushenko, A. *Best Practice: RANS Turbulence Modeling in Ansys CFD*; ANSYS Rep.; ANSYS, Inc.: Canonsburg, PA, USA, 2021.

113. ANSYS. *ANSYS Fluent Theory Guide*; ANSYS, Inc.: Canonsburg, PA, USA, 2023.

114. Rumsey, C.; Smith, B.; Huang, G. *NASA Turbulence Modeling Resource*; NASA, Langley Research Center: Hampton, VA, USA, 2023. Available online: https://turbmodels.larc.nasa.gov (accessed on 11 November 2023).

115. Frank-Ito, D.O.; Garcia, G. Clinical Implications of Nasal Airflow Simulations. In *Clinical and Biomedical Engineering in the Human Nose*; Biological and Medical Physics, Biomedical Engineering; Inthavong, K., Singh, N., Wong, E., Tu, J., Eds.; Springer: Singapore, 2021; pp. 157–192. ISBN 9789811567155.

116. Kimbell, J.S.; Garcia, G.J.M.; Frank, D.O.; Cannon, D.E.; Pawar, S.S.; Rhee, J.S. Computed Nasal Resistance Compared with Patient-Reported Symptoms in Surgically Treated Nasal Airway Passages: A Preliminary Report. *Am. J. Rhinol. Allergy* **2012**, *26*, 94–98. [CrossRef]

117. Cherobin, G.B.; Voegels, R.L.; Pinna, F.R.; Gebrim, E.M.M.S.; Bailey, R.S.; Garcia, G.J.M. Rhinomanometry Versus Computational Fluid Dynamics: Correlated, but Different Techniques. *Am. J. Rhinol. Allergy* **2021**, *35*, 245–255. [CrossRef] [PubMed]

118. Zachow, S.; Muigg, P.; Hildebrandt, T.; Doleisch, H.; Hege, H.-C. Visual Exploration of Nasal Airflow. *IEEE Trans. Vis. Comput. Graph.* **2009**, *15*, 1407–1414. [CrossRef]

119. Hildebrandt, T. Das Konzept der Rhinorespiratorischen Homöostase—Ein Neuer Theoretischer Ansatz für die Diskussion Physiologischer und Physikalischer Zusammenhänge bei der Nasenatmung. Ph.D. Thesis, Albert-Ludwigs-Universität, Freiburg im Breisgau, Germany, 2011. Available online: https://freidok.uni-freiburg.de/data/8177 (accessed on 1 January 2024).

120. Hildebrandt, T.; (Charité–Universitätsmedizin Berlin, Berlin, Germany). Personal communication, 2020.

121. Schmidt, N.; Behrbohm, H.; Goubergrits, L.; Hildebrandt, T.; Brüning, J. Comparison of Rhinomanometric and Computational Fluid Dynamic Assessment of Nasal Resistance with Respect to Measurement Accuracy. *Int. J. Comput. Assist. Radiol. Surg.* **2022**, *17*, 1519–1529. [CrossRef]

122. Dong, J.; Sun, Q.; Shang, Y.; Zhang, Y.; Tian, L.; Tu, J. Numerical Comparison of Inspiratory Airflow Patterns in Human Nasal Cavities with Distinct Age Differences. *Int. J. Numer. Methods Biomed. Eng.* **2022**, *38*, e3565. [CrossRef] [PubMed]

123. Hemtiwakorn, K.; Mahasitthiwat, V.; Tungjitkusolmun, S.; Hamamoto, K.; Pintavirooj, C. Patient-Specific Aided Surgery Approach of Deviated Nasal Septum Using Computational Fluid Dynamics. *IEEJ Trans. Electr. Electron. Eng.* **2015**, *10*, 274–286. [CrossRef]

124. Osman, J.; Großmann, F.; Brosien, K.; Kertzscher, U.; Goubergrits, L.; Hildebrandt, T. Assessment of Nasal Resistance Using Computational Fluid Dynamics. *Curr. Dir. Biomed. Eng.* **2016**, *2*, 617–621. [CrossRef]

125. Berger, M.; Giotakis, A.I.; Pillei, M.; Mehrle, A.; Kraxner, M.; Kral, F.; Recheis, W.; Riechelmann, H.; Freysinger, W. Agreement between Rhinomanometry and Computed Tomography-Based Computational Fluid Dynamics. *Int. J. Comput. Assist. Radiol. Surg.* **2021**, *16*, 629–638. [CrossRef]

126. Kaneda, S.; Goto, F.; Okami, K.; Mitsutani, R.; Takakura, Y. Numerical Simulation of Nasal Resistance Using Three-Dimensional Models of the Nasal Cavity and Paranasal Sinus. *Tokai J. Exp. Clin. Med.* **2023**, *48*, 56–61.

127. Hebbink, R.H.J.; Wessels, B.J.; Hagmeijer, R.; Jain, K. Computational Analysis of Human Upper Airway Aerodynamics. *Med. Biol. Eng. Comput.* **2023**, *61*, 541–553. [CrossRef]

128. Guide: Guide for the Verification and Validation of Computational Fluid Dynamics Simulations (AIAA G-077-1998(2002)). In *Guide: Guide for the Verification and Validation of Computational Fluid Dynamics Simulations (AIAA G-077-1998(2002))*; Computational Fluid Dynamics Committee (Ed.) American Institute of Aeronautics and Astronautics, Inc.: Washington, DC, USA, 1998; ISBN 978-1-56347-285-5. [CrossRef]

129. Pawade, A.R. Computational Modeling of Airflow in a Human Nasal Cavity. Master's Thesis, Chalmers University of Technology, Gothenburg, Sweden, 2021.

130. Reid, A.W.N.; Chen, D.H.; Wen, H.; Li, H.; Wang, Z.; Hu, Y.; Zhang, F.; Bele, E.; Tan, P.J.; East, C. The Virtual Nose: Assessment of Static Nasal Airway Obstruction Using Computational Simulations and 3D-Printed Models. *Facial Plast. Surg. Aesthetic Med.* **2022**, *24*, 20–26. [CrossRef]

131. Patel, R.G.; Garcia, G.J.M.; Frank-Ito, D.O.; Kimbell, J.S.; Rhee, J.S. Simulating the Nasal Cycle with Computational Fluid Dynamics. *Otolaryngol. Neck Surg.* **2015**, *152*, 353–360. [CrossRef] [PubMed]

132. Susaman, N.; Cingi, C.; Mullol, J. Is the Nasal Cycle Real? How Important Is It? In *Challenges in Rhinology*; Cingi, C., Bayar Muluk, N., Scadding, G.K., Mladina, R., Eds.; Springer International Publishing: Cham, Switzerland, 2021; pp. 1–8. ISBN 978-3-030-50898-2.

133. Hasegawa, M. Nasal Cycle and Postural Variations in Nasal Resistance. *Ann. Otol. Rhinol. Laryngol.* **1982**, *91*, 112–114. [CrossRef] [PubMed]

134. Haight, J.S.; Cole, P. Unilateral Nasal Resistance and Asymmetrical Body Pressure. *J. Otolaryngol. Suppl.* **1986**, *16*, 1–31.

135. Fodil, R.; Brugel-Ribere, L.; Croce, C.; Sbirlea-Apiou, G.; Larger, C.; Papon, J.-F.; Delclaux, C.; Coste, A.; Isabey, D.; Louis, B. Inspiratory Flow in the Nose: A Model Coupling Flow and Vasoerectile Tissue Distensibility. *J. Appl. Physiol.* **2005**, *98*, 288–295. [CrossRef]
136. Bailie, N.; Hanna, B.; Watterson, J.; Gallagher, G. An Overview of Numerical Modelling of Nasal Airflow. *Rhinology* **2006**, *44*, 53–57.
137. Akmenkalne, L.; Prill, M.; Vogt, K. Nasal Valve Elastography: Quantitative Determination of the Mobility of the Nasal Valve. *Rhinol. Online* **2019**, *2*, 81–86. [CrossRef]
138. Vogt, K.; Zhang, L. Airway Assessment by Four-Phase Rhinomanometry in Septal Surgery. *Curr. Opin. Otolaryngol. Head Neck Surg.* **2012**, *20*, 33–39. [CrossRef] [PubMed]
139. Vogt, K.; Wernecke, K.-D.; Behrbohm, H.; Gubisch, W.; Argale, M. Four-Phase Rhinomanometry: A Multicentric Retrospective Analysis of 36,563 Clinical Measurements. *Eur. Arch. Otorhinolaryngol.* **2016**, *273*, 1185–1198. [CrossRef] [PubMed]
140. Groß, T.F.; Peters, F. A Fluid Mechanical Interpretation of Hysteresis in Rhinomanometry. *ISRN Otolaryngol.* **2011**, *2011*, 126520. [CrossRef] [PubMed]
141. Bui, N.L.; Ong, S.H.; Foong, K.W.C. Automatic Segmentation of the Nasal Cavity and Paranasal Sinuses from Cone-Beam CT Images. *Int. J. Comput. Assist. Radiol. Surg.* **2015**, *10*, 1269–1277. [CrossRef] [PubMed]
142. Aanderaa, L.Ø. Experimental Investigation of Rhinomanometry. Master's Thesis, NTNU, Trondheim, Norway, 2021.
143. Silkoff, P.E.; Chakravorty, S.; Chapnik, J.; Cole, P.; Zamel, N. Reproducibility of Acoustic Rhinometry and Rhinomanometry in Normal Subjects. *Am. J. Rhinol.* **1999**, *13*, 131–136. [CrossRef] [PubMed]
144. Carney, A.S.; Bateman, N.D.; Jones, N.S. Reliable and Reproducible Anterior Active Rhinomanometry for the Assessment of Unilateral Nasal Resistance. *Clin. Otolaryngol. Allied Sci.* **2000**, *25*, 499–503. [CrossRef] [PubMed]
145. Thulesius, H.L.; Cervin, A.; Jessen, M. Can We Always Trust Rhinomanometry? *Rhinol. J.* **2011**, *49*, 46–52. [CrossRef] [PubMed]
146. ERCOFTAC. *Best Practice Guidelines, Industrial Computational Fluid Dynamics of Single-Phase Flows*; ERCOFTAC: London, UK, 2000; ISBN 978-0-9955779-2-3.
147. Andersson, B.; Andersson, R.; Håkansson, L.; Mortensen, M.; Sudiyo, R.; Van Wachem, B. Best Practice Guidelines. In *Computational Fluid Dynamics for Engineers*; Cambridge University Press: Cambridge, UK, 2011; pp. 174–180. ISBN 978-1-107-01895-2.
148. Schillaci, A.; Quadrio, M. Importance of the Numerical Schemes in the CFD of the Human Nose. *J. Biomech.* **2022**, *138*, 111100. [CrossRef]
149. Proetz, A.W. Air Currents in the Upper Respiratory Tract and Their Clinical Importance. *Ann. Otol. Rhinol. Laryngol.* **1951**, *60*, 439–467. [CrossRef]
150. Sullivan, K.J.; Chang, H.K. Steady and Oscillatory Transnasal Pressure-Flow Relationships in Healthy Adults. *J. Appl. Physiol.* **1991**, *71*, 983–992. [CrossRef]
151. Hahn, I.; Scherer, P.W.; Mozell, M.M. Velocity Profiles Measured for Airflow through a Large-Scale Model of the Human Nasal Cavity. *J. Appl. Physiol.* **1993**, *75*, 2273–2287. [CrossRef]
152. Keyhani, K.; Scherer, P.W.; Mozell, M.M. Numerical Simulation of Airflow in the Human Nasal Cavity. *J. Biomech. Eng.* **1995**, *117*, 429–441. [CrossRef] [PubMed]
153. Shi, H.; Kleinstreuer, C.; Zhang, Z. Laminar Airflow and Nanoparticle or Vapor Deposition in a Human Nasal Cavity Model. *J. Biomech. Eng.* **2006**, *128*, 697–706. [CrossRef] [PubMed]
154. Hörschler, I.; Schröder, W.; Meinke, M. On the Assumption of Steadiness of Nasal Cavity Flow. *J. Biomech.* **2010**, *43*, 1081–1085. [CrossRef] [PubMed]
155. Womersley, J.R. Method for the Calculation of Velocity, Rate of Flow and Viscous Drag in Arteries When the Pressure Gradient Is Known. *J. Physiol.* **1955**, *127*, 553–563. [CrossRef] [PubMed]
156. Loudon, C.; Tordesillas, A. The Use of the Dimensionless Womersley Number to Characterize the Unsteady Nature of Internal Flow. *J. Theor. Biol.* **1998**, *191*, 63–78. [CrossRef]
157. Doorly, D.J.; Taylor, D.J.; Schroter, R.C. Mechanics of Airflow in the Human Nasal Airways. *Respir. Physiol. Neurobiol.* **2008**, *163*, 100–110. [CrossRef]
158. Bosykh, L.Y.; Ganimedov, V.L.; Muchnaya, M.I.; Sadovskii, A.S. Influence of the Respiratory Cycle Structure on the Flow Field in Human Nasal Cavity at a Fixed Level of Breath Depth. *AIP Conf. Proc.* **2016**, *1770*, 030085.
159. Bradshaw, K.; Warfield-McAlpine, P.; Vahaji, S.; Emmerling, J.; Salati, H.; Sacks, R.; Fletcher, D.F.; Singh, N.; Inthavong, K. New Insights into the Breathing Physiology from Transient Respiratory Nasal Simulation. *Phys. Fluids* **2022**, *34*, 115103. [CrossRef]
160. Tretiakow, D.; Tesch, K.; Markiet, K.; Przewoźny, T.; Kusiak, A.; Cichońska, D.; Skorek, A. Numerical Analysis of the Ostiomeatal Complex Aeration Using the CFD Method. *Sci. Rep.* **2023**, *13*, 3980. [CrossRef]
161. Koeppen, B.M.; Stanton, B.A. (Eds.) *Berne & Levy Physiology*, 7th ed.; Elsevier: Philadelphia, PA, USA, 2018; ISBN 978-0-323-39394-2.
162. He, S.; Seddighi, M. Transition of Transient Channel Flow after a Change in Reynolds Number. *J. Fluid Mech.* **2015**, *764*, 395–427. [CrossRef]
163. Guerrero, B.; Lambert, M.F.; Chin, R.C. Transient Dynamics of Accelerating Turbulent Pipe Flow. *J. Fluid Mech.* **2021**, *917*, A43. [CrossRef]
164. Guerrero, B.; Lambert, M.F.; Chin, R.C. Transient Behaviour of Decelerating Turbulent Pipe Flows. *J. Fluid Mech.* **2023**, *962*, A44. [CrossRef]
165. Yellin, E.L. Laminar-Turbulent Transition Process in Pulsatile Flow. *Circ. Res.* **1966**, *19*, 791–804. [CrossRef] [PubMed]

166. Gündoğdu, M.Y.; Çarpinlioğlu, M.Ö. Present State of Art on Pulsatile Flow Theory. Part 1. Laminar and Transitional Flow Regimes. *JSME Int. J. Ser. B* **1999**, *42*, 384–397. [CrossRef]
167. Gündoğdu, M.Y.; Çarpinlioğlu, M.Ö. Present State of Art on Pulsatile Flow Theory. Part 2. Turbulent Flow Regime. *JSME Int. J. Ser. B* **1999**, *42*, 398–410. [CrossRef]
168. Xu, D.; Warnecke, S.; Song, B.; Ma, X.; Hof, B. Transition to Turbulence in Pulsating Pipe Flow. *J. Fluid Mech.* **2017**, *831*, 418–432. [CrossRef]
169. Sayadi, T.; Moin, P. Large Eddy Simulation of Controlled Transition to Turbulence. *Phys. Fluids* **2012**, *24*, 114103. [CrossRef]
170. Zang, H.; Liu, Y.; Han, D.; Zhang, L.; Wang, T.; Sun, X.; Li, L. Airflow and Temperature Distribution inside the Maxillary Sinus: A Computational Fluid Dynamics Simulation. *Acta Otolaryngol. (Stockh.)* **2012**, *132*, 637–644. [CrossRef]
171. Kim, S.K.; Na, Y.; Kim, J.-I.; Chung, S.-K. Patient Specific CFD Models of Nasal Airflow: Overview of Methods and Challenges. *J. Biomech.* **2013**, *46*, 299–306. [CrossRef]
172. Paz, C.; Suarez, E.; Concheiro, M.; Porteiro, J. CFD Transient Simulation of a Breathing Cycle in an Oral-Nasal Extrathoracic Model. *J. Appl. Fluid Mech.* **2017**, *10*, 777–784. [CrossRef]
173. Chen, S.; Wang, J.; Liu, D.; Lei, L.; Wu, W.; Liu, Z.; Lee, C. Open Oral Cavity Has Little Effects on Upper Airway Aerodynamics in Children with Obstructive Sleep Apnea Syndrome: A Computational Fluid Dynamics Study Based on Patient-Specific Models. *J. Biomech.* **2021**, *121*, 110383. [CrossRef] [PubMed]
174. Stoddard, D.G.; Pallanch, J.F.; Hamilton, G.S. The Effect of Vibrissae on Subjective and Objective Measures of Nasal Obstruction. *Am. J. Rhinol. Allergy* **2015**, *29*, 373–377. [CrossRef] [PubMed]
175. Lee, C.F.; Arifin Ahmad, K.; Ismail, R.; Abdul Hamid, S. Numerical Study of Mucous Layer Effects on Nasal Airflow. *Biomed. Eng. Appl. Basis Commun.* **2012**, *24*, 327–332. [CrossRef]
176. Frank-Ito, D.O.; Wofford, M.; Schroeter, J.D.; Kimbell, J.S. Influence of Mesh Density on Airflow and Particle Deposition in Sinonasal Airway Modeling. *J. Aerosol Med. Pulm. Drug Deliv.* **2016**, *29*, 46–56. [CrossRef] [PubMed]
177. Launder, B.E.; Spalding, D.B. *Lectures in Mathematical Models of Turbulence*; Academic Press: London, UK; New York, NY, USA, 1972; ISBN 978-0-12-438050-9.
178. Courant, R.; Friedrichs, K.; Lewy, H. Über die partiellen Differenzengleichungen der mathematischen Physik. *Math. Ann.* **1928**, *100*, 32–74. [CrossRef]
179. Courant, R.; Friedrichs, K.; Lewy, H. On the Partial Difference Equations of Mathematical Physics. *IBM J. Res. Dev.* **1967**, *11*, 215–234. [CrossRef]
180. Taylor, D.J.; Doorly, D.J.; Schroter, R.C. Inflow Boundary Profile Prescription for Numerical Simulation of Nasal Airflow. *J. R. Soc. Interface* **2010**, *7*, 515–527. [CrossRef]
181. Wu, H.; Wang, M.; Wang, J.; An, Y.; Wang, H.; Huang, Y. Direct Visualizations of Air Flow in the Human Upper Airway Using In-Vitro Models. *Sci. China Life Sci.* **2019**, *62*, 235–243. [CrossRef] [PubMed]
182. Han, J.K.; Stringer, S.P.; Rosenfeld, R.M.; Archer, S.M.; Baker, D.P.; Brown, S.M.; Edelstein, D.R.; Gray, S.T.; Lian, T.S.; Ross, E.J.; et al. Clinical Consensus Statement: Septoplasty with or without Inferior Turbinate Reduction. *Otolaryngol. Neck Surg.* **2015**, *153*, 708–720. [CrossRef]
183. Puccia, R.; Pawar, S.S. Key Points on Functional Rhinoplasty Patient Evaluation. *Curr. Otorhinolaryngol. Rep.* **2022**, *10*, 127–133. [CrossRef]
184. Urban, M.J.; Friedman, J.J.; Husain, I.; LoSavio, P.S. Overview of Recent Advances in Surgical Treatments for OSA. *Curr. Sleep Med. Rep.* **2020**, *6*, 199–207. [CrossRef]
185. Vicory, J.; Garcia, G.J.M.; Rhee, J.S.; Enquobahrie, A. Toward Automatic Atlas-Based Surgical Planning for Septoplasty. *Int. J. Comput. Assist. Radiol. Surg.* **2022**, *17*, 403–411. [CrossRef] [PubMed]
186. Lintermann, A. Application of Computational Fluid Dynamics Methods to Understand Nasal Cavity Flows. In *All Around the Nose*; Cingi, C., Bayar Muluk, N., Eds.; Springer International Publishing: Cham, Switzerland, 2020; pp. 75–84. ISBN 978-3-030-21216-2.
187. Tu, J.; Inthavong, K.; Ahmadi, G. *Computational Fluid and Particle Dynamics in the Human Respiratory System*; Biological and Medical Physics, Biomedical Engineering; Springer: Dordrecht, The Netherlands; New York, NY, USA, 2012; ISBN 978-94-007-4487-5. [CrossRef]
188. Inthavong, K.; Singh, N.; Wong, E.; Tu, J. (Eds.) *Clinical and Biomedical Engineering in the Human Nose: A Computational Fluid Dynamics Approach*; Biological and Medical Physics, Biomedical Engineering; Springer: Singapore, 2021; ISBN 9789811567155.
189. Leong, S.C. A Review of the Implications of Computational Fluid Dynamic Studies on Nasal Airflow and Physiology. *Rhinol. J.* **2010**, *48*, 139–145. [CrossRef]
190. Radulesco, T.; Meister, L.; Bouchet, G.; Giordano, J.; Dessi, P.; Perrier, P.; Michel, J. Functional Relevance of Computational Fluid Dynamics in the Field of Nasal Obstruction: A Literature Review. *Clin. Otolaryngol.* **2019**, *44*, 801–809. [CrossRef]
191. Faizal, W.M.; Ghazali, N.N.N.; Khor, C.Y.; Badruddin, I.A.; Zainon, M.Z.; Yazid, A.A.; Ibrahim, N.B.; Razi, R.M. Computational Fluid Dynamics Modelling of Human Upper Airway: A Review. *Comput. Methods Programs Biomed.* **2020**, *196*, 105627. [CrossRef]
192. Burgos, M.A.; Sanmiguel-Rojas, E.; del Pino, C.; Sevilla-García, M.A.; Esteban-Ortega, F. New CFD Tools to Evaluate Nasal Airflow. *Eur. Arch. Otorhinolaryngol.* **2017**, *274*, 3121–3128. [CrossRef]
193. Pugachev, A.; Arnold, M.; Burgmann, S.; Janoske, U.; Bicsák, Á.; Abel, D.; Linssen, J.; Bonitz, L. Application of Patient-specific Simulation Workflow for Obstructive Sleep Apnea Diagnosis and Treatment with a Mandibular Advancement Device. *Int. J. Numer. Methods Biomed. Eng.* **2020**, *36*, e3350. [CrossRef] [PubMed]

194. Tretiakow, D.; Tesch, K.; Meyer-Szary, J.; Markiet, K.; Skorek, A. Three-Dimensional Modeling and Automatic Analysis of the Human Nasal Cavity and Paranasal Sinuses Using the Computational Fluid Dynamics Method. *Eur. Arch. Otorhinolaryngol.* **2021**, *278*, 1443–1453. [CrossRef]

195. Waldmann, M.; Grosch, A.; Witzler, C.; Lehner, M.; Benda, O.; Koch, W.; Vogt, K.; Kohn, C.; Schröder, W.; Göbbert, J.H.; et al. An Effective Simulation- and Measurement-Based Workflow for Enhanced Diagnostics in Rhinology. *Med. Biol. Eng. Comput.* **2022**, *60*, 365–391. [CrossRef] [PubMed]

196. Quadrio, M.; Pipolo, C.; Corti, S.; Lenzi, R.; Messina, F.; Pesci, C.; Felisati, G. Review of Computational Fluid Dynamics in the Assessment of Nasal Air Flow and Analysis of Its Limitations. *Eur. Arch. Otorhinolaryngol.* **2014**, *271*, 2349–2354. [CrossRef]

197. Le, T.B.; Moghaddam, M.G.; Woodson, B.T.; Garcia, G.J.M. Airflow Limitation in a Collapsible Model of the Human Pharynx: Physical Mechanisms Studied with Fluid-structure Interaction Simulations and Experiments. *Physiol. Rep.* **2019**, *7*, e14099. [CrossRef] [PubMed]

198. Xi, J.; Wang, J.; Si, X.A.; Dong, H. Direct Numerical Simulations and Flow-Pressure Acoustic Analyses of Flapping-Uvula-Induced Flow Evolutions within Normal and Constricted Pharynx. *Theor. Comput. Fluid Dyn.* **2023**, *37*, 131–149. [CrossRef]

199. Larimi, M.M.; Babamiri, A.; Biglarian, M.; Ramiar, A.; Tabe, R.; Inthavong, K.; Farnoud, A. Numerical and Experimental Analysis of Drug Inhalation in Realistic Human Upper Airway Model. *Pharmaceuticals* **2023**, *16*, 406. [CrossRef]

200. Stuiver, M. Biophysics of the Sense of Smell. Ph.D. Thesis, University of Groningen, Groningen, The Netherlands, 1958.

201. Masing, H. Investigations about the Course of Flow in the Nose Model. *Arch Klin Exp Ohr Nas Kehlkopf* **1967**, *189*, 371–381. [CrossRef]

202. Hornung, D.E.; Leopold, D.A.; Youngentob, S.L.; Sheehe, P.R.; Gagne, G.M.; Thomas, F.D.; Mozell, M.M. Airflow Patterns in a Human Nasal Model. *Arch. Otolaryngol. Head Neck Surg.* **1987**, *113*, 169–172. [CrossRef]

203. Kelly, J.T.; Prasad, A.K.; Wexler, A.S. Detailed Flow Patterns in the Nasal Cavity. *J. Appl. Physiol.* **2000**, *89*, 323–337. [CrossRef]

204. Mylavarapu, G.; Murugappan, S.; Mihaescu, M.; Kalra, M.; Khosla, S.; Gutmark, E. Validation of Computational Fluid Dynamics Methodology Used for Human Upper Airway Flow Simulations. *J. Biomech.* **2009**, *42*, 1553–1559. [CrossRef] [PubMed]

205. Ormiskangas, J.; Valtonen, O.; Kivekäs, I.; Dean, M.; Poe, D.; Järnstedt, J.; Lekkala, J.; Harju, T.; Saarenrinne, P.; Rautiainen, M. Assessment of PIV Performance in Validating CFD Models from Nasal Cavity CBCT Scans. *Respir. Physiol. Neurobiol.* **2020**, *282*, 103508. [CrossRef]

206. Van Strien, J.; Shrestha, K.; Gabriel, S.; Lappas, P.; Fletcher, D.F.; Singh, N.; Inthavong, K. Pressure Distribution and Flow Dynamics in a Nasal Airway Using a Scale Resolving Simulation. *Phys. Fluids* **2021**, *33*, 011907. [CrossRef]

207. Zhao, K.; Jiang, J. What Is Normal Nasal Airflow? A Computational Study of 22 Healthy Adults: Normal Human Nasal Airflow. *Int. Forum Allergy Rhinol.* **2014**, *4*, 435–446. [CrossRef]

208. Ramprasad, V.H.; Frank-Ito, D.O. A Computational Analysis of Nasal Vestibule Morphologic Variabilities on Nasal Function. *J. Biomech.* **2016**, *49*, 450–457. [CrossRef]

209. Sanmiguel-Rojas, E.; Burgos, M.A.; del Pino, C.; Sevilla-García, M.A.; Esteban-Ortega, F. Robust Nondimensional Estimators to Assess the Nasal Airflow in Health and Disease: Mathematical Estimators to Assess Nasal Airflow. *Int. J. Numer. Methods Biomed. Eng.* **2018**, *34*, e2906. [CrossRef]

210. Li, L.; Han, D.; Zang, H.; London, N.R. Aerodynamics Analysis of the Impact of Nasal Surgery on Patients with Obstructive Sleep Apnea and Nasal Obstruction. *ORL* **2022**, *84*, 62–69. [CrossRef]

211. Ormiskangas, J.; Valtonen, O.; Harju, T.; Rautiainen, M.; Kivekäs, I. Computational Fluid Dynamics Calculations in Inferior Turbinate Surgery: A Cohort Study. *Eur. Arch. Otorhinolaryngol.* **2023**, *280*, 4923–4931. [CrossRef]

212. Sagandykova, N.S.; Fakhradiyev, I.R.; Sajjala, S.R.; Taukeleva, S.A.; Shemetova, D.E.; Saliev, T.M.; Tanabayeva, S.B.; Zhao, Y. Patient-Specific CFD Simulation of Aerodynamics for Nasal Pathology: A Combined Computational and Experimental Study. *Comput. Methods Biomech. Biomed. Eng. Imaging Vis.* **2021**, *9*, 470–479. [CrossRef]

213. Corda, J.V.; Shenoy, B.S.; Lewis, L.; Khader, S.M.A.; Ahmad, K.A.; Zuber, M. Nasal Airflow Patterns in a Patient with Septal Deviation and Comparison with a Healthy Nasal Cavity Using Computational Fluid Dynamics. *Front. Mech. Eng.* **2022**, *8*, 1009640. [CrossRef]

214. Swift, D.L.; Proctor, D.F. Access of Air to the Respiratory Tract. In *Respiratory Defense Mechanisms, Part I*; Brain, J.D., Proctor, D.F., Reid, L.M., Eds.; Marcel Dekker: New York, NY, USA, 1977; pp. 63–95.

215. Schreck, S.; Sullivan, K.J.; Ho, C.M.; Chang, H.K. Correlations between Flow Resistance and Geometry in a Model of the Human Nose. *J. Appl. Physiol.* **1993**, *75*, 1767–1775. [CrossRef] [PubMed]

216. Simmen, D.; Scherrer, J.L.; Moe, K.; Heinz, B. A Dynamic and Direct Visualization Model for the Study of Nasal Airflow. *Arch. Otolaryngol. Neck Surg.* **1999**, *125*, 1015. [CrossRef] [PubMed]

217. Xi, J.; Kim, J.; Si, X.A. Effects of Nostril Orientation on Airflow Dynamics, Heat Exchange, and Particle Depositions in Human Noses. *Eur. J. Mech. BFluids* **2016**, *55*, 215–228. [CrossRef]

218. Bates, A.J. Fundamentals of Fluid Dynamics. In *Clinical and Biomedical Engineering in the Human Nose*; Biological and Medical Physics, Biomedical Engineering; Inthavong, K., Singh, N., Wong, E., Tu, J., Eds.; Springer: Singapore, 2021; pp. 117–156. ISBN 9789811567155.

219. Segal, R.A.; Kepler, G.M.; Kimbell, J.S. Effects of Differences in Nasal Anatomy on Airflow Distribution: A Comparison of Four Individuals at Rest. *Ann. Biomed. Eng.* **2008**, *36*, 1870–1882. [CrossRef] [PubMed]

220. Phuong, N.L.; Ito, K. Investigation of Flow Pattern in Upper Human Airway Including Oral and Nasal Inhalation by PIV and CFD. *Build. Environ.* **2015**, *94*, 504–515. [CrossRef]

221. Cozzi, F.; Felisati, G.; Quadrio, M. Velocity Measurements in Nasal Cavities by Means of Stereoscopic PIV—Preliminary Tests. *J. Phys. Conf. Ser.* **2017**, *882*, 012010. [CrossRef]

222. Chung, S.-K.; Son, Y.R.; Shin, S.J.; Kim, S.-K. Nasal Airflow during Respiratory Cycle. *Am. J. Rhinol.* **2006**, *20*, 379–384. [CrossRef]

223. Van Hove, S.C.; Storey, J.; Adams, C.; Dey, K.; Geoghegan, P.H.; Kabaliuk, N.; Oldfield, S.D.; Spence, C.J.T.; Jermy, M.C.; Suresh, V.; et al. An Experimental and Numerical Investigation of CO_2 Distribution in the Upper Airways During Nasal High Flow Therapy. *Ann. Biomed. Eng.* **2016**, *44*, 3007–3019. [CrossRef]

224. Kleven, M.; Singh, N.P.; Messina, J.C.; Djupesland, P.G.; Inthavong, K. Development of Computational Fluid Dynamics Methodology for Characterization of Exhalation Delivery System Performance in a Nasal Airway with Draf-III Surgery. *J. Aerosol Sci.* **2023**, *169*, 106121. [CrossRef]

225. Subramaniam, R.P.; Richardson, R.B.; Morgan, K.T.; Kimbell, J.S.; Guilmette, R.A. Computational Fluid Dynamics Simulations of Inspiratory Airflow in the Human Nose and Nasopharynx. *Inhal. Toxicol.* **1998**, *10*, 473–502. [CrossRef]

226. Zhao, K.; Dalton, P.; Yang, G.C.; Scherer, P.W. Numerical Modeling of Turbulent and Laminar Airflow and Odorant Transport during Sniffing in the Human and Rat Nose. *Chem. Senses* **2006**, *31*, 107–118. [CrossRef]

227. Rhee, J.S. Role of Virtual Surgery in Preoperative Planning: Assessing the Individual Components of Functional Nasal Airway Surgery. *Arch. Facial Plast. Surg.* **2012**, *14*, 354–359. [CrossRef]

228. Hildebrandt, T.; Goubergrits, L.; Heppt, W.; Bessler, S.; Zachow, S. Evaluation of the Intranasal Flow Field through Computational Fluid Dynamics. *Facial Plast. Surg.* **2013**, *29*, 093–098. [CrossRef]

229. Zhu, J.H.; Lee, H.P.; Lim, K.M.; Lee, S.J.; Teo Li San, L.; Wang, D.Y. Inspirational Airflow Patterns in Deviated Noses: A Numerical Study. *Comput. Methods Biomech. Biomed. Engin.* **2013**, *16*, 1298–1306. [CrossRef]

230. Cheng, G.C.; Koomullil, R.P.; Ito, Y.; Shih, A.M.; Sittitavornwong, S.; Waite, P.D. Assessment of Surgical Effects on Patients with Obstructive Sleep Apnea Syndrome Using Computational Fluid Dynamics Simulations. *Math. Comput. Simul.* **2014**, *106*, 44–59. [CrossRef] [PubMed]

231. Garcia, G.J.M.; Hariri, B.M.; Patel, R.G.; Rhee, J.S. The Relationship between Nasal Resistance to Airflow and the Airspace Minimal Cross-Sectional Area. *J. Biomech.* **2016**, *49*, 1670–1678. [CrossRef] [PubMed]

232. Inthavong, K.; Ma, J.; Shang, Y.; Dong, J.; Chetty, A.S.R.; Tu, J.; Frank-Ito, D. Geometry and Airflow Dynamics Analysis in the Nasal Cavity during Inhalation. *Clin. Biomech.* **2019**, *66*, 97–106. [CrossRef]

233. Brüning, J.; Hildebrandt, T.; Heppt, W.; Schmidt, N.; Lamecker, H.; Szengel, A.; Amiridze, N.; Ramm, H.; Bindernagel, M.; Zachow, S.; et al. Characterization of the Airflow within an Average Geometry of the Healthy Human Nasal Cavity. *Sci. Rep.* **2020**, *10*, 3755. [CrossRef]

234. Ormiskangas, J.; Valtonen, O.; Harju, T.; Rautiainen, M.; Kivekäs, I. Computational Fluid Dynamics Assessed Changes of Nasal Airflow after Inferior Turbinate Surgery. *Respir. Physiol. Neurobiol.* **2022**, *302*, 103917. [CrossRef]

235. Siu, J.; Inthavong, K.; Dong, J.; Shang, Y.; Douglas, R.G. Nasal Air Conditioning Following Total Inferior Turbinectomy Compared to Inferior Turbinoplasty—A Computational Fluid Dynamics Study. *Clin. Biomech.* **2021**, *81*, 105237. [CrossRef]

236. Corda, J.V.; Shenoy, B.S.; Ahmad, K.A.; Lewis, L.; Khader, S.M.A.; Zuber, M. Nasal Airflow Comparison in Neonates, Infant and Adult Nasal Cavities Using Computational Fluid Dynamics. *Comput. Methods Programs Biomed.* **2022**, *214*, 106538. [CrossRef]

237. Mihaescu, M.; Murugappan, S.; Kalra, M.; Khosla, S.; Gutmark, E. Large Eddy Simulation and Reynolds-Averaged Navier–Stokes Modeling of Flow in a Realistic Pharyngeal Airway Model: An Investigation of Obstructive Sleep Apnea. *J. Biomech.* **2008**, *41*, 2279–2288. [CrossRef]

238. Mihaescu, M.; Murugappan, S.; Gutmark, E.; Donnelly, L.F.; Kalra, M. Computational Modeling of Upper Airway Before and After Adenotonsillectomy for Obstructive Sleep Apnea. *Laryngoscope* **2008**, *118*, 360–362. [CrossRef]

239. Ito, Y.; Cheng, G.C.; Shih, A.M.; Koomullil, R.P.; Soni, B.K.; Sittitavornwong, S.; Waite, P.D. Patient-Specific Geometry Modeling and Mesh Generation for Simulating Obstructive Sleep Apnea Syndrome Cases by Maxillomandibular Advancement. *Math. Comput. Simul.* **2011**, *81*, 1876–1891. [CrossRef]

240. De Backer, J.W.; Vanderveken, O.M.; Vos, W.G.; Devolder, A.; Verhulst, S.L.; Verbraecken, J.A.; Parizel, P.M.; Braem, M.J.; Van de Heyning, P.H.; De Backer, W.A. Functional Imaging Using Computational Fluid Dynamics to Predict Treatment Success of Mandibular Advancement Devices in Sleep-Disordered Breathing. *J. Biomech.* **2007**, *40*, 3708–3714. [CrossRef]

241. Chen, X.B.; Lee, H.P.; Hin Chong, V.F.; Wang, D.Y. Assessment of Septal Deviation Effects on Nasal Air Flow: A Computational Fluid Dynamics Model. *Laryngoscope* **2009**, *119*, 1730–1736. [CrossRef]

242. Gökcan, M.K.; Wanyonyi, S.N.; Kurtuluş, D.F. Computational Fluid Dynamics: Analysis of a Real Nasal Airway. In *Challenges in Rhinology*; Cingi, C., Bayar Muluk, N., Scadding, G.K., Mladina, R., Eds.; Springer International Publishing: Cham, Switzerland, 2021; pp. 501–517. ISBN 978-3-030-50898-2.

243. Mylavarapu, G.; Mihaescu, M.; Fuchs, L.; Papatziamos, G.; Gutmark, E. Planning Human Upper Airway Surgery Using Computational Fluid Dynamics. *J. Biomech.* **2013**, *46*, 1979–1986. [CrossRef]

244. Karbowski, K.; Kopiczak, B.; Chrzan, R.; Gawlik, J.; Szaleniec, J. Accuracy of Virtual Rhinomanometry. *Pol. J. Med. Phys. Eng.* **2023**, *29*, 59–72. [CrossRef]

245. Faizal, W.M.; Khor, C.Y.; Yaakob, M.N.C.; Ghazali, N.N.N.; Zainon, M.Z.; Ibrahim, N.B.; Razi, R.M. Turbulent Kinetic Energy of Flow during Inhale and Exhale to Characterize the Severity of Obstructive Sleep Apnea Patient. *Comput. Model. Eng. Sci.* **2023**, *136*, 43–61. [CrossRef]

246. Wakayama, T.; Suzuki, M.; Tanuma, T. Effect of Nasal Obstruction on Continuous Positive Airway Pressure Treatment: Computational Fluid Dynamics Analyses. *PLoS ONE* **2016**, *11*, e0150951. [CrossRef]

247. Ghahramani, E.; Abouali, O.; Emdad, H.; Ahmadi, G. Numerical Investigation of Turbulent Airflow and Microparticle Deposition in a Realistic Model of Human Upper Airway Using LES. *Comput. Fluids* **2017**, *157*, 43–54. [CrossRef]

248. Lu, M.Z.; Liu, Y.; Ye, J.Y.; Luo, H.Y. Large Eddy Simulation of Flow in Realistic Human Upper Airways with Obstructive Sleep. *Procedia Comput. Sci.* **2014**, *29*, 557–564. [CrossRef]

249. Calmet, H.; Gambaruto, A.M.; Bates, A.J.; Vázquez, M.; Houzeaux, G.; Doorly, D.J. Large-Scale CFD Simulations of the Transitional and Turbulent Regime for the Large Human Airways during Rapid Inhalation. *Comput. Biol. Med.* **2016**, *69*, 166–180. [CrossRef] [PubMed]

250. Calmet, H.; Inthavong, K.; Owen, H.; Dosimont, D.; Lehmkuhl, O.; Houzeaux, G.; Vázquez, M. Computational Modelling of Nasal Respiratory Flow. *Comput. Methods Biomech. Biomed. Engin.* **2021**, *24*, 440–458. [CrossRef] [PubMed]

251. Wang, Y.; Elghobashi, S. On Locating the Obstruction in the Upper Airway via Numerical Simulation. *Respir. Physiol. Neurobiol.* **2014**, *193*, 1–10. [CrossRef]

252. Lintermann, A.; Meinke, M.; Schröder, W. Fluid Mechanics Based Classification of the Respiratory Efficiency of Several Nasal Cavities. *Comput. Biol. Med.* **2013**, *43*, 1833–1852. [CrossRef]

253. Aljawad, H.; Rüttgers, M.; Lintermann, A.; Schröder, W.; Lee, K.C. Effects of the Nasal Cavity Complexity on the Pharyngeal Airway Fluid Mechanics: A Computational Study. *J. Digit. Imaging* **2021**, *34*, 1120–1133. [CrossRef]

254. Otopront. Available online: https://www.otopront.de (accessed on 1 January 2024).

255. Yushkevich, P.A.; Piven, J.; Hazlett, H.C.; Smith, R.G.; Ho, S.; Gee, J.C.; Gerig, G. User-Guided 3D Active Contour Segmentation of Anatomical Structures: Significantly Improved Efficiency and Reliability. *NeuroImage* **2006**, *31*, 1116–1128. [CrossRef]

256. ANSYS. Available online: https://www.ansys.com (accessed on 1 January 2024).

257. ANSYS. *ANSYS Fluent User's Guide*; ANSYS, Inc.: Canonsburg, PA, USA, 2023.

258. Shih, T.-H.; Liou, W.W.; Shabbir, A.; Yang, Z.; Zhu, J. A New K-ϵ Eddy Viscosity Model for High Reynolds Number Turbulent Flows. *Comput. Fluids* **1995**, *24*, 227–238. [CrossRef]

259. Nicoud, F.; Ducros, F. Subgrid-Scale Stress Modelling Based on the Square of the Velocity Gradient Tensor. *Flow Turbul. Combust.* **1999**, *62*, 183–200. [CrossRef]

260. Weickert, M.; Teike, G.; Schmidt, O.; Sommerfeld, M. Investigation of the LES WALE Turbulence Model within the Lattice Boltzmann Framework. *Comput. Math. Appl.* **2010**, *59*, 2200–2214. [CrossRef]

261. Patton, K.T.; Thibodeau, G.A. *Anatomy & Physiology*, 8th ed.; Mosby/Elsevier: St. Louis, MO, USA, 2013; ISBN 978-0-323-08357-7.

262. Yeung, P.K.; Sreenivasan, K.R.; Pope, S.B. Effects of Finite Spatial and Temporal Resolution in Direct Numerical Simulations of Incompressible Isotropic Turbulence. *Phys. Rev. Fluids* **2018**, *3*, 064603. [CrossRef]

263. Kolmogorov, A.N. The Local Structure of Turbulence in Incompressible Viscous Fluid for Very Large Reynolds Numbers. *Proc. R. Soc. Lond.* **1991**, *434*, 9–13. [CrossRef]

264. Rubinstein, R.; Clark, T.T. "Equilibrium" and "Non-Equilibrium" Turbulence. *Theor. Appl. Mech. Lett.* **2017**, *7*, 301–305. [CrossRef]

265. De Bonilla, J.S.-D.; McCaffrey, T.V.; Kern, E.B. The Nasal Valve: A Rhinomanometric Evaluation of Maximum Nasal Inspiratory Flow and Pressure Curves. *Ann. Otol. Rhinol. Laryngol.* **1986**, *95*, 229–232. [CrossRef] [PubMed]

266. Schumacher, M.J. Nasal Dyspnea: The Place of Rhinomanometry in Its Objective Assessment. *Am. J. Rhinol.* **2004**, *18*, 41–46. [CrossRef] [PubMed]

267. Haghnegahdar, A.; Bharadwaj, R.; Feng, Y. Exploring the Role of Nasal Hair in Inhaled Airflow and Coarse Dust Particle Dynamics in a Nasal Cavity: A CFD-DEM Study. *Powder Technol.* **2023**, *427*, 118710. [CrossRef]

268. OSASMOD. PI: Müller, B. Grant Title: Modeling of Obstructive Sleep Apnea by Fluid-Structure Interaction in the Upper Airways. Grant Number: 231741. Funding Body: The Research Council of Norway. Start-End Dates: 2014–2018. 2014. Available online: https://prosjektbanken.forskningsradet.no/project/FORISS/231741 (accessed on 1 January 2024).

269. VIRTUOSA. PI: Skallerud, B.H. Grant Title: Virtual Surgery in the Upper Airways—New Solutions to Obstructive Sleep Apnea Treatment. Grant Number: 303218. Funding Body: The Research Council of Norway. Start-End Dates: 2020–2024. 2020. Available online: https://prosjektbanken.forskningsradet.no/project/FORISS/303218 (accessed on 1 January 2024).

270. SINTEF. Available online: http://www.sintef.no (accessed on 1 January 2024).

271. Sigma2—The National Infrastructure for High Performance Computing and Data Storage in Norway. Available online: https://www.sigma2.no (accessed on 1 January 2024).

272. Norwegian Research Infrastructure Services (NRIS). NIRD Research Data Archive. Available online: https://archive.sigma2.no/ (accessed on 1 January 2024).

Article

Impact of Variation in Commissural Angle between Fused Leaflets in the Functionally Bicuspid Aortic Valve on Hemodynamics and Tissue Biomechanics

Elias Sundström [1],* and Justin T. Tretter [2]

[1] Department of Engineering Mechanics, Flow Research Center, KTH Royal Institute of Technology, Teknikringen 8, 100 44 Stockholm, Sweden
[2] Congenital Valve Procedural Planning Center, Department of Pediatric Cardiology and Division of Pediatric Cardiac Surgery, Cleveland Clinic Children's, and The Heart, Vascular, and Thoracic Institute, Cleveland Clinic, Cleveland, OH 44195, USA
* Correspondence: elias@kth.se

Abstract: In subjects with functionally bicuspid aortic valves (BAVs) with fusion between the coronary leaflets, there is a natural variation of the commissural angle. What is not fully understood is how this variation influences the hemodynamics and tissue biomechanics. These variables may influence valvar durability and function, both in the native valve and following repair, and influence ongoing aortic dilation. A 3D aortic valvar model was reconstructed from a patient with a normal trileaflet aortic valve using cardiac magnetic resonance (CMR) imaging. Fluid–structure interaction (FSI) simulations were used to compare the effects of the varying commissural angles between the non-coronary with its adjacent coronary leaflet. The results showed that the BAV with very asymmetric commissures (120° degree commissural angle) reduces the aortic opening area during peak systole and with a jet that impacts on the right posterior wall proximally of the ascending aorta, giving rise to elevated wall shear stress. This manifests in a shear layer with a retrograde flow and strong swirling towards the fused leaflet side. In contrast, a more symmetrical commissural angle (180° degree commissural angle) reduces the jet impact on the posterior wall and leads to a linear decrease in stress and strain levels in the non-fused non-coronary leaflet. These findings highlight the importance of considering the commissural angle in the progression of aortic valvar stenosis, the regional distribution of stresses and strain levels experienced by the leaflets which may predispose to valvar deterioration, and progression in thoracic aortic dilation in patients with functionally bicuspid aortic valves. Understanding the hemodynamics and biomechanics of the functionally bicuspid aortic valve and its variation in structure may provide insight into predicting the risk of aortic valve dysfunction and thoracic aortic dilation, which could inform clinical decision making and potentially lead to improved aortic valvar surgical outcomes.

Keywords: bicuspid aortic valve; commissural angle; fluid–structure interaction; magnetic resonance

Citation: Sundström, E.; Tretter, J.T. Impact of Variation in Commissural Angle between Fused Leaflets in the Functionally Bicuspid Aortic Valve on Hemodynamics and Tissue Biomechanics. *Bioengineering* **2023**, *10*, 1219. https://doi.org/10.3390/bioengineering10101219

Academic Editor: Chiara Giulia Fontanella

Received: 14 September 2023
Revised: 12 October 2023
Accepted: 13 October 2023
Published: 18 October 2023

1. Introduction

A functionally bileaflet or bicuspid aortic valve (BAV) with a trisinuate aortic root is a common congenital heart condition that presents challenges in diagnosis and treatment due to its variable progression of valvar dysfunction and thoracic aortic dilation [1,2]. The severity of aortic valvar stenosis is known to be related to the size of the valvar opening. Proper leaflet coaptation is influenced by the relative dimensions of the leaflets in relation to the dimensions of the planes throughout the aortic root, from the virtual basal ring to the sinutubular junction. However, the role of variation in the position of the two commissures present in this most common form of a bicuspid aortic valve, or commissural angle, remains unclear in its impact on valvar function [3–5].

Recently, a study using cardiac magnetic resonance (CMR) imaging and fluid–structure interaction (FSI) simulations investigated the impact of the height of the interleaflet triangle or commissural height on the degree of aortic valvar stenosis, hemodynamics, and tissue biomechanics. The study found that a larger zone of fusion in the functionally BAV with an inversely reduced interleaflet triangle height resulted in a linear rise in wall shear stress, peak velocity, pressure gradient, and strain levels, forming more asymmetric vortex systems and the recirculation of flow toward the side of leaflet fusion within the trisinuate aortic root. The study's findings highlight the importance of considering the interleaflet triangle height as a crucial factor in the development of thoracic aortic dilation in patients with BAV, along with potential considerations related to durable aortic valvar repair [6].

Flow MRI has also been used to investigate how the degree of asymmetry of the BAV, with commissural angles between 120 and 180°, affects the outflow jet as compared to the normal trileaflet aortic valve (TAV) [7]. The blood flow was considered during peak systole and its implications for aortopathy (aortic disease). The results showed that asymmetric BAVs had eccentric outflow jets that affected specific regions of the aortic wall based on the position of the smaller leaflet. In contrast, symmetric BAVs had more centered outflow jets that did not impact the aortic wall. The symmetry of the BAV and the position of the smaller leaflet were key factors influencing the outflow jet characteristics. However, there was no quantification of stresses and strain levels in the leaflet material that may predispose to aortic valvar dysfunction.

This current study aims to build upon previous research [6,8–10] and investigate the resulting hemodynamic and tissue biomechanical impact related to variation in the commissural angle found in the functionally bileaflet aortic valves with a trisinuate aortic root (Figure 1). Like the variation in the degree of fusion between the two fused leaflets in this common form of a bicuspid aortic valve, the variation described in the commissural angle is believed to affect the systolic valvar opening area and the interplay between hemodynamics and biomechanical responses in the thoracic aorta. Therefore, this study will complement the previous assessments of the effects of normal variation in the rotation position of the aortic root relative to the base of the left ventricle and the degree of leaflet fusion [10], as well as the variation of the interleaflet triangle height [6].

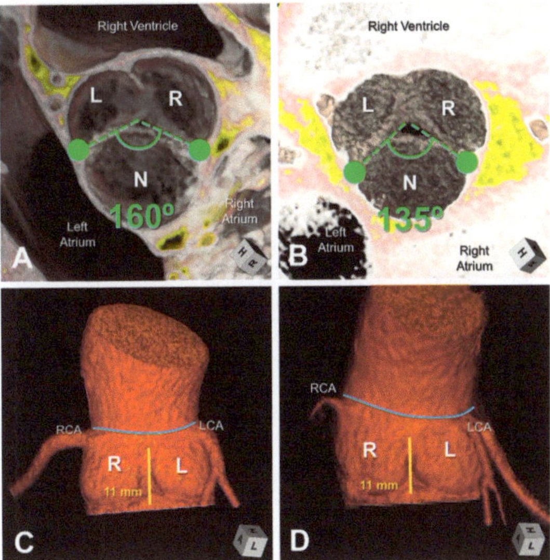

Figure 1. Computed tomographic 3D reconstructions of two patients with functionally bileaflet aortic valves with fusion between the coronary leaflets and trisinuate aortic roots are demonstrated (Patient

1 = Panels (**A,C**); Patient 2 = Panels (**B,D**)). Panels (**A,B**) demonstrate a short axis view of the aortic valve with the angle between the two commissures measured to be symmetrical and very asymmetrical, respectively. Prominent raphes are visualized at the zone of fusion between the coronary leaflets for both valves. Panels (**C,D**) demonstrate the blood-filled trisinuate aortic roots, both with similar commissural heights (11 mm) of the hypoplastic interleaflet triangle under the zone of fusion between the coronary leaflets. The sinutubular junction is marked with a blue line. L, left coronary sinus; LCA, left coronary artery; N, non-coronary sinus; R, right coronary artery; RCA, right coronary artery.

2. Method

2.1. CMR

Cardiac magnetic resonance (CMR) scans were performed at Cincinnati Children's Hospital Medical Center (CCHMC) on an adolescent subject who had normal cardiovascular anatomy and function, see [9,10]. All demographic information, such as age, weight, gender, and diagnosis, was de-identified. The study was therefore deemed exempt from the ethical review and approval by the Cincinnati Children's Institutional Review Board.

The CMR acquisition protocol used a 1.5 T CMR machine (Ingenia, Philips Healthcare, Best, the Netherlands) equipped with a phased-array coil. The protocol incorporated various imaging sequences, including the axial aortic root cine stack, phase-contrast velocity sequence, and non-contrast 3D mDixon angiogram. The short-axis aortic root cine stack was obtained using a steady-state free precession pulse sequence. This sequence employed a repetition time of 7.8 ms, an echo time of 4.7 ms, a flip angle of 15 degrees, and sequential 2 mm slices with no interslice gap. To capture the phase-contrast velocity information, a gradient-echo sequence was employed with a repetition time of 4.3 ms, an echo time of 2.7 ms, a flip angle of 12 degrees, and a slice thickness of 6 mm. The encoding velocity used was 1.5 m/s. The scanning protocol achieved a mean time resolution of 30–40 ms, resulting in 30 phases per cardiac cycle. In the coronal plane, a non-contrast 3D mDixon angiogram was acquired, using a repetition time of 5.3 ms, a flip angle of 15 degrees, and a spatial resolution of 1 mm. The details of the aortic root were assessed by analyzing the information obtained from the aortic root cine steady-state free precession sequence, see [9,10]. In addition, the protocol for the long-axis sagittal stack 4D Flow MRI was acquired using a velocity encoding of 2 m/s, a repetition time of 3.5 ms, an echo time of 1.9 ms, and a flip angle of 8°, see reference [8].

2.2. Aortic Reconstruction

Geometry reconstruction of the subject's aortic valve and thoracic aortic anatomy was performed using the segmentation software 3D Slicer 5.2.2 [11]. The outlines of the aortic fluid domain were defined by specifying a threshold intensity on the 3D mDixon angiogram CMR dataset. The resulting control volume encompassed the aortic sinuses, thoracic aorta, and the head and neck vessels, including the right brachiocephalic, left common carotid, and left subclavian arteries. High-curvature features were captured using fine, high-quality tessellation applied to the control surfaces, which was followed by a smoothing protocol to reduce irregularities.

A short distance upstream of the aortic root, the inlet surface was extruded, and a uniform blood flow velocity was specified using the flow rate shown in Figure 2 [12]. This allowed for the development of a velocity profile with a boundary layer inferiorly to the aortic valve.

Leaflet attachment lines and commissures observed in the CMR images were used to reconstruct anatomically accurate semilunar contact between the leaflets and the walls of the aortic sinuses. The spatial resolution of the CMR images facilitated the identification of the leaflets' semilunar features. The reconstructed leaflet surface was then inflated to a thickness of 0.7 mm, representing the lower end of a multi-ethnic population dataset [13], thus forming the solid domain of the leaflet tissue. However, due to the limited spatial resolution of the CMR, some high-curvature features of the aortic leaflets were not captured.

To aid the reconstruction process, the remaining leaflet attachment lines, which were not visible in the CMR images, were specified using ratios derived from an averaged homograft dataset from a population with normal aortic valves [14].

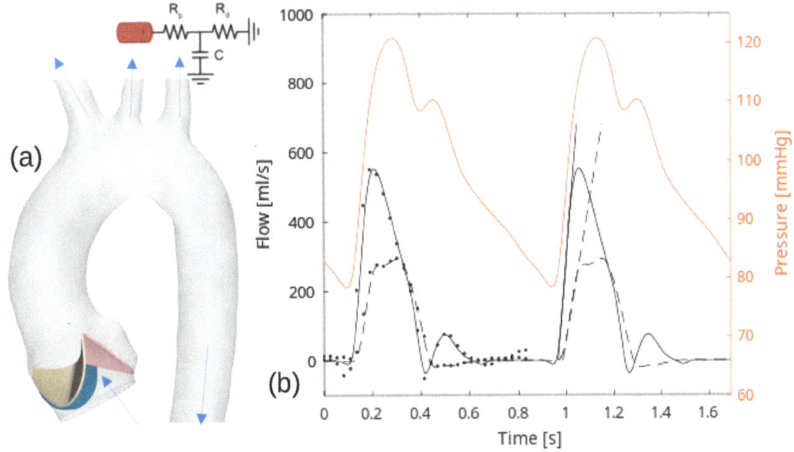

Figure 2. (**a**) Reconstructed geometry of the thoracic aorta is shown to the left, complete with specified boundary conditions (transparent grey color), and with the aortic valvar leaflets: left coronary (L), non-coronary (N), and right coronary (R). (**b**) Graphs in this section depict the flow rate during the cardiac cycle, both at the inlet and outlet (in the descending aorta). The pressure boundary conditions at the head and neck vessels across the cardiac cycle are established using the three-parameter Windkessel model. The solid surfaces of the leaflets that intersect with the aortic root are fixed, whereas all other solid surfaces are free.

The most common type of bicuspid aortic valve is characterized by fused leaflets, resulting in a functional bileaflet valve within a trisinuate aortic root. There is a variable degree of a raphe at the zone of fusion. This form is present in 90–95% of individuals with a bicuspid aortic valve [4,5]. The functionally bileaflet phenotypes include right–left leaflet fusion, right–non-leaflet fusion, and rarely left–non-leaflet fusion. Figure 3 illustrates long-axis and short-axis views of the complex 3D structure of the aortic root and its leaflets, similar to 3D CT and MRI. In all phenotypes shown, three well-defined aortic sinuses are evident, with variability in the degree of fusion between the coronary leaflets.

The functionally bileaflet aortic valve is further characterized by variation in the angle between its two commissures. The short-axis view in Figure 3 displays different angles of the commissures. The functionally bileaflet aortic valve with commissural symmetry is defined by a commissural angle of 160–180 degrees; that with asymmetrical commissures exhibits an angle of 140–159 degrees; and that with very asymmetrical commissures exhibits an angle of 120–139 degrees.

2.3. Computational Models

Similar to previous studies [6,8–10], we took into account the conservation of mass and momentum laws to simulate the transport of blood flow through the aorta. We described the non-Newtonian behavior of blood using a mass transport equation for the volume fraction between red blood cells (RBCs) and blood plasma. The density of the blood was modeled as a linear combination of the densities of RBCs and blood plasma. The diffusion rate of RBCs was governed by a parameter that depends on the viscosity and mass diffusivity. To incorporate rheology effects, we used a modified power law approach based on the Ostwald–de Waele rheology model. This modification aimed to adjust the viscosity to match the viscosity of blood plasma in scenarios with zero shear rate and zero RBC volume fraction. Empirical correlations were used to determine the model parameters [15].

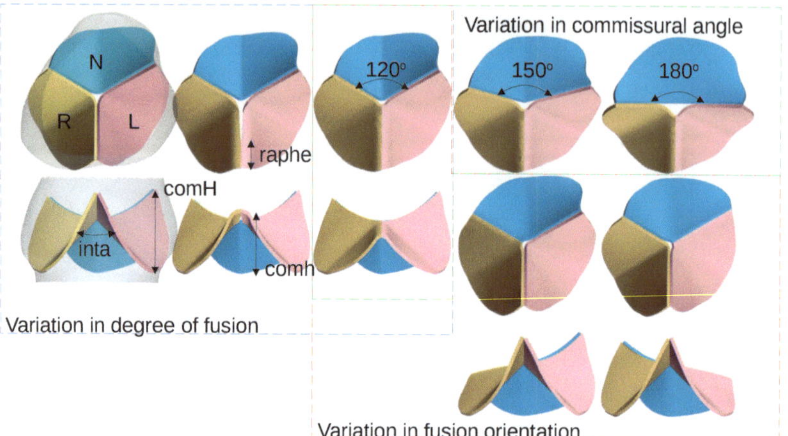

Figure 3. Within the blue hashed box, we have short-axis and long-axis reconstructed geometry of the normal trileaflet aortic valve, and with two variations in the extent of the zone of fusion, one with 2/3rd commissural height (comH) and partial fusion between the coronary leaflets, and the other with 1/3rd commissural height and complete fusion between the coronary leaflets. With decreasing commissural height, there is increased fusion and angle of the apex of the interleaflet triangle (inta). Within the green hashed box, there is a variation of the commissural angle, i.e., 120°, 150°, and 180° deg. This increases the surface area of the unfused non-coronary leaflet. Within the red hashed box, there is a variation in the fusion orientation with the RN and rare LN leaflet fusion phenotypes.

The flow variation at the inlet boundary was measured with CMR phase contrast data during the cardiac cycle, see Figure 2. For the head and neck vessels, we employed a Windkessel three-element circuit function to connect the flow rate and pressure. We used a numerical scheme to obtain the pressure level at each time step, considering the physiological pressure ratios. The proximal resistance (Rp), distal resistance (Rd), and compliance (C) in the Windkessel model were tuned for physiological pressures of 120 mmHg at peak systole and 80 mmHg at the end of diastole. The governing equation and constitutive relations were solved using the finite volume method, with a Rhie and Chow-type velocity coupling and a semi-implicit method for pressure-linked equations [16–19].

The solid tissue was modeled as a nearly incompressible hyperelastic material using the Ogden model [20,21]. The elastic properties were determined through an inverse optimization-based process to match the displacement of the leaflet edge with CMR data. The model was constrained by fixed support at the interface between the leaflets and the aortic root, whereas all other surfaces were free to move. The exchange of information between the fluid and solid domains was governed by a fluid–solid contact interface. The stiffness matrix was updated using Newton iteration methods to handle nonlinear material specifications and large deformations. Four wedge layers discretized the leaflets, whereas polyhedral cell and prism layers discretized the fluid domain. Grid convergence studies were performed in previous studies to ensure accurate results [10]. The FSI model used in previous studies was validated by comparing the velocity distribution between CMR data and numerical results for the baseline case [9,10]. The computed flow field showed good agreement with the literature [12,22–24], with a difference of about 15%. The simulations were performed in parallel on 112 cores (Intel Xeon E5-2680), each case taking approximately six hours for one cardiac cycle.

3. Result

Figure 4 presents a comprehensive depiction of blood flow through the aorta and the aortic valve during various stages of the cardiac cycle for the normal trileaflet aortic valve

(TAV) and the functionally bileaflet aortic valve (BAV). During early systole ($t/T = 0.04$), the aortic valve initiates its opening, leading to the ejection of blood from the left ventricle. In the narrower section of the valve, local flow velocity begins to accelerate, and a jet of flow forms through the valve in both TAV and BAV cases with approximately stagnant flow in the ascending aorta. In the next time instant ($t/T = 0.07$), the aortic valve is now half open, where the BAV cases show severe stenosis on the side with the fully fused left and right leaflets, limiting the displacement as compared to the normal TAV case. The streamlines in Figure 5 in both TAV and BAV cases during early systole are directed towards the convex side of the ascending aorta, which is consistent with the curvature of the ascending aorta.

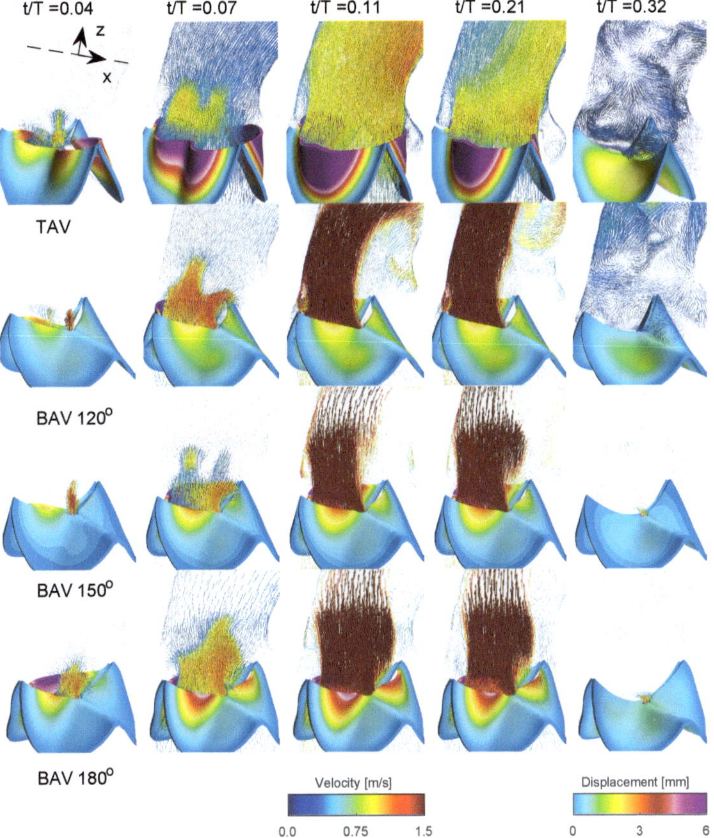

Figure 4. Blood flow visualization using velocity vectors at different time instants during the cardiac cycle for the TAV case and the BAV cases with fully fused coronary leaflets. The velocity vectors are presented on a vertical plane, coronally oriented.

Around the time of peak systole (between t/T values of 0.1 and 0.2), the velocities in the ascending aorta increase, corresponding to the pulse wave velocity assessment and the time delay for the pulse wave to propagate through the aorta, see Figure 2. In the TAV case, the velocity vectors remain relatively aligned with the aorta. However, in the BAV cases, there is a higher peak velocity that impacts more towards the convex tissue wall near the sinotubular junction and proximally in the ascending aorta compared to the TAV case. The velocity field depicts a strong shear layer with a recirculating flow towards the concave side of the ascending aorta, which is in good agreement with a previous blood speckle imaging study [25]. The shear layer intensifies, and during post-peak systole (between t/T values of 0.2 and 0.3), it occupies a significant portion of the region, extending from the aortic root to the proximal aortic arch and its right brachiocephalic and the left common carotid arteries. The streamlines in the short-axis cut of the ascending aorta at peak systole for the TAV case show two counter-rotating vortices, see Figure 5. These are Dean-like vortices that develop due to the curvature of the ascending aorta. In the BAV cases, there are also counter-rotating vortices but with an asymmetry compared to the TAV case [26]. Towards post-peak systole, the TAV case depicts three coherent vortices that coincide with the apexes of the three commissures at the level of the sinutubular junction. In the BAV case, towards post-peak systole, the streamlines depict a swirling flow with local incoherent vortices located between the shear-layer and circulation zone.

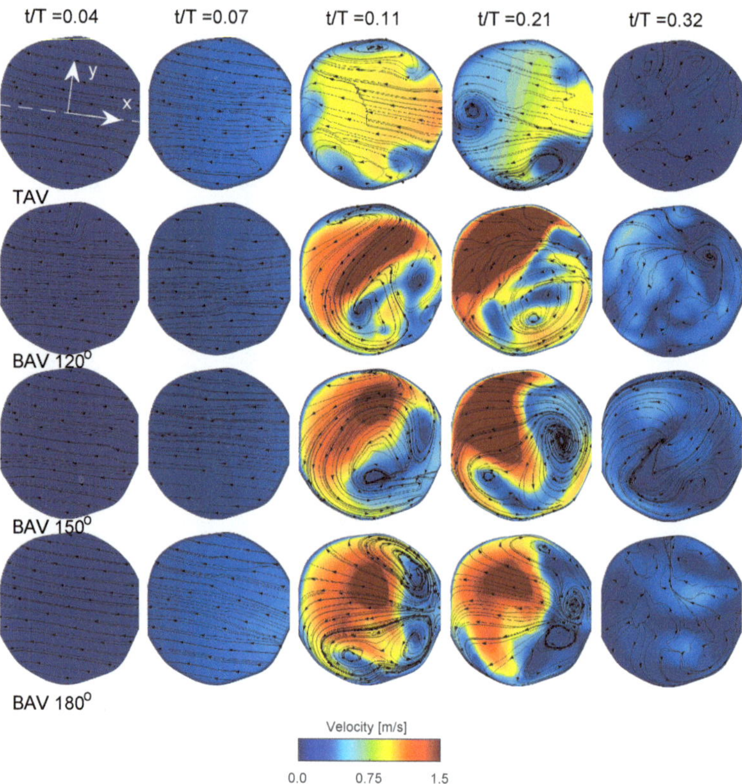

Figure 5. Velocity and streamline distribution shown on the short-axis cut plane. The plane is located $1.5 \times$ comH proximally of the sinutubular junction as shown with the white dashed line in Figure 4. The keys are the same as those in Figure 4.

As the systolic phase concludes (between t/T values of 0.2 and 0.3), the valve closes, leading to a decrease in blood flow concurrent with a reduction in aortic pressure during diastole, see last time instant in Figures 4 and 5. As time progresses towards diastole, the flow will settle down due to diffusion, promoting the filling of the left ventricle with fresh blood in preparation for another systole. The streamlines on the short-axis cut for the BAV case exhibit residual flow with a small swirl, indicating a longer diffusion time compared to the TAV case.

Figure 6 quantifies the streamwise and cross-flow velocity profiles on the short-axis cut plane, c.f. Figure 5 for the location of the profile. For the TAV case, the streamwise velocity (Figure 6a) indicates a top hat distribution with a slight slope towards the convex side of the ascending aorta. There is a fair degree of agreement with the 4D Flow MRI data, where the difference in the peak velocities is within 15%. The BAV cases show higher velocities towards the convex side of the ascending aorta. There is a general trend of an increasing velocity gradient with a reduced commissural angle. On the opposite side, there is a retrograde flow, i.e., towards the side with the RL leaflet fusion. It is observed that the shear layer shifts towards the fusion side, and the magnitude of the flow reversal reduces with an increasing commissural angle. The cross-flow component (Figure 6b) for the TAV case indicates little to no swirl, which is in fair agreement with the 4D flow MRI. However, all BAV cases indicate a notable cross flow that correlates with a strong swirling component, c.f. Figure 5. There is a general trend that a reduced commissural angle increases the tangential velocity gradient on the convex side of the ascending aorta. It is also observed that the magnitude or the cross flow reduces with an increased commissural angle.

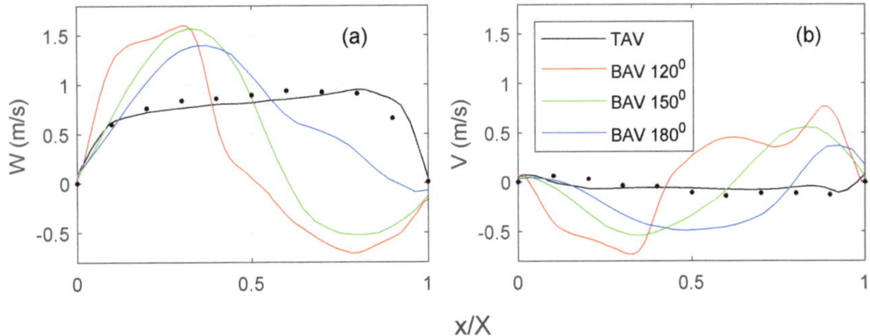

Figure 6. (a) Streamwise velocity profile (W) and (b) cross-flow velocity profile (V) distribution along the white dashed line that is located proximal to the sinutubular junction, see annotated white dashed line in Figure 5. The x-axis of the profile is normalized with the total length, where $x/X = 0$ is towards the convex side and $x/X = 1$ is towards the concave side of the ascending aorta. All cases are for peak systole. The 4D Flow MRI data for the normal TAV cases are shown with black dots.

Figure 7a shows the aortic opening area for the TAV case compared with the BAV case with fully fused coronary leaflets for three different commissural angles ($120°$, $150°$, $180°$). Both the TAV and BAV cases open around $t/T = 0.05$. The opening is nearly linear until the peak opening, around $t/T = 0.1$. The TAV case shows a normal aortic opening area around 4 cm^2, whereas the BAV cases are stenoic with an aortic opening area below 2 cm^2. There are only trivial to no oscillation wiggles, indicating critical damping. All cases show a non-symmetric top-hat-like distribution with more rapid opening than closing. In addition, the TAV case closes earlier than the BAV cases, which correlates with a higher LV pressure to produce the same flow rate.

Figure 7b,c depict the radial displacement of the non-coronary and right coronary leaflets, i.e., the free edge's mid-point. The TAV case shows an asymmetric opening where the right coronary leaflet exhibits a larger radial displacement than the non-coronary leaflet, which is probably due to the normal minimal asymmetries present in the geometry of the valvar leaflets. For the BAV cases, there is a linear trend with increasing radial displacement of the unfused non-coronary leaflet as a function of the commissural angle with a decreasing commissural angle. Specifically, the BAV 120° deg case has a radial displacement of the non-coronary leaflet that is twice as large compared to that of the BAV 180° case. It is also observed that the BAV 180° case opens up earlier than the BAV 120° case. When considering the radial displacement of the fused right leaflet there is also a similar linear trend but reversed, where the radial displacement increases with an increased commissural angle. However, this trend is not as clear compared to the radial displacement of the unfused non-coronary leaflet.

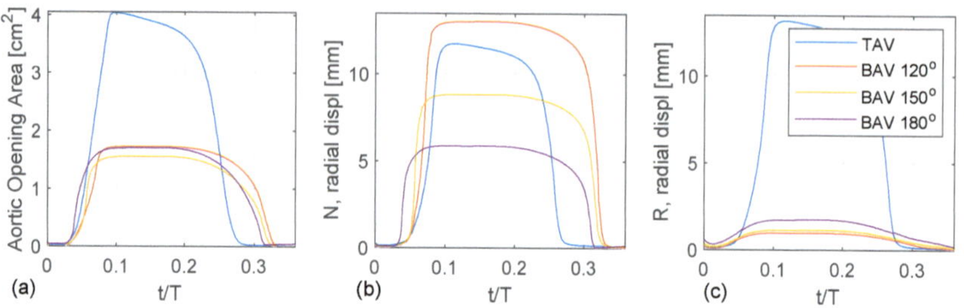

Figure 7. (**a**) Aortic opening area as a function of the cardiac cycle for the TAV and BAV cases. (**b**) Radial displacement of the non-coronary leaflet at the mid-point location of the free edge. (**c**) Radial displacement of the right leaflet at the mid-point location of the free edge.

Figure 8 shows the von Mises stress distribution for the TAV and BAV cases during different stages of the cardiac cycle. At the beginning of the valve opening $t/T = 0.04$, all cases shows stress levels below 0.04 MPa. Between $t/T = 0.07$ and $t/T = 0.021$, the stress level gradually increases, where elevated stress levels are present at the interface of the leaflets with the aortic root and near the leaflet free edge. Upon the valve closure, the entire valve is pushed downwards due to the adverse pressure gradient, and the stress level gradually increases with concentrated levels at the interface of the leaflets with the aortic root but also in the mid-portion of the leaflet surface.

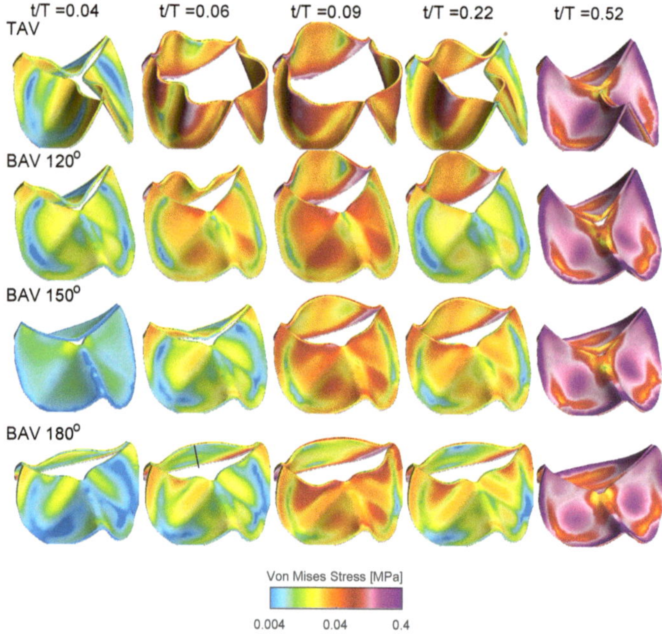

Figure 8. Isometric views showing the von Mises stress distribution during opening and closing for the TAV case and the BAV cases during the cardiac cycle.

Figure 9 shows the Frobenius norm of the strain tensor for the same cases and time instant as in Figure 8. At the beginning of the opening $t/T = 0.04$, all cases shows low strain levels, indicating that the valve is close to its neutral stress-free configuration. Between time instants $t/T = 0.07$ and $t/T = 0.021$, the strain level increases, with the TAV case showing larger strain due its larger displacement and aortic opening area, c.f. Figure 7. Similar to the quantification of the radial displacement of the unfused non-coronary leaflet, i.e., Figure 7b, there is a reduced strain as a function of the commissural angle. However, the strain on the fused side of the BAV does not show a linear variation as a function of the commissural angle. Instead, the fused RL leaflet for the BAV 180° depicts a growing strain concentrated along the fold that starts close to the location of the free edge of the zone of fusion and its raphe, and extending across the midline of both fused leaflets towards their respective nadirs. As time evolves to time instant $t/T = 0.52$, the valve closes, and the strain level gradually increases in the mid-portion of the leaflets due to the increasing adverse pressured gradient during diastole.

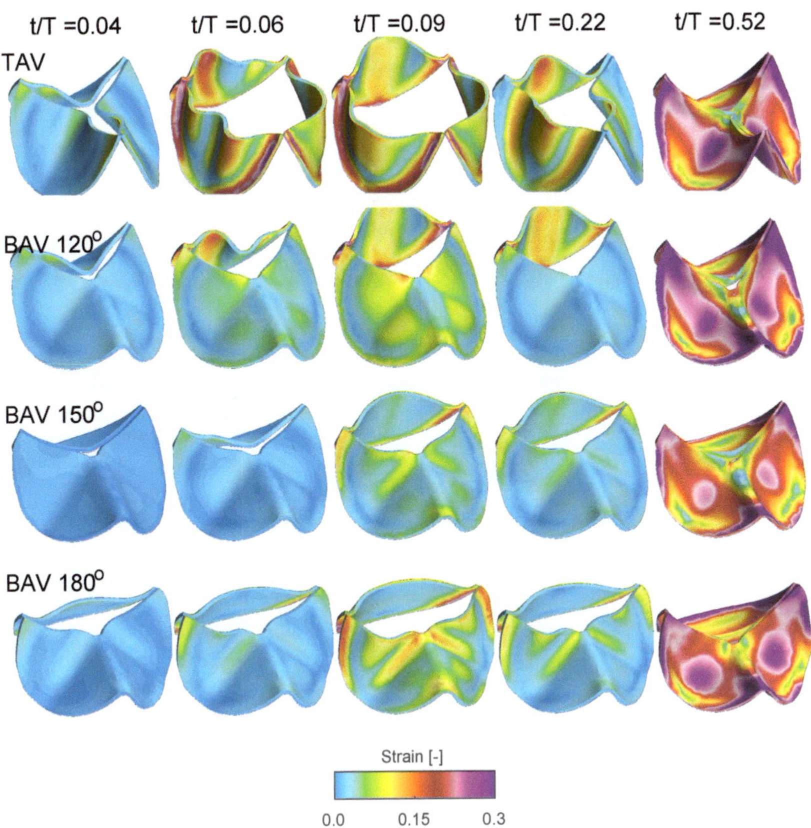

Figure 9. Isometric views showing the Frobenius norm of the strain tensor during opening and closing for the TAV case and the BAV cases with full coronary leaflet fusion during the cardiac cycle. Same keys as in Figure 8.

The stress and strain levels shown previously are now further quantified along the vertical intersection of the non-coronary leaflet, see Figure 10 (see black line annotations in Figure 8 for the location of the vertical intersection). At peak systole ($t/T = 0.1$), the stress and strain levels are similarly distributed along the non-coronary leaflet for the TAV and the BAV 120° deg cases, see Figure 10a,b. The stress shows a minimum at a location around 20–30% of the normalized effective leaflet height s and a maximum at the intersection of the leaflet with the aortic root ($s = 0$). As the commissural angle increases, in the BAV 150° deg and 180° deg cases, both stress and strain levels reduce.

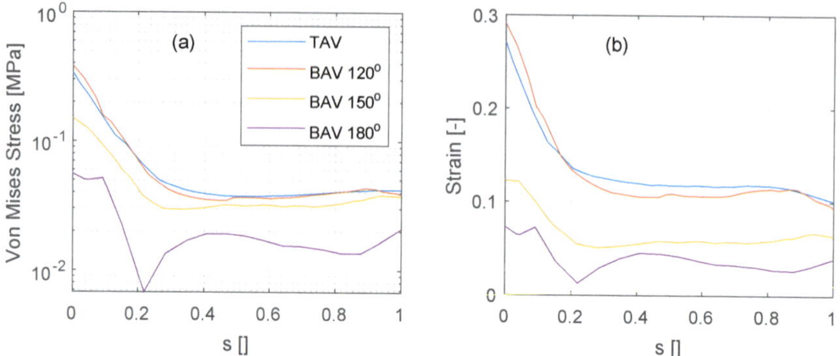

Figure 10. Variation of (**a**) von Mises stress and (**b**) Frobenius norm of the strain tensor along the mid of the non-coronary leaflet between the intersection with the aortic root ($s = 0$) up to the free edge ($s = 1$). Data are presented for the TAV case and the BAV cases.

4. Discussion

This FSI analysis examines the influence of variation in the commissural angle in the most common BAV, the functionally bileaflet valve with trisinuate aortic root with a fusion between the left and right coronary leaflets. The study reveals the significance of this morphological feature in determining the resulting hemodynamics and tissue biomechanics. The decreasing aortic valve opening area seen in the spectrum from TAV to BAV with partial fusion, and to BAV with full fusion leads to increasing elevation in the blood flow velocity and pressure gradient, with an increasing wall shear stress seen along the convex surface of the proximal ascending aorta. It was found that the commissural angle significantly influences the aortic valvar outflow jet, where a highly asymmetric BAV leads to increased velocity gradient and wall shear stress on the posterior convex side of the ascending aorta. It is also evident that a BAV with very asymmetric commissures introduces a strong swirling flow during peak systole. These results suggest that the behavior of the aortic valvar outflow jet is influenced by both the symmetry and position of the smaller leaflet in BAVs, which is in good agreement with a previous 4D flow MRI study [7]. Conversely, BAVs with symmetrically positioned commissures showed mildly eccentric aortic valvar outflow jets at peak systole that did not impinge on the aortic wall. In BAVs with right and left coronary leaflet fusion, increasing the commissural angle will correspond to an increase in the size of the non-fused non-coronary leaflet. This correlates with decreasing stress and strain levels on the non-coronary leaflet (recall Figure 10). This result is not evident from previous studies relying on only 4D Flow MRI. With increasing the commissural angle, and hence decreasing the size of the fused coronary leaflets, there is not a clear trend in the stress and strain levels placed on the fused coronary leaflets.

The augmentation in the peak systolic velocity is intuitively linked to the valvar opening area, which corresponds to the degree of left and right coronary leaflet fusion. The alteration in stress, strain, and flow patterns is also influenced by the commissural angle. In this analysis, the commissural angle primarily affects the unfused non-coronary leaflet. Increasing the commissural angle leads to an increased surface area of the non-coronary leaflet but also reduced curvature at the intersection with the aortic root. This directs the aortic valvar opening area towards the center reducing the wall shear stress on the proximal ascending aorta.

Compared to the less common true BAV that is bileaflet with a bisinuate root, the functionally BAV and trisinuate aortic root assessed in this study is more commonly associated with aortic valvar stenosis and aortic dilation [27]. In the latter type, two functional commissures are present, with two normal underlying interleaflet triangles, both extending to the level of the sinutubular junction (Figure 3). In contrast, the interleaflet triangle inferior to the zone of fusion is hypoplastic, with no functional commissure, with its apex falling short of the sinutubular junction. These data support the idea that, in the functionally bicuspid aortic valve, variation in the commissural angle may impact the regional distribution of leaflet stresses and strains, which may predispose to ongoing leaflet thickening, sclerosis, and eventual calcification. This may in turn impact the progression of aortic valvar stenosis. Moreover, the study shows how changes in the commissural angle directly relate to disruptions in hemodynamics and tissue biomechanics within the aortic root and subsequent thoracic aorta. This understanding may further illuminate risk factors for the progression of aortic dilation.

These findings carry implications for evaluating congenitally malformed valves and guiding surgical repairs. To best achieve durable repairs, it was demonstrated that the commissural angle, whether symmetric versus very asymmetric, may provide guidance in maintaining a functionally bileaflet valve versus reconfiguration to a trileaflet valve, respectively [22,28]. While this general approach has been scientifically validated [22], the hemodynamic mechanisms underlying these successes have not been elucidated. Specifically, our results demonstrate that the non-fused leaflet in the functionally BAV has decreased stress and strain when larger, in the setting of symmetrical commissures. This may support why an approach leaving this variation functionally bileaflet leads to a more durable repair. Similarly, the non-fused leaflet experiences varying stress and strain when smaller in the setting of very asymmetric commissures in a functionally BAV. This suggests why converting this type to a trileaflet valve may be the preferred surgical approach. Prior investigations into computational fluid dynamics in BAV have predominantly focused on altered hemodynamics arising from the leaflet orientation and valve opening area [23]. Although these impact the resultant hemodynamics and tissue biomechanics, they merely scratch the surface of the intricate three-dimensional variation of the aortic root and its valve. This study adds to existing knowledge by demonstrating the substantial effects of the variation seen in the commissural angle of the most common phenotype of the functionally BAV.

The study findings suggest that clinical assessment of the affected commissural angles could enhance the understanding of the projected progression of aortic valvar stenosis and thoracic aortic dilation. In addition, an improved understanding of the impact of the commissural angle on the resulting hemodynamics and tissue biomechanics may help better fine-tune surgical repair approaches in the functional BAV. These findings require clinical validation to understand the progression of aortic valvar stenosis and aortic root and ascending aortic dilation, along with the impact of valvar repair durability.

5. Limitations

The study used data from a normal TAV subject and subsequently simulated variations in the commissural angle in a BAV with fusion between the right and the left leaflets. This approach offers a notable advantage for a small-scale study by isolating the studied variable and additional variables introduced by dissimilarities observed in the geometry and dimensions of the thoracic aorta in both normal and congenitally malformed aortic roots.

This investigation centered on the functionally BAV with fusion involving the coronary leaflets. Future research is required to investigate other phenotypes of both functionally BAVs.

In the simulation, the inlet and outlet stations of the thoracic aorta were considered fixed supports, and the contact between the leaflets and the thoracic aorta was modeled as solid. While the motion of the heart between systole and diastole induces cyclical displacement at these aortic stations, the diastolic displacement of the aortic root and velocity were documented at around 1 cm and 10 cm/s, respectively, using echocardiography and tissue Doppler velocimetry [29]. Similar assessments of aortic root motion have been conducted using CMR [30]. Given that the thoracic aorta's length and the rapidity of jet development during valvar opening differ by an order of magnitude, aortic root motion might not significantly impact stress levels in the ascending aorta. However, challenges persist due to the low signal-to-noise ratio and image quality of temporally varying 3D echocardiography and CMR, which hamper the accurate quantification of the aortic root. This aspect is currently the subject of intense research and will be addressed in subsequent studies.

Acknowledging certain assumptions, the FSI analysis in this study encompassed (a) a uniform aortic leaflet thickness, (b) uniform material properties of the aorta, (c) fixed spatial support of the aorta, and (d) the absence of aortic root pull and twist during the cardiac cycle. Naturally, these parameters exhibit variations based on age and gender. Yet, refining these parameters through CMR techniques remains challenging due to the inherent limitations of the low signal-to-noise ratio and image quality.

6. Conclusions

Variation in the commissural angle of the functionally BAV impacts the stress and strain of both the non-fused and fused leaflets. This variation also influences the hemodynamics and tissue biomechanics experienced in the subsequent thoracic aorta. This may influence the progression of aortic valvar stenosis and aortic root and ascending aortic dilation, along with the durability of valvar repair strategies. Clinical studies are warranted to validate these findings and determine the utility of assessing and surgically manipulating this variation in commissural angle in the functionally BAV.

Author Contributions: Conceptualization, E.S. and J.T.T.; formal analysis, E.S. and J.T.T.; writing—review and editing, E.S. and J.T.T.; funding acquisition, E.S. All authors have read and agreed to the published version of the manuscript.

Funding: This research was funded by the Swedish Research Council (grant number 2021-04894).

Institutional Review Board Statement: Ethical review and approval were waived by the Cincinnati Children's Institutional Review Board because the study was a retrospective analysis using de-identified data from fewer than 4 subjects

Informed Consent Statement: Informed consent was obtained from all subjects involved in the study. Written informed consent has been obtained from the patient(s) to publish this paper.

Data Availability Statement: Data available on request due to restrictions in repository access.

Conflicts of Interest: The authors declare no conflict of interest.

References

1. Nistri, S.; Basso, C.; Marzari, C.; Mormino, P.; Thiene, G. Frequency of bicuspid aortic valve in young male conscripts by echocardiogram. *Am. J. Cardiol.* **2005**, *96*, 718–721. [CrossRef]
2. Basso, C.; Boschello, M.; Perrone, C.; Mecenero, A.; Cera, A.; Bicego, D.; Thiene, G.; De Dominicis, E. An echocardiographic survey of primary school children for bicuspid aortic valve. *Am. J. Cardiol.* **2004**, *93*, 661–663. [CrossRef]
3. Tretter, J.T.; Spicer, D.E.; Mori, S.; Chikkabyrappa, S.; Redington, A.N.; Anderson, R.H. The significance of the interleaflet triangles in determining the morphology of congenitally abnormal aortic valves: Implications for noninvasive imaging and surgical management. *J. Am. Soc. Echocardiogr.* **2016**, *29*, 1131–1143.
4. Bonow, R.O.; Carabello, B.A.; Chatterjee, K.; De Leon, A.C.; Faxon, D.P.; Freed, M.D.; Gaasch, W.H.; Lytle, B.W.; Nishimura, R.A.; O'Gara, P.T.; et al. ACC/AHA 2006 guidelines for the management of patients with valvular heart disease: a report of the American College of Cardiology/American Heart Association Task Force on Practice Guidelines (writing Committee to Revise the 1998 guidelines for the management of patients with valvular heart disease) developed in collaboration with the Society of Cardiovascular Anesthesiologists endorsed by the Society for Cardiovascular Angiography and Interventions and the Society of Thoracic Surgeons. *J. Am. Coll. Cardiol.* **2006**, *48*, e1–e148. [PubMed]
5. Cohen, L.S.; Friedman, W.F.; Braunwald, E. Natural history of mild congenital aortic stenosis elucidated by serial hemodynamic studies. *Am. J. Cardiol.* **1972**, *30*, 1–5. [CrossRef] [PubMed]
6. Sundström, E.; Tretter, J.T. Impact of Variation in Interleaflet Triangle Height Between Fused Leaflets in the Functionally Bicuspid Aortic Valve on Hemodynamics and Tissue Biomechanics. *J. Eng. Sci. Med. Diagn. Ther.* **2022**, *5*, 031102. [CrossRef]
7. Hattori, K.; Nakama, N.; Takada, J.; Nishimura, G.; Moriwaki, R.; Kawasaki, E.; Nagao, M.; Goto, Y.; Niinami, H.; Iwasaki, K. Bicuspid aortic valve morphology and aortic valvular outflow jets: an experimental analysis using an MRI-compatible pulsatile flow circulation system. *Sci. Rep.* **2021**, *11*, 2066.
8. Jonnagiri, R.; Sundström, E.; Gutmark, E.; Anderson, S.; Pednekar, A.; Taylor, M.; Tretter, J.; Gutmark-Little, I. Influence of aortic valve morphology on vortical structures and wall shear stress. *Med. Biol. Eng. Comput.* **2023**, *61*, 1489–1506. [CrossRef] [PubMed]

9. Sundström, E.; Jonnagiri, R.; Gutmark-Little, I.; Gutmark, E.; Critser, P.; Taylor, M.; Tretter, J. Hemodynamics and tissue biomechanics of the thoracic aorta with a trileaflet aortic valve at different phases of valve opening. *Int. J. Numer. Methods Biomed. Eng.* **2020**, *36*, e3345. [CrossRef]

10. Sundström, E.; Jonnagiri, R.; Gutmark-Little, I.; Gutmark, E.; Critser, P.; Taylor, M.; Tretter, J. Effects of Normal Variation in the Rotational Position of the Aortic Root on Hemodynamics and Tissue Biomechanics of the Thoracic Aorta. *Cardiovasc. Eng. Technol.* **2020**, *11*, 47–58. [CrossRef]

11. Fedorov, A.; Beichel, R.; Kalpathy-Cramer, J.; Finet, J.; Fillion-Robin, J.C.; Pujol, S.; Bauer, C.; Jennings, D.; Fennessy, F.; Sonka, M.; et al. 3D Slicer as an image computing platform for the Quantitative Imaging Network. *Magn. Reson. Imaging* **2012**, *30*, 1323–1341. [PubMed]

12. Nestola, M.G.C.; Faggiano, E.; Vergara, C.; Lancellotti, R.M.; Ippolito, S.; Antona, C.; Filippi, S.; Quarteroni, A.; Scrofani, R. Computational comparison of aortic root stresses in presence of stentless and stented aortic valve bio-prostheses. *Comput. Methods Biomech. Biomed. Eng.* **2017**, *20*, 171–181. [CrossRef] [PubMed]

13. Rosero, E.B.; Peshock, R.M.; Khera, A.; Clagett, P.; Lo, H.; Timaran, C.H. Sex, race, and age distributions of mean aortic wall thickness in a multiethnic population-based sample. *J. Vasc. Surg.* **2011**, *53*, 950–957. [PubMed]

14. De Kerchove, L.; Momeni, M.; Aphram, G.; Watremez, C.; Bollen, X.; Jashari, R.; Boodhwani, M.; Astarci, P.; Noirhomme, P.; El Khoury, G. Free margin length and coaptation surface area in normal tricuspid aortic valve: An anatomical study. *Eur. J. Cardio-Thorac. Surg.* **2018**, *53*, 1040–1048.

15. Van Wyk, S.; Wittberg, L.P.; Fuchs, L. Wall shear stress variations and unsteadiness of pulsatile blood-like flows in 90-degree bifurcations. *Comput. Biol. Med.* **2013**, *43*, 1025–1036.

16. Sundström, E.; Oren, L.; Farbos de Luzan, C.; Gutmark, E.; Khosla, S. Fluid-Structure Interaction Analysis of Aerodynamic and Elasticity Forces During Vocal Fold Vibration. *J. Voice* **2022**. [CrossRef]

17. Sundström, E.; Oren, L. Pharyngeal flow simulations during sibilant sound in a patient-specific model with velopharyngeal insufficiency. *J. Acoust. Soc. Am.* **2019**, *145*, 3137–3145. [CrossRef]

18. Sundström, E.; Oren, L. Sound production mechanisms of audible nasal emission during the sibilant /s/. *J. Acoust. Soc. Am.* **2019**, *146*, 4199–4210. [CrossRef]

19. Sundström, E.; Boyce, S.; Oren, L. Effects of velopharyngeal openings on flow characteristics of nasal emission. *Biomech. Model. Mechanobiol.* **2020**, *19*, 1447–1459. [CrossRef]

20. Holzapfel, G.A.; Gasser, T.C.; Ogden, R.W. A new constitutive framework for arterial wall mechanics and a comparative study of material models. *J. Elast. Phys. Sci. Solids* **2000**, *61*, 1–48.

21. Simo, J.C.; Taylor, R.L. Quasi-incompressible finite elasticity in principal stretches. Continuum basis and numerical algorithms. *Comput. Methods Appl. Mech. Eng.* **1991**, *85*, 273–310. [CrossRef]

22. Schäfers, H.J. The 10 Commandments for Aortic Valve Repair. *Innovations* **2019**, *14* 188–198. [CrossRef] [PubMed]

23. Mei, S.; de Souza Júnior, F.S.; Kuan, M.Y.; Green, N.C.; Espino, D.M. Hemodynamics through the congenitally bicuspid aortic valve: A computational fluid dynamics comparison of opening orifice area and leaflet orientation. *Perfusion* **2016**, *31*, 683–690. [CrossRef] [PubMed]

24. Pasta, S.; Gentile, G.; Raffa, G.; Bellavia, D.; Chiarello, G.; Liotta, R.; Luca, A.; Scardulla, C.; Pilato, M. In silico shear and intramural stresses are linked to aortic valve morphology in dilated ascending aorta. *Eur. J. Vasc. Endovasc. Surg.* **2017**, *54*, 254–263. [CrossRef] [PubMed]

25. Sundström, E.; Michael, J.; Najm, H.K.; Tretter, J.T. Blood Speckle Imaging: An Emerging Method for Perioperative Evaluation of Subaortic and Aortic Valvar Repair. *Bioengineering* **2023**, *10*, 1183. [CrossRef]

26. Sundström, E.; Laudato, M. Machine Learning-Based Segmentation of the Thoracic Aorta with Congenital Valve Disease Using MRI. *Bioengineering* **2023**, *10*, 1216. [CrossRef]

27. Evangelista, A.; Gallego, P.; Calvo-Iglesias, F.; Bermejo, J.; Robledo-Carmona, J.; Sánchez, V.; Saura, D.; Arnold, R.; Carro, A.; Maldonado, G.; et al. Anatomical and clinical predictors of valve dysfunction and aortic dilation in bicuspid aortic valve disease. *Heart* **2018**, *104*, 566–573. [CrossRef]

28. Tretter, J.T.; Izawa, Y.; Spicer, D.E.; Okada, K.; Anderson, R.H.; Quintessenza, J.A.; Mori, S. Understanding the aortic root using computed tomographic assessment: a potential pathway to improved customized surgical repair. *Circ. Cardiovasc. Imaging* **2021**, *14*, e013134. [CrossRef]

29. Karwat, P.; Klimonda, Z.; Styczyński, G.; Szmigielski, C.; Litniewski, J. Aortic root movement correlation with the function of the left ventricle. *Sci. Rep.* **2021**, *11*, 4473. [CrossRef]

30. Plonek, T.; Berezowski, M.; Kurcz, J.; Podgorski, P.; Sąsiadek, M.; Rylski, B.; Mysiak, A.; Jasinski, M. The evaluation of the aortic annulus displacement during cardiac cycle using magnetic resonance imaging. *BMC Cardiovasc. Disord.* **2018**, *18*, 1–6. [CrossRef]

 *bioengineering*

Article

Machine Learning-Based Segmentation of the Thoracic Aorta with Congenital Valve Disease Using MRI

Elias Sundström [1,*] and Marco Laudato [1,2]

[1] Department of Engineering Mechanics, FLOW Research Center, KTH Royal Institute of Technology, Teknikringen 8, 10044 Stockholm, Sweden

[2] Department of Engineering Mechanics, The Marcus Wallenberg Laboratory for Sound and Vibration Research, KTH Royal Institute of Technology, Teknikringen 8, 10044 Stockholm, Sweden

* Correspondence: elias@kth.se

Abstract: Subjects with bicuspid aortic valves (BAV) are at risk of developing valve dysfunction and need regular clinical imaging surveillance. Management of BAV involves manual and time-consuming segmentation of the aorta for assessing left ventricular function, jet velocity, gradient, shear stress, and valve area with aortic valve stenosis. This paper aims to employ machine learning-based (ML) segmentation as a potential for improved BAV assessment and reducing manual bias. The focus is on quantifying the relationship between valve morphology and vortical structures, and analyzing how valve morphology influences the aorta's susceptibility to shear stress that may lead to valve incompetence. The ML-based segmentation that is employed is trained on whole-body Computed Tomography (CT). Magnetic Resonance Imaging (MRI) is acquired from six subjects, three with tricuspid aortic valves (TAV) and three functionally BAV, with right–left leaflet fusion. These are used for segmentation of the cardiovascular system and delineation of four-dimensional phase-contrast magnetic resonance imaging (4D-PCMRI) for quantification of vortical structures and wall shear stress. The ML-based segmentation model exhibits a high Dice score (0.86) for the heart organ, indicating a robust segmentation. However, the Dice score for the thoracic aorta is comparatively poor (0.72). It is found that wall shear stress is predominantly symmetric in TAVs. BAVs exhibit highly asymmetric wall shear stress, with the region opposite the fused coronary leaflets experiencing elevated tangential wall shear stress. This is due to the higher tangential velocity explained by helical flow, proximally of the sinutubal junction of the ascending aorta. ML-based segmentation not only reduces the runtime of assessing the hemodynamic effectiveness, but also identifies the significance of the tangential wall shear stress in addition to the axial wall shear stress that may lead to the progression of valve incompetence in BAVs, which could guide potential adjustments in surgical interventions.

Keywords: machine learning segmentation; 4D-PCMRI; aortic valve disease

check for
updates

Citation: Sundström, E.; Laudato, M. Machine Learning-Based Segmentation of the Thoracic Aorta with Congenital Valve Disease Using MRI. *Bioengineering* **2023**, *10*, 1216. https://doi.org/10.3390/bioengineering10101216

Academic Editors: Enrique Berjano and Mario Petretta

Received: 16 August 2023
Revised: 21 September 2023
Accepted: 12 October 2023
Published: 18 October 2023

1. Introduction

The aorta is the main artery in the human body, with the life-sustaining role of distributing oxygenated blood to all parts of the body via systemic circulation. With its complex morphology, it has been extensively researched to explore the space of parameters that can impact the biomechanical function of the aorta [1]. The bicuspid aortic valve (BAV) is a congenital aortic valvar disease present in 1–2 percent of the population. It is characterized by a fusion of the right and left coronary leaflets, but can also show other geometric variations on the raphe length, interleaflet triangle, and the rotational position of the aortic valve. Such rotational position has been hypothesized to influence wall shear stress (WSS) and helicity of blood flow in the ascending aorta [2–5]. This is believed not only to affect the valve competency, but also increase the risk for aortic dilation [6]. Additionally, the calcification of aortic valvar leaflets can lead to complications such as aortic regurgitation, increased shear stress, and pressure loss [7].

Meierhofer et al. [8] utilized three-dimensional time-resolved phase-contrast magnetic resonance imaging (4D-PCMRI) to compare blood flow patterns between individuals with BAV and those with normal tricuspid aortic valves (TAV). They found that non-stenotic BAVs exhibit higher tangential shear stress but lower axial shear stress compared to TAV, which has been confirmed through in vitro studies [9]. Furthermore, qualitative analysis has shown that vortical flow structures, which potentially contribute to aortic dilation, are more prevalent in BAVs. However, the scales and strength of these vortical structures were not quantified [8]. Similarly, Dux-Santoy et al. [10] quantified higher WSS magnitude in non-stenotic BAV compared to healthy volunteers but found no correlation with the pathogenesis of aortic dilation. In two related 4D-PCMRI studies, it was observed that the jet angle emanating from the valve differs in TAV and BAV subjects [11,12]. This variation was also reported in the assessment of 4D-PCMRI data by Barker [13], who observed a correlation of the jet impingement on the aortic wall in functional BAVs to coincide with the opposite side of the fused leaflets. The use of 4D-PCMRI has also increased the understanding of aortic stiffness associated with aortic disease, degenerative aneurysms, and chronic dissections [14,15]. Hope et al. [16] investigated 4D-PCMRI data and hypothesized that counter-rotating helices in the ascending aorta play a role in reducing velocity fluctuations and lessening the wall shear stress. MRI has also been used in retrospective studies of non-atherosclerotic aortic arch pathologies (NA-AAPs), including conditions like bicuspid aortic valve, inflammatory diseases, and heritable connective tissue disorders. Certain aortic arch variations, like bovine arch and vascular rings, can lead to aortic wall stiffening, promoting atherosclerosis growth and aneurysm formation, highlighting arterial stiffness as a significant risk factor for cardiovascular outcomes [17].

The analysis of 4D-PCMRI faces challenges due to its relatively long lead times, requiring manual segmentation and the process of identification of anatomical landmarks for localizing flow characteristics. To address these limitations, machine learning (ML) algorithms have been employed to fully automate the cardiac imaging workflow [18,19]. Previous studies have used ML algorithms to identify vascular anatomical landmarks in various imaging modalities. It has also been used in emergency clinical scenarios to facilitate accurate diagnosis of different thoracic aortic pathologies [20]. Notably, the 3D U-net convolutional neural network (see next section for details) has shown promise in aortic segmentation using PCMRI [21]. However, due to the data requirements of ML algorithms and the scarcity of large 4D-PCMRI datasets with consistent annotations, limited research exists on the application of ML algorithms for analyzing 4D-PCMRI data.

The study aims to test ML-based segmentation using MRI data to analyze vortical structures and wall shear stress in subjects with BAV and contrast it against normal TAV. With the use of ML-based segmentation, this study aims to provide a physics-based understanding of the biomechanical characteristics of the aorta and provide valuable insights for clinical practice.

2. Method

2.1. MRI Acquisition

The present study used cardiac MRI data sets acquired at Cincinnati Children's Hospital Medical Center (CCHMC). All demographic information was anonymized (i.e., age, sex, etc.) and CCHMC Institutional Review Board deemed the study to be exempt from any ethical inquiries [2–4]. These data sets were acquired from six subjects, three individuals with non-stenotic tricuspid aortic valves (TAV) and trisinuate aortic roots, as well as three individuals with non-stenotic functionally bicuspid aortic valves (BAV) featuring right–left coronary leaflet fusion (specifically type 1 fusion according to Sievers et al. [22]).

All MRI scans were performed using a 1.5 Tesla clinical MRI scanner (Ingenia, Philips Healthcare; Best, Netherlands) using a phased-array coil. The study of the protocol for the long-axis sagittal stack 4D phase-contrast magnetic resonance imaging (4D-PCMRI), the short-axis aortic root cine stack, and the aortic root phase-contrast velocity sequence was carried out, in addition to the noncontrast 3D mDixon angiogram for analysis. The short-axis aortic root cine stack was obtained using a steady-state free precession pulse sequence with specific parameters: a repetition time of 7.8 ms, an echo time of 4.7 ms, a flip angle of 15°, and sequential 2 mm slices without an interslice gap. For the short-axis aortic root phase-contrast velocity encoded sequence, a gradient echo sequence was employed with a repetition time of 4.3 ms, an echo time of 2.7 ms, a flip angle of 12°, and a slice thickness of 6 mm. The encoding velocity for this sequence was set at 1.5 m/s. Each cardiac cycle was represented by 30 phases for both sequences, resulting in a mean temporal resolution of 30–40 ms. The non-contrast coronal 3D mDixon angiogram was acquired using a repetition time of 5.3 ms, a flip angle of 15°, and 1 mm slices with no interslice gap. The 4D-PCMRI sagittal stack was obtained using a velocity encoding of 2 m/s, s repetition time of 3.5 ms, an echo time of 1.9 ms, and a flip angle of 8° [2–4].

2.2. ML-Based Segmentation

Semantic segmentation is one of the oldest problems in computer vision [23]. It is defined as the ability to label every pixel of an image, even when the object under analysis is completely unknown [24]. In this perspective, segmentation represents a more complex task than object recognition. The latter, indeed, is limited to classifying objects in an image within a set of a-priori specified labels. Segmentation, on the other hand, is a more general problem as it requires the computer to identify and isolate unknown objects.

It is possible to approximately classify the many instances of ML implementations for semantic segmentation problems in three main groups: weakly supervised methods [25], region-based semantic segmentation [26], and fully convolutional network (FCN)-based segmentation [27]. While weakly supervised methods have the advantage of not requiring any labeling of the training data set, they show poor performances in terms of object localization [28]. The other methods can be framed as supervised learning implementation and are based on Convolutional Neural Networks (CNN). One relevant example is U-Net [29]. Its architecture is based on FCN, but is characterized by the presence of multiple up-sampling layers. Essentially, the first half (contractive path) of a U-Net implementation can be seen as a classical contracting CNN, while the second half (expansive path) is symmetrically growing again using up-sampling operators (see Figure 1).

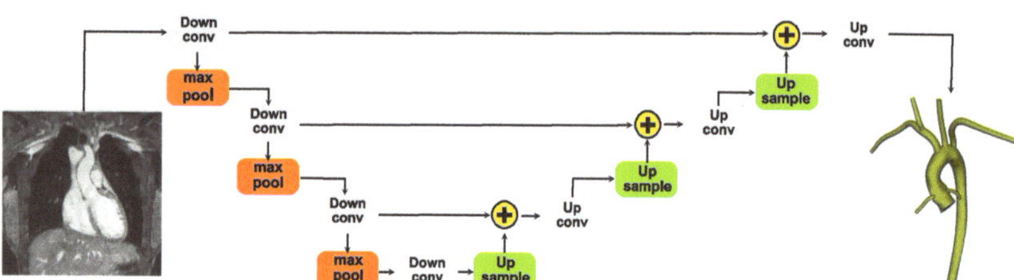

Figure 1. Schematics of a U-Net architecture. The raw MRI data are first down-sampled via convolution and max pool operations. The resulting latent space is up-sampled again via convolution and concatenation (yellow plus signs in the figure). The output is the segmented 3D domain.

A particularly successful U-Net based implementation specialized in biomedical images is the so-called nnU-Net [30]. Its main feature is the possibility to automatically configure all the hyperparameters of the network by modeling them in terms of fixed parameters, interdependent heuristic rules, and empirical decisions. Such configuration does not require any manual intervention and it can provide a highly accurate segmentation on the 23 public data sets usually employed in biomedical segmentation competitions [31]. A particularly convenient implementation available is called TotalSegmentator [18], which can segment 104 anatomical structures in the human body with a Dice similarity coefficient score of 0.943. The training data set consists of 1368 CT images, manually labeled. The architecture follows the original encoder–decoder nnU-Net scheme (see Figure 1) with the following minor modifications. The activation function employed for the network is the leaky RELUs which has a negative slope of 0.01. Moreover, instance normalization in place of standard normalization is employed, as the batch size is relatively small. Strided convolution is implemented for the down-sampling, whereas the up-sampling is obtained via convolution transposed. The training runs over 1000 epochs, where one epoch includes 250 randomly chosen mini-batches. The algorithm employed in the training is a stochastic gradient descent with an initial learning rate of 0.01, which is then dynamically modified during the training. The loss function is cross-entropy summed with the Dice score. The images are normalized, re-sampled, and then processed by the neural network using a sliding window. The re-sampling, in particular, is a crucial step, as often in the medical domain the information is arranged on nonhomogeneous grids. The information is therefore arranged on a homogeneous grid using 3-spline-based interpolation. A fine feature of nnU-Net is that it automatically adapts its topology to the GPU memory budget. In particular, the algorithm seeks the largest sustainable patch size, which is in turn connected to more contextual information. Finally, the default convolutional kernel size is $3 \times 3 \times 3$. However, as medical data often show a different resolution along one axis, the network is able to automatically set the kernel dimension in that direction to 1.

The main advantage of TotalSegmentator is that it can segment a wide range of clinical data (also on pathological cases) with superior performances concerning other publicly available algorithms [32–34]. Clearly, the Dice score alone is not enough to provide a full measure of the accuracy of the segmentation. Typical failure cases are the missing small parts of anatomical structures and the mixing of neighboring parts. However, the main limitation is that the training data set consists only of CT data. Consequently, the performance of the segmentation for any other kind of clinical data needs to be investigated. One of the goals of this work is to test the TotalSegmentator's performance on MRI data to study the cardiovascular assessment of vortical structures and wall shear stress connected to valve incompetence. As discussed in detail in the next section, although the Dice score is comparatively worse (0.8) on MRI data, such an ML-based segmentation is able to reduce the runtime of the cardiac assessment under analysis and allow for the subsequent flow analysis.

3. Result

3.1. Segmentation Evaluation

The ML-based segmentation algorithm identified most of the larger cardiovascular structures: left atrium, left ventricle, myocardium, right atrium, right ventricle, pulmonary artery, and aorta; see Figure 2. The axial cut through the major chambers of the heart shows a good qualitative agreement with the background MRI. The segmented aorta (light green) was broadly underestimated compared to the manual segmentation (olive green). Details of the aortic root, the leaflets, coronary arteries, as well as the head and neck arteries, i.e., brachiocephalic artery, left common carotid artery and left subclavian artery, were not identified in the dataset. However, a small fraction towards the descending aorta was identified in both TAV and BAV cases. The segmented cross-sectional areas in Figure 2 (bottom row) are quantified in Table 1 compared to the ground truth (in parentheses) and including the Dice score (in square brackets). The Dice score of the left ventricle segmentation was 0.86 ± 0.06 for the TAV cases and 0.89 ± 0.06 for the BAV cases. The Dice score of the segmented aorta, i.e., of the region that was identified, was 0.72 ± 0.12 for the TAV cases and 0.82 ± 0.06 for the BAV cases.

Table 1. Segmented cross-sectional areas of the left ventricle and the aorta measured on the axial cut shown in Figure 2. The ground truth values (in parentheses) of the left ventricle and the aorta were obtained by manually tracing a closed loop around the region of interest. The mean μ and standard deviation σ of the Dice score (in square brackets) are given for both TAV and BAV cases.

Parameter	Aorta (cm^2)	Left Ventricle (cm^2)
TAV1	1.4 (2.0) [0.82]	17.4 (20.9) [0.91]
TAV2	1.3 (1.9) [0.81]	11.1 (14.2) [0.88]
TAV3	0.9 (2.1) [0.60]	18.4 (22.5) [0.90]
μ and σ of Dice score	[0.72 ± 0.12]	[0.86 ± 0.06]
BAV1	2.9 (2.2) [0.86]	24.1 (27.1) [0.94]
BAV2	2.2 (3.5) [0.75]	20.8 (29.3) [0.83]
BAV3	2.7 (3.7) [0.84]	14.3 (17.8) [0.89]
μ and σ of Dice score	[0.82 ± 0.06]	[0.89 ± 0.06]

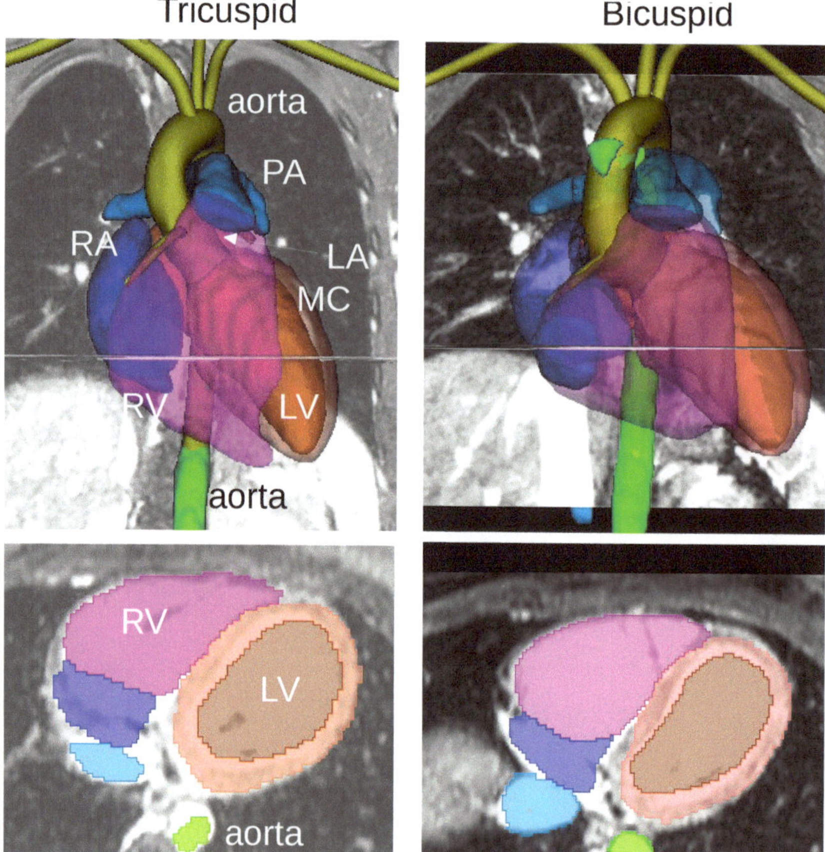

Figure 2. Segmentation of cardiovascular structures of the TAV (**top left panel**) and BAV (**top right panel**) subject by the ML-based segmentation. The segmented structures on the top coronal view are: the left ventricle (LV), myocardium (MC), left atrium (LA), right ventricle (RV), right atrium (RA), pulmonary artery (PA), aorta. The axial plane at the mid-position of the LV is shown in the (**bottom panels**).

3.2. Segmentation Runtime

Table 2 provides a summary of the runtime, RAM (random access memory), and GPU (graphics processing unit) memory requirements for the MRI resolution analysis of both TAV and BAV cases. Both cases cover the thorax and abdomen, with a voxel size of $320 \times 320 \times 100$ in the TAV and $400 \times 400 \times 100$ voxels in the BAV. The runtime and RAM and GPU memory requirements were monitored on a Linux workstation with an Intel Core i9 5.2 GHz CPU and an Nvidia GeForce GTX 1050 Ti. Overall, the runtime of the whole heart segmentation clocked in at about 2 min, which is significantly faster than the manual segmentation that takes in the order of a day (c.f., olive green colored aorta in Figure 1).

Table 2. Image size, runtime, RAM, and GPU memory requirements of the TAV and BAV cases.

Case	Size (Voxels) (mm)	Runtime	RAM	GPU Mem
TAV	$(320 \times 320 \times 100)$ $(0.9 \times 0.9 \times 2.4$ mm$)$	1 min 48 s	5.1 GB	3.0 GB
BAV	$(400 \times 400 \times 100)$ $(0.8 \times 0.8 \times 2.8$ mm$)$	2 min 2 s	5.4 GB	3.2 GB

3.3. Flow Rate, Pulse Wave Velocity, and Arterial Distensibility

Figure 3a compares the measured flow rates between the TAV and BAV cases. Notably, the BAV case exhibits a higher peak flow rate compared to the TAV case. When considering the normalized timescale, the TAV case demonstrates a larger fraction between the systolic and diastolic phases than the BAV case. Both cases exhibit minimal regurgitant fraction, and the net flow fraction between the descending and ascending flow is larger in the BAV case. Table 3 provides an overview of the cardiac output.

The calculation of pulse wave velocity (PWV) involves two crucial parameters: the Aortic Length measurement and the time interval between the upslopes of the flow curves. The time interval is determined by measuring the temporal distance from the point where the tangent of the aortic ascending flow curve reaches zero to the point where the tangent of the descending aorta curve also reaches zero. The calculated PWV values are 3.2 m/s for the TAV case and 3.3 m/s for the BAV case, indicating similar arterial stiffness. The distensibility was assessed using the relation $PWV = \sqrt{1/\rho D}$. The change in area $(\Delta A/A)$ is measured in the ascending aorta between peak systolic and the end of diastole, as shown in Figure 3b. The pressure drop is similar for both cases and is determined by the fraction of the area change over the distensibility.

Figure 3c shows the velocity in the ascending aorta, where the error bar represents the standard deviation over time. Both the TAV and BAV cases indicate similar velocity at peak systole (around 60 cm/s), which is due to the larger cross-sectional area in the TAV case. Although there is a slightly larger velocity in the BAV case at peak systole that results in a higher peak kinetic energy compared to the TAV case; see Figure 3d.

Table 3. Different parameters of the cardiac output calculated from the aortic ascending and descending flow.

Parameter	TAV	BAV
Heart rate (bpm)	70	47
Net volume (mL)	67	108
Ascending flow (L/min)	4.7	5.1
Regurgitant fraction (%)	0.3	1.5
Descending flow (L/min)	3.4	3.2
Aortic length (mm)	112	156
PWV time to foot (m/s)	3.9	3.8
Distensibility (1/mmHg)	0.008	0.009
$\Delta A/A$	0.3	0.27
ΔP (mmHg)	36	31

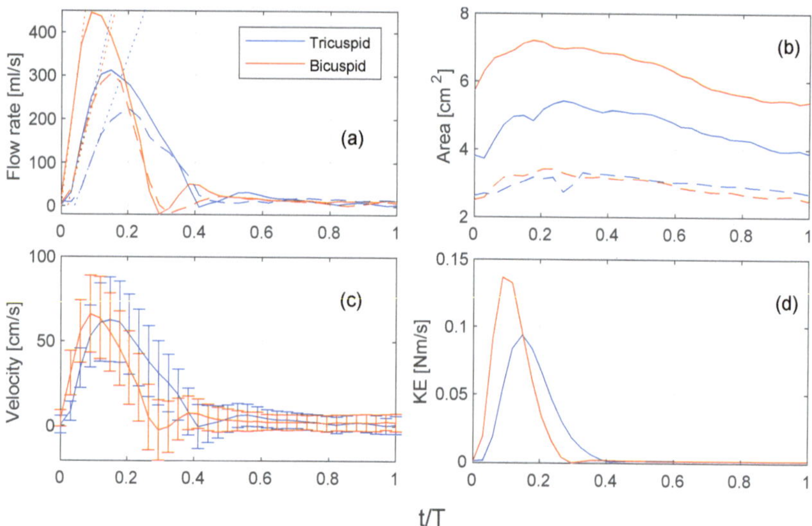

Figure 3. Comparison of different flow quantities for the TAV (blue) and BAV (red) cases: (**a**) ascending (full line) and descending (dashed line) flow, (**b**) area, (**c**) velocity, (**d**) kinetic energy. The dotted lines in (**a**) indicate the up slopes of the flow curves that are used for determining the time interval of the PWV.

3.4. Vortical Structures

Figure 4 provides an overview of the blood flow through the left atrium (LA), left ventricle (LV), and aorta during different stages of the cardiac cycle for both the TAV and BAV cases. In the early systole (at approximately $t/T = 0.05$), the aortic valve opens, initiating the ejection of blood from the LV. At the narrower section of the valve, the local flow velocity starts to accelerate, and the streamlines indicate a smooth and streamlined flow in both the TAV and BAV cases. The flow through the valve forms a jet, with the BAV case exhibiting a higher peak velocity that impacts more on the convex tissue wall near the sinotubular junction compared to the TAV case.

Around the time of peak systole (between t/T values of 0.1 and 0.3), the velocities in the descending aorta increase, which corresponds to the pulse wave velocity assessment and the time delay for the pulse wave to propagate from the ascending to the descending aorta. In the TAV case, the streamlines remain relatively aligned with the aorta. However, in the BAV case, some streamlines curl around a strong counterclockwise rotating vortex towards the concave side of the ascending aorta. This vortex qualitatively grows in size, and during post-peak systole (between t/T values of 0.2 and 0.3), it occupies a significant portion of the region proximal to the aortic root and interacts with upper arterial branches, i.e., the brachiocephalic artery and the left common carotid artery. Beyond the head and neck vessels, the flow aligns with the proximal thoracic descending aorta.

As the systolic phase concludes, the valve closes, and the blood flow diminishes, coinciding with a decrease in aortic pressure during diastole. As time progresses to approximately $t/T = 0.7$, in mid-diastole, the mitral valve opens, allowing blood flow from the left atrium to the left ventricle. This is accompanied by an increased flow velocity, facilitating the filling of the LV with fresh blood in preparation for another systole.

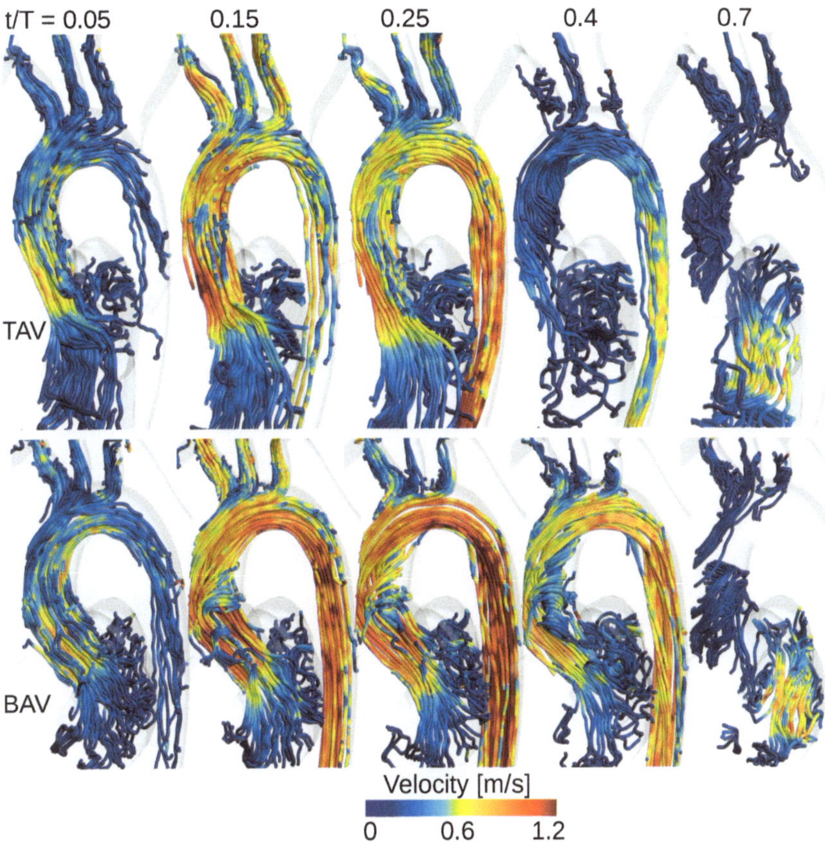

Figure 4. Streamlines colored by the velocity magnitude at different instances during the cardiac cycle for the TAV (**top row**) and BAV (**bottom row**) subjects. The streamlines show flow structures in the aorta, left atrium, and the left ventricle.

Figures 5 and 6 provide additional information on the blood flow through a short axis view, complementing the streamlines shown in the previous Figure 4. The former illustrates the velocity magnitude, while the latter displays the axial vorticity. These figures depict the same time instances of the cardiac cycle, starting from systole when the valve opens.

Around peak systole (between t/T values of 0.1 and 0.3), both the TAV and BAV cases exhibit vortical flow patterns. In the TAV case, the streamlines are directed towards the convex side of the aorta, aligning with the curvature of the ascending aorta. Simultaneously, two secondary Dean-like counter-rotating vortices form along the perpendicular axis to the curvature. However, these vortices are not perfectly symmetric, with slightly higher vorticity observed in the vortex towards the negative y-axis, corresponding to the left cusp of the aortic valve.

In the BAV case, the flow generates a robust swirling motion characterized by positive axial vorticity, indicating counter-clockwise rotation. The streamlines near the center of the vortex converge between t/T values of 0.1 and 0.2, indicating an increase in vortex strength. Beyond peak systole, the kinetic energy of the flow begins to diminish, resulting in reduced magnitudes of both velocity and axial vorticity.

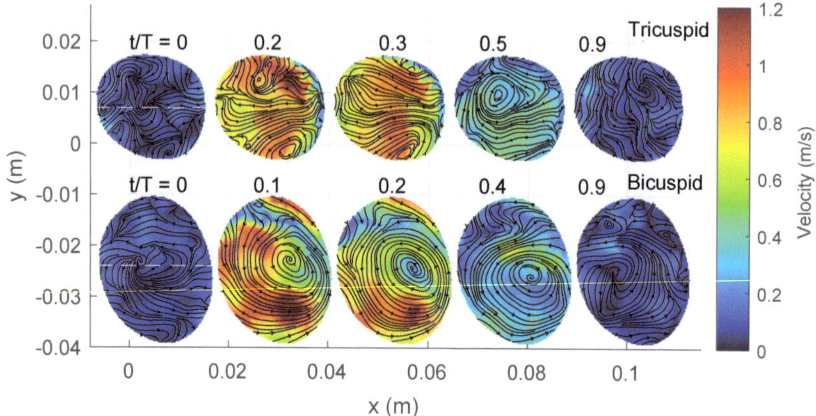

Figure 5. Short-axis view showing velocity magnitude with streamlines at the location proximally of the sinutubular junction in the ascending aorta (c.f. Figure 4 TAV (top row) and BAV (bottom row).

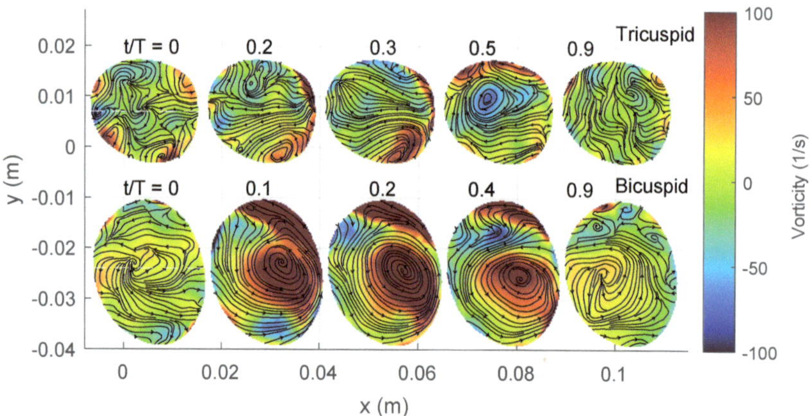

Figure 6. Short-axis view showing the axial vorticity with streamlines. Same keys as Figure 5.

3.5. Wall Shear Stress

The blood flow depicted in Figures 4–6 is further analyzed along a horizontal profile indicated by a white dashed line in Figure 5. In the TAV case, the axial velocity exhibits small magnitudes at the beginning of systole, progressively increasing towards peak systole and forming a symmetric top-hat profile, see Figure 7a. This behavior suggests the development of a boundary layer with a steeper gradient near the endothelium. Consequently, the axial shear stress component (Figure 7c) shows lower magnitudes at the center and gradually increases towards the endothelial wall. This result is consistent for all considered cases, which can be seen in the evaluation of the mean and standard deviation of the velocity and shear stresses at peak systole; see Figure 8.

In the BAV case, the axial velocity profile also evolves, but with greater asymmetry, featuring a slope towards the convex side (in the direction of the raphe between the left and right coronary leaflets). The velocity gradient is larger on this side compared to the TAV case, resulting in increased axial wall shear stress (Figure 7d). However, on the opposite side ($x/X = 1$), there is a small gradient with low axial velocity, leading to a lower axial wall shear stress compared to the TAV case. This is also supported in the statistical comparison of the 3 TAV and 3 BAV cases (Figure 8).

The cross-flow velocity component (y-axis) in the BAV case exhibits more significant levels compared to the TAV case, as illustrated in Figure 7e,f, as well as in the statistical assessment in Figure 8b. This observation aligns with the presence of a strong swirl and higher axial vorticity, as shown earlier in Figure 6. Consequently, the steeper gradient in the cross-flow velocity manifests as elevated tangential wall shear stress on the endothelium, particularly near the center of the vortex at $x/X = 0.7$.

In the BAV cases, the axial velocity changes signify transitioning from positive to negative flow, which coincides with the location of the vortex core at $x/X = 0.7$. This change in direction results in a non-zero oscillatory shear index (OSI) in the axial flow direction. A similar behavior is observed in the TAV case, although the magnitude of flow reversal is lower. On the other hand, when considering the cross-flow velocity component, it consistently maintains a positive sign, indicating that the oscillatory shear index (OSI) in the tangential flow direction is close to zero in both cases.

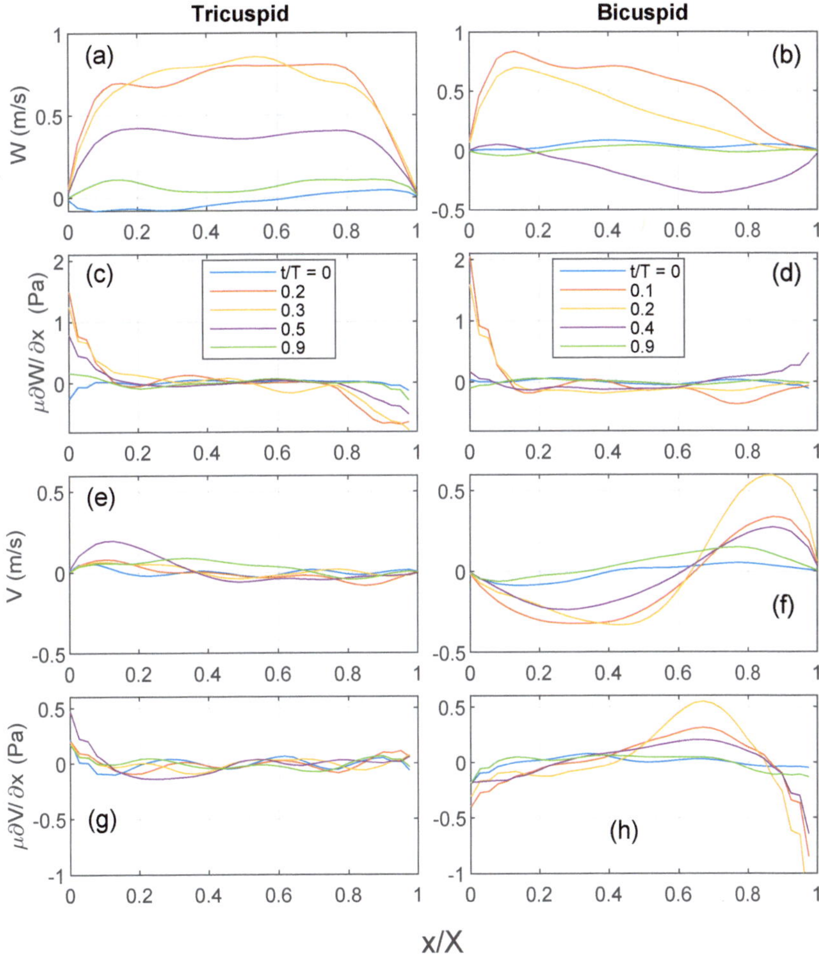

Figure 7. Quantification of velocity and shear stress along the dashed line annotated in Figure 5 that is located proximally of the sinutubular. The x-axis of the profile is normalized with the total length where $x/X = 0$ is towards the convex side and $x/X = 1$ is towards the concave side of the ascending aorta. (**a,b**) Streamwise velocity, (**c,d**) streamwise shear stress, (**e,f**) cross-flow velocity, (**g,h**) shear-stress in cross-flow direction. TAV (left column) and BAV (right column).

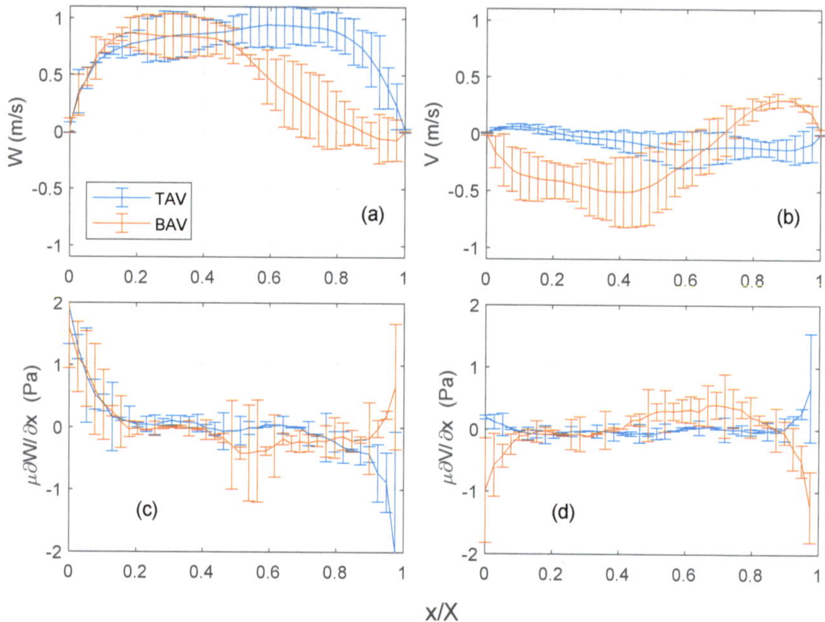

Figure 8. Error bar showing the mean and standard deviation of the velocity and shear stress, along the dashed line annotated in Figure 5 for all considered cases, i.e., 3 TAV and 3 BAV cases. The same keys as in Figure 7, but here the data are during peak systole.

4. Discussion

In this study, we employed machine learning (ML)-based segmentation techniques on an MRI dataset to segment the anatomical structures of the cardiovascular system. The ML-based segmentation approach, trained on CT datasets, exhibited a high level of accuracy, with a Dice score of 0.86 of the main chambers of the heart, which is in good agreement with segmentation using CT data [18]. However, its performance was relatively poorer when segmenting the aorta, achieving a Dice score of 0.72 in that particular region, which agrees with other studies that report better performance with CT compared to MRI [17]. Therefore, using MRI datasets on ML-based segmentation that is pre-trained on CT could affect the Dice score. In addition, the leaflet thickness of the aortic valve and the diameter of the coronary arteries are two features with dimensions below the resolution of 4D-PCMRI, which may also explain the segmentation failure and lower Dice score of the aorta.

The ML-based segmentation method demonstrated robustness when applied to clinical MRI data, effectively reducing the time required for the entire heart segmentation compared to manual segmentation. The runtime for the segmentation process was less than approximately 2 min, and it demanded about 5 GB of RAM and GPU memory. As a result, this approach can be executed on a workstation, making it practical and feasible for routine usage.

The functionally bicuspid aortic valve and trisinuate aortic root discussed in this study are more frequently linked to aortic valve narrowing and aortic enlargement [35]. In this latter type, two functional commissures exist with two normal interleaflet triangles, extending to the sinutubular junction. On the contrary, the interleaflet triangle below the fusion zone is underdeveloped without a functional commissure, as its tip does not reach the sinutubular junction. These data support the notion that fusion between the right and left leaflets leads to an asymmetric wall shear stress distribution, with increased circumferential wall shear stress compared to TAVs [36]. This can potentially lead to leaflet thickening, sclerosis, and eventual calcification, influencing aortic valve narrowing progression. Additionally, the study highlights that the BAV aortopathy directly impacts hemodynamics in the thoracic aorta, shedding light on potential risk factors for aortic enlargement progression [37].

These findings have important implications for assessing congenitally malformed valves and providing insights for surgical interventions. Studies have shown that the fusion orientation and the level of hypoplasty of the commissural apex can guide decisions between preserving a functional bicuspid valve or re-configuring it to a tricuspid valve, optimizing the potential for long-lasting repairs [1,38].

5. Limitation

The sub-optimal Dice score obtained for the aorta resulted in segmentation failure of the head and neck vessels, coronary arteries, and the fine details of the valve leaflets. In cases where segmentation failure occurs, it may be necessary to resort to manual segmentation, although this would incur additional turnaround time for the complete segmentation of the heart.

One limitation of the ML-based segmentation method employed in this study is that it was trained on ECG-gated CT datasets. Consequently, the study was constrained to utilizing 3D mDixon data that were ECG-gated. Therefore, the physiological displacement of the heart throughout the cardiac cycle could impact the delineation of the 4D-PCMRI data, which evolve both spatially and temporally. This influences the flow quantification in the left ventricle, as it experiences significant volume changes during systolic contraction. However, for vessels that undergo moderate displacement and area changes, this factor is less significant. Nonetheless, including 4D-PCMRI data in the training datasets for the ML-based segmentation would help overcome this limitation and enhance the segmentation performance. This will be the objective of a future study.

The study's scope is constrained by the limited number of subjects, focusing on the functionally bicuspid aortic valve (BAV) with a trisinuate aortic root when compared to the normal trisinuate aortic valve (TAV). Nonetheless, the cohort size in this study is consistent with other studies [39]. Subsequent studies will employ this methodology for more intricate analysis of flow structures, incorporating a more extensive cohort that matches age, sex, and valve anatomy. This expanded dataset will include normal aortic valves as well as various forms of bicuspid and unicuspid aortic valves. These future investigations will aim to establish connections between these observations and the likelihood of aortic dilation [40].

6. Conclusions

The examined ML-based segmentation method demonstrated precise measurement of left ventricular volume and accurate identification of the thoracic aorta region, regardless of aortic pathology. These findings suggest that this method holds significant promise as a valuable tool for promptly assessing aortic pathologies like the bicuspid aortic valve in a clinical setting.

The quantification of flow structures in TAV and BAV morphologies was accomplished using ML-based segmentation to delineate the 4D-PCMRI flow measurements. Comparative analysis revealed that BAVs exhibited a more pronounced impingement of the jet towards the convex side of the ascending aorta, in contrast to TAVs. At peak systole, the axial vorticity data in the ascending aorta demonstrated that BAV subjects displayed a significant vortex structure with a counterclockwise swirl, which was more prominent than in the normal TAV cases. This is in agreement with previous observations [4,8,16,36]. This vortex structure led to increased tangential wall shear stress in BAV compared to TAV, concluding the significance of quantifying the tangential component of wall shear stress in addition to the axial wall shear stress.

The quantitative assessment of vortical flow structures contributes to a deeper understanding of the relationship between hemodynamics in the proximal thoracic aorta and differences in aortic valvar morphology. The Computational Fluid Dynamics model in ideal [41,42] and patient-specific [43–45] geometries will be implemented to compare with the observed behavior. Further investigations involving larger cohort sizes will be conducted to determine the potential clinical utility of these findings, particularly concerning the propensity for aortic dilation in individuals with BAV. However, it is anticipated that the main findings will remain consistent even with a larger cohort.

Author Contributions: Conceptualization, E.S. and M.L.; formal analysis, E.S. and M.L.; writing—review and editing, E.S. and M.L.; funding acquisition, E.S. All authors have read and agreed to the published version of the manuscript.

Funding: This research was funded by the Swedish Research Council (grant number 2021-04894).

Institutional Review Board Statement: Ethical review and approval were waived by the Cincinnati Children's Institutional Review Board because the study was a retrospective analysis using de-identified data from a limited number of subjects.

Informed Consent Statement: Informed consent was obtained from all subjects involved in the study. Written informed consent has been obtained from the patient(s) to publish this paper.

Data Availability Statement: Data available on request due to restrictions in repository access.

Acknowledgments: The authors acknowledge the support from KTH Engineering Mechanics in the thematic areas of Biomechanics, Health, and Biotechnology (BHB).

Conflicts of Interest: The authors declare no conflict of interest.

References

1. Tretter, J.T.; Spicer, D.E.; Mori, S.; Chikkabyrappa, S.; Redington, A.N.; Anderson, R.H. The significance of the interleaflet triangles in determining the morphology of congenitally abnormal aortic valves: Implications for noninvasive imaging and surgical management. *J. Am. Soc. Echocardiogr.* **2016**, *29*, 1131–1143. [CrossRef] [PubMed]
2. Sundström, E.; Jonnagiri, R.; Gutmark-Little, I.; Gutmark, E.; Critser, P.; Taylor, M.; Tretter, J. Effects of Normal Variation in the Rotational Position of the Aortic Root on Hemodynamics and Tissue Biomechanics of the Thoracic Aorta. *Cardiovasc. Eng. Technol.* **2020**, *11*, 47–58. [CrossRef] [PubMed]
3. Sundström, E.; Jonnagiri, R.; Gutmark-Little, I.; Gutmark, E.; Critser, P.; Taylor, M.; Tretter, J. Hemodynamics and tissue biomechanics of the thoracic aorta with a trileaflet aortic valve at different phases of valve opening. *Int. J. Numer. Methods Biomed. Eng.* **2020**, *36*, e3345. [CrossRef] [PubMed]
4. Jonnagiri, R.; Sundström, E.; Gutmark, E.; Anderson, S.; Pednekar, A.; Taylor, M.; Tretter, J.; Gutmark-Little, I. Influence of aortic valve morphology on vortical structures and wall shear stress. *Med. Biol. Eng. Comput.* **2023**, *61*, 1489–1506. [CrossRef]
5. Sundström, E.; Tretter, J.T. Impact of Variation in Interleaflet Triangle Height between Fused Leaflets in the Functionally Bicuspid Aortic Valve on Hemodynamics and Tissue Biomechanics. *J. Eng. Sci. Med. Diagn. Ther.* **2022**, *5*, 031102. [CrossRef]
6. Guala, A.; Dux-Santoy, L.; Teixido-Tura, G.; Ruiz-Muñoz, A.; Galian-Gay, L.; Servato, M.L.; Valente, F.; Gutiérrez, L.; González-Alujas, T.; Johnson, K.M.; et al. Wall shear stress predicts aortic dilation in patients with bicuspid aortic valve. *Cardiovasc. Imaging* **2022**, *15*, 46–56. [CrossRef] [PubMed]
7. Bech-Hanssen, O.; Svensson, F.; Polte, C.L.; Johnsson, Å.A.; Gao, S.A.; Lagerstrand, K.M. Characterization of complex flow patterns in the ascending aorta in patients with aortic regurgitation using conventional phase-contrast velocity MRI. *Int. J. Cardiovasc. Imaging* **2018**, *34*, 419–429. [CrossRef] [PubMed]
8. Meierhofer, C.; Schneider, E.P.; Lyko, C.; Hutter, A.; Martinoff, S.; Markl, M.; Hager, A.; Hess, J.; Stern, H.; Fratz, S. Wall shear stress and flow patterns in the ascending aorta in patients with bicuspid aortic valves differ significantly from tricuspid aortic valves: A prospective study. *Eur. Heart J.-Imaging* **2013**, *14*, 797–804. [CrossRef]
9. Saikrishnan, N.; Mirabella, L.; Yoganathan, A.P. Bicuspid aortic valves are associated with increased wall and turbulence shear stress levels compared to trileaflet aortic valves. *Biomech. Model. Mechanobiol.* **2015**, *14*, 577–588. [CrossRef]
10. Dux-Santoy, L.; Guala, A.; Sotelo, J.; Uribe, S.; Teixidó-Turà, G.; Ruiz-Muñoz, A.; Hurtado, D.E.; Valente, F.; Galian-Gay, L.; Gutiérrez, L.; et al. Low and oscillatory wall shear stress is not related to aortic dilation in patients with bicuspid aortic valve: A time-resolved 3-dimensional phase-contrast magnetic resonance imaging study. *Arterioscler. Thromb. Vasc. Biol.* **2020**, *40*, e10–e20. [CrossRef]
11. Mahadevia, R.; Barker, A.J.; Schnell, S.; Entezari, P.; Kansal, P.; Fedak, P.W.; Malaisrie, S.C.; McCarthy, P.; Collins, J.; Carr, J.; et al. Bicuspid aortic cusp fusion morphology alters aortic three-dimensional outflow patterns, wall shear stress, and expression of aortopathy. *Circulation* **2014**, *129*, 673–682. [CrossRef] [PubMed]
12. Mirabella, L.; Barker, A.J.; Saikrishnan, N.; Coco, E.R.; Mangiameli, D.J.; Markl, M.; Yoganathan, A.P. MRI-based protocol to characterize the relationship between bicuspid aortic valve morphology and hemodynamics. *Ann. Biomed. Eng.* **2015**, *43*, 1815–1827. [CrossRef] [PubMed]

13. Barker, A.J.; Markl, M.; Bürk, J.; Lorenz, R.; Bock, J.; Bauer, S.; Schulz-Menger, J.; von Knobelsdorff-Brenkenhoff, F. Bicuspid aortic valve is associated with altered wall shear stress in the ascending aorta. *Circ. Cardiovasc. Imaging* **2012**, *5*, 457–466. [CrossRef] [PubMed]

14. Takahashi, K.; Sekine, T.; Ando, T.; Ishii, Y.; Kumita, S. Utility of 4D flow MRI in thoracic aortic diseases: A Literature Review of Clinical Applications and Current Evidence. *Magn. Reson. Med. Sci.* **2022**, *21*, 327–339. [CrossRef]

15. Adriaans, B.P.; Wildberger, J.E.; Westenberg, J.J.; Lamb, H.J.; Schalla, S. Predictive imaging for thoracic aortic dissection and rupture: Moving beyond diameters. *Eur. Radiol.* **2019**, *29*, 6396–6404. [CrossRef]

16. Hope, T.A.; Markl, M.; Wigström, L.; Alley, M.T.; Miller, D.C.; Herfkens, R.J. Comparison of flow patterns in ascending aortic aneurysms and volunteers using four-dimensional magnetic resonance velocity mapping. *J. Magn. Reson. Imaging Off. J. Int. Soc. Magn. Reson. Med.* **2007**, *26*, 1471–1479. [CrossRef]

17. Wawak, M.; Tekieli, Ł.; Badacz, R.; Pieniążek, P.; Maciejewski, D.; Trystuła, M.; Przewłocki, T.; Kabłak-Ziembicka, A. Clinical Characteristics and Outcomes of Aortic Arch Emergencies: Takayasu Disease, Fibromuscular Dysplasia, and Aortic Arch Pathologies: A Retrospective Study and Review of the Literature. *Biomedicines* **2023**, *11*, 2207.

18. Wasserthal, J.; Meyer, M.; Breit, H.C.; Cyriac, J.; Yang, S.; Segeroth, M. TotalSegmentator: Robust segmentation of 104 anatomical structures in CT images. *arXiv* **2022**, arXiv:2208.05868.

19. Berhane, H.; Scott, M.; Elbaz, M.; Jarvis, K.; McCarthy, P.; Carr, J.; Malaisrie, C.; Avery, R.; Barker, A.J.; Robinson, J.D.; et al. Fully automated 3D aortic segmentation of 4D flow MRI for hemodynamic analysis using deep learning. *Magn. Reson. Med.* **2020**, *84*, 2204–2218. [CrossRef]

20. Artzner, C.; Bongers, M.N.; Kärgel, R.; Faby, S.; Hefferman, G.; Herrmann, J.; Nopper, S.L.; Perl, R.M.; Walter, S.S. Assessing the Accuracy of an Artificial Intelligence-Based Segmentation Algorithm for the Thoracic Aorta in Computed Tomography Applications. *Diagnostics* **2022**, *12*, 1790. [CrossRef]

21. Fujiwara, T.; Berhane, H.; Scott, M.B.; Englund, E.K.; Schäfer, M.; Fonseca, B.; Berthusen, A.; Robinson, J.D.; Rigsby, C.K.; Browne, L.P.; et al. Segmentation of the Aorta and Pulmonary Arteries Based on 4D Flow MRI in the Pediatric Setting Using Fully Automated Multi-Site, Multi-Vendor, and Multi-Label Dense U-Net. *J. Magn. Reson. Imaging* **2022**, *55*, 1666–1680. [CrossRef]

22. Sievers, H.H.; Schmidtke, C. A classification system for the bicuspid aortic valve from 304 surgical specimens. *J. Thorac. Cardiovasc. Surg.* **2007**, *133*, 1226–1233. [CrossRef] [PubMed]

23. Guo, Y.; Liu, Y.; Georgiou, T.; Lew, M.S. A review of semantic segmentation using deep neural networks. *Int. J. Multimed. Inf. Retr.* **2018**, *7*, 87–93. [CrossRef]

24. Weinland, D.; Ronfard, R.; Boyer, E. A survey of vision-based methods for action representation, segmentation and recognition. *Comput. Vis. Image Underst.* **2011**, *115*, 224–241. [CrossRef]

25. Papandreou, G.; Chen, L.C.; Murphy, K.P.; Yuille, A.L. Weakly-and semi-supervised learning of a deep convolutional network for semantic image segmentation. In Proceedings of the IEEE International Conference on Computer Vision, Santiago, Chile, 7–13 December 2015; pp. 1742–1750. [CrossRef]

26. Caesar, H.; Uijlings, J.; Ferrari, V. Region-based semantic segmentation with end-to-end training. In Proceedings of the Computer Vision—ECCV 2016: 14th European Conference, Amsterdam, The Netherlands, 11–14 October 2016; pp. 381–397.

27. Long, J.; Shelhamer, E.; Darrell, T. Fully convolutional networks for semantic segmentation. In Proceedings of the IEEE Conference on Computer Vision and Pattern Recognition, Boston, MA, USA, 7–12 June 2015; pp. 3431–3440.

28. Bearman, A.; Russakovsky, O.; Ferrari, V.; Fei-Fei, L. What's the point: Semantic segmentation with point supervision. In Proceedings of the European Conference on Computer Vision, Amsterdam, The Netherlands, 11–14 October 2016; pp. 549–565.

29. Ronneberger, O.; Fischer, P.; Brox, T. U-net: Convolutional networks for biomedical image segmentation. In Proceedings of the Medical Image Computing and Computer-Assisted Intervention—MICCAI 2015: 18th International Conference, Munich, Germany, 5–9 October 2015; pp. 234–241.

30. Isensee, F.; Jaeger, P.F.; Kohl, S.A.; Petersen, J.; Maier-Hein, K.H. nnU-Net: A self-configuring method for deep learning-based biomedical image segmentation. *Nat. Methods* **2021**, *18*, 203–211. [CrossRef]

31. Bernard, O.; Lalande, A.; Zotti, C.; Cervenansky, F.; Yang, X.; Heng, P.A.; Cetin, I.; Lekadir, K.; Camara, O.; Ballester, M.A.G.; et al. Deep learning techniques for automatic MRI cardiac multi-structures segmentation and diagnosis: Is the problem solved? *IEEE Trans. Med. Imaging* **2018**, *37*, 2514–2525. [CrossRef]

32. Chen, X.; Sun, S.; Bai, N.; Han, K.; Liu, Q.; Yao, S.; Tang, H.; Zhang, C.; Lu, Z.; Huang, Q.; et al. A deep learning-based auto-segmentation system for organs-at-risk on whole-body computed tomography images for radiation therapy. *Radiother. Oncol.* **2021**, *160*, 175–184. [CrossRef]

33. Sundar, L.K.S.; Yu, J.; Muzik, O.; Kulterer, O.C.; Fueger, B.; Kifjak, D.; Nakuz, T.; Shin, H.M.; Sima, A.K.; Kitzmantl, D.; et al. Fully automated, semantic segmentation of whole-body 18F-FDG PET/CT images based on data-centric artificial intelligence. *J. Nucl. Med.* **2022**, *63*, 1941–1948. [CrossRef]

34. Trägårdh, E.; Borrelli, P.; Kaboteh, R.; Gillberg, T.; Ulén, J.; Enqvist, O.; Edenbrandt, L. RECOMIA—A cloud-based platform for artificial intelligence research in nuclear medicine and radiology. *EJNMMI Phys.* **2020**, *7*, 51. [CrossRef]

35. Evangelista, A.; Gallego, P.; Calvo-Iglesias, F.; Bermejo, J.; Robledo-Carmona, J.; Sánchez, V.; Saura, D.; Arnold, R.; Carro, A.; Maldonado, G.; et al. Anatomical and clinical predictors of valve dysfunction and aortic dilation in bicuspid aortic valve disease. *Heart* **2018**, *104*, 566–573. [CrossRef]

36. Sundström, E.; Tretter, T. J. Impact of Variation in Commissural Angle between Fused Leaflets in the Functionally Bicuspid Aortic Valve on Hemodynamics and Tissue Biomechanics. *Bioengineering* 2023, *submitted (unpublished)*.

37. Sundström, E.; Michael, J.; Najm, H.K.; Tretter, J.T. Blood Speckle Imaging: An Emerging Method for Perioperative Evaluation of Subaortic and Aortic Valvar Repair. *Bioengineering* **2023**, *10*, 1183. [CrossRef]

38. Schäfers, H.J. The 10 commandments for aortic valve repair. *Innovations* **2019**, *14*, 188–198. [CrossRef] [PubMed]

39. Faggiano, E.; Antiga, L.; Puppini, G.; Quarteroni, A.; Luciani, G.B.; Vergara, C. Helical flows and asymmetry of blood jet in dilated ascending aorta with normally functioning bicuspid valve. *Biomech. Model. Mechanobiol.* **2013**, *12*, 801–813. [CrossRef]

40. Bissell, M.M.; Hess, A.T.; Biasiolli, L.; Glaze, S.J.; Loudon, M.; Pitcher, A.; Davis, A.; Prendergast, B.; Markl, M.; Barker, A.J.; et al. Aortic dilation in bicuspid aortic valve disease: Flow pattern is a major contributor and differs with valve fusion type. *Circ. Cardiovasc. Imaging* **2013**, *6*, 499–507. [CrossRef]

41. Laudato, M.; Mosca, R.; Mihaescu, M. Buckling critical pressures in collapsible tubes relevant for biomedical flows. *Sci. Rep.* **2023**, *13*, 9298. [CrossRef]

42. Sundström, E.; Oren, L.; Farbos de Luzan, C.; Gutmark, E.; Khosla, S. Fluid-Structure Interaction Analysis of Aerodynamic and Elasticity Forces During Vocal Fold Vibration. *J. Voice* **2022**. [CrossRef] [PubMed]

43. Sundström, E.; Oren, L. Change in aeroacoustic sound mechanism during sibilant sound with different velopharyngeal opening sizes. *Med. Biol. Eng. Comput.* **2021**, *59*, 937–945. [CrossRef]

44. Sundström, E.; Boyce, S.; Oren, L. Effects of velopharyngeal openings on flow characteristics of nasal emission. *Biomech. Model. Mechanobiol.* **2020**, *19*, 1447–1459. [CrossRef]

45. Sundström, E.; Oren, L. Sound production mechanisms of audible nasal emission during the sibilant/s/. *J. Acoust. Soc. Am.* **2019**, *146*, 4199–4210. [CrossRef]

bioengineering

Article

Blood Speckle Imaging: An Emerging Method for Perioperative Evaluation of Subaortic and Aortic Valvar Repair

Elias Sundström [1,*], Michael Jiang [2], Hani K. Najm [3,4] and Justin T. Tretter [3,4]

[1] Department of Engineering Mechanics, FLOW Research Center, KTH Royal Institute of Technology, Teknikringen 8, 100 44 Stockholm, Sweden
[2] Department of Pediatric Cardiology, Cleveland Clinic, Cleveland, OH 44195, USA
[3] Congenital Valve Procedural Planning Center, Department of Pediatric Cardiology, Cleveland, OH 44195, USA
[4] Division of Pediatric Cardiac Surgery, and the Heart, Vascular, and Thoracic Institute, Cleveland Clinic, Cleveland, OH 44195, USA
* Correspondence: elias@kth.se; Tel.: +46-(0)70-449-5063

Abstract: Background: This article presents the use of blood speckle Imaging (BSI) as an echocardiographic approach for the pre- and post-operative evaluation of subaortic membrane resection and aortic valve repair. Method: BSI, employing block-matching algorithms, provided detailed visualization of flow patterns and quantification of parameters from ultrasound data. The 9-year-old patient underwent subaortic membrane resection and peeling extensions of the membrane from under the ventricular-facing surface of all three aortic valve leaflets. Result: Post-operatively, BSI demonstrated improvements in hemodynamic patterns, where quantified changes in flow velocities showed no signs of stenosis and trivial regurgitation. The asymmetric jet with a shear layer and flow reversal on the posterior aspect of the aorta was corrected resulting in reduced wall shear stress on the anterior aspect and reduced oscillatory shear index, which is considered a contributing element in cellular alterations in the structure of the aortic wall. Conclusion: This proof-of-concept study demonstrates the potential of BSI as an emerging echocardiographic approach for evaluating subaortic and aortic valvar repair. BSI enhances the quantitative evaluation of the left ventricular outflow tract of immediate surgical outcomes beyond traditional echocardiographic parameters and aids in post-operative decision-making. However, larger studies are needed to validate these findings and establish standardized protocols for clinical implementation.

Keywords: blood speckle imaging; aortic valve repair; aortic stenosis

Citation: Sundström, E.; Jiang, M.; Najm, H.K.; Tretter, J.T. Blood Speckle Imaging: An Emerging Method for Perioperative Evaluation of Subaortic and Aortic Valvar Repair. *Bioengineering* **2023**, *10*, 1183. https://doi.org/10.3390/bioengineering10101183

Academic Editor: Chung-Hao Lee

Received: 29 August 2023
Revised: 3 October 2023
Accepted: 6 October 2023
Published: 12 October 2023

1. Introduction

Aortic valvar repair is increasingly pursued for treating aortic valvar pathologies when feasible, aiming to restore valvar function and avoid the need for valvar replacement. This, however, requires a thorough understanding of the complex anatomy of the aortic root and its valve, and technical proficiency which is only obtained with experience and time spent on this steep learning curve [1]. Accurate evaluation of the repaired valve is crucial for assessing immediate surgical outcomes and optimizing post-operative management. While traditional echocardiographic techniques offer valuable information, they may have limitations in capturing detailed hemodynamic changes associated with aortic valvar repair [2]. On many occasions, we lack understanding of the recurrence of the subaortic obstruction, despite a good initial gradient drop. This lack of proper assessment of flow dynamics may explain the reason for the recurrence of the subaortic membrane and moderate progressive subaortic stenosis.

Continuous-wave Doppler echocardiography is a standard clinical procedure for evaluating the extent of aortic stenosis. It involves measuring the peak velocity of blood flow as it passes through the aortic valve during systole. By applying the simplified

Bernoulli equation, clinicians can estimate the transvalvar pressure gradient [3]. This non-invasive technique is preferred over cardiac catheterization due to its accessibility, affordability, and minimal invasiveness [4].

The application of continuous-wave Doppler echocardiography has been found to be limited to the estimation of peak velocity and trasvalvar pressure gradient when contrasted with the equation that factors in the complete hemodynamic profile at the point of greatest constriction [5,6]. Relying solely on peak velocity measurements neglects the momentum of blood flow spanning the entire vascular cross-section, a crucial element in accurately estimating the velocity profile and Wall Shear Stress (WSS). Moreover, the estimation of peak velocity using Doppler echocardiography heavily relies on the skills of the operator. Any misalignment between the angle of insonation and the direction of blood flow can result in inaccuracies of the maximum velocity [7]. Numerous non-invasive alternatives have been investigated, although they have not yet been incorporated into clinical practice [6,8].

Blood speckle imaging (BSI) has emerged as a recent alternative approach for evaluating the severity of aortic stenosis [9,10]. This technique involves the direct measurement and visualization of blood vector velocity patterns, captured at ultra-high frame rates in the kilohertz range [9,11,12]. BSI holds promise in mitigating the limitations of conventional Doppler echocardiography, such as angle dependence and the reliance on acquiring single peak velocities [5].

BSI relies on existing technology for tissue speckle tracking, applied to assess myocardial deformation [13]. The methodology entails defining a small image kernel within the initial vessel image, then tracking the same speckle pattern in subsequent frames using a "best match" search algorithm. This process is iterated across a grid of measurements to quantify both the velocity and direction of blood flow [9,14]. This approach to obtaining blood flow velocity data offers an advantage, potentially enabling the calculation of shear stress from velocity information across a cross-sectional profile, as opposed to relying on a single streamline as seen in traditional Doppler echocardiography.

This proof-of-concept study explores the retrospective application of BSI as a novel echocardiographic approach for pre- and post-operative evaluation of aortic valve repair in one 9-year-old patient with recurrent subaortic membrane. Tissue and hemodynamics data were acquired with the 9VT-D transesophageal probe by GE HealthCare (GE Vingmed Ultrasound AS, Horten, Norway), which is the first 4D transesohpageal echocardiography transducer small enough for use in young children. To the authors' knowledge, this is the first paper describing its use with BSI. This echocardiographic imaging mode acquires 2D images at an exceptionally fast temporal frequency and employs a block-matching algorithm [15–20] to track the speckle movement across the images to produce 2D vector fields. The flow field visualization allows for further characterization of the flow dynamics including turbulence and shear force calculation [21–24]. We sought to utilize BSI data in this clinical case to retrospectively analyze the flow patterns and derived properties to visually assess and quantify the hemodynamic changes from surgical subaortic membrane resection and aortic valvar repair.

2. Methods

2.1. Surgical Repair

The 9-year-old patient with recurrent subaortic membranes underwent surgical resection in accordance with standard clinical practice. Deliberate interrogation of the leaflet revealed the non-obvious extension of the subaortic membrane onto the leaflets. This membranous extension was delicately peeled off, uncovering a thin pliable leaflet under the inferior (ventricle-facing) surface. Standard intra-operative transesophageal imaging demonstrated satisfactory resection of subaortic membrane without significant aortic valvar regurgitation or stenosis.

2.2. Quantification of the Velocity Field from BSI

The Vivid E95 ultrasound system and 9VT-D transesophageal probe by GE HealthCare (GE Vingmed Ultrasound AS, Horten, Norway) were used for the acquisition of the BSI images. The BSI images consisted of a cine of standard 2D sonographic images, blood velocity data, confidence level, and metadata. The tissue data consisted of single channel data, where the pixel value represents the image brightness, see Figure 1a. The velocity data were multichannel and were combined into a velocity field. A second single channel was used for quantification of the confidence level in the blood velocity, with a range of 0 as low and 1 as high confidence [25]. A confidence level of 1 corresponded to a high fidelity of the block-matching protocol that tracks the displacement of blood speckles. Low confidence levels indicate less reliable velocity estimates and could be caused by low image quality or blood speckles moving out of plane [22–24].

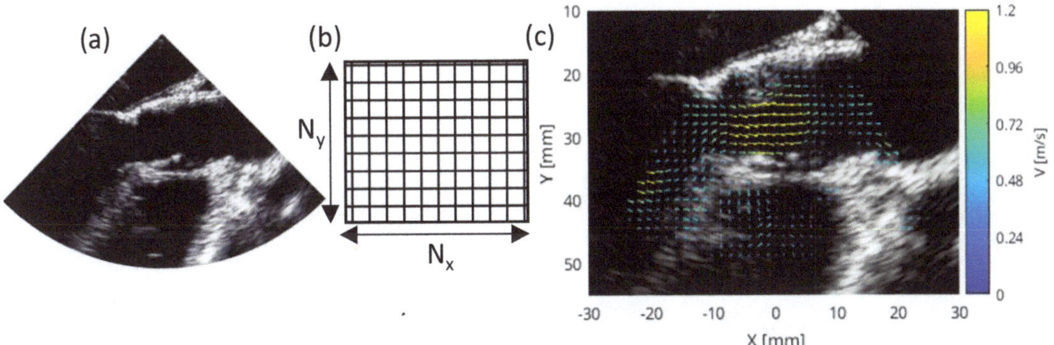

Figure 1. Blood speckle imaging (BSI) data (**a**) consisting of tissue (single channel) and velocity (multi-channel). The multichannel data were combined into a velocity field and then mapped onto the 2D grid (**b**) for visualization and quantification of the flow (**c**).

The physical distance between pixels in the BSI images defines the spatial extent of the image and allows determining the size of each pixel, which was 0.0825 mm/pixel. The number of pixels in the x- and y-directions were obtained from the tissue data, which were subsequently used for creating a 2D grid spanning the region of interest. The velocity field was then mapped on this 2D grid to facilitate visualization and flow quantification; see Figure 1b,c. The BSI images were acquired with a pulse repetition frequency (PRF) of 6 kHz and with a maximum velocity of 2 m/s of the tracking in both pre- and post-operative cases.

2.3. Quantification of Wall Shear Stress Indicators

Wall shear stress (WSS) is an important parameter to characterize the force per unit area that the blood flow exerts on the endothelial wall in the aorta. WSS varies in space and time depending on the blood flow conditions and geometry of the thoracic aorta and the aortic valve. To characterize WSS in the preoperative and postoperative cases, the study used two common parameters. $TAWSS$, or time-averaged wall shear stress, represents the localized, time-averaged value of wall shear stress (WSS) [26,27].

$$TAWSS = \frac{1}{T} \int_0^T |WSS_i| dt. \tag{1}$$

Spatial variations in WSS provide insights into the magnitude and the non-uniformity of WSS, but they do not convey information about the temporal variations in WSS. Therefore, the study also uses the oscillatory shear index (OSI) for quantification of frequency fluctuations in the wall shear stress (WSS) [27], which is defined as

$$OSI = \frac{1}{2}\left(1 - \frac{\left|\int_0^T WSS_i dt\right|}{\int_0^T |WSS_i| dt}\right).$$

(2)

OSI is a parameter that ranges from 0 to 0.5. When WSS remains consistently positive (whether oscillatory or non-oscillatory), OSI equals 0. When the WSS changes sign to the extent that the integral of the positive and negative sequences becomes equal, OSI reaches the value of 0.5.

3. Results

3.1. Echocardiographic Improvement after Subaortic Membrane Resection and Aortic Valvar Repair

During the pre-operative transesophageal echocardiogram (Figure 2), BSI depicted abnormal flow patterns and high peak velocity associated with the left ventricular outflow obstruction and restricted valvar leaflets. Figure 3 demonstrates significant improvements in flow patterns, with quantified changes in flow velocities indicating the absence of stenosis and minimal regurgitant volumes.

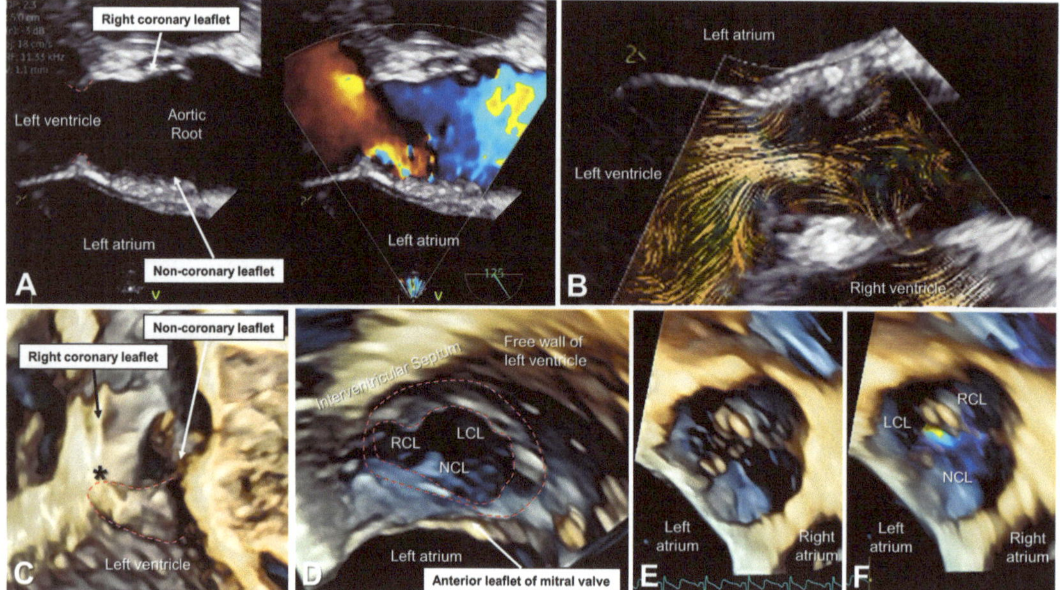

Figure 2. Pre-operative transesophageal echocardiographic evaluation. (**A**) Color Doppler comparison in the systole demonstrates flow acceleration starting at the level of the subaortic membrane (the membrane is outlined with red hashed lines in this and the other panels). (**B**) BSI in the systole demonstrates narrowing of the outflow jet at the level of the subaortic membrane with subsequent turbulence. (**C**) 3D long-axis imaging in the systole shows the prominent subaortic membrane which has an attachment to the nadir of the right coronary leaflet (black asterisk) with mild thickening of the leaflet extending from the leaflet hinge to the tip. (**D**) 3D imaging in the systole viewing in the short axis from the apex of the ventricle looking up towards the left ventricular outflow tract demonstrates the circumferential membrane which is most prominent under the coronary leaflets. (**E**) 3D short axis of the aortic valve in the diastole (**F**) with a 3D color Doppler demonstrates the trileaflet aortic valve with notable thickening of the right coronary leaflet (RCL) with mild to moderate central regurgitation. LCL, left coronary leaflet; NCL, non-coronary leaflet.

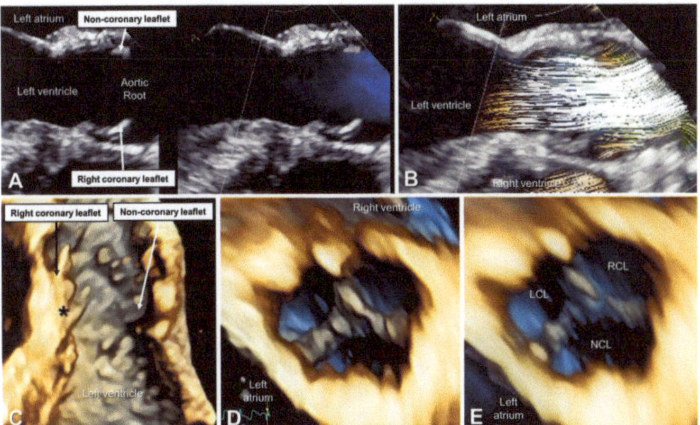

Figure 3. Post-operative transesophageal echocardiographic evaluation. (**A**) Color Doppler comparison in the systole demonstrates no significant residual subaortic membrane with laminar flow across the left ventricular outflow tract and the aortic valve. (**B**) BSI in the systole confirms laminar flow. (**C**) 3D long-axis imaging in the systole shows a very trivial residual subaortic membrane near the nadir of the right coronary leaflet (black asterisk) with a noticeably thinner right coronary leaflet. (**D**) 3D short axis of the aortic valve in the diastole (**E**) with a 3D color Doppler demonstrates notable thinning of the right coronary leaflet (RCL) tip with improve coaptation and no significant residual regurgitation. LCL, left coronary leaflet; NCL, non-coronary leaflet.

3.2. BSI Confidence Level

Figure 4 shows the confidence level for the pre- and post-operative cases during the cardiac cycle. The post-operative case shows high confidence levels in the region around the aortic root and proximally of the sinutubular junction between peak systole $t/T = 0.2$ and towards the end of systole $t/T = 0.4$, where t is time and T is the period of the cardiac cycle. The pre-operative case shows less uniformity with local spots of low confidence in the aortic root, especially around the stenotic left ventricular outflow tract and the aortic valve. Both cases indicate lower confidence in the left atrium (LA), which is at a depth further from the TEE transducer.

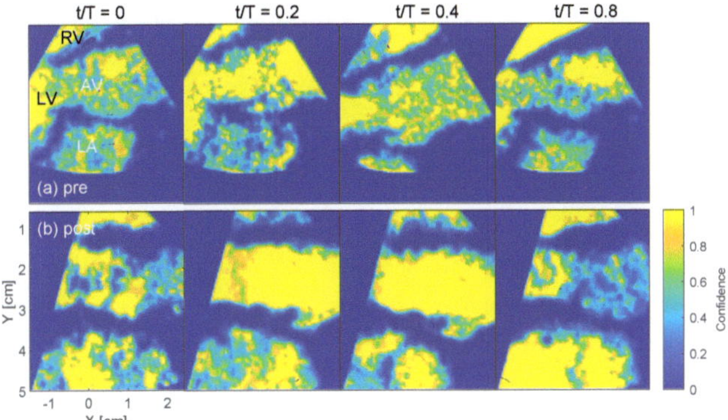

Figure 4. Parasternal long-axis view of the left ventricular (LV) outflow tract and the aortic valve (AV) showing the BSI confidence level, with range 0 as low and 1 as high confidence. The TEE probe has the same orientation as in Figure 1 and the confidence level is shown at the beginning of systole $t/T = 0$, peak systole $t/T = 0.2$, towards the end of systole $t/T = 0.4$, and during diastole $t/T = 0.8$. Pre-operative images are displayed across the top panels and post-operative images across the bottom panels.

3.3. BSI Flow Field

Figure 5 shows the blood flow through a parasternal long-axis view of the left ventricular outflow tract and aortic valve during different stages of the cardiac cycle for both the pre- and post-operative cases. In the early systole or isovolumetric contraction $t/T = 0$, the aortic valve is closed with minimal flow velocity. At early and late systole $t/T = 0.2$ and 0.4, respectively, the valve is open, and the velocity vector field shows flow acceleration across the aortic valve. This flow acceleration in the pre-operative case appears more prominent and is directed slightly anteriorly (upwards image). In late systole, a large recirculation zone forms in the posterior region of the aortic root (downstream from the stenotic aortic valve). In contrast, the post-operative vector field demonstrates a smoother flow profile throughout the systole.

At the end of the systolic phase, the valve closes, and the velocity magnitude of the blood flow reduces, indicating a decrease in aortic pressure during diastole. At the time of mid-diastole $t/T = 0.8$, following isovolumetric relaxation, the mitral valve opens, discharging blood flow from the left atrium to the left ventricle, i.e., filling of the left ventricle to prepare for a new systole.

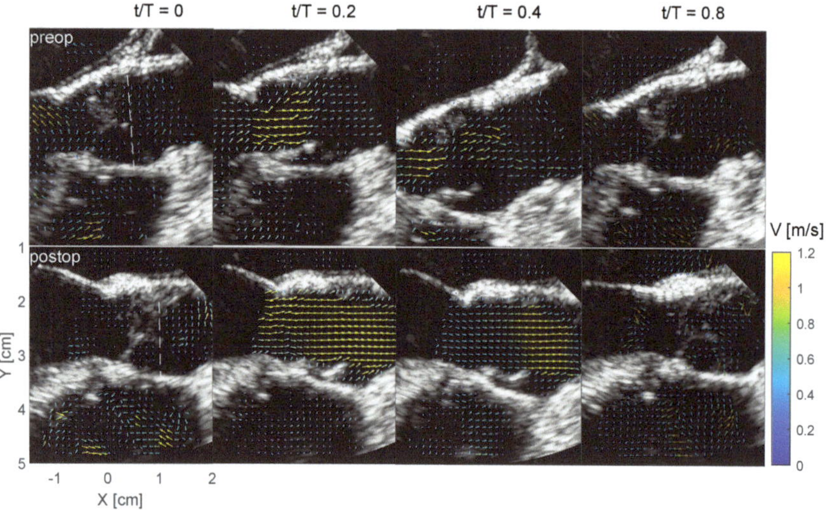

Figure 5. Parasternal long-axis view of the left ventricular (LV) outflow tract and the aortic valve (AV) showing the velocity field at different time instants during the cardiac cycle, where the vector color indicates the velocity magnitude. The TEE probe has the same orientation as in Figure 1. Pre-operative images are displayed across the top panels and post-operative images are displayed across the bottom panels.

3.4. BSI Peak Velocity Compared with CW Doppler Measurements

Figure 6 shows the continuous-wave (CW) Doppler assessment for the pre-operative and the post-operative transesophageal echocardiographic studies surrounding the surgical procedure. There is a peak velocity of approximately 3.5 m/s at peak systole for the pre-operative case, indicating moderate left ventricular outflow tract and aortic valvar stenosis. In the post-operative case, the maximum velocity at peak systole is around 1.4 m/s, indicating resolution of the left ventricular outflow tract and aortic valve obstruction.

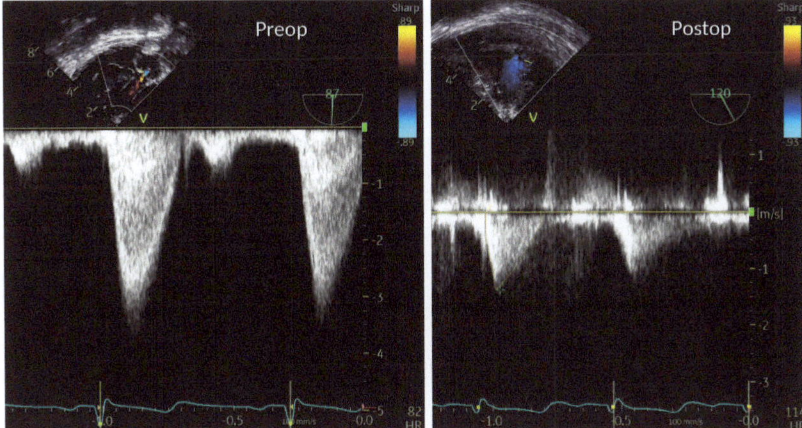

Figure 6. Continuous-wave Doppler assessment from the pre-operative (**left panel**) and post-operative (**right panel**) transesophageal echocardiographic studies.

The maximum velocity at the narrow section of the valve as obtained with BSI and CW Doppler for both pre-operative and post-operative cases are compared in Figure 7a. In the pre-operative case, there is a larger difference in maximum velocity at peak systole, where BSI indicates lower levels. This coincides with the lower BSI confidence level in the pre-operative case; see Figure 7b. In the post-operative case, there is a good agreement, both in terms of peak velocity and its variation during the cardiac cycle.

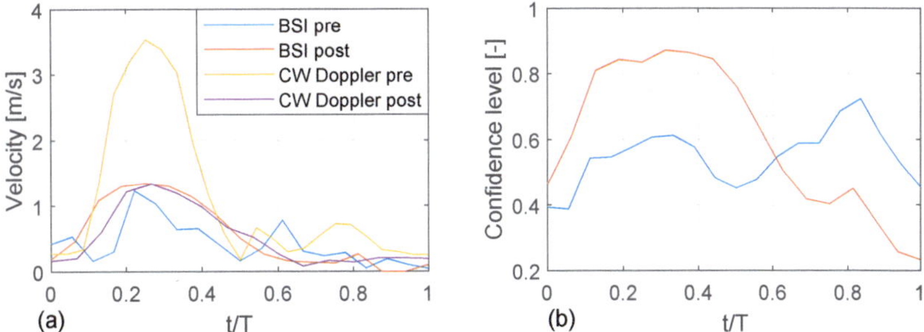

Figure 7. (a) Maximum velocity at peak systole as a function of the cardiac cycle for pre- and post-operative cases, for BSI and continuous-wave (CW) Doppler data, respectively. The maximum velocity is evaluated as the maximum valve on the white dashed lines in Figure 5. (b) Confidence level as a function of the cardiac cycle for the pre- and post-operative cases (evaluated as the mean value on the white dashed lines in Figure 5).

3.5. BSI Velocity and Shear Stress Quantification

The blood flow presented in Figure 5 is further quantified along the vertical profile (white dashed line in Figure 5). In the pre-operative case, the long-axis velocity component indicates low magnitudes at early systole $t/T = 0$, which progress towards peak systole $t/T = 0.2$ and forming an anteriorly directed jet with flow reversal on the posterior aspect of the aortic root. There is a larger velocity gradient at the shear layer of the jet, which results in increased shear stress (bottom left panel). Towards post-systole $t/T = 0.4$, the velocity magnitude decreases and the center of the jet is displaced approximately 5 mm in the short axis direction. In the post-operative case (right top panel in Figure 8), a relatively symmetric top-hat velocity profile develops at peak systole, with only minor recirculation towards the endothelial walls. In this case, there is a steeper gradient in the jet's shear layer on the side posteriorly of the probe, resulting in increased shear stress level (bottom right panel in Figure 8). There is also less displacement along the short axis of the jet's center in the post-operative case compared with the pre-operative case.

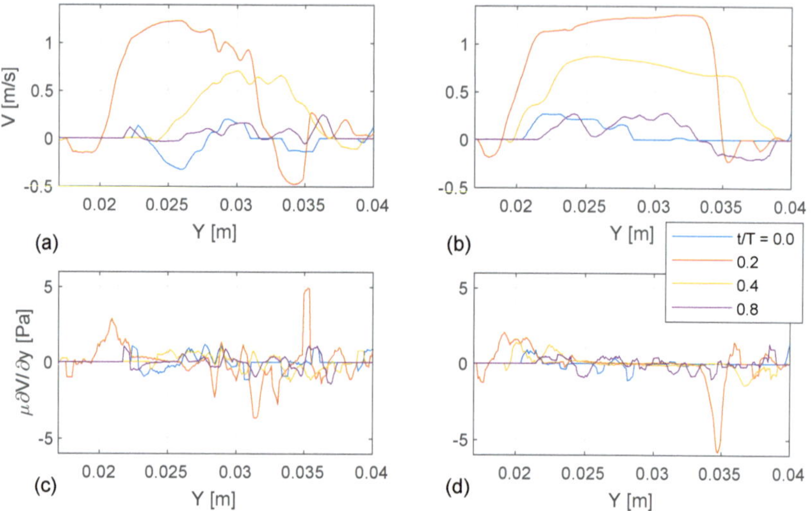

Figure 8. The long-axis velocity component (V) is given as a function of the short-axis (Y) for (**a**) pre-operative and (**b**) post-operative cases. The long-axis shear stress ($\mu\partial V/\partial y$) component is given on the bottom row for (**c**) pre-operative and (**d**) post-operative cases. Both the velocity and the shear stress are given at different instances of the cardiac cycle ($t/T = 0, 0.2, 0.4, 0.8$). The location of the profiles is along the white dashed line, annotated in Figure 5. The y-direction corresponds with the short axis, where a lower y is towards the anterior wall (higher up in the screen and closer to the probe, c.f. Figures 2 and 3) and a higher y is towards the posterior wall (more distal to the probe).

In both pre-operative and post-operative cases, the axial velocity undergoes a change in sign, transitioning from positive to negative flow, which coincides with the location of the shear layer. In the preoperative case, the retrograde flow towards the posterior wall is stronger than in the postoperative case. Nevertheless, integrating over the cardiac cycle shows that the peaks of the time average wall shear stress ($TAWSS$) are similar, i.e., between 1.5 and 1.6 for both cases; see Table 1. However, a retrograde flow with a change in flow direction results in a non-zero oscillatory shear index in the axial flow direction [26,27]. This is indicated in Table 1 where the OSI is significantly larger in the pre-operative case. Time durations with significant wall shear stress in combination with non-zero OSI have been suggested as a risk factor for cell-driven changes in the aortic wall structure [28], including development and progression of atherosclerosis [29,30].

Table 1. Peak values of WSS indicators between pre-operative and post-operative cases.

Case	Pre-Operative	Post-Operative
TAWSS	1.52	1.58
OSI	0.24	0.01

4. Discussion

In this study, we used blood speckle Imaging for pre- and post-operative evaluation of subaortic membrane resection and aortic valvar repair in a young pediatric patient utilizing the newly available mini-3D transesophageal probe. The post-operative case showed high BSI confidence levels in the region of the aortic root, whereas the pre-operative case showed relatively lower and less uniform BSI confidence levels in this region. The suboptimal BSI confidence level obtained pre-operatively resulted in an underprediction of peak velocities compared to the CW Doppler measurement. In cases with suboptimal BSI confidence levels, it would be necessary to resort to CW Doppler assessment of peak velocity.

Despite the limitation with BSI and its measurement of peak velocity in the pre-operative case, it is possible to quantify its velocity profile and compare it with the post-operative case, which is not possible with CW Doppler. This comparison revealed that the pre-operative case exhibits a more anteriorly directed jet with a large recirculation zone on the posterior side of the aortic root. In the post-operative case at peak systole, the axial velocity data in the ascending aorta demonstrated a more symmetric top-hat profile, which is in good agreement with previous observations [31–35]. The steeper velocity gradient led to a higher axial shear stress in the shear layer in the post-operative case. However, this is most likely due to the underestimation of the peak velocities with BSI in the stenotic pre-operative case.

Both pre-operative and post-operative cases indicate a non-zero oscillator shear index (OSI), since the axial velocity profile undergoes a change in sign, i.e., alternating from positive to negative flow during the cardiac cycle [26,27]. This is more significant in the pre-operative case due to its stronger recirculation zone that is located towards the posterior side of the aortic valve. In the study by Kiema et al. (2022) [28], it was observed that in cases of a situation of non-zero OSI together with significant WSS, there is elevated risk for the incidence of aortic wall damage, including progression of atherosclerosis [29,30].

The quantitative assessment of axial velocity and shear stress contributes to a deeper understanding of the relationship between hemodynamics in the proximal thoracic aorta and differences in aortic valvar morphology. The Computational Fluid Dynamics model in patient-specific geometries [36–38] will be performed to compare with the observed behavior. Despite BSI's current capabilities constraining velocity magnitude to 2 m/s and within 2D imaging planes, it may serve as a powerful tool to validate any simulation results. In areas of laminar blood flow without extreme velocities, such as the primarily regurgitant aortic valve, the 3D vector field generated from BSI may serve as an inlet boundary condition for computational models. Furthermore, even in the aortic valve with significant stenosis or subvalvar obstruction, post-operative assessment following repair may help to assess immediate hemodynamic results and predict long-term valvar repair durability. While the recurrence of a subaortic membrane may be inevitable in certain cases, this technique may provide a better understanding of the completeness of the surgical intervention when compared to simple Doppler and color Doppler assessment of the residual gradient or leak. In addition, this may add some early indications for the substrate for possible recurrence in the future. Further investigations involving larger cohort sizes will be conducted to determine the potential clinical utility of these findings, particularly in relation to assessing for favorable hemodynamics following left ventricular outflow tract and aortic valvar repair.

5. Limitations

In the pre-operative case, for a considerable duration of systole, the flow acceleration through the stenotic subaortic membrane and the valve exceeds the upper limit of BSI of 2 m/s. The BSI block-matching algorithm is unable to track the speckle pattern beyond the proportional maximum search distance, resulting in unpredictable underestimation of the flow velocity. In the post-operative case, with velocities never exceeding the 2 m/s limit, the agreement as compared to CW Doppler is good, both in terms of magnitude and its variation during the cardiac cycle. Thus, BSI demonstrates robustness when applied to the aortic flow with moderate flow velocities.

Another limitation of the BSI employed in this study is that it is limited to 2D planar assessment. The velocity component orthogonal to the imaging plane is not available. Capturing the maximum velocity can be optimized by aligning it with the long axis of the LVOT and aorta. However, we lose the vorticity around this long-axis, as well as any jet or turbulent flow which may deviate outside the 2D imaging plane. The 3D physiological displacement of the heart throughout the cardiac cycle, which evolves both spatially and temporally, also is not fully captured. This influences the flow quantification of relative velocities to the walls in the left ventricle, as it experiences significant volume changes during systolic contraction. However, for vessels that undergo moderate displacement and area changes, this factor is less significant. Nonetheless, 4D BSI would help overcome this limitation.

There is a need to validate both BSI and CW Doppler techniques in the setting of multilevel obstruction. Previous studies have measured the pressure drop in silicon phantom models of the aortic valve using PendoTech pressure sensors to assess the accuracy of CW Doppler and BSI [6]. However, so far, it has been challenging to perform these pressure measurements in vivo.

6. Conclusions

For the postoperative non-stenotic condition, blood speckle imaging (BSI) demonstrated good agreement with the continuous wave Doppler echocardiography estimation of peak velocity. This demonstrates the potential for assessing shear stresses within the aortic valve region. However, BSI shows a tendency to underestimate peak velocities for the preoperative condition, with high stenotic burden, likely due to limitations in tracking higher flow velocities and managing speckle decorrelation. Despite similar TAWSS in both cases, it was observed that the OSI level reduced significantly postoperatively, indicating a lower risk factor for the ongoing development and progression of adjacent aortic wall and leaflet damage, including atherosclerosis.

While BSI presents several advantages over traditional Doppler methods, its use for shear stress assessments in aortic stenosis needs further capability. Extended clinical investigations are imperative before BSI can be considered a viable means to enhance the accuracy of wall shear stress estimations in the clinical assessment of aortic stenosis.

Author Contributions: Conceptualization, E.S., M.J., H.K.N. and J.T.T.; formal analysis, E.S., M.J. and J.T.T.; writing—review and editing, E.S., M.J., H.K.N. and J.T.T.; funding acquisition, E.S. All authors have read and agreed to the published version of the manuscript.

Funding: This research was funded by the Swedish Research Council (grant number 2021-04894).

Institutional Review Board Statement: Ethical review and approval were waived by the Cleveland Clinic Institutional Review Board because the study was retrospective using de-identified data from fewer than 4 subjects.

Informed Consent Statement: Informed consent was obtained from all subjects involved in the study. Written informed consent has been obtained from the patient(s) to publish this paper.

Data Availability Statement: Data available on request due to restrictions in repository access.

Conflicts of Interest: The authors declare no conflict of interest.

References

1. Tretter, J.T.; Spicer, D.E.; Franklin, R.C.G.; Béland, M.J.; Aiello, V.D.; Cook, A.C.; Crucean, A.; Loomba, R.S.; Yoo, S.J.; Quintessenza, J.A.; et al. Expert Consensus Statement: Anatomy, Imaging, and Nomenclature of Congenital Aortic Root Malformations. *Ann. Thorac. Surg.* **2023**, *116*, 6–16. [CrossRef]
2. Tretter, J.T.; Spicer, D.E.; Mori, S.; Chikkabyrappa, S.; Redington, A.N.; Anderson, R.H. The significance of the interleaflet triangles in determining the morphology of congenitally abnormal aortic valves: Implications for noninvasive imaging and surgical management. *J. Am. Soc. Echocardiogr.* **2016**, *29*, 1131–1143. [CrossRef] [PubMed]
3. Currie, P.J.; Seward, J.; Reeder, G.; Vlietstra, R.; Bresnahan, D.; Bresnahan, J.; Smith, H.; Hagler, D.; Tajik, A. Continuous-wave Doppler echocardiographic assessment of severity of calcific aortic stenosis: A simultaneous Doppler-catheter correlative study in 100 adult patients. *Circulation* **1985**, *71*, 1162–1169. [CrossRef] [PubMed]
4. Nishimura, R.A.; Carabello, B.A. Hemodynamics in the cardiac catheterization laboratory of the 21st century. *Circulation* **2012**, *125*, 2138–2150. [CrossRef] [PubMed]

5. Donati, F.; Myerson, S.; Bissell, M.M.; Smith, N.P.; Neubauer, S.; Monaghan, M.J.; Nordsletten, D.A.; Lamata, P. Beyond Bernoulli: Improving the accuracy and precision of noninvasive estimation of peak pressure drops. *Circ. Cardiovasc. Imaging* **2017**, *10*, e005207. [CrossRef] [PubMed]

6. Dockerill, C.; Gill, H.; Fernandes, J.F.; Nio, A.Q.; Rajani, R.; Lamata, P. Blood speckle imaging compared with conventional Doppler ultrasound for transvalvular pressure drop estimation in an aortic flow phantom. *Cardiovasc. Ultrasound* **2022**, *20*, 18. [CrossRef] [PubMed]

7. Ring, L.; Shah, B.N.; Bhattacharyya, S.; Harkness, A.; Belham, M.; Oxborough, D.; Pearce, K.; Rana, B.S.; Augustine, D.X.; Robinson, S.; et al. Echocardiographic assessment of aortic stenosis: A practical guideline from the British Society of Echocardiography. *Echo Res. Pract.* **2021**, *8*, G19–G59. [CrossRef] [PubMed]

8. Gill, H.; Fernandes, J.; Chehab, O.; Prendergast, B.; Redwood, S.; Chiribiri, A.; Nordsletten, D.; Rajani, R.; Lamata, P. Evaluation of aortic stenosis: From Bernoulli and Doppler to Navier-Stokes. *Trends Cardiovasc. Med.* **2023**, *33*, 32–43. [CrossRef]

9. Lasse Løvstakken, L.; Lie, G.R. BSI (Blood Speckle Imaging). GE Healthcare. 2017. Available online: https://www.gehealthcare.com/-/media/28568e93cbe741d48a10114e399ebd91.pdf (accessed on 12 December 2022).

10. Trahey, G.E.; Allison, J.W.; Von Ramm, O.T. Angle independent ultrasonic detection of blood flow. *IEEE Trans. Biomed. Eng.* **1987**, *BME-34*, 965–967. [CrossRef] [PubMed]

11. Hansen, K.L.; Nielsen, M.B.; Jensen, J.A. Vector velocity estimation of blood flow—A new application in medical ultrasound. *Ultrasound* **2017**, *25*, 189–199. [CrossRef] [PubMed]

12. Nyrnes, S.A.; Fadnes, S.; Wigen, M.S.; Mertens, L.; Lovstakken, L. Blood speckle-tracking based on high–frame rate ultrasound imaging in pediatric cardiology. *J. Am. Soc. Echocardiogr.* **2020**, *33*, 493–503. [CrossRef] [PubMed]

13. Geyer, H.; Caracciolo, G.; Abe, H.; Wilansky, S.; Carerj, S.; Gentile, F.; Nesser, H.J.; Khandheria, B.; Narula, J.; Sengupta, P.P. Assessment of myocardial mechanics using speckle tracking echocardiography: Fundamentals and clinical applications. *J. Am. Soc. Echocardiogr.* **2010**, *23*, 351–369. [CrossRef] [PubMed]

14. Fadnes, S.; Nyrnes, S.A.; Torp, H.; Lovstakken, L. Shunt flow evaluation in congenital heart disease based on two-dimensional speckle tracking. *Ultrasound Med. Biol.* **2014**, *40*, 2379–2391. [CrossRef] [PubMed]

15. Je, C.; Park, H.M. Optimized hierarchical block matching for fast and accurate image registration. *Signal Process. Image Commun.* **2013**, *28*, 779–791. [CrossRef]

16. Li, R.; Zeng, B.; Liou, M.L. A new three-step search algorithm for block motion estimation. *IEEE Trans. Circuits Syst. Video Technol.* **1994**, *4*, 438–442.

17. Lu, J.; Liou, M.L. A simple and efficient search algorithm for block-matching motion estimation. *IEEE Trans. Circuits Syst. Video Technol.* **1997**, *7*, 429–433.

18. Po, L.M.; Ma, W.C. A novel four-step search algorithm for fast block motion estimation. *IEEE Trans. Circuits Syst. Video Technol.* **1996**, *6*, 313–317.

19. Zhu, S.; Ma, K.K. A new diamond search algorithm for fast block-matching motion estimation. *IEEE Trans. Image Process.* **2000**, *9*, 287–290. [CrossRef] [PubMed]

20. Nie, Y.; Ma, K.K. Adaptive rood pattern search for fast block-matching motion estimation. *IEEE Trans. Image Process.* **2002**, *11*, 1442–1449. [CrossRef]

21. Stefani, L.; De Luca, A.; Maffulli, N.; Mercuri, R.; Innocenti, G.; Suliman, I.; Toncelli, L.; Vono, M.C.; Cappelli, B.; Pedri, S.; et al. Speckle tracking for left ventricle performance in young athletes with bicuspid aortic valve and mild aortic regurgitation. *Eur. J. Echocardiogr.* **2009**, *10*, 527–531. [CrossRef]

22. Swillens, A.; Segers, P.; Lovstakken, L. Two-dimensional flow imaging in the carotid bifurcation using a combined speckle tracking and phase-shift estimator: A study based on ultrasound simulations and in vivo analysis. *Ultrasound Med. Biol.* **2010**, *36*, 1722–1735. [CrossRef]

23. Swillens, A.; Segers, P.; Torp, H.; Lovstakken, L. Two-dimensional blood velocity estimation with ultrasound: Speckle tracking versus crossed-beam vector Doppler based on flow simulations in a carotid bifurcation model. *IEEE Trans. Ultrason. Ferroelectr. Freq. Control* **2010**, *57*, 327–339. [CrossRef] [PubMed]

24. Ekroll, I.K.; Swillens, A.; Segers, P.; Dahl, T.; Torp, H.; Lovstakken, L. Simultaneous quantification of flow and tissue velocities based on multi-angle plane wave imaging. *IEEE Trans. Ultrason. Ferroelectr. Freq. Control* **2013**, *60*, 727–738. [CrossRef] [PubMed]

25. Faragallah, O.S.; El-Hoseny, H.; El-Shafai, W.; Abd El-Rahman, W.; El-Sayed, H.S.; El-Rabaie, E.S.M.; Abd El-Samie, F.E.; Geweid, G.G. A comprehensive survey analysis for present solutions of medical image fusion and future directions. *IEEE Access* **2020**, *9*, 11358–11371. [CrossRef]

26. Suo, J.; Oshinski, J.N.; Giddens, D.P. Blood flow patterns in the proximal human coronary arteries: Relationship to atherosclerotic plaque occurrence. *Mol. Cell. Biomech.* **2008**, *5*, 9.

27. Chen, X.; Gao, Y.; Lu, B.; Jia, X.; Zhong, L.; Kassab, G.S.; Tan, W.; Huo, Y. Hemodynamics in coronary arterial tree of serial stenoses. *PLoS ONE* **2016**, *11*, e0163715. [CrossRef] [PubMed]

28. Kiema, M.; Sarin, J.K.; Kauhanen, S.P.; Torniainen, J.; Matikka, H.; Luoto, E.S.; Jaakkola, P.; Saari, P.; Liimatainen, T.; Vanninen, R.; et al. Wall shear stress predicts media degeneration and biomechanical changes in thoracic aorta. *Front. Physiol.* **2022**, *13*, 934941. [CrossRef] [PubMed]

29. Zarins, C.K.; Giddens, D.P.; Bharadvaj, B.; Sottiurai, V.S.; Mabon, R.F.; Glagov, S. Carotid bifurcation atherosclerosis. Quantitative correlation of plaque localization with flow velocity profiles and wall shear stress. *Circ. Res.* **1983**, *53*, 502–514. [CrossRef] [PubMed]

30. LaDisa, J.F., Jr.; Dholakia, R.J.; Figueroa, C.A.; Vignon-Clementel, I.E.; Chan, F.P.; Samyn, M.M.; Cava, J.R.; Taylor, C.A.; Feinstein, J.A. Computational simulations demonstrate altered wall shear stress in aortic coarctation patients treated by resection with end-to-end anastomosis. *Congenit. Heart Dis.* **2011**, *6*, 432–443. [CrossRef] [PubMed]

31. Jonnagiri, R.; Sundström, E.; Gutmark, E.; Anderson, S.; Pednekar, A.; Taylor, M.; Tretter, J.; Gutmark-Little, I. Influence of aortic valve morphology on vortical structures and wall shear stress. *Med. Biol. Eng. Comput.* **2023**, *61*, 1489–1506. [CrossRef] [PubMed]

32. Hope, T.A.; Markl, M.; Wigström, L.; Alley, M.T.; Miller, D.C.; Herfkens, R.J. Comparison of flow patterns in ascending aortic aneurysms and volunteers using four-dimensional magnetic resonance velocity mapping. *J. Magn. Reson. Imaging: Off. J. Int. Soc. Magn. Reson. Med.* **2007**, *26*, 1471–1479. [CrossRef]

33. Meierhofer, C.; Schneider, E.P.; Lyko, C.; Hutter, A.; Martinoff, S.; Markl, M.; Hager, A.; Hess, J.; Stern, H.; Fratz, S. Wall shear stress and flow patterns in the ascending aorta in patients with bicuspid aortic valves differ significantly from tricuspid aortic valves: A prospective study. *Eur. Heart-J. Cardiovasc. Imaging* **2013**, *14*, 797–804. [CrossRef]

34. Chambers, J.B.; Myerson, S.G.; Rajani, R.; Morgan-Hughes, G.J.; Dweck, M.R. Multimodality imaging in heart valve disease. *Open Heart* **2016**, *3*, e000330. [CrossRef] [PubMed]

35. Oechtering, T.H.; Sieren, M.M.; Hunold, P.; Hennemuth, A.; Huellebrand, M.; Scharfschwerdt, M.; Richardt, D.; Sievers, H.H.; Barkhausen, J.; Frydrychowicz, A. Time-resolved 3-dimensional magnetic resonance phase contrast imaging (4D Flow MRI) reveals altered blood flow patterns in the ascending aorta of patients with valve-sparing aortic root replacement. *J. Thorac. Cardiovasc. Surg.* **2020**, *159*, 798–810. [CrossRef] [PubMed]

36. Sundström, E.; Jonnagiri, R.; Gutmark-Little, I.; Gutmark, E.; Critser, P.; Taylor, M.; Tretter, J. Hemodynamics and tissue biomechanics of the thoracic aorta with a trileaflet aortic valve at different phases of valve opening. *Int. J. Numer. Methods Biomed. Eng.* **2020**, *36*. [CrossRef] [PubMed]

37. Sundström, E.; Tretter, J.T. Impact of Variation in Interleaflet Triangle Height Between Fused Leaflets in the Functionally Bicuspid Aortic Valve on Hemodynamics and Tissue Biomechanics. *J. Eng. Sci. Med. Diagn. Ther.* **2022**, *5*, 031102. [CrossRef]

38. Sundström, E.; Jonnagiri, R.; Gutmark-Little, I.; Gutmark, E.; Critser, P.; Taylor, M.; Tretter, J. Effects of Normal Variation in the Rotational Position of the Aortic Root on Hemodynamics and Tissue Biomechanics of the Thoracic Aorta. *Cardiovasc. Eng. Technol.* **2020**, *11*, 47–58. [CrossRef] [PubMed]

 *bioengineering*

Article

Vortical Structures Promote Atheroprotective Wall Shear Stress Distributions in a Carotid Artery Bifurcation Model

Nora C. Wild [1], Kartik V. Bulusu [1] and Michael W. Plesniak [1,2,*]

1 Department of Mechanical and Aerospace Engineering, The George Washington University, 800 22nd Street NW, Science & Engineering Hall, Suite 3000, Washington, DC 20052, USA; caroline_zalud@gwu.edu (N.C.W.); bulusu@gwu.edu (K.V.B.)

2 Department of Biomedical Engineering, The George Washington University, 800 22nd Street NW, Science & Engineering Hall, Suite 3000, Washington, DC 20052, USA

* Correspondence: plesniak@gwu.edu; Tel.: +1-(202)-994-9803

Abstract: Carotid artery diseases, such as atherosclerosis, are a major cause of death in the United States. Wall shear stresses are known to prompt plaque formation, but there is limited understanding of the complex flow structures underlying these stresses and how they differ in a pre-disposed high-risk patient cohort. A 'healthy' and a novel 'pre-disposed' carotid artery bifurcation model was determined based on patient-averaged clinical data, where the 'pre-disposed' model represents a pathological anatomy. Computational fluid dynamic simulations were performed using a physiological flow based on healthy human subjects. A main hairpin vortical structure in the internal carotid artery sinus was observed, which locally increased instantaneous wall shear stress. In the pre-disposed geometry, this vortical structure starts at an earlier instance in the cardiac flow cycle and persists over a much shorter period, where the second half of the cardiac cycle is dominated by perturbed secondary flow structures and vortices. This coincides with weaker favorable axial pressure gradient peaks over the sinus for the 'pre-disposed' geometry. The findings reveal a strong correlation between vortical structures and wall shear stress and imply that an intact internal carotid artery sinus hairpin vortical structure has a physiologically beneficial role by increasing local wall shear stresses. The deterioration of this beneficial vortical structure is expected to play a significant role in atherosclerotic plaque formation.

Keywords: physiological pulsatile flow; vortical structures; wall shear stresses; healthy and pre-disposed geometry; cardiovascular disease

 check for **updates**

Citation: Wild, N.C.; Bulusu, K.V.; Plesniak, M.W. Vortical Structures Promote Atheroprotective Wall Shear Stress Distributions in a Carotid Artery Bifurcation Model. *Bioengineering* **2023**, *10*, 1036. https://doi.org/10.3390/bioengineering10091036

Academic Editors: Ephraim Gutmark and Iris Little

Received: 21 June 2023
Revised: 4 August 2023
Accepted: 24 August 2023
Published: 3 September 2023

1. Introduction

In the United States, about 50% of deaths are caused by cardiovascular diseases such as strokes and heart attacks [1]. Strokes alone affect over 15 million people every year [2], and up to 30% of strokes are estimated to be caused by carotid artery disease [2,3]. Carotid artery disease is often located at the carotid artery bifurcation (CAB), which is situated in the neck. The common carotid artery (CCA) branches off from the aorta and in turn branches into the internal (ICA) and external (ECA) carotid arteries, where the CAB is the bifurcation from CCA into ICA and ECA. The ICA and ECA transport blood to the brain and face, respectively [1].

Carotid artery disease often involves the formation of atherosclerotic plaques. Plaque growth and/or rupture can block blood supply to the downstream brain vasculature. Atherosclerosis describes the local deposit of cholesterol and lipids on the arterial wall and mainly occurs at locations of complex vessel geometry, such as bifurcations. It mostly affects large- and medium-sized blood vessels [1–6]. The formation of a stenosis, which requires the implantation of a stent, is a common consequence of atherosclerotic plaque formation, which poses an additional risk for thrombus formation [7].

The innermost layer in the vessel lumen, the endothelium, is exposed to wall shear stresses, which are present in the entire cardiovascular system [2,3]. These shear stresses on the endothelial cells are caused by blood flow over the vessel lumen; they are required to sustain healthy endothelial cells [8]. Atherosclerosis is caused when pathological flow and consequently abnormal shear stresses are present, when they vary in magnitude or have strong temporal variations [1,9]. These pathological mechanical forces are sensed and transmitted by the endothelial cells, resulting in a diseased cell through a mechanotransduction process [10,11]. In atherosclerotic regions, the flow is disturbed, the shear stress is lower than normal [1,12], and the spatial shear stress gradients are large [13–16]. Due to their crucial role in the origin of cardiovascular diseases, shear stresses triggering plaque formation in the carotid artery bifurcation are of great clinical interest and are investigated using computational fluid dynamic (CFD) simulations.

The CFD studies of patient-specific artery geometries [17], as well as simplified artery models, are both commonly reported in the literature, incorporating 2D [18] and 3D [19–22] blood vessel geometries. The use of simplified and patient-averaged models allows for a broader understanding of general flow phenomena in the carotid artery bifurcation and their influence on atherosclerosis, and thus to decouple it from specific patient anatomies [23]. In addition to geometric simplification, a steady flow assumption is used broadly [24–26]. More realistically, physiological pulsatile flow can be applied by using a pulsatile inflow waveform at the CCA, which is mostly performed in combination with patient-specific geometries [27]. Typically, 'healthy' vessel geometries are modeled [28] that are based on general patient averages without any emphasis on the patient cohorts' actual risk for future plaque development. Atherosclerotic vessel geometries are typically modeled under diseased conditions [17,29], incorporating a local narrowing of the vessels' diameter in the form of a plaque built-up (stenosis). This paper takes a novel approach by emphasizing a patient cohort's predisposition toward plaque formation by modeling a nondiseased pre-disposed CAB. We are unaware of any published study investigating flow fields and vortical structures and their impact on the wall shear stress distribution in carotid artery bifurcation geometries with a different predisposition toward the formation of atherosclerotic plaques.

Flow separation in the ICA sinus is observed and correlated to increased atherosclerosis risk [1,19–22,26,30,31]. Thus, it is hypothesized that the sinus geometry is an important indicator of risk. This includes the bifurcation angle, as a larger bifurcation angle causes larger areas of nonaxial shear stress at the branches' outer walls [23], and an asymmetric bifurcation is linked to increased atherosclerosis risk [32]. Instantaneous, as well as time-averaged, wall shear stresses are frequently used metrics in determining proatherogenic regions within the geometry [17,19–24,27,29,30].

Shear stress as a mechanical stimulation of atherosclerosis is widely studied. There are, however, only a few studies of the complex flow within the carotid artery bifurcation affecting these shear stresses. Thus, our study aims to increase the understanding of the complex flow, including secondary flow and vortical structures, within the CAB and its impact on the wall shear stress distribution. Such an understanding is necessary for the early detection of atherosclerosis by allowing for the identification of patients with pre-disposed increased risk before the onset of the disease and symptoms.

A vortex is defined as a specific region of fluid circulating around an axis. The fluid velocity is typically greatest at this axis and decreases with increasing distance to this axis. The vortex core is furthermore a local pressure minimum. A common example of a vortical structure is a vortex ring (Figure 1a). Secondary flows are flows with flow direction in the plane perpendicular to the main flow direction. Those in-plane secondary velocities can develop in pipe flow (similar to a circular cross-sectional vessel) when the pipe or vessel has a curvature. These secondary flows result in Dean vortices, which are a pair of counter-rotating vortices (Figure 1b).

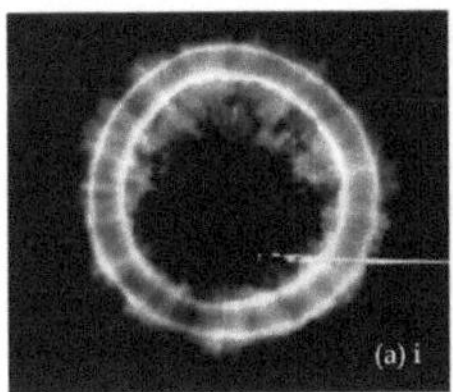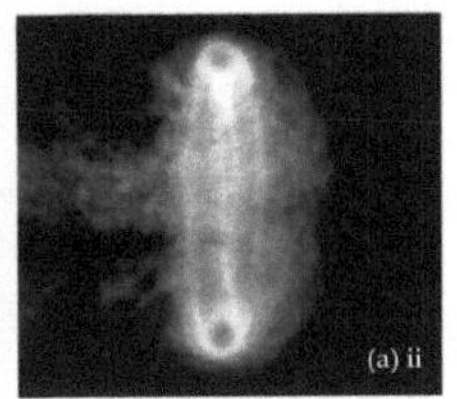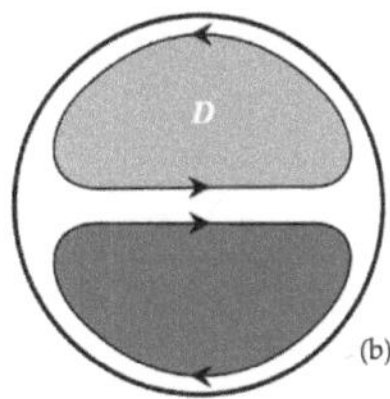

Figure 1. Examples of vortical structures: (**a**) vortex ring front (**i**) and side (**ii**) view [33]; (**b**) schematic of a Dean (D) vortex pair viewed from upstream in a curved pipe; clockwise and counter-clockwise rotation of the secondary flow is indicated by the arrows [34].

Secondary flows emerge in branching vessels because the axial main flow has to follow two curved paths. The formation of secondary flow structures leads to the formation of vortices, which alter the wall shear stress distribution [35–37], its magnitude, and multidirectionality [38–40]. Thus, the characteristics of these vortices and secondary flow patterns play a crucial role in the onset of cardiovascular diseases [41]. Very few studies investigate vortices in the flow [17] despite their significant influence on the flow and shear stresses. Rindt et al. found that, at the inlet of the ICA, reversed flows have the same magnitude as the mainstream velocity [18]. Gijsen et al. observed the formation of Dean vortices after the bifurcation under steady flow [26]. These vortices originate from the curvature due to the transition from CCA into ICA, causing the fluid close to the branching dividing wall to be transported toward the nondivider wall. In addition to the increase in cross-sectional area in the ICA sinus, this leads to flow reversal close to the nondivider ICA wall [26]. Kumar et al. reported the helical structure of velocity streamlines downstream of the bifurcation [31]. Morbiducci et al. investigated vorticity in 2D planes in healthy patient-specific geometries under pulsatile flow [21], like Perktold et al. describing the formation of secondary flow structures in 2D planes for a healthy carotid artery bifurcation model under pulsatile flow [30]. The main limitation of these previous studies is the 2D analysis of secondary flows without detailed description or 3D analysis of vortical structures. Nagargoje and Gupta [25] performed a pulsatile computational study in a very simplified carotid artery bifurcation model and found that secondary flows are dependent on sinus size and that a larger sinus increases the low WSS recirculation area, without further specifying secondary flow structures or their 3D structure. One of the few studies to examine 3D vortical structures in a carotid artery bifurcation was published by Chen et al. in 2020 [42]. They investigated patient-specific geometries, with unspecified atherosclerosis risk, finding differences in the time duration of dominant reverse flow and the duration of dominant secondary flows based on varying carotid artery bifurcation geometries. They found that secondary flows are dominant in the deceleration phase at end-systole and that a larger ICA bifurcation angle results in stronger (having higher maximum magnitude) secondary flow during the inflow acceleration throughout the pulsatile cardiac cycle and found the formation of Dean vortices at the systolic peak for geometries having an ICA bifurcation angle around 30°. Complex bifurcation geometry according to Chen et al. breaks the symmetry of the Dean vortices. Furthermore, they observed the formation of straight vortex tubes at the systolic peak. They found that vessels having a higher flare (A_{max}/A_{CCA}) have larger regions of reversed flow and are more dominated by secondary flows.

Despite the wide range of studies of flow and wall shear stress distribution in carotid artery bifurcation geometries, there is a need for a greater understanding of the evolution and characterization of three-dimensional vortical structures and their role in vascular flow and wall shear stress distribution. The few studies of three-dimensional vortical structures in cardiovascular flows, specifically CAB flow, use them to visualize the three-dimensional distribution of vorticity and helicity. The study by Chen et al. uses classical three-dimensional vortex criteria to highlight secondary flows and flow separation zones. Our study aims to address secondary flow by characterizing vorticial structures using established vortex identification methods in a carotid artery bifurcation model under physiological pulsatile flow. Our objective is to increase the understanding of physiological flow patterns, specifically the role of vortices, in the human carotid artery. The modeling of a representative 'pre-disposed' CAB geometry and its hemodynamics using advanced CFD simulations aims to identify pathological flow and vortex patterns. Thus, this study aims not only to increase the understanding of the role of vortices in disease development but also intends to inform the early detection of disease with the focus on identifying patients at high risk that are pre-disposed and likely to form an atherosclerotic plaque—before the plaque develops. Modern biomedical imaging techniques, such as ultrasonography [43] or 4D flow MRI [44,45], allow for the precise visualization of CAB flow in patients, and more emerging techniques such as vector flow imaging allow for the intuitive and quantitative imaging of vortical structures [46]. Helical flow patterns in arteries and vortical flow structures, such as vortex rings in the heart [45], are identifiable in the state of the art of medical imaging and are therefore accessible to clinicians [47]. The clinical analysis of vortices, such as the left-ventricle vortex ring, is often performed by acquiring the flow field using 4D flow MRI [47], extracting vortical structures using the λ_2 criterion [44], and finding its vortex core position [45]. The altered development of a main vortex is associated with pathological cardiovascular flow and function in other parts of the cardiovascular system, such as the diastolic vortex formation in the left ventricle of the human heart [45]. Thus, there is a high potential to transfer this vortex-based clinical evaluation to carotid bifurcation atherosclerosis risk assessment. To achieve this goal, this study focuses on the identification of coherent vortical structure in the CAB and vortex characteristics to distinguish between a physiological or a pathological vortical structure.

There is a lack of knowledge on the impact of three-dimensional vortices in the CAB on possibly proatherogenic shear stress distributions. In order to fill this knowledge gap, we investigate three-dimensional vortical structures in the ICA sinus and their influence on wall shear stress for a representative 'healthy' and a proatherogenic 'pre-disposed' vessel geometry. This will facilitate a fundamental understanding of vortex evolution over the cardiac cycle and allow for the characterization of the differences between physiological and pathological vortical structures. The accessibility of vortex visualization using medical imaging presents a promising tool to identify high-risk CAB atherosclerosis patients through pathological vortex formation in the future. It can be hypothesized that specific geometrical features affect the vortical structures and thus result in a proatherogenic shear stress distribution. Hemodynamics in a 'healthy' and 'pre-disposed' CAB geometry are calculated using a three-dimensional CFD simulation with physiological pulsatile flow.

This study will highlight the potential role of three-dimensional vortical structures in atherosclerosis by linking internal vortical structures to the wall shear stress distribution known to critically contribute to disease onset. Furthermore, the results will correlate a statistically 'pre-disposed' carotid bifurcation geometry to differing vortex characteristics, aiming to inform the future clinical early detection of CAB atherosclerosis.

2. Materials and Methods

The implemented methods will be discussed in the following sections. First, the modeling of statistically 'healthy' and 'pre-disposed' carotid artery bifurcation geometries is explicated. Second, the three-dimensional computational fluid dynamic simulation and the associated physiological boundary conditions are described.

2.1. Carotid Artery Bifurcation Geometry

The average person is not affected by clinically relevant carotid stenosis (over 50% severity) as stenoses appear in only about 2–8% of the population [48]). Among others, a factor increasing the atherosclerosis risk is the vessel's geometry, i.e., its anatomical features. Clinically, two main geometrical risk factors have been identified: (i) larger bifurcation angle, especially larger ICA angles, and (ii) a small ICA/CCA diameter ratio [49–51]). To study the influence of these geometrical risk factors, computed flow fields of two carotid artery bifurcation models were compared. The first is based on vessel measurements from healthy volunteers without any specific increased atherosclerosis risk. Our 'healthy' geometry model's dimensions were based on these healthy patient averages [52–54]. The CCA had an inlet diameter of 6.5mm and branched into the ICA and ECA with a symmetric 60° angle (Figure 2a). Secondly, a 'pre-disposed' geometry was modeled, incorporating both known geometrical risk factors. An asymmetric branching angle was realized by increasing the ICA angle to 45° [49], and the ICA/CCA diameter ratio was reduced to 0.5 by reducing the ICA diameter [34,50] (Figure 2b). The ICA sinus was included in both geometries as it is a strong proatherogenic region. Entry and exit lengths were added to the ICA and ECA to improve numerical convergence, prevent nonphysical reverse flow at the outlet, and achieve a developing inflow.

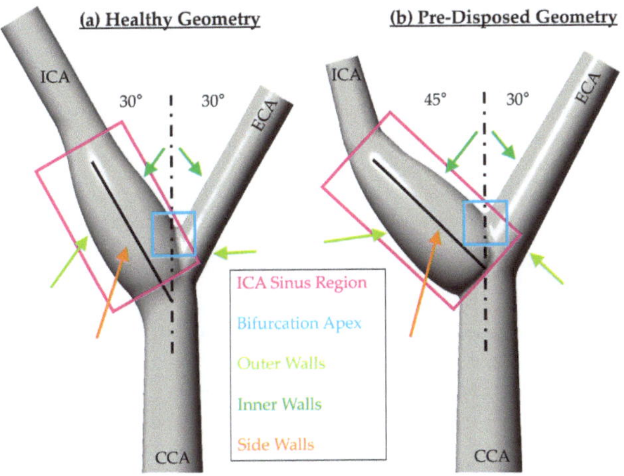

Figure 2. 'Healthy' (**a**, left) and 'pre-disposed' (**b**, right) model geometry of the carotid artery bifurcation: (**a**) symmetric total branching angle with 30° ICA branching angle, characteristic of a 'healthy' geometry; (**b**) asymmetric total branching angle with 45° ICA branching angle, characteristic of a 'pre-disposed' geometry. Regions of interest are highlighted.

2.1.1. Healthy Geometry

The modeled 'healthy' geometry dimensions are based on the physiological average of a healthy volunteer population [52–54]. The geometry had a circular 6.5 mm CCA inlet diameter and the CCA diameter tapered continuously toward the symmetric bifurcation. As it is a crucial region for atherosclerotic plaque formation, the ICA sinus was incorporated in the model geometry according to Figure 2.

2.1.2. Pre-Disposed Geometry

This novel pre-disposed geometry represents a patient cohort with an increased risk for atherosclerotic plaque formation and is based on this patient population average. Based on clinical data, the two main geometrical risk factors clinically known to be associated with increased atherosclerosis risk—(i) an increased ICA angle, resulting in an asymmetric total CAB bifurcation angle, and (ii) a reduced ICA/CCA diameter ratio—were incorporated into this 'pre-disposed' geometry. The model is a variant of the healthy geometry, where CCA and ECA specifications remain unchanged. First, the ICA angle was increased to the typical 45° [49], resulting in an asymmetric and larger total bifurcation angle. Secondly, the ICA/CCA diameter ratio was reduced to 0.5 [50] between the inlet and ICA downstream of the sinus by reducing the ICA diameter (Table 1). The development of this pre-disposed geometry allowed us to study how the flow field differs between a healthy and pre-disposed population and how differences in forming vortical structures might influence atherosclerosis, triggering wall shear stress distributions.

Table 1. Main anatomical risk factors in CAB showing geometrical differences between 'healthy' and 'pre-disposed' model geometry.

Geometrical Risk Factors	'Healthy' Model	'Pre-Disposed' Model
Total CAB Branching Angle	60°	75°
ICA Branching Angle	30°	45°
Diameter Ratio ICA/CCA [1]	0.7	0.5

[1] CCA diameter measured at the inlet; ICA diameter measured downstream of the sinus.

2.2. Computational Fluid Dynamics (CFD)

Computational fluid dynamics (CFD) was used to compute three-dimensional time-resolved flow fields in both geometries. ANSYS® meshing tool was used to discretize both geometries. ANSYS® FLUENT Academic Research Mechanical Release 2021 R1 was used to specify the physics of the Navier–Stokes equations using a finite volume method and a pressure-based solver under time-dependent laminar conditions. A CFD solver was chosen due to the rigid wall approximation, which allowed for the comparison of three-dimensional flow structures between the two fixed geometries. A high-performance computing cluster (Pegasus, GWU) was used to solve the Navier–Stokes equations. ANSYS® FLUENT 2021 R2 and Python JUPYTER Notebook 6.4.8 were used to post-process the results.

2.2.1. Spatial and Time Discretization

Both domains were discretized using tetrahedral core meshing and boundary layer inflation meshing at the vessel's lumen wall. A mesh independence study was performed by varying both mesh variables and observing the change in local velocity as well as wall shear stresses over the wall. High accuracy, and thus mesh independence, were achieved and validated, with a total of 2,080,502 mesh elements in the domain. Several cycles of the physiological inflow waveform were run until independence from the initial conditions and cycle-to-cycle repeatability was achieved. Data from the third physiological cycle were analyzed and are reported herein. The fixed time step was set to ensure a maximum Courant–Friedrichs–Lewy (CFL) number of less than 1 for all timesteps.

2.2.2. Boundary Conditions

The carotid artery bifurcation model was subjected to physiological pulsatile flow, which was provided at the CCA through a time-varying uniform velocity inlet condition. Mass outflow was controlled at the ICA and ECA outlet, and the rigid walls' no-slip boundary conditions were applied. Medium and large-sized vessels, such as the carotid artery bifurcation, experience negligible non-Newtonian effects [55], so a Newtonian fluid assumption was used. To facilitate interdisciplinary comparison with mechanotransduction-related cell studies, a constant kinematic viscosity of $6.95 \cdot 10^{-7}$ m^2/sec was used. The flow is Reynolds number matched to physiological blood flow in this vessel, where the CCA inflow Reynolds number is defined as CCA inflow velocity, multiplied by the CCA diameter and divided by the kinematic viscosity. The CCA inlet Reynolds number (Re) waveform peaks at Re = 1530, with a mean Re of 385. This inflow waveform is based on patient-averaged flow in the CCA [56] based on a healthy patient cohort and has a physiological period (T) of 1 s, corresponding to a heartbeat frequency of 60 bpm. For the user-defined velocity function at the inlet, a digitized waveform was used. The digitized model was devel-oped in our lab. The original data (which we digitized) is from Holdsworth et al. [57–60] (Figure 3).

At the ICA and ECA outlets, the mass outflow was controlled with a resistive outlet, implemented through a pressure outlet boundary condition. This resistive outlet condition was chosen due to its numerical robustness and nearly identical behavior to traditional three-element Windkessel models [61]. Resistive outlet conditions allowed for the control of the flow split between the ICA and ECA outlets and matched physiological clinical data. They are defined through a pressure outlet as they follow $\Delta p = R \cdot Q$, where Δp is the pressure drop in Pascals, R is the vascular resistance in $[\text{Pa·s}/\text{m}^3]$, and Q is the volume flow rate in $[\text{m}^3/\text{s}]$. Using resistive outlet conditions in the CFD simulation, resistance values (R) for both branches were iteratively changed until the mass outflow of the ICA matched patient-averaged clinical data [62]. To ensure continuity, ECA outflow was defined as CCA inflow minus ICA outflow. The values of vascular resistances iteratively determined and used in the 'healthy' as well as 'pre-disposed' geometry were $R_{\text{ICA}} = 1.5{\cdot}10^6$ Pa·s·m^{-3} and $R_{\text{ECA}} = 11{\cdot}10^6$ Pa·s·m^{-3} for the ICA and ECA, respectively, for both geometries. Thus, the outlet pressure boundary conditions were time-dependent functions due to their linear dependence on the physiological inflow waveform and their respective resistances.

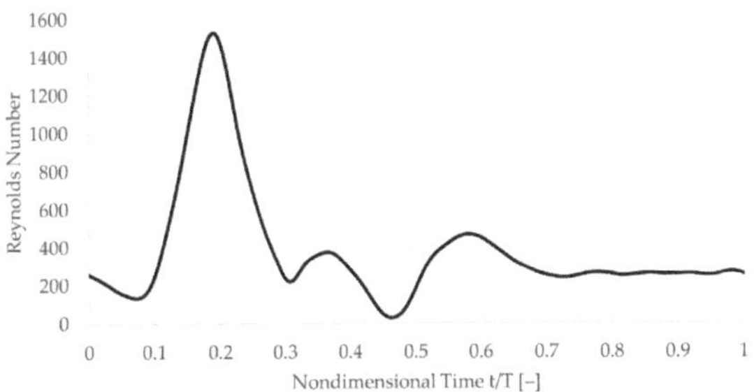

Figure 3. Physiological pulsatile inflow into the common carotid artery inlet over one cardiac cycle. The diagram represents the time dependence of the total inflow velocity magnitude at the inlet boundary. Waveform is the patient-averaged CCA flow of a healthy patient population [56].

2.3. Time-Averaged Wall Shear Stress and Vortex Identification

Wall shear stresses (WSS) are a main driver of atherosclerotic plaque formation, and thus WSS metrics provide an indication of expected cell response and disease onset. By contrast, time-averaged-WSS (TAWSS) is a commonly used metric, and this study focused on instantaneous WSS, as in our prior study, the observed WSS differences in the 'pre-disposed' geometry resulted in a proatherogenic TAWSS distribution [63]. Furthermore, this allowed for an increased understanding of the instantaneous flows and stresses underlying an unfavorable TAWSS distribution. WSS and TAWSS are based on Equations (1) and (2), respectively, where μ is the dynamic viscosity, $\vec{u}$ the local flow velocity, and T the period of the cardiac cycle.

$$\text{WSS} = \vec{\tau_w} = \mu\cdot\left(\frac{\partial \vec{u}}{\partial y}\right)_{wall} \tag{1}$$

$$\text{TAWSS} = \frac{1}{\text{T}}\cdot\int_0^{\text{T}}\left|\vec{\tau_w}\right|\,\text{dt} \tag{2}$$

Flow structures such as vortices can significantly impact the WSS distribution. Currently, there are only few investigations of vortical structures in bifurcating vessels, specifically the CAB, and rarely are these investigated with respect to their influence on the WSS distribution. The resulting vortical structures arise from secondary flows due to the vessels' curvature when branching from the CCA into the ICA or ECA, respectively. Secondary flow structures are expected to resemble the features observed in curved pipes under physiological pulsatile inflow, similar to the results reported by Cox et al. [41]. In curved pipe flow, a counter-rotating 'Dean' vortex pair is present, which forms in the flow acceleration phase. Around the inflow peak, this counter-rotating vortex pair typically gets deformed, and a second vortex pair, rotating in the opposite direction, will form. During the inflow deceleration phase, a separation or split of the deformed main counter-rotating vortex pair (Dean vortex) can be observed in pipe flow [41]. This study revealed similar secondary flow patterns in the bifurcating vessel as previously observed in a single curved pipe [64,65].

The identification of three-dimensional vortical structures in the flow field was performed using the λ_2 criterion, one of the most accepted vortex identification techniques [66], which is widely used in cardiovascular fluid dynamics. Furthermore, its advantage is that it is a quantitative method as it is based on the physical definition of the vortical structure and does not depend on a visualization technique [67]. The λ_2 criterion is an Eulerian and eigenvalue-based vortex identification method [68]. The three-dimensional velocity field functions as input, where the velocity field is described as $\vec{u} = (u, v, w)$. From this, the velocity gradient tensor J is computed using Equation (3).

$$J = \begin{bmatrix} \frac{\partial u}{\partial x} & \frac{\partial u}{\partial y} & \frac{\partial u}{\partial z} \\ \frac{\partial v}{\partial x} & \frac{\partial v}{\partial y} & \frac{\partial v}{\partial z} \\ \frac{\partial w}{\partial x} & \frac{\partial w}{\partial y} & \frac{\partial w}{\partial z} \end{bmatrix} \tag{3}$$

This velocity gradient tensor is decomposed into the strain deformation tensor S (Equation (4)), which is its symmetric part, and the rotation rate tensor Ω (Equation (5)), its asymmetric part [68].

$$S = \frac{J + J^T}{2} \tag{4}$$

$$\Omega = \frac{J - J^T}{2} \tag{5}$$

The eigenvalues of $\left(S^2 + \Omega^2\right)$ are computed and labeled such that $\lambda_1 \geq \lambda_2 \geq \lambda_3$. A voxel is detected to be part of a vortex core if it has two negative eigenvalues; thus, if $\lambda_2 < 0$. The vortex region is defined as a connected region of $\lambda_2 < 0$ voxels [67,69–71]. To highlight strong vs. weak vortical structures, an additional threshold of λ_2 can be applied [67]. When analyzing λ_2, we visualized, the three-dimensional iso-surfaces of the chosen λ_2 threshold.

To isolate and track key vortices originating in the wall boundary layer a PYTHON code was developed for post-processing. A vortex core was defined as a local pressure minimum, and thus the exact location of the vortex origin at the wall was identified by tracking a local pressure minimum at the vessel wall. This allowed for the analysis of the moment in time and location for vortex formation and its evolution over the cardiac cycle. The vortical structure of interest was first identified using the λ_2 criterion defining the broad location of vortex origin, e.g., the sinus side wall. To guarantee that the pressure minimum location of the vortex of interest was captured, a three-dimensional reduced field of view or search window was created and kept stationary or moved with the bulk flow velocity downstream where applicable. The location of the vortex core within this search window was identified based on the position of the pressure minimum on the three-dimensional surface in the search window for each time step. The position was provided in a translated and rotated coordinate system, where the origin lay in the bifurcation point and the axial downstream direction followed the ICA sinus center axis (following a black solid line in the sinus, as illustrated in Figure 2). The path traced by the pressure minimum over time was provided for one side of the vessel due to its symmetry around the z-axis. The following results will present the location, velocity, and acceleration of a main vortex core's origin at the sinus side wall.

3. Results

The results show WSS, secondary flow fields, and three-dimensional vortical structures for a typical 'healthy' and 'pre-disposed' carotid artery bifurcation geometry. The analysis focused on the ICA sinus as it is the prevalent region for atherosclerotic plaque formation in the CAB. The results highlight the impact that large vortical structures have on the WSS distribution, hinting at their crucial role in atherosclerosis. This is followed by the analysis of the development of a main hairpin vortical structure in the ICA sinus, specifically, its occurrence time window within the cardiac cycle and its evolution. The results highlight the importance of three-dimensional vortical structures within the CAB to increase the understanding of hemodynamic effects underlying proatherogenic WSS distributions.

3.1. Influence of Three-Dimensional Vortical Structures on WSS

Wall shear stress is widely known to be one of the key stresses affecting endothelial cells and plays a major role in atherosclerosis. Thus, the flow fields within the CAB significantly determine whether a physiological or pathological shear stress distribution emerges. There is a gap in the literature on the study of complex flow structures such as vortices, present in the CAB flow, and how they impact the WSS distribution. In this section, vortical structures and concomitant instantaneous WSS are evaluated to gain deeper insights into the flows related to carotid artery bifurcation.

Three-dimensional vortices are found to strongly influence the wall shear stress distribution. This can be seen in Figure 4, where the 'healthy' geometry is shown on the left (framed in blue), and the 'pre-disposed' geometry is shown on the right (framed in red). Slightly transparent wall shear stress contours are shown on the CAB walls. Vortices present in the fluid volume are visualized in a gray solid color using the λ_2 criterion. The middle row inset indicates the moment in time of the instantaneous result with dots on the physiological waveform in blue and red for the 'healthy' and 'pre-disposed', respectively. For both geometries, a strong, hairpin-shaped vortical structure forms in the ICA sinus. This main hairpin vortical structure starts to form at an earlier instance in the cardiac cycle in the case of the 'pre-disposed' geometry and then for the 'healthy' geometry.

Starting with the 'healthy' geometry (left, in blue), it is observed that this hairpin vortical structure changes its orientation over the cardiac cycle. The vortical structure starts forming in the inflow deceleration phase ($t/T = 0.23$) at an upstream sinus position around the CAB branching point. Initially, the vortical structure is oriented in extension to the CCA, so its imaginary center axis is almost parallel to the CCA center axis. By the end of the inflow deceleration phase ($t/T = 0.31$), the main hairpin vortical structure's center axis is almost perpendicular to the ICA center axis. Thus, throughout the inflow deceleration, this main hairpin vortical structure rotates, orienting its 'head' in the axial ICA flow direction. The 'pre-disposed' geometry (right, framed red) shows a similar phenomenon, forming one main hairpin vortical structure that changes its orientation in a similar pattern. The formation of this main hairpin vortical structure starts earlier in the cardiac cycle than in the 'healthy' case. In addition, by comparing the vortical structure's orientation over time, a time shift in the vortical structure's orientation can be observed, $t/T=0.05$ earlier means: 5% of the total cardiac cycle length earlier AS WELL AS 13% earlier relative to the total at-tached lifespan of the 'pre-disposed' hairpin vortical structure. Previous studies have established that wall shear stress is an essential mechanism that drives the biochemical mechanotransduction process related to atherosclerotic plaque formation [8,10,11]. Typically, TAWSS is studied to capture effects over the entire cardiac cycle, and it is known that low TAWSS are proatherogenic. Herein, the instantaneous time-resolved WSS results show that, throughout the cardiac cycle, various locations of the ICA sinus experience low WSS for both geometries. Both geometries, however, show a high-WSS band across the sinus, and furthermore, the position of this high-WSS band aligns with the internal position of the main hairpin vortical structure. This clearly shows the strong influence of vortical structures on the wall shear stress distribution. From these results, we anticipate a positive physiological impact of this hairpin vortical structure as it increases WSS in the generally low-WSS flow separation zone at the ICA sinus side and outer walls. This vortical structure's behavior will be discussed in more detail in Section 3.3.1. Thus, this study increases current knowledge by determining vortical structures as a hypothesized cause for a proatherogenic wall shear stress distribution.

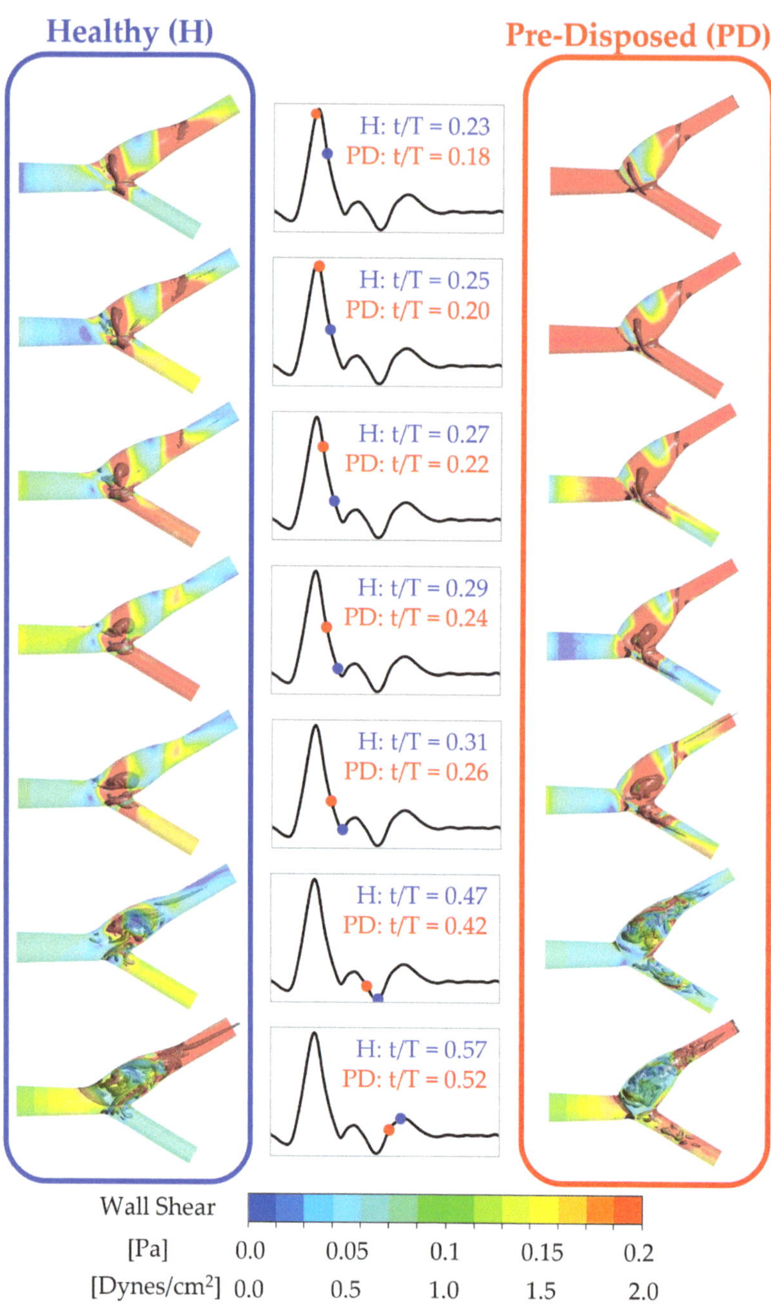

Figure 4. Wall shear stress distribution and the underlying three-dimensional vortices. The wall shear stress distribution is shown as slightly transparent (color bar on the bottom), and vortices in the vessel volume are shown in gray using a λ_2 threshold. The 'healthy' geometry is shown in the left column (famed in blue), and the 'pre-disposed' geometry is shown in the right column (framed in red). The middle column insets show an instant in the cycle for the 'healthy' and 'pre-disposed' geometry with a blue and red dot on the pulsatile inflow waveform, respectively.

Thus, in summary, both geometries, 'healthy' and 'pre-disposed', develop a main hairpin vortical structure in the ICA sinus. These results demonstrate the strong influence of this observed hairpin vortical structure on the WSS distribution. Its internal location is linked to an increased WSS in an otherwise low-WSS separation zone. Thus, this hairpin vortical structure could have a positive physiological impact by increasing WSS to nonatherogenic values in the critical sinus region. Therefore, it is important to deepen our knowledge of vortical structure formation in the CAB to understand its possibly crucial role in atherosclerosis and plaque formation.

3.2. Secondary Flows in the ICA Sinus

The CAB's branching results in a curvature within the vessel's axial direction and thus secondary flows are presented. The previous section showed the impact of three-dimensional vortices on WSS. As secondary flows facilitate vortical structure formation, the evolution of secondary flows over the cardiac cycle is investigated and compared between a 'healthy' and a 'pre-disposed' geometry in this section. The focus again lies on the ICA sinus as it is the location mostly affected by atherosclerosis.

The secondary velocity magnitude is shown at six planes perpendicular to the ICA branching centerline path (color bar) (see Figure 5). Planes are labeled 1–6, with 1 being most upstream, close to the CCA branching point, and 6 most downstream closer to the ICA outlet (as labeled in Figure 5a). The secondary velocity streamlines are added in black over the magnitude contours, and the instantaneous axial bulk inflow is represented with the black axial vectors in the CCA. Three-dimensional vortical structures using a λ_2 threshold are shown in gray. The 'healthy' geometry is shown on the left (a), and the 'pre-disposed' geometry is presented on the right (b). The time instances of data are indicated with the orange diamonds on the pulsatile inflow waveform in the middle inset.

Toward the end of the inflow deceleration phase at $t/T = 0.27$ (Figure 5a,b), both geometries exhibit a pronounced hairpin vortical structure in the ICA sinus. In the 'healthy' geometry (Figure 5a), it is located around plane 1. The vortical structure's intersection with the imaginary plane can be seen in the curved high-density streamlines at its location. Thus, a counter-rotating vortex pair is visible in plane 1. This location also exhibits a high secondary velocity magnitude that is concentrated at the sinus side walls. The secondary velocity magnitude weakens downstream, where no three-dimensional vortical structures with comparable strength are present. Similarly, counter-rotating vortex pairs can be observed further downstream, in planes 5 and 6. The 'pre-disposed' geometry generally experiences much higher secondary velocities in that phase (Figure 5b). At plane 1, a similar counter-rotating vortex pair is observed where the three-dimensional vortical structure intersects. This vortex pair is located closer to the inner sinus wall and can similarly be found in plane 2. In contrast to the 'healthy' geometry, no clear symmetric counter-rotating vortex pairs can be observed in planes 5 and 6. Thus, the vortical structure remains more coherent and persists further downstream in the healthy geometry.

Shortly after the strong inflow deceleration phase at $t/T = 0.37$, the three-dimensional hairpin vortices experience significant changes in both geometries. In the 'healthy' geometry very dense streamlines and high secondary velocity magnitudes around the vortical structure legs are observed, where its head turns toward the inner sinus wall (Figure 5c). The highest secondary velocities are still concentrated in the first upstream half of the ICA sinus. In contrast, high secondary velocities occupy the full sinus region in the 'pre-disposed' geometry (Figure 5d). Circular streamlines clearly indicate the location of three-dimensional vortices.

At the inflow minimum at $t/T = 0.47$, high secondary velocity is still concentrated in the sinus region, mostly in the upstream regions on planes 1 and 2 for the 'healthy' geometry (Figure 5e). Circular streamlines in plane 3 indicate that the observed bonelike vortical structure passes through the hairpin vortical structure loop, and plane 2 continues at lower λ_2 values downstream. Very dense circular streamlines are visible in all planes, indicating the presence of long 'leg-like' three-dimensional vortical structures through the whole

ICA. The 'pre-disposed' geometry has very different secondary flow patterns (Figure 5f). Here, higher secondary velocities are present in the downstream half of the sinus. Similar to the 'healthy' geometry, three-dimensional structures and counter-rotating vortex pairs are mostly located close to the sinus outer wall. But the upstream sinus part (around planes 1 and 2) contains many small, very disordered vortical structures, disorganized streamlines, and only small disconnected areas of high secondary velocity.

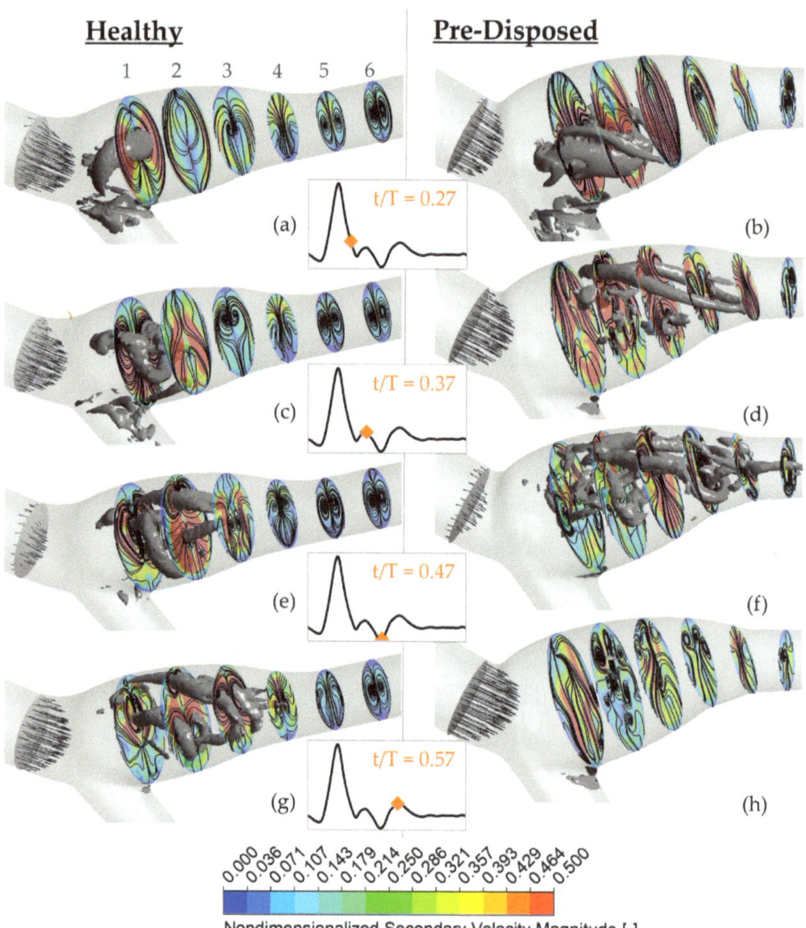

Figure 5. Secondary velocities in the ICA sinus over the cardiac cycle. 'Healthy' geometry is on the left (**a,c,e,g**), and 'pre-disposed' geometry is on the right (**b,d,f,h**). The middle insets show the analyzed instant in time, which is highlighted with an orange diamond on the pulsatile inflow waveform. The axial inflow in the CAB just upstream of the bifurcation is indicated with black axial velocity vectors. Velocity results are presented on 6 planes perpendicular to the ICA center axis; planes are labeled 1-6 where 1 is most upstream near the CCA and 6 is the most downstream (labeled in **a**). Nondimensionalized secondary velocity magnitude (u/U) (secondary velocity magnitude divided by U = U (t/T = 0.27), where U is the mean velocity magnitude at the CCA inlet) is defined with the color bar, and streamlines are added in black. Three-dimensional vortical structures are determined using a constant λ_2 criterion over all time instances t/T and shown in gray.

Finally, secondary flows and vortical structures exhibit significantly different behavior between the two geometries at $t/T = 0.57$, when the inflow starts to level off in the 'diastolic' phase. In the 'healthy' geometry (Figure 5g), one clear counter-rotating vortex pair is identifiable in each plane, and the large three-dimensional structures are still connected and mainly concentrated in the sinus. Beyond the sinus, the flow keeps its organized structure of one clearly defined vortex pair (see planes 5 and 6) throughout the complete time range. For the 'pre-disposed' geometry, no three-dimensional structures of comparable strength exist in this late stage in the cardiac cycle (Figure 5h). The secondary velocity magnitude is generally low with only small disconnected local increased secondary velocities. No clear counter-rotating vortex pairs are observed in the 'pre-disposed' geometry, neither downstream of the sinus (plane 6) nor in the sinus. Furthermore, when analyzing streamlines, a very disordered secondary flow can be observed, especially mid-sinus as is visible by many small circular streamline islands in planes 2–4 located toward the outer sinus wall.

In summary, a difference in the secondary flow magnitude over the full cardiac cycle is observed between the two geometries. The 'pre-disposed' geometry experiences larger regions of strong secondary flows in the inflow deceleration phase and smaller regions in the 'diastolic' phase compared with the 'healthy' geometry. The 'healthy' geometry exhibits a distinct counter-rotating vortex pair throughout the cardiac cycle—as expected for a curved vessel. The 'pre-disposed' geometry, on the other hand, lacks those characteristic counter-rotating vortex pairs throughout most of the vessel over most of the cycle. The 'pre-disposed' geometry has disturbed flow, with small disordered vortices and small disconnected areas of high and low secondary flow velocities.

3.3. Characterization of the ICA Sinus Vortical Structure

The three-dimensional structures significantly influence the shear stress distributions, especially in the strongly affected ICA sinus area. Thus, this section describes vortical structure formation and evolution throughout the cardiac cycle in the ICA sinus. It focuses on the main hairpin vortical structure due to its high impact on the wall shear stress. This analysis aims to explain the flow characteristics underlying a pathological proatherogenic WSS distribution for the 'pre-disposed' geometry versus a physiological WSS distribution for the 'healthy' geometry. An increased understanding of the role of vortices' association with atherosclerosis has the potential to become an important tool in early detection, treatment, and intervention planning.

3.3.1. Instant of Formation, Duration, and Rotation of Sinus Vortical Structure

The occurrence and orientation of the main hairpin vortical structure in the ICA sinus (described in Section 3.1) were investigated further. While this pronounced hairpin vortical structure forms in both cases, the temporal starting point of its formation and its duration vary between the cases. The hairpin vortical structure is visualized by choosing a λ_2 threshold to filter out vortices of smaller scale and strength. The existence window of the hairpin vortical structure is defined as its presence using this constant threshold. This lifespan of the vortical structure is plotted in Figure 6 over the pulsatile inflow waveform in blue and red for the 'healthy' and 'pre-disposed' cases, respectively. Several differences are evident. As was found from the three-dimensional vortical structure contours analyzed in Section 3.1, the formation of the main hairpin vortical structure starts earlier for the 'pre-disposed' geometry ($t/T = 0.05$ earlier). Notably, its formation starts in the inflow acceleration phase, whereas for the 'healthy' geometry, the hairpin vortical structure only starts to develop shortly after the inflow peak, in the inflow deceleration phase. In the 'pre-disposed' geometry, the hairpin vortical structure loses its dominance compared with the other vorticity in the sinus, where this endpoint of its predominance is observed at $t/T = 0.42$, shortly after the main inflow deceleration. The hairpin vortical structure in the 'healthy' geometry persists much longer than for the 'pre-disposed' case, almost to the end of the cardiac cycle ($t/T = 0.98$). These results highlight two main differences between the distinct hairpin vortical structure in the ICA sinus. In a 'healthy' geometry, the vortical structure starts forming only in the inflow deceleration and persists almost three times as long as that in 'pre-disposed' geometry.

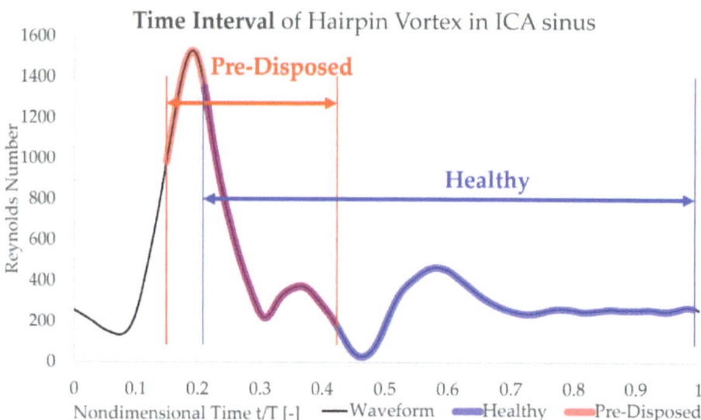

Figure 6. The life span (wall attachment) of the main hairpin vortical structure is shown on the pulsatile inflow waveform (black) for the 'healthy' geometry in blue and for the 'pre-disposed' geometry in red. The hairpin vortical structure is observed earlier for the 'pre-disposed' case and also stops to be detected significantly earlier.

3.3.2. Axial Pressure Gradient over ICA Sinus

The following section focuses on explaining a possible cause for the different hairpin vortical structure characteristics between a 'healthy' and a 'pre-disposed' geometry analyzed before. We argue that the differences in the axial pressure gradient over the ICA sinus—caused by its differing geometrical resistance—might cause discrepancies in vortical structure behavior. A favorable flow-driving pressure gradient is defined to be negative ($\Delta P / \Delta \zeta_{ICA} < 0$), whereas an adverse pressure gradient is positive ($\Delta P / \Delta \zeta_{ICA} > 0$). The axial direction is defined as the direction of the ICA center axis. For the presented analysis, $\Delta \zeta_{ICA}$ spans over the length of the ICA sinus. The results for the axial ICA pressure gradient over the cardiac cycle are shown in Figure 7, for the 'healthy' and 'pre-disposed' geometry in blue and red, respectively. Local pressure gradient minima and maxima are indicated by the lines, where the magenta, yellow, and black lines highlight favorable pressure gradient extrema, and the cyan and green lines indicate adverse pressure gradient extrema.

Significantly large differences are observed for the favorable pressure gradient extrema. The first flow-driving pressure gradient is the strongest pressure gradient over the entire cardiac cycle, experiences a very steep slope, and has the largest peak difference between the geometries (Figure 7, magenta line). The pre-disposed geometry experiences a 55% lower pressure gradient at this first peak. The maximum adverse pressure gradient is very similar for the two geometries (Figure 7, cyan line). The 'pre-disposed' geometry experiences an almost equal adverse pressure gradient to the 'healthy' geometry in the first peak (cyan line). Progressing in time, the 'healthy' geometry experiences another strong, favorable pressure gradient (yellow line); this pressure gradient peak is about 75% lower for the pre-disposed geometry. In the later stages of the cardiac cycle, the hairpin vortical structure is detected only for the 'healthy' geometry. As indicated by the green line (Figure 7), the 'healthy' geometry again experiences a stronger adverse pressure gradient. The last significant favorable pressure peak, before the pressure gradient levels off at a low value for both geometries, is located around t/T = 0.5, indicated by the black dashed line in Figure 7.

In summary, the adverse axial pressure gradient time trace within the ICA sinus has a very similar shape for the two geometries, especially in the time window where a vortical structure is detected for both. Large differences are, however, observed in the favorable pressure gradient magnitude, where the 'healthy' geometry experiences much stronger driving pressure gradients. These low favorable pressure gradients in a pre-disposed geometry are hypothesized to promote a proatherogenic vortical structure formation, resulting in a pathological wall shear stress distribution.

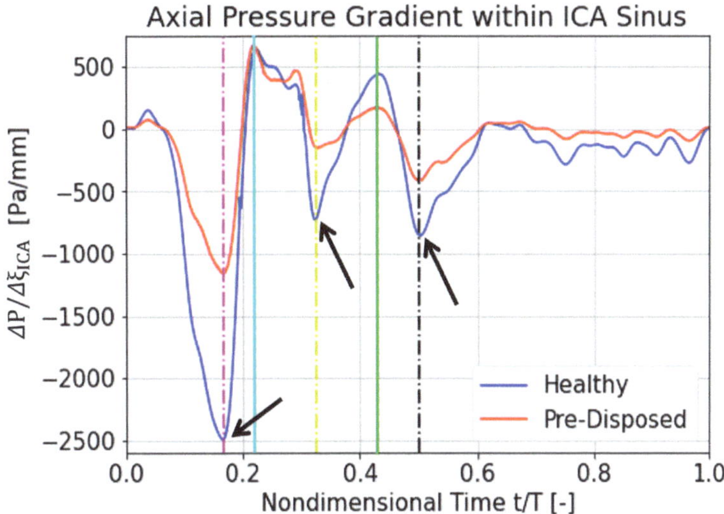

Figure 7. Axial pressure gradient over the ICA sinus through the cardiac cycle. 'Healthy' geometry is in blue, and 'pre-disposed' is illustrated in red. The indicated moments in time are shown with dashed lines: t/T = 0.165 (magenta), t/T = 0.22 (cyan), t/T = 0.325 (yellow), t/T = 0.43 (green), and t/T = 0.50 (black). Lines indicating favorable pressure gradients are marked with dash dots, and lines indicating adverse pressure gradients are solid. Large differences in the favorable pressure gradient are observed, which are higher for the 'healthy' geometry (highlighted with the arrows). Smaller differences are evident in the case of an adverse pressure gradient.

3.3.3. Streamwise Motion of Vortex Core

In this section, the streamwise motion of the main hairpin vortical structure is presented. The investigation of its motion characterizes the main hairpin vortical structure and identifies the differences between a 'healthy' and 'pre-disposed' geometry. Furthermore, it provides insights into the location and motion of the high-WSS band in the sinus.

As previously observed, the vortical structure forms around the bifurcation point and tilts into the ICA. In what follows, 'axial motion' is used to describe the vortices' motion along the ICA axial flow direction in the translated and rotated coordinate system, as indicated in Section 2.3. Figure 8 shows the three-dimensional vortical structures in the λ_2 criterion in gray in the vessel and the nondimensional pressure is shown with slightly transparent wall contours. Both geometries are shown after the inflow deceleration phase at t/T = 0.325, with the 'healthy' geometry framed in blue (Figure 8a) and the 'pre-disposed' geometry framed in red (Figure 8b), respectively. The core of a three-dimensional vortical structure is defined with a pressure minimum condition. Pressure contour plots allow for the visualization of the pressure origin. It is evident that the main hairpin vortical structure originates from the ICA sinus side walls, as indicated by the local pressure minimum on the sinus walls for both geometries (Figure 8) that matches the underlying hairpin vortical structure. In the following section, the motion of the pressure minimum at the vortical structure origin on the ICA sinus side wall will be tracked to indicate the position and motion of the hairpin vortical structure.

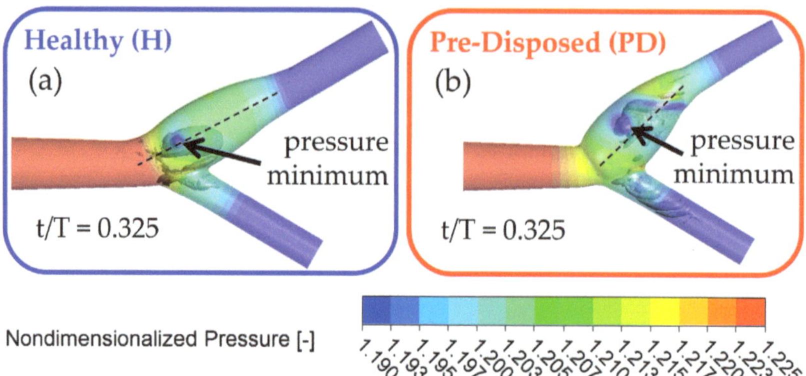

Figure 8. Nondimensionalized static pressure on artery wall (transparent, color bar) with underlying three-dimensional vortical structure using λ_2 vortex identification. Pressure is nondimensionalized by dividing by P0, where P0 = P(t/T = 0.191) at the CCA inlet center for the entire cardiac cycle. The illustrated instantaneous moment is t/T = 0.325. The pressure minimum on the sinus side wall shows the position of the vortical structure core of the main hairpin vortical structure: (**a**) blue shows the 'healthy' geometry; (**b**) red indicates the 'pre-disposed' geometry.

The position of the vortical structure origin was determined by finding the axial location of the pressure minimum in the vortical structure core. The axial position of the vortical structure and velocity are plotted in Figure 9a,b, respectively, over the cardiac cycle, where the 'healthy' geometry is displayed in blue, and the 'pre-disposed' geometry in red. In what follows, the 'vortex structure core' will be used to describe the pressure minimum at the vortical structure's wall origin. Instances of time representing axial pressure minima acting over the vortex core (as discussed in Section 3.3.2) are highlighted with dashed lines and will be elucidated in this and the following section.

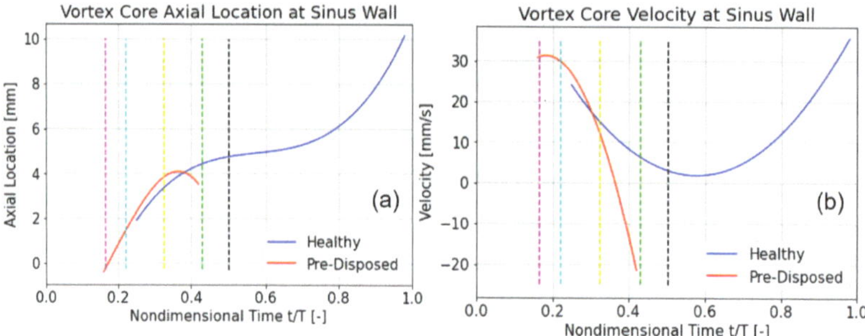

Figure 9. Vortical structure motion over the cardiac cycle. 'Healthy' geometry is in blue, and 'pre-disposed' is shown in red: (**a**) location of vortex core along axial ICA direction; (**b**) velocity of vortex core. The indicated moments in time are shown with the dashed lines: t/T = 0.165 (magenta), t/T = 0.22 (cyan), t/T = 0.325 (yellow), t/T = 0.43 (green), and t/T = 0.50 (black).

The location of the vortex core minimum pressure is plotted in Figure 9a using a translated and rotated coordinate system with the origin at the CAB bifurcation point and streamwise direction aligned with the ICA center axis (as shown in Figure 2 and detailed in Section 2.3). The distinct time shift of vortex pressure minima at the wall can be confirmed. The much earlier occurrence of this wall pressure minimum is clear in the case of the 'pre-disposed' geometry, as well as its significantly shorter duration of persistence. For both geometries, the vortex core translates downstream initially. Approximately t/T = 0.2 after onset, the downstream translation (or convection) of the vortex core slows significantly for both geometries. Afterward, a relatively stationary period between t/T = 0.5 and 0.7, the vortex core of the 'healthy' geometry again translates downstream at a much higher rate. In the 'pre-disposed' geometry, however, the vortex core does not recover from the slowed-down downstream motion, translates upstream very shortly, and then quickly deteriorates completely.

The changes in the downstream motion are further elucidated in the velocity plots (Figure 9b). It can be observed that the velocity of the vortex core initially decreases for both geometries. In the 'healthy' geometry, the velocity decreases slowly and nearly reaches a standstill. This very low vortex core translation velocity ('almost stillstand') occurs around t/T = 0.6. The vortex core does not completely slow and increases its velocity again, resulting in faster downstream motion. In the case of the 'pre-disposed' geometry, the velocity of the vortex core decreases at a nearly constant rate. In this case, the vortex core reaches a stillstand (velocity = 0 mm/s), at which time it is located at the most downstream location (the maximum axial location). The vortex core reaches a negative velocity early after formation, which continues to decrease almost linearly. This results in an upstream vortex core motion, soon followed by its deterioration.

In summary, the hairpin vortical structure initially translates downstream, for both geometries. Both cases then experience a slowed downstream motion, but the vortical structure in the 'healthy' case recovers around mid-cycle, avoids stagnation, and continues its downstream motion at a rapid rate. In the 'pre-disposed' geometry, on the other hand, the vortical structure starts decelerating early, stagnates, moves upstream, and deteriorates shortly after.

The axial pressure gradient peaks (as described in Section 3.3.2) acting on the hairpin vortical structure are hypothesized to be the driving factor of the vortical structure translation phenomena. Thus, the vortex core behavior will be linked to the acting axial pressure gradient in what follows. The first favorable pressure gradient peak (Figure 7, magenta line) occurs approximately at the time when both hairpin vortices start to form and intersect very closely with the starting point of the vortical structure in the 'pre-disposed' geometry (compare the magenta line in Figures 7 and 9a). The 'healthy' geometry experiences a 200% stronger initial favorable pressure gradient than the 'pre-disposed' geometry.

For the 'healthy' geometry, the first adverse pressure gradient peak occurs shortly before the hairpin vortical structure starts to form (compare the cyan line in Figures 7 and 9a). In contrast to the 'healthy' geometry, the 'pre-disposed' geometry has two distinct peaks in this adverse pressure gradient period, where the second one lies between the cyan and yellow lines in Figure 7. Comparing the second adverse pressure gradient peak (Figure 7, green line) to the occurrences in the vortex core's motion (Figure 9a,b, green dashed line), we find that it correlates with its slowed downstream position and translation (Figure 9a). In the 'healthy' geometry, the final favorable pressure peak (Figure 7, black dashed line) shows an interesting possible effect on the hairpin vortical structure characteristics. The moment of this axial pressure gradient peak is indicated by the black dashed line in Figure 9a,b. This is possibly preventing the vortical structure's stillstand (Figure 9b) and slowly accelerating the vortical structure's downstream motion again (Figure 9a,b).

In summary, these results strongly support the hypothesis that the differences in the axial pressure gradient significantly affect vortical structure evolution in the sinus. A favorable pressure gradient has a greater impact than an adverse pressure gradient on the main hairpin vortical structure's evolution.

4. Discussion

This study presents a novel analysis of complex flows, including three-dimensional vortical structures, and wall shear stress distribution in the artery model. In addition to this analysis, the study includes a novel patient-averaged 'pre-disposed' CAB geometry, representative of a patient cohort that is clinically at higher risk for CAB atherosclerotic plaque formation but has not yet developed stenosis. The comparison of this 'pre-disposed' geometry with a 'healthy' geometry elucidates how certain vortical structures promote a proatherogenic WSS distribution. A dominant vortical structure influencing the ICA sinus WSS was identified and characterized over the entire cardiac cycle. This expands the state-of-the-art research, which observed three-dimensional flow structures in carotid artery bifurcation [26,30,31,42] by exploring their dependence on geometrical risk factors and their atheroprotective impact on the wall shear stress distribution.

It was found that the presence of a vortical structure strongly affects the wall shear stress distribution; specifically, a highly influential hairpin vortex was found in the ICA sinus. The ICA sinus is the CAB location most commonly affected by atherosclerosis through typically very low WSS in the sinus flow separation zone [19,30,31]. Interestingly, the hairpin vortical structure increases the WSS significantly. Thus, this finding suggests that the main sinus hairpin vortical structure has a positive physiological impact by increasing ICA sinus WSS to atheroprotective levels.

Secondary flows facilitate vortical structure formation [30,35,37], and significant differences between 'healthy' and 'pre-disposed' secondary flow fields were detected, related to the observed differences in the vortical structure. During the vortical structure characterization, the main hairpin vortical structure in the 'pre-disposed' geometry forms earlier and deteriorates significantly earlier than in the 'healthy' case. Similarly, the 'pre-disposed' geometry has high secondary velocities earlier in the cardiac cycle but lower secondary velocities in the 'diastolic' phase of the cycle in which its hairpin vortical structure deteriorates. Furthermore, the 'pre-disposed' case contains more small distributed vortices in the 'diastolic' phase in the ICA sinus and does not contain the classical counter-rotating vortex pair expected for a curved vessel. The earlier transition into a disturbed secondary flow and vortical structures in the sinus coincides with the reduced peaks of favorable axial pressure gradients in the ICA. Furthermore, the hairpin vortical structure in the 'healthy' geometry, experiencing a higher driving favorable pressure gradient, never slows to zero velocity in its downstream translation, whereas it does under the lower favorable pressure gradient for the 'pre-disposed' case. This standstill and the subsequent upstream motion are shortly followed by the deterioration of the vortical structure in the 'pre-disposed' case. This strongly reduced favorable pressure gradient might have a strong influence on the vortical structure behavior, slow it down, and result in its deterioration, ultimately causing proatherogenic wall shear stress distributions. The differences present in the 'pre-disposed' geometry reduce favorable pressure gradients in the ICA sinus, which is the hypothesized driving force for a long-lasting stable vortical structure. We propose that the downstream motion of a persistent hairpin vortical structure, which increases WSS, is physiologically beneficial. We relate the lower pathological WSS found in the ICA sinus of a 'pre-disposed' geometry to weaker distorted secondary flows, mitigating the persistence and stability of the hairpin vortical structure caused by a weaker favorable axial pressure gradient.

The findings support our hypothesis that three-dimensional vortical structures impact WSS distribution, and their premature dissolution might play a significant role in the development of atherosclerotic plaques. This is further supported by the observed differences in vortical structures between a 'healthy' and a 'pre-disposed' geometry, where the latter experiences a proatherogenic WSS distribution.

Limitations

Clinically reported patient-averaged geometric anatomical features were used to develop the model vessels, in order to reduce complexity and to account for inter-patient variability. While this allowed us to elucidate the significant impact of three-dimensional vortical structures on a proatherogenic wall shear stress distribution, a further expansion of our study will allow us to augment the findings. Thus, it will be valuable to expand the study to larger datasets of model geometries, allowing for a large parametric study, and thus further investigate a variety of healthy and pathological patient-specific geometries. Future work will also deepen the investigation into the characterization of the main hairpin vortical structure and its behavior under different pressure gradients corresponding to the axial pressure gradient within the ICA sinus. A prospective goal will be to define the clear structural and time-evolution parameter differences of this vortical structure to allow for a clear distinction between a physiological and a pathological CAB vortical structure formation, thus informing the clinical early identification of affected patients through medical imaging in the future.

5. Conclusions

Physiological flow in carotid artery bifurcation models was studied to increase the understanding of how vortical flow structures may impose an atheroprotective wall shear stress distribution. The CFD simulations of a 'healthy' and a 'pre-disposed' CAB model geometry revealed substantial differences in the behavior of the main three-dimensional vortical structures in the ICA sinus. This study revealed a strong correlation between a long-lasting, stably persisting main vortical structure and the related increased WSS, known to have a positive physiological impact on the endothelium. This supports our hypothesis that the formation and persistence of a vortical structure is physiologically beneficial and that the deterioration of this structure would play a significant role in the development of atherosclerosis. The difference in vortical structure behavior not only increases our understanding of why certain vessel geometries are more prone to develop atherosclerosis but also could be observable in medical imaging, leading to a potential clinical diagnostic. The characterization of the main ICA sinus vortical structure is a precursor to developing a promising clinical indicator for atherosclerosis risk. Thus, the advances made in this study could be translated to inform the clinical early detection of patients at higher risk for carotid bifurcation atherosclerosis in the future.

Author Contributions: Conceptualization, N.C.W., K.V.B. and M.W.P.; methodology, N.C.W., K.V.B. and M.W.P.; software, N.C.W.; validation, N.C.W.; formal analysis, N.C.W.; investigation, N.C.W.; resources, N.C.W. and M.W.P.; data curation, N.C.W.; writing—original draft preparation, N.C.W.; writing—review and editing, K.V.B. and M.W.P.; visualization, N.C.W.; supervision, K.V.B. and M.W.P.; project administration, N.C.W. and M.W.P.; funding acquisition, M.W.P. All authors have read and agreed to the published version of the manuscript.

Funding: This research was supported by the National Science Foundation, Biomechanics & Mechanobiology (BMMB) Program, under grant CMMI-1854415. The first author was further supported by the George Washington University graduate research assistantship and the Michael K. Myers Merit Scholarship. This study was completed in part with resources provided by the High-Performance Computing Cluster at The George Washington University, Information Technology, Research Technology Services.

Institutional Review Board Statement: Not applicable.

Informed Consent Statement: Not applicable.

Data Availability Statement: Data are available upon request due to restrictions. The data presented in this study are available upon request from the corresponding author. The data are not publicly available due to the use of this data in ongoing research and planned future publication by this research group.

Acknowledgments: We thank Elizabeth Gregorio (The George Washington University) and Megan C. Leftwich (The George Washington University) for their support and expertise.

Conflicts of Interest: The authors declare no conflict of interest. The funders had no role in the design of the study; in the collection, analyses, or interpretation of data; in the writing of the manuscript; or in the decision to publish the results.

References

1. Chandran, K.B.; Rittgers, S.E.; Yoganathan, A.P. *Biofluid Mechanics—The Human Circulation*, 1st ed.; CRC Press: Boca Raton, FL, USA, 2006.
2. Advanced Vascular Surgery. TCAR (Carotid Revascularization). 2022. Available online: https://www.avsurgery.com/services/tcar-carotid-revascularization/ (accessed on 22 April 2022).
3. Mayfield Brain & Spine. Carotid Stenosis (Carotid Artery Disease). 2022. Available online: https://mayfieldclinic.com/pe-carotidstenosis.htm (accessed on 22 April 2022).

4. Dai, G.; Kaazempur-Mofrad, M.R.; Natarajan, S.; Zhang, Y.; Vaughn, S.; Blackman, B.R.; Kamm, R.D.; García-Cardeña, G.; Gimbrone, M.A., Jr. Distinct endothelial phenotypes evoked by arterial waveforms derived from atherosclerosis-susceptible and-resistant regions of human vasculature. *Proc. Natl. Acad. Sci. USA* **2004**, *101*, 14871–14876. Available online: https://www.pnas.org/doi/pdf/10.1073/pnas.0406073101 (accessed on 1 March 2022). [PubMed]
5. Fung, Y. *Biomechanics: Circulation*, 2nd ed.; Springer Science & Business Media: Dordrecht, The Netherlands, 2013.
6. Chiu, J.-J.; Wang, D.L.; Chien, S.; Skalak, R.; Usami, S. Effects of disturbed flow on endothelial cells. *J. Biomech. Eng.* **1998**, *120*, 2–8. [PubMed]
7. Polanczyk, A.; Podyma, M.; Stefanczyk, L.; Szubert, W.; Zbicinski, I. A 3D model of thrombus formation in a stent-graft after implantation in the abdominal aorta. *J. Biomech.* **2015**, *48*, 425–431. [CrossRef] [PubMed]
8. Hann, S.Y.; Cui, H.; Zalud, N.C.; Esworthy, T.; Bulusu, K.; Shen, Y.-L.; Plesniak, M.W.; Zhang, L.G. An in vitro analysis of the effect of geometry-induced flows on endothelial cell behavior in 3D printed small-diameter blood vessels. *Biomater. Adv.* **2022**, *137*, 212832. [CrossRef]
9. Zalud, N.C.; Bulusu, K.V.; Plesniak, M.W. Simulation of a Pro-Atherogenic High-Risk Carotid Artery Bifurcation Geometry. *Bull. Am. Phys. Soc.* **2022**. Available online: https://meetings.aps.org/Meeting/DFD22/Session/U05.3 (accessed on 1 December 2022).
10. French, A.S. Mechanotransduction. *Annu. Rev. Physiol.* **1992**, *54*, 135–152. [CrossRef] [PubMed]
11. Tarbell, J.M.; Weinbaum, S.; Kamm, R.D. Cellular fluid mechanics and mechanotransduction. *Ann. Biomed. Eng.* **2005**, *33*, 1719–1723. [CrossRef]
12. Himburg, H.A.; Grzybowski, D.M.; Hazel, A.L.; LaMack, J.A.; Li, X.M.; Friedman, M.H. Spatial comparison between wall shear stress measures and porcine arterialendothelial permeability. *Am. J. Physiol. Heart Circ. Physiol.* **2004**, *286*, H1916–H1922. [CrossRef]
13. Barber, K.M.; Pinero, A.; Truskey, G.A. Effects of recirculating flow on U-937 cell adhesion to human umbilical vein endothelial cells. *Am. J. Physiol. Heart Circ. Physiol.* **1998**, *275*, H591–H599.
14. Ravensbergen, J.; Ravensbergen, J.W.; Krijger, J.K.B.; Hillen, B.; Hoogstraten, H.W. Localizing role of hemodynamics in atherosclerosis in several human vertebrobasilar junction geometries. *Arter. Thromb Vasc. Biol.* **1998**, *18*, 708–716. Available online: http://ahajournals.org (accessed on 1 May 2023).
15. Conway, D.E.; Williams, M.R.; Eskin, S.G.; McIntire, L.V. Endothelial cell responses to atheroprone flow are driven by two separate flow components: Low time-average shear stress and fluid flow reversal. *Am. J. Physiol. Heart Circ. Physiol.* **2009**, *298*, H367–H374. [CrossRef] [PubMed]
16. Rouleau, L.; Farcas, M.; Tardif, J.C.; Mongrain, R.; Leask, R.L. Endothelial cell morphologic response to asymmetric stenosis hemodynamics: Effects of spatial wall shear stress gradients. *J. Biomech. Eng.* **2010**, *132*, 081013. [CrossRef]
17. Lopes, D.; Puga, H.; Teixeira, J.; Lima, R. Blood flow simulations in patient-specific geometries of the carotid artery: A systematic review. *J. Biomech.* **2020**, *111*, 110019. [CrossRef] [PubMed]
18. Rindt, C.; van de Vosse, F.N.; Steenhovw, A.A.; Jamsen, J.D.; Reneman, R.S. A numerical and experimental analysis of the flow field in a two-dimensional model of the human carotid artery bifurcation. *J. Biomech.* **1987**, *20*, 499–509. [CrossRef]
19. Li, C.H.; Gao, B.L.; Wang, J.W.; Liu, J.F.; Li, H.; Yang, S.T. Hemodynamic factors affecting carotid sinus atherosclerotic stenosis. *World Neurosurg.* **2019**, *121*, e262–e276. [CrossRef] [PubMed]
20. Augst, A.D.; Ariff, B.; Thom, S.A.G.M.; Xu, X.Y.; Hughes, A.D. Analysis of complex flow and the relationship between blood pressure, wall shear stress, and intima-media thickness in the human carotid artery. *Am. J. Physiol. Heart Circ. Physiol.* **2007**, *293*, 1031–1037. [CrossRef]
21. Morbiducci, U.; Gallo, D.; Massai, D.; Ponzini, R.; Deriu, M.A.; Antiga, L.; Redaelli, A.; Montevecchi, F.M. On the importance of blood rheology for bulk flow in hemodynamic models of the carotid bifurcation. *J. Biomech.* **2011**, *44*, 2427–2438. [CrossRef]
22. Lopes, D.; Puga, H.; Teixeira, J.C.; Teixeira, S.F. Influence of arterial mechanical properties on carotid blood flow: Comparison of CFD and FSI studies. *Int. J. Mech. Sci.* **2019**, *160*, 209–218. [CrossRef]
23. Ku, D.N.; Giddens, D.P.; Zarins, C.K.; Glagov, S. Pulsatile flow and atherosclerosis in the human carotid bifurcation positive correlation between plaque location and low and oscillating shear stress. *Arterioscler. Off. J. Am. Heart Assoc. Inc.* **1985**, *5*, 293–302. Available online: http://ahajournals.org (accessed on 1 May 2023). [CrossRef]
24. Bharadvaj, B.K.; Mabon, R.F.; Giddens, D.P. Steady flow in a model of the human carotid bifurcation. Part II—Laser-Doppler anemometer measurements. *J. Biomech.* **1982**, *15*, 363–378. [CrossRef]
25. Nagargoje, M.; Gupta, R. Effect of sinus size and position on hemodynamics during pulsatile flow in a carotid artery bifurcation. *Comput. Methods Programs Biomed.* **2020**, *192*, 105440. [CrossRef] [PubMed]
26. Gijsen, F.J.H.; Van De Vosse, F.N.; Janssen, J.D. The influence of the non-Newtonian properties of blood on the flow in large arteries: Steady flow in a carotid bifurcation model. *J. Biomech.* **1999**, *32*, 601–608. [CrossRef]
27. Stroud, J.S.; Berger, S.A.; Saloner, D. Numerical analysis of flow through a severely stenotic carotid artery bifurcation. *J. Biomech. Eng.* **2002**, *124*, 9–20. [CrossRef]
28. Zhou, X.; Yin, L.; Xu, L.; Liang, F. Non-periodicity of blood flow and its influence on wall shear stress in the carotid artery bifurcation: An in vivo measurement-based computational study. *J. Biomech.* **2020**, *101*, 109617. [CrossRef]
29. Bouteloup, H.; de Oliveira Marinho, J.G.; Chatpun, S.; Espino, D.M. Computational analysis to predict the effect of pre-bifurcation stenosis on the hemodynamics of the internal and external carotid arteries. *J. Mech. Eng. Sci.* **2020**, *14*, 7029–7039. [CrossRef]

30. Perktold, K.; Resch, M.; Peter, R.O. Three-dimensional numerical analysis of pulsatile flow and wall shear stress in the carotid artery bifurcation. *J. Biomech.* **1991**, *24*, 409–420. [CrossRef] [PubMed]
31. Kumar, N.; Khader, S.M.A.; Pai, R.; Khan, S.H.; Kyriacou, P.A. Fluid structure interaction study of stenosed carotid artery considering the effects of blood pressure. *Int. J. Eng. Sci.* **2020**, *154*, 103341. [CrossRef]
32. Nguyen, K.T.; Clark, C.D.; Chancellor, T.J.; Papavassiliou, D.V. Carotid geometry effects on blood flow and on risk for vascular disease. *J. Biomech.* **2008**, *41*, 11–19. [CrossRef]
33. Akhmetov, D.G. *Vortex Rings*; Springer Science & Business Media: Dordrecht, The Netherlands, 2009.
34. Cox, C. Development of a High-Order Navier-Stokes Solver Using Flux Reconstruction to Simulate Three-Dimensional Vortex Structures in a Curved Artery Model. Ph.D. Thesis, The George Washington University, Washington, DC, USA, 2018.
35. Dewey, C.F. Dynamics of arterial flow. *Adv. Exp. Med. Biol.* **1979**, *115*, 55–89.
36. Giddens, D.P.; Zarins, C.K.; Glagov, S. Response of arteries to near-wall fluid dynamic behavior. *Appl. Mech. Rev.* **1990**, *43*, S98–S102. [CrossRef]
37. Nerem, R.M. Vascular Fluid Mechanics, the Arterial Wall, and Atherosclerosis. *J. Biomech. Eng.* **1992**, *114*, 274–282. [CrossRef] [PubMed]
38. Peiffer, V.; Sherwin, S.J.; Weinberg, P.D. Computation in the rabbit aorta of a new metric–The transverse wall shear stress—To quantify the multidirectional character of disturbed blood flow. *J. Biomech.* **2013**, *46*, 2651–2658. [CrossRef] [PubMed]
39. Mantha, A.; Karmonik, C.; Benndorf, G.; Strother, C.; Metcalfe, R. Hemodynamics in a Cerebral Artery before and after the Formation of an Aneurysm. *Am. J. Neuroradiol.* **2006**, *27*, 1113–1118. Available online: http://ahajournals.org (accessed on 1 May 2023). [PubMed]
40. Chakraborty, A.; Chakraborty, S.; Jala, V.R.; Haribabu, B.; Sharp, M.K.; Berson, R.E. Effects of biaxial oscillatory shear stress on endothelial cell proliferation and morphology. *Biotechnol. Bioeng.* **2011**, *109*, 695–707. [CrossRef]
41. Cox, C.; Najjari, M.R.; Plesniak, M.W. Three-dimensional vortical structures and wall shear stress in a curved artery model. *Phys. Fluids* **2019**, *31*, 121903. [CrossRef]
42. Chen, Y.; Yang, X.; Iskander, A.J.; Wang, P. On the flow characteristics in different carotid arteries. *Phys. Fluids* **2020**, *32*, 101902. [CrossRef]
43. Goddi, A.; Fanizza, M.; Bortolotto, C.; Raciti, M.V.; Fiorina, I.; He, X.; Du, Y.; Calliada, F. Vector flow imaging techniques: An innovative ultrasonographic technique for the study of blood flow. *J. Clin. Ultrasound* **2017**, *45*, 582–588. [CrossRef]
44. Elbaz, M.S.; van der Geest, R.J.; Calkoen, E.E.; de Roos, A.; Lelieveldt, B.P.; Roest, A.A.; Westenberg, J.J. Assessment of viscous energy loss and the association with three-dimensional vortex ring formation in left ventricular inflow: In vivo evaluation using four-dimensional flow MRI. *Magn. Reson. Med.* **2016**, *77*, 794–805. [CrossRef]
45. Elbaz, M.S.M.; E Calkoen, E.; Westenberg, J.J.M.; Lelieveldt, B.P.F.; Roest, A.A.W.; van der Geest, R.J. Vortex flow during early and late left ventricular filling in normal subjects: Quantitative characterization using retrospectively-gated 4D flow cardiovascular magnetic resonance and three-dimensional vortex core analysis. *J. Cardiovasc. Magn. Reson.* **2014**, *16*, 78. [CrossRef]
46. Goddi, A.; Bortolotto, C.; Raciti, M.V.; Fiorina, I.; Aiani, L.; Magistretti, G.; Sacchi, A.; Tinelli, C.; Calliada, F. High-Frame Rate Vector Flow Imaging of the Carotid Bifurcation in Healthy Adults: Comparison with Color Doppler Imaging. *J. Ultrasound Med.* **2018**, *37*, 2263–2275. [CrossRef]
47. Youn, S.W.; Lee, J. From 2D to 4D Phase-Contrast MRI in the Neurovascular System: Will It Be a Quantum Jump or a Fancy Decoration? *J. Magn. Reson. Imaging* **2020**, *55*, 347–372. [CrossRef] [PubMed]
48. Park, J.H.; Razuk, A.; Saad, P.F.; Telles, G.J.P.; Karakhanian, W.K.; Fioranelli, A.; Rodrigues, A.C.; Volpiani, G.G.; Campos, P.; Yamada, R.M.; et al. Carotid stenosis: What is the high-risk population? *Clinics* **2012**, *67*, 865–870. [CrossRef] [PubMed]
49. Phan, T.G.; Beare, R.J.; Jolley, D.; Das, G.; Ren, M.; Wong, K.; Chong, W.; Sinnott, M.D.; Hilton, J.E.; Srikanth, V.; et al. Carotid Artery Anatomy and Geometry as Risk Factors for Carotid Atherosclerotic Disease. *Stroke* **2012**, *43*, 1596–1601. [CrossRef] [PubMed]
50. Spanos, K.; Petrocheilou, G.; Karathanos, C.; Labropoulos, N.; Mikhailidis, D.; Giannoukas, A. Carotid Bifurcation Geometry and Atherosclerosis. *Angiology* **2016**, *68*, 757–764. [CrossRef]
51. Thomas, J.B.; Antiga, L.; Che, S.L.; Milner, J.S.; Steinman, D.A.H.; Spence, J.D.; Rutt, B.K.; Steinman, D.A. Variation in the Carotid Bifurcation Geometry of Young Versus Older Adults. *Stroke* **2005**, *36*, 2450–2456. [CrossRef] [PubMed]
52. Tada, S.; Tarbell, J.M. A Computational Study of Flow in a Compliant Carotid Bifurcation–Stress Phase Angle Correlation with Shear Stress. *Ann. Biomed. Eng.* **2005**, *33*, 1202–1212. [CrossRef]
53. Marshall, I.; Papathanasopoulou, P.; Wartolowska, K. Carotid flow rates and flow division at the bifurcation in healthy volunteers. *Physiol. Meas.* **2004**, *25*, 691–697. [CrossRef]
54. Goubergrits, L.; Affeld, K.; Fernandez-Britto, J.; Falcon, L. Geometry of the human common carotid artery. A vessel cast study of 86 specimens. *Pathol. Res. Pract.* **2002**, *198*, 543–551. [CrossRef]
55. Boyd, J.; Buick, J.M. Comparison of Newtonian and non-Newtonian flows in a two-dimensional carotid artery model using the lattice Boltzmann method. *Phys. Med. Biol.* **2007**, *52*, 6215–6228. [CrossRef]
56. Holdsworth, D.W.; Norley CJ, D.; Frayne, R.; Steinman, D.A.; Rutt, B.K. Characterization of common carotid artery blood-flow waveforms in normal human subjects. *Physiol. Meas.* **1999**, *20*, 219. [CrossRef]
57. Bulusu, K.V.; Plesniak, M.W. Secondary flow morphologies due to model stent-induced perturbations in a 180° curved tube during systolic deceleration. *Exp. Fluids* **2013**, *54*, 1493. [CrossRef]

58. Glenn, A.L.; Bulusu, K.V.; Shu, F.; Plesniak, M.W. Secondary flow structures under stent-induced perturbations for cardiovascular flow in a curved artery model. *Int. J. Heat Fluid Flow* **2012**, *35*, 76–83. [CrossRef]

59. Najjari, M.R.; Plesniak, M.W. Evolution of vortical structures in a curved artery model with non-Newtonian blood-analog fluid under pulsatile inflow conditions. *Exp. Fluids* **2016**, *57*, 100. [CrossRef]

60. Peterson, S.D.; Plesniak, M.W. The influence of inlet velocity profile and secondary flow on pulsatile flow in a model artery with stenosis. *J. Fluid Mech.* **2008**, *616*, 263–301. [CrossRef]

61. Capuano, F.; Loke, Y.-H.; Balaras, E. Blood Flow Dynamics at the Pulmonary Artery Bifurcation. *Fluids* **2019**, *4*, 190. [CrossRef]

62. Ford, M.D.; Alperin, N.; Lee, S.H.; Holdsworth, D.W.; Steinman, D.A. Characterization of volumetric flow rate waveforms in the normal internal carotid and vertebral arteries. *Physiol. Meas.* **2005**, *26*, 477–488. [CrossRef]

63. Zalud, N.C.; Bulusu, K.V.; Plesniak, M.W. Shear stress metrics associated with pro-atherogenic high-risk anatomical features in a carotid artery bifurcation model. *Clin. Biomech.* **2023**, *105*, 105956. [CrossRef]

64. Plesniak, M.W.; Bulusu, K.V. Morphology of Secondary Flows in a Curved Pipe with Pulsatile Inflow. *J. Fluids Eng.* **2016**, *138*, 101203. [CrossRef]

65. Bulusu, K.V.; Plesniak, M.W. Insights on arterial secondary flow structures and vortex dynamics gained using the MRV technique. *Int. J. Heat Fluid Flow* **2018**, *73*, 143–153. [CrossRef]

66. Kheradvar, A.; Pedrizzetti, G. *Vortex Formation in the Cardiovascular System*, 1st ed.; Springer: London, UK, 2012. [CrossRef]

67. ElBaz, M.S.; Lelieveldt, B.P.; Westenberg, J.J.; Van Der Geest, R.J. Automatic extraction of the 3D left ventricular diastolic transmitral vortex ring from 3D whole-heart phase contrast MRI using Laplace-Beltrami signatures. In *Statistical Atlases and Computational Models of the Heart. Imaging and Modeling Challenges, Proceedings of the 4th International Workshop, STACOM 2013, Held in Conjunction with MICCAI 2013, Nagoya, Japan, 26 September 2013*; Springer: Berlin/Heidelberg, Germany, 2014; pp. 204–211.

68. Gao, Y.; Liu, C. Rortex and comparison with eigenvalue-based vortex identification criteria. *Phys. Fluids* **2018**, *30*, 085107. [CrossRef]

69. Chen, Q.; Zhong, Q.; Qi, M.; Wang, X. Comparison of vortex identification criteria for planar velocity fields in wall turbulence. *Phys. Fluids* **2015**, *27*, 085101. [CrossRef]

70. Jeong, J.; Hussain, F. On the identification of a vortex. *J. Fluid Mech.* **1995**, *285*, 69–94. [CrossRef]

71. Jiang, M.; Machiraju, R.; Thompson, D. Detection and visualization of cortices. In *The Visualization Handbook*; Hansen, C.D., Johnson, C.R., Eds.; Elsevier Academic Press: Amsterdam, The Netherlands, 2005; p. 295.

 *bioengineering*

Article

Assessment of Rheological Models Applied to Blood Flow in Human Thoracic Aorta

Alexander Fuchs [1,2,3], Niclas Berg [3], Laszlo Fuchs [3] and Lisa Prahl Wittberg [3,*]

1 Department of Radiology in Linköping, Linköping University, 581 83 Linköping, Sweden
2 Department of Health, Medicine and Caring Sciences, Linköping University, 581 83 Linköping, Sweden
3 FLOW, Department of Engineering Mechanics, Royal Institute of Technology (KTH),
100 44 Stockholm, Sweden; lf@mech.kth.se (L.F.)
* Correspondence: prahl@kth.se

Abstract: Purpose: The purpose of this study is to assess the importance of non-Newtonian rheological models on blood flow in the human thoracic aorta. Methods: The pulsatile flow in the aorta is simulated using the models of Casson, Quemada and Walburn–Schneck in addition to a case of fixed (Newtonian) viscosity. The impact of the four rheological models (using constant hematocrit) was assessed with respect to (i) magnitude and deviation of the viscosity relative to a reference value (the Newtonian case); (ii) wall shear stress (WSS) and its time derivative; (iii) common WSS-related indicators, OSI, TAWSS and RRT; (iv) relative volume and surface-based retrograde flow; and (v) the impact of rheological models on the transport of small particles in the thoracic aorta. Results: The time-dependent flow in the thoracic aorta implies relatively large variations in the instantaneous WSS, due to variations in the instantaneous viscosity by as much as an order of magnitude. The largest effect was observed for low shear rates (tens s^{-1}). The different viscosity models had a small impact in terms of time- and spaced-averaged quantities. The significance of the rheological models was clearly demonstrated in the instantaneous WSS, for the space-averaged WSS (about 10%) and the corresponding temporal derivative of WSS (up to 20%). The longer-term accumulated effect of the rheological model was observed for the transport of spherical particles of 2 mm and 2 mm in diameter (density of 1200 kg/m^3). Large particles' total residence time in the brachiocephalic artery was 60% longer compared to the smaller particles. For the left common carotid artery, the opposite was observed: the smaller particles resided considerably longer than their larger counterparts. Conclusions: The dependence on the non-Newtonian properties of blood is mostly important at low shear regions (near walls, stagnation regions). Time- and space-averaging parameters of interest reduce the impact of the rheological model and may thereby lead to under-estimation of viscous effects. The rheological model affects the local WSS and its temporal derivative. In addition, the transport of small particles includes the accumulated effect of the blood rheological model as the several forces (e.g., drag, added mass and lift) acting on the particles are viscosity dependent. Mass transport is an essential factor for the development of pathologies in the arterial wall, implying that rheological models are important for assessing such risks.

Keywords: thoracic aorta flow; rheological models for blood; effects of blood viscosity

1. Introduction

Blood is a non-Newtonian fluid mainly composed of water but also containing a wide range of cells, micelles, and molecules of widely different sizes. The multiple functionalities of blood and its components include transporting substances needed by the different organs, being a major player in maintaining an optimal environment (e.g., pH, temperature) in the body as well as defending the body against microorganisms and stopping bleeding. Hence, blood composition may change in terms of numbers and types of cells, micelles and molecules as a response to needs. This adaptive behavior can be rather quick, implying that for all of us, blood rheological properties may vary over the day. In addition, pathological variations of the blood constituents affect blood viscosity. For example, red blood cell (RBC) size distribution (or RBC distribution width (RDW)) has been found to be a predictor of morbidity and mortality (cf Lippi et al. [1], Danese et al. [2], Ananthaseshan et al. [3]). Non-uniform distribution of RBCs within the circulatory system contributes to the difficulty in determining whole blood viscosity (WBV). Blood rheology has been the subject of many studies found in the literature. Often, WBV measurements have been carried out using standard (shear) rheometers (cf Cowan et al. [4]), implying that the flow is laminar and that the shear rate is well defined and solely dependent on the rotation rate (cf Agarwal et al. [5]). Yamamoto et al. [6] used a compact-sized falling needle rheometer on fresh blood samples to measure the relationship between the shear stress (τ) and shear rate (γ). The study identified three typical "regions": the "Casson" region for a low shear rate range (below 140 s^{-1}), the transition region (up to about 160 s^{-1}) and the Newtonian fluid region for higher shear rates (above 160 s^{-1}). The range of human blood viscosity was found to be in the range of 5.5 to 6.4 mPa s, and 4.5 to 5.3 mPa s for males and females, respectively. Moreover, Wang et al. [7] measured the WBV for a group of healthy individuals showing that both inter- and intra-individual variations were higher in the morning than later in the day for all shear rates. In addition to blood cell content, the types and number of lipoproteins also influence WBV [8].

Over past decades, different types of whole blood viscosity models based on rheological data have been proposed, where power-law-based models were suggested due to the resemblance between blood and other shear-thinning fluids. A recent review of most existing models of whole blood viscosity was presented by Hund et al. [9]. All models include several parameters determined by fitting the models to measurements. For example, Marcinkowska-Gapinska et al. [10] measured the viscosity of 100 whole blood and plasma samples over a range of shear rates (between 0.01 and 100 s^{-1}). Applying the measured data to the models of Casson, Ree–Eyring and Quemada, the Quemada model was found to provide the best fit. Gallagher et al. [11] considered models of Bird, Carreau [12], Cross and Yasuda. The aim of the study was to address the problem of inferring model parameters by fitting them to experiments. By refitting published data, families of parameter sets capturing the data equally well were identified. The fitted parameters yielded almost indistinguishable fits to experimental data, but the different parameter sets predicted very different flow profiles. This finding shows the difficulties in representing the fluid physical properties well. To assess the sensitivity of the models, random perturbations were added to the measured data from which new model parameters were derived, showing that the problem was inherent to the models considered. The effects of the Casson and Carreau–Yasuda models on a steady and oscillatory 2D flow in a straight and curved pipe geometry were studied by Boyd et al. [13] using the Lattice Boltzmann method. Differences in velocity and shear were found at low Reynolds and Womersley numbers, although were rather small in terms of velocity profiles. More recently, attempts have been made to improve the low shear stress behavior of rheological models. Jedrzejczak et al. [14] proposed a population balance-based model of blood including hemolysis. The proposed model was compared to the characteristic viscosity of the Carreau Yasuda model for viscosimetric conditions. The model predicted a smoother transition between high and low viscosity zones in a constricted pipe. The model was verified through results from ex vivo experiments. Arzani [15] proposed a hybrid Newtonian and non-Newtonian rheology model by switching to a traditional Carreau–Yasuda model when the residence time exceeded a threshold value. Lagrangian particle tracking was used to detect stagnant regions with increased rouleaux formation likelihood.

Blood flow simulations of clinical interest require considering blood flow in patient-relevant settings. Several studies have assessed and compared different rheological models, providing results displaying discrepancies among the conclusions. Akherat et al. [16] simulated blood flow in arteriovenous (AV) fistula using reconstructed 3D geometries and the models of Quemada and Casson. The results displayed no major differences in the flow field and the flow characteristics. Instead, the shape of the geometry was found to be far more important for the WSS distribution as compared to the effect due to the rheological model. Investigating the flow in a patient-specific aorta model, Karimi et al. [17] applied nine rheological models (three Casson model variants, Carreau, Carreau–Yasuda, Cross, Power-law, Modified Power-law, and Generalized Power-law parameters), focusing on WSS and the deviation of the computed viscosity as compared to a reference viscosity (3.45×10^{-3} Pa s). The largest differences in WSS were located near the branches and found more pronounced at a low flow rate (diastole) where the viscosity was considerably larger than the value used in the Newtonian computations. Karimi et al. [17] concluded that various rheological models may yield equally good results apart from the Cross model. A similar conclusion was presented by Johnston et al. [18], who numerically investigated the effects of six rheological models (including a Newtonian) on WSS during the cardiac cycle. The results were comparable for all models under steady flow conditions and at unsteady mid-range flow velocities (around 0.2 m/s). It was reported that a Newtonian model was a good approximation in regions of mid-range to high shear, whereas for low shear rates, the Power-law model was found more appropriate. However, in the numerical study of a pipe flow by Jahangiri et al. [19], the rheological models of Carreau, Carreau–Yasuda, modified Casson, Power-law, Generalized Power-Law and Walburn–Schneck, except the Generalized Power-law model, resulted in graphically the same axial velocity profile and WSS behavior. In contrast, Apostolidis et al. [20], considering the flow in a coronary artery setup, found as much as 50% change in WSS (instantaneous and local) when considering the effect of the Casson model. Mendieta et al. [21] studied the importance of blood rheology in patient-specific simulations of stenotic carotid arteries. Four rheological models (Newtonian and four non-Newtonian models (Carreau, Cross, Quemada and Power-law) were considered. Averaged quantities related to WSS descriptors (such as those used here, Section 3 of the results) were compared. The main conclusion was that the assumption of a Newtonian model is reasonable in terms of the overall flow pattern or the mean values, but a non-Newtonian model is necessary when the low TAWSS region and strongly stagnant flow regions are in focus. Further patient-specific-related geometries were studied by Liepsch et al. [22] to assess the impact of non-Newtonian viscosity models on the hemodynamics of a cerebral aneurysm. The flow in the aneurysm was computed using the Newtonian, Power-law, Bird–Carreau, Casson and Local viscosity models. Although similar flow patterns were observed both for Newtonian and non-Newtonian models, a quantitative comparison performed on a group of monitoring points revealed an average difference between the Newtonian and non-Newtonian models of about 12%. Furthermore, in low-speed regions, the differences were even larger (20% to 63%). Skiadopoulos et al. [23] considered Newtonian, Quemada and Casson blood viscosity models for simulating pulsatile flow in patient-specific geometry of the iliac bifurcation. The effect of the rheological models was monitored through the WSS distribution, magnitude and oscillations and viscosity behavior as a function of the shear rate. In addition to the commonly used WSS-related indicators (OSI and TAWSS), the study considered the (wall) area averaged WSS and the corresponding area averaged shear rate. The magnitude of the WSS and its oscillations were found to depend on the shear rate and the rheological model. The main conclusion of Skiadopoulos et al. [23] was that the Newtonian approximation is mostly applicable for high shear and flow rates. The Newtonian model was found to overestimate the possibility of the formation of atherosclerotic lesions in regions with oscillatory WSS. In a recent paper, Mendieta et al. [21] compared the effect of rheological models on the WSS-related parameters in stenotic carotid flows. Four rheological models were considered (Carreau, Cross, Quemada and Power-law). The largest differences between the Newtonian and non-Newtonian models were noted for an OSI of about 12% (in terms of maximum and mean values). Regarding the TAWSS, the difference was less than 6%, except for the Quemada model where the difference was as much as 26%. The authors concluded that the assumption of a Newtonian model can be reasonable; however, non-Newtonian models were found necessary in low TAWSS regions. Mirza and Ramaswamy [8] observed that the WSS using Newtonian and non-Newtonian models differ noticeably.

Most of the above cases considered laminar and/or transitional flow regimes. In contrast, Molla and Paul [24] studied a turbulent flow in a channel with a constriction, computed by Large Eddy Simulations. Five rheological models (Power-law, Carreau, Quemada, Cross and a modified Casson) were compared in terms of peak shear rate, mean shear stress and pressure, re-circulation zones and turbulent kinetic energy. The main finding was that the non-Newtonian viscosity models extended the post-stenotic re-circulation region and reduced the turbulent kinetic energy downstream of the stenosis. In the simulations, the shear rate was limited to lower than 100 s^{-1}, i.e., in the range where the viscosity differs strongly from the high shear rate range which in turn implies higher levels of viscosity and a lower level of turbulent kinetic energy (TKE). In terms of TKE, the differences between the rheological models were rather modest.

Rheological blood models are calibrated to in vitro measured data. However, when applied to patient-specific simulations, the main goal of the models is to be able to account for important fluid physics and capture relevant clinical observations. Despite the consensus that the importance of non-Newtonian rheological effects is more pronounced for low shear rates, the conclusions with respect to the importance of modeling the non-Newtonian effect may differ depending on the application. Hence, the aim of the current study was to provide some general principles for when to employ Newtonian or non-Newtonian rheological models for blood flow simulation of clinical interest. We show how several of the commonly used parameters to quantify blood flow characteristics depend on the chosen rheological model and the origin of these dependencies.

2. Materials and Methods

The geometry of the thoracic aorta was derived from a computed tomography angiography (CTA) study of a healthy patient (Figure 1a). The computational domain consisted of the ascending aorta, the aortic arch, the descending aorta and the three main branching arteries: the brachiocephalic (BC) artery, the left common carotid artery (LC) and the left subclavian artery (LS)). The blood was assumed to be incompressible with constant bulk density and fixed concentration of red blood cells (RBC). The effects of other blood components were neglected. Thus, the blood was modeled as a non-homogenous mixture satisfying conservation of mass (Equation (1a)) and momentum (Equation (1b)).

$$\frac{\partial \rho u_i}{\partial t} + \frac{\partial (\rho u_i u_j)}{\partial x_j} = -\frac{\partial p}{\partial x_i} + \frac{\partial}{\partial x_j} \mu \frac{\partial u_i}{\partial x_j} \tag{1a}$$

$$\frac{\partial \rho}{\partial t} + \frac{\partial \rho u_i}{\partial x_i} = 0 \tag{1b}$$

where and μ are the density and viscosity of the mixture, respectively, p is the pressure and u_i is the Cartesian velocity component in the i:th direction. The bulk viscosity depends on the local shear rate, accounted for by applying the following rheological models: Newtonian fluid (viscosity, $\mu = 3.35$ mPa s and density $\rho = 1102$ kg/m^3) and three non-Newtonian models of Walburn and Schneck [25], Casson [26] and Quemada [27,28] to account for the local mixture viscosity, $\mu = \mu_{eff}$. The hematocrit value was set to a fixed value of $\alpha = 0.45$.

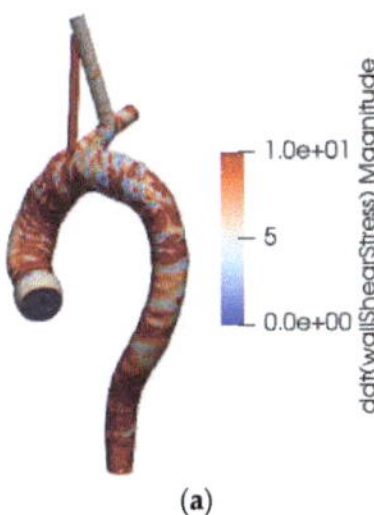
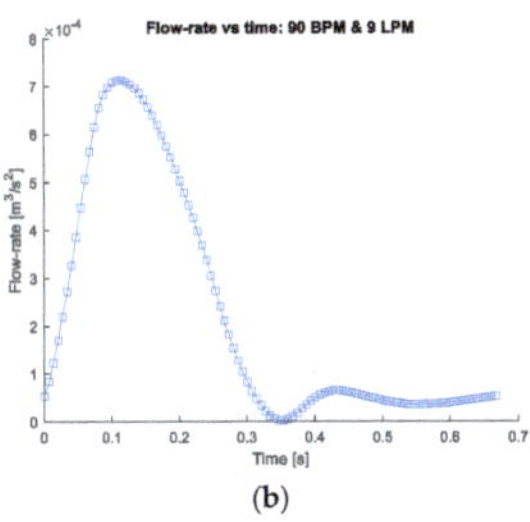

(a)   (b)

Figure 1. (**a**) The computational geometry of the thoracic aorta. The figures depict the absolute values of time derivative of the wall shear stress (WSS) at the end of systole (time = 0.333 on the right frame). (**b**) The Cardiac output (flow rate) as function of the cardiac cycle. The cardiac output corresponds to 9 L/min (LPM) and heart-rate of 90 beats/min (BPM).

2.1. Mixture Viscosity Models

The simplest non-Newtonian model, used in multiple publications and available in several common pieces of software, is the Power-law model of Ostwald de Waele [6]. In its simplest form, the model directly relates effective viscosity (m_{eff}) to the shear rate $\mu_{eff} = k\gamma^{n-1}$, where γ is the magnitude of the shear rate tensor $\gamma_{ij} = \frac{1}{2}\left(\frac{\partial u_i}{\partial x_j} + \frac{\partial u_j}{\partial x_i}\right)$. The power of the shear rate is negative for shear-thinning fluids ($n < 1$). The three considered non-Newtonian models were as follows:

- The Walburn–Schneck [25] model for the effective viscosity (μ_{eff}) of the blood (mixture), which is an extension of the Power-law model. In the current work, using $\alpha = 0.45$, the effective viscosity is modeled by:

$$\mu_{eff} = max\left(0.0034, \gamma^{-0.0225}\right) \tag{2}$$

- The Casson [26] model, based on a calibrated Power-law concept. The following form and parameters were used:

$$\mu_{eff} = min\left(\mu_{max}, \frac{\left[\sqrt{k_C(\alpha)\gamma} + \sqrt{\tau_y(\alpha)}\right]^2}{\gamma}\right) \tag{3}$$

The model parameters were derived for a hematocrit of $\alpha = 0.45$ (Cokelet et al. [29]; Perktold et al. [30]).

- The Quemada model [27,28]. The model may include two variables: the hematocrit (α) and the shear rate (γ). Here, a constant $\alpha = 0.45$ was used leading to a simpler formulation:

$$\mu_{eff} = \mu_p(1 - 0.225k(\gamma))^{-2} \tag{4}$$

with plasma viscosity $\mu_p = 1.32$ mPa s. k is an expression of γ (further details are found in Fuchs et al. [31].

2.2. Boundary Conditions

Regarding boundary conditions, no-slip conditions were set on the (rigid) walls of the thoracic aorta (Figure 1a). The outlet sections were extended to avoid the effects of the outlet conditions on the domain of interest. The exit planes were the brachiocephalic artery (BC), left common carotid artery (LC), left subclavian artery (LC) and the exit plane of the thoracic aorta (EXT). At the inlet, a time-dependent flow rate profile derived from a measured human cardiac profile was imposed (Figure 1b), representing 90 heartbeats per minute (BPM) and 9 L per minute (LPM). This inflow condition was chosen since the flow profile induces temporal and spatial gradients. The axial inlet velocity vector was uniformly distributed across the inlet plane. The other components of the velocity vector were set to 0. The main branches were extended so that no flow recirculation occurred at any time during the cardiac cycle. The flow rate in the main branches, BC, LC and LS was set to 15%, 7.5% and 7.5%, respectively (Benim et al. [32]). The flow out from the thoracic aorta (EXT plane) was set to 70%. The computational domain was discretized by about 5 million computational cells and found to yield adequately accurate results.

The governing equation was discretized using a formally second-order finite-volume scheme. The discrete equations were advanced in time using an implicit solver (OpenFoam 5.0). The results were processed using MATLAB, Paraview, VTK and our own python scripts.

2.3. Viscosity Models and WSS Sensitivity Indicators
2.3.1. Monitoring Effects of Non-Newtonian Viscosity

The different non-Newtonian models lead to blood mixture viscosity coefficients varying in space and time. To quantify the variation of the viscosity, the following indices were used (Johnston et al. [18]): the relative viscosity, I_L, and the non-Newtonian importance factors, $I_{g\text{-}space}$ and $I_{g\text{-}time}$ capturing the space and time effects, respectively.

$$I_L = \frac{\mu}{\mu_{ref}} \tag{5}$$

$$I_{g-space}(t) = \frac{1}{N} \frac{\left[\sum_{i=1}^{N} \mu(x_i,t) - \mu_{ref})^2\right]^{1/2}}{\mu_{ref}} \; ; \; I_{g-time}(x) = \frac{1}{M} \frac{\left[\sum_{j=1}^{M} \mu(x,t_j) - \mu_{ref})^2\right]^{1/2}}{\mu_{ref}} \tag{6}$$

where m_{ref} is the reference viscosity (μ_{ref} = 3.5 mPa s, used for the Newtonian case).

2.3.2. WSS-related Indicators

The WSS plays a major role in the processes in the arterial wall and depends directly on the near-wall viscosity. To assess the impact of the rheological models, different WSS indicators were used.

- WSS(x,t) and TD_WSS(x,t)

$$\text{WSS}_i(x,t) = \left|n_j \tau_{ij}\right| \; \tau_{ij} = \mu \gamma_{ij} \; \text{TD\_WSS}(x,t) = \left|\frac{\partial \text{WSS}}{\partial t}\right| \tag{7}$$

with n_j being the wall j-th component of the wall normal vector.

- Time average-based expression of WSS: TAWSS, OSI and RRT.

Time-averaged Wall Shear Stress (TAWSS) cf. Suo et al. [33] and Chen et al. [34], a local time-averaged WSS:

$$\text{TAWSS} = \frac{1}{T} \int_0^T |\text{WSS}_i| dt \tag{8}$$

Spatial variation of WSS provides an indication of the WSS level and its spatial non-uniformity (i.e., WSS gradient), but does not contain any information about WSS temporal variation.

Oscillatory Shear Index (OSI) (cf He et al. [35], Chen et al. [34]), a measure for temporal sign change in WSS vector. The OSI is defined as

$$\text{OSI} = \frac{1}{2}\left(1 - \frac{\left|\int_0^T \text{WSS}_i dt\right|}{\int_0^T |\text{WSS}_i| dt}\right) \tag{9}$$

Hence, the OSI varies between 0 and 0.5. When WSS_i has a constant sign (possibly oscillatory but maintaining the same sign of the WSS vector), OSI = 0. When WSS_i changes signs such that the integral of the positive and negative sequences are equal, the OSI gets the value 0.5. Values of 0 < OSI < 0.5 indicate a sign oscillatory WSS.

The Relative Residence Time (RRT) has been formulated (cf Rikhtegar et al. [36], Gallo et al. [37]) as follows:

$$\text{RRT} = \frac{1}{(1 - 2 \cdot \text{OSI}) \cdot \frac{1}{T}\int_0^T \frac{|\text{WSS}_i|}{\mu}dt} \tag{10}$$

3. Results

To elucidate the differences and similarities of the flow and shear stress characteristics due to the different rheological models, the results are presented in terms of the following parameters within the lumen or near the aortic wall:

- The relative size of the viscosity coefficient, I_L, and the so-called non-Newtonian importance factor, I_g (space and time).
- The impact of the rheological model on
  - The WSS and its time derivative (TD_WSS)
  - WSS-related indicators: OSI, TAWSS and RRT. The results are presented in the form of spatial and/or temporal distribution of the different parameters
  - The extent of retrograde flow particle transport in the thoracic aorta, expressed in terms of residence times.
- The extent of shear stress magnitude below 100.

3.1. Viscosity Coefficient

As the hematocrit is kept constant, the considered rheological models depend only on the shear rate. Figure 2a depicts the mixture viscosity of the four rheological models. Behaving as Newtonian fluids at large shear rates and differing mainly at low shear rates, differences between the rheological models are expected for local shear rates below the order of $100 \, \text{s}^{-1}$. A direct comparison of the contributions of the different models is displayed in terms of I_L and I_g, Equation (5), and shown in Figure 2a,b. The space averaging of the normalized viscosity (I_L) and the non-Newtonian importance factor enables the assessment of the temporal variations of the rheological models during the cardiac cycle. Obviously, for the Newtonian model $I_L = 1$ and $I_g = 0$ and not shown. The Walburn–Schneck model yields the largest values as compared to the other two models. The Casson model presents a mean increase in viscosity only by a factor less than 1.4 whereas the corresponding value for the Quemada model is about 1.8. All three models have the largest viscosity in early systole, decreasing during systole and increasing at late diastole. This behavior reflects the shear-thinning property. In systole and early diastole, the shear rates are largest, leading to lower viscosity whereas, with a lower shear rate, the viscosity increases. The importance factor (I_g) also shows that the deviation from the Newtonian reference value is smallest for the Casson model and largest for Walburn–Schneck. The local minima are found around 0.1 s and 0.45 s, Figure 2c, and correspond to the instantaneous peaks of flow rate and largest shear (Figure 1b). The local peak at about 0.2 s is linked to the strongest flow rate deceleration inducing retrograde flow and temporarily decreasing shear rates in parts of the domain. A similar effect is observed in the late diastolic phase.

An insight into the behavior of the rheological models for the aortic flow under consideration is obtained by considering the probability distribution (PDF) of I_L and $I_{g\text{-space}}$. The PDFs for the rheological models of Quemada, Casson and Walburn–Schneck are depicted in Figure 3. The peak probability of the relative time-averaged viscosity is about 1.6, 1.25 and 1.7 for the Quemada, Casson and Walburn–Schneck models, respectively. The corresponding highest probability for $I_{g\text{-space}}$ is at about 0.5, 0.2 and 0.5, respectively. I_L values close to unity and correspondingly $I_{g\text{-space}}$ close to 0 imply values close to the reference (Newtonian) viscosity. The peaks shown in Figure 3a indicate that during the cardiac cycle, there are regions where the viscosity is lower than that of the reference value due to the Quemada model. At low shear rates, the models yield considerably larger viscosity values as compared to the reference viscosity. This leads to $I_L > 1$ and larger $I_{g\text{-space}}$ values.

The largest viscosity values, although associated with low probability, are noted for the Walburn–Schneck and Quemada models. The I_L distributions show that the Casson model has a smaller "signature" relative to the reference value, though the level of fluctuation is relatively large. The Walburn–Schneck distribution indicates that the flow under consideration has a relatively large volume (number of computational cells) with low shear. This effect is less observable in parameters based on time- and space-averaged parameters.

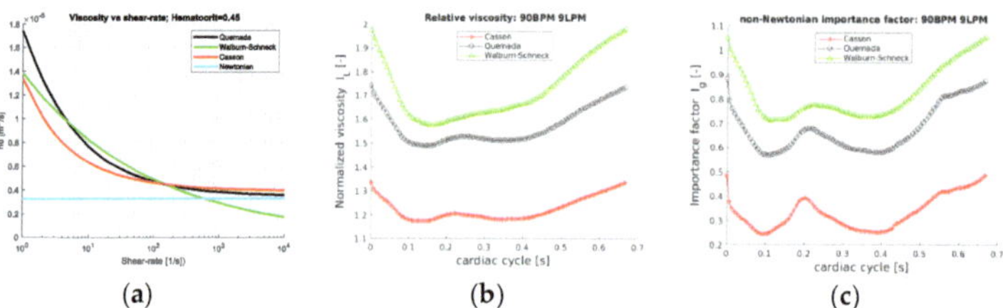

Figure 2. (**a**) Mixture viscosity versus shear rate for the three non-Newtonian models (Equations (2)–(4)), for a = 0.45. (**b**) The space-averaged normalized viscosity $I_L(t)$ and (**c**) the importance factor ($I_{g\text{-}space}(t)$), plotted over the cardiac cycle for 90 BPM and 9 LPM.

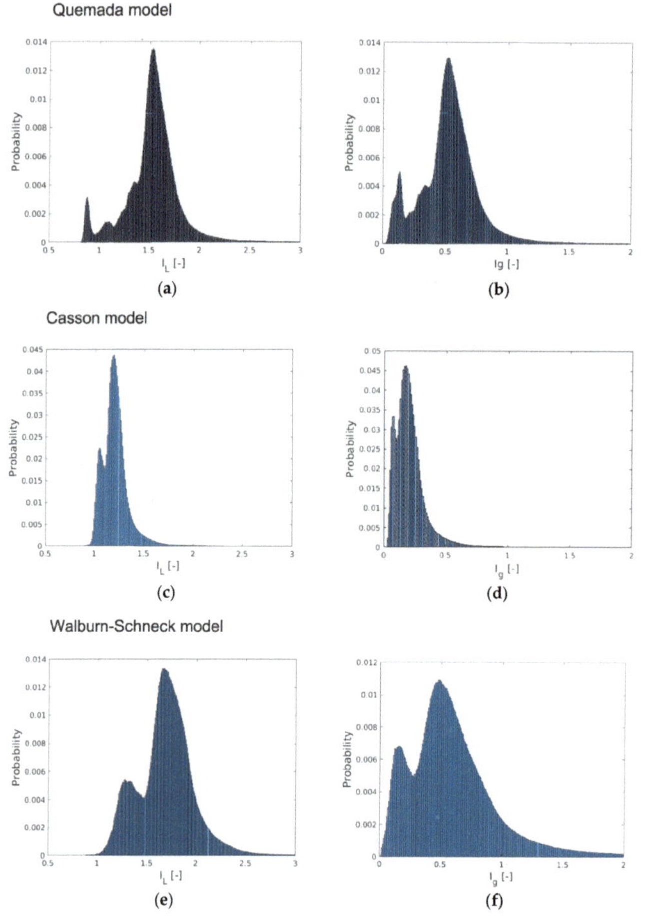

Figure 3. The probability distribution of the relative viscosity (I_L) and importance factor ($I_{g\text{-}time}$) computed by (**a,b**) the Quemada model, (**c,d**) Casson and (**e,f**) Walburn–Schneck.

A quantitative but less detailed comparison of the two viscosity parameters I_L and I_g, is given in Table 1. The table shows the mean, standard deviation (std), peak and minimal values of these parameters. The largest peak viscosity value was found with the Walburn–Schneck model. The Casson model shows the smallest mean values but a relatively large peak value.

Table 1. Time and space statistical values for I_L and I_g, with the three non-Newtonian models.

Model	Quemada		Casson		Walburn–Schneck	
	I_L	I_g	I_L	I_g	I_L	I_g
Mean	1.536	0.538	1.215	0.215	1.690	0.631
Std	0.288	0.266	0.166	0.158	0.290	0.535
Max	5.755	4.745	6.002	4.991	4.480	14.352
Min	0.817	0.011	0.906	0.017	0.876	0.009

3.2. Model Impact on WSS-Related Quantities

Atherosclerosis is an arterial wall process. Viscous effects are important near the walls and the wall shear stress (WSS) is known to have a significant role in the formation of wall pathologies. The time-averaged probability distributions of the WSS and its temporal derivative (Equation (7)) show similar probability distributions with the different rheological models. Some small differences in terms of the peaks are noted in the WSS (about 2 Pa) and its time derivative (about 20 Pa/s). A common observation is that most of the aortic wall is subjected to a low WSS and only a smaller portion of the wall is subjected to a large and oscillatory WSS. The statistics of the WSS and its time derivative are provided in Table 2. Comparing the models, the mean values of WSS, about 2 Pa, and spatial WSS fluctuation (noted in terms of root mean square (RMS)) are similar. The time derivative of the WSS shows larger differences, where the Walburn–Schneck model displays the largest mean and RMS values. The near-wall behavior of the model is a possible explanation of the larger dissipation generated by this model during systole. The peak of the time derivative of the WSS is smallest for the Casson and the Quemada models, also reflected in Table 2.

Table 2. The mean and RMS of spatial and temporal statistics of WSS and its time derivative.

Model	| WSS | (Pa)	$\left\| \frac{\partial \text{WSS}}{\partial t} \right\|$ (Pa/s)
Newtonian	2.122 ± 3.294	36.349 ± 115.49
Quemada	2.044 ± 2.678	32.251 ± 90.47
Casson	2.045 ± 2.994	35.058 ± 112.17
Walburn–Schneck	2.049 ± 3.145	46.360 ± 171.82

The instantaneous WSS in the thoracic aorta is depicted in Figure 1a. The TAWSS (Equation (8)), OSI (Equation (9)) and RRT (Equation (10)) have been associated with the formation of atherosclerosis. Implicitly, these markers (parameters) include blood viscosity and are thus affected by the rheological models. The results show strong similarity in the markers for all four rheological models. The OSI has mostly values below about 0.1 apart from some well-localized regions. Larger OSI values (close to 0.5, indicating sign oscillations of the WSS vector) are seen in the inner wall of the ascending aorta, in the aortic sinus and near the branches of the aortic arch. The RRT indicates analogous behavior with the largest values at the same locations as the OSI. The TAWSS displays stronger values at the junction between the aortic sinus and the ascending aorta and proximal parts of the brachiocephalic artery. In terms of the averaged parameters, the differences between the models are rather small and require close examination of the non-averaged data for making a qualitative assessment.

Visualizations of wall indicators have commonly been adopted in the literature. Here, a different way of assessing the differences between the models is proposed, considering the probability distribution of the OSI, RRT and TAWSS. The OSI probability is largest for a low OSI, where the distribution of probability clearly depends on the rheological model. Lower OSI values indicate the change of sign in the WSS. The Newtonian and the Walburn–Schneck models yield a larger probability for an OSI < 0.1 as compared to the other two models. Thus, the Newtonian and Walburn–Schneck models have larger rates of sign change in the resulting WSS. For a larger OSI (i.e., >0.3), the probability level is similar for the different models, an effect occurring in surface regions without retrograde flow. The TAWSS and the RRT probabilities are insensitive to the viscosity model used, both in terms of values and distribution.

Table 3 summarizes the quantitative statistical data for the three WSS-based markers. The mean and RMS values of the OSI, TAWSS and RRT are close to each other for the different models. It is evident that the specific rheological models are less important for these markers. As the markers are averaged in space and time, the markers may not correctly reflect the impact of the WSS on the arterial processes. Therefore, more sensitive markers must be developed to explain and monitor the formation and progress of arterial wall remodeling.

Table 3. The space-time averages of three WSS indicators demonstrate the marginal effects of the rheological models on these parameters.

Model	OSI (-)	TAWSS (Pa)	RRT (s)
Newtonian	0.112 ± 0.100	3.146 ± 4.889	0.0027 ± 0.0049
Quemada	0.105 ± 0.100	3.023 ± 3.975	0.0020 ± 0.0033
Casson	0.113 ± 0.100	3.031 ± 4.443	0.0026 ± 0.0049
Walburn–Schneck	0.112 ± 0.100	3.032 ± 4.667	0.0026 ± 0.0046

3.3. Impact of Viscosity on Retrograde Flow

A less common measure of the impact of blood rheology on the flow is using the volume of retrograde flow in the thoracic aorta. Two such measures can be defined: (1) based on the direction of the axial velocity compared to the direction of the aortic centerline and (2) based on the (negative) direction of the wall shear stress (i.e., wall parallel component) relative to the centerline direction. The volume and negative WSS of retrograde flow are normalized by the total aortic volume and aortic surface, respectively. The average and maximal values of these quantities are given in Table 4. The time- and space-averaged values show that the amount of retrograde flow is relatively large, about 10% (mean) and about 35% (peak). However, comparing the different models, the values are similar, only differing by about 2% for the averaged and less than 1% in terms of peak relative retrograde volume. The corresponding values for the WSS < 0 are about 7% and 2%, respectively.

Table 4. The time- and space-averaged retrograde flow volume and the corresponding negative WSS.

Model	Relative Retrograde Flow Volume		Relative WSS < 0 Area	
	Average	Peak	Average	Peak
Newtonian	0.10240	0.36405	0.21834	0.82772
Quemada	0.10476	0.36308	0.20845	0.82053
Casson	0.10487	0.36605	0.22388	0.83970
Walburn–Schneck	0.10445	0.36578	0.22001	0.83248

Figure 4 depicts the temporal variation of the relative retrograde flow volume and the corresponding relative negative WSS. The peak values of retrograde flow, with both measures, occur at the end of systole (about 0.35 s). The second peak occurs at the flow maximum during diastole (about 0.55 s). The reason for this effect is the same as retrograde flow in cyclic pipe flow (Womersley effect) and is explained in more detail in Fuchs et al. [31,38]. As seen, the differences between the four rheological models are rather limited. One may also note that most differences are found in late diastole.

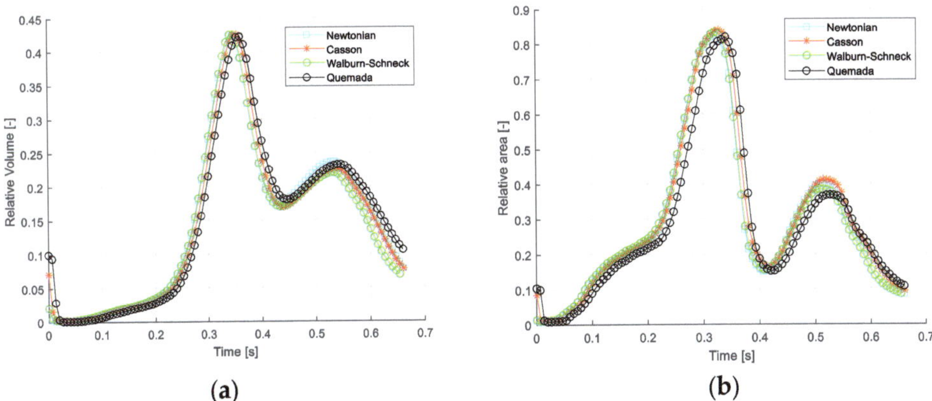

Figure 4. Relative retrograde flow in the thoracic aorta vs. cardiac cycle: (**a**) relative volumetric flow and (**b**) negative WSS area.

3.4. Shear Rate (SR) and Rheology Model Extent during Cardiac Cycle

The extent of low shear rate (<100 s^{-1}) within the thoracic aorta as a function of the cardiac cycle is required to enable the understanding of the consequences of the rheological modeling on, for example, particle transport (discussed in the next section). Figure 5 depicts the probability distributions (counts) for the range of shear rates (SR, 0 s^{-1} to 1000 s^{-1}) and the kinematic viscosity (2×10^{-6} to 1×10^{-5} m^2/s) for the four rheological models and three time instances in the cardiac cycles: T = 0.12 s (peak systole), 0.33 s (end-systole) and 0.5 s (mid-diastole). Maximal counts in all four cases are observed at low shear rates (below 100 s^{-1}). For the three rheological models, it is noted that the width of the distribution is wider for a SR above 200 s^{-1} to 1000 s^{-1}. At end-systole, the distribution SRs for the three rheological models show the presence of very low values of SR (about 15 s^{-1}) and a second peak at about 65 s^{-1}. At mid-diastole, the three rheological models have a narrower distribution showing larger counts at a low SR. In contrast, in the distributions of the Newtonian case, the SR distribution is narrow at systole and wide in mid-diastole. The wider distribution of the SR at T = 0.12 s for the three models leads to a peak in kinematic viscosity near 4×10^{-6} m^2/s due to the presence of a low SR. The second peak and the tail in the kinematic viscosity (at T = 0.12 s) is due to the low SR values (peaks at around 35 s^{-1}) for the three rheological models. The main differences between the three models are observed in the distribution widths of the kinematic viscosity: they are widest (3×10^{-6} to 8×10^{-6} m^2/s) for the Walburn–Schneck during diastole (T > 0.33 s), whereas the Casson model has a narrower distribution (3×10^{-6} to 5×10^{-6} m^2/s) and the Quemada model is in between the two others.

T = 0.12s   T = 0.33s   T = 0.50s

Newtonian Model

Quemada Model

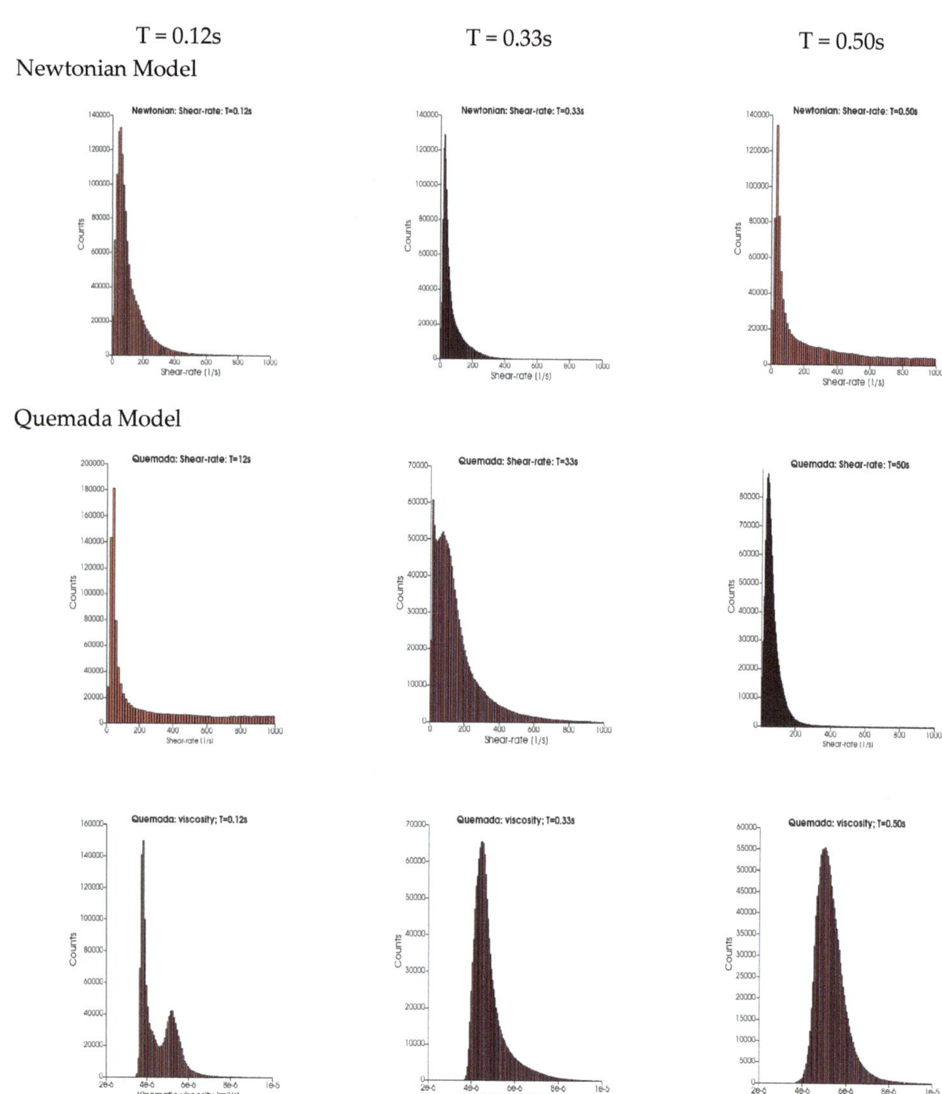

Figure 5. *Cont.*

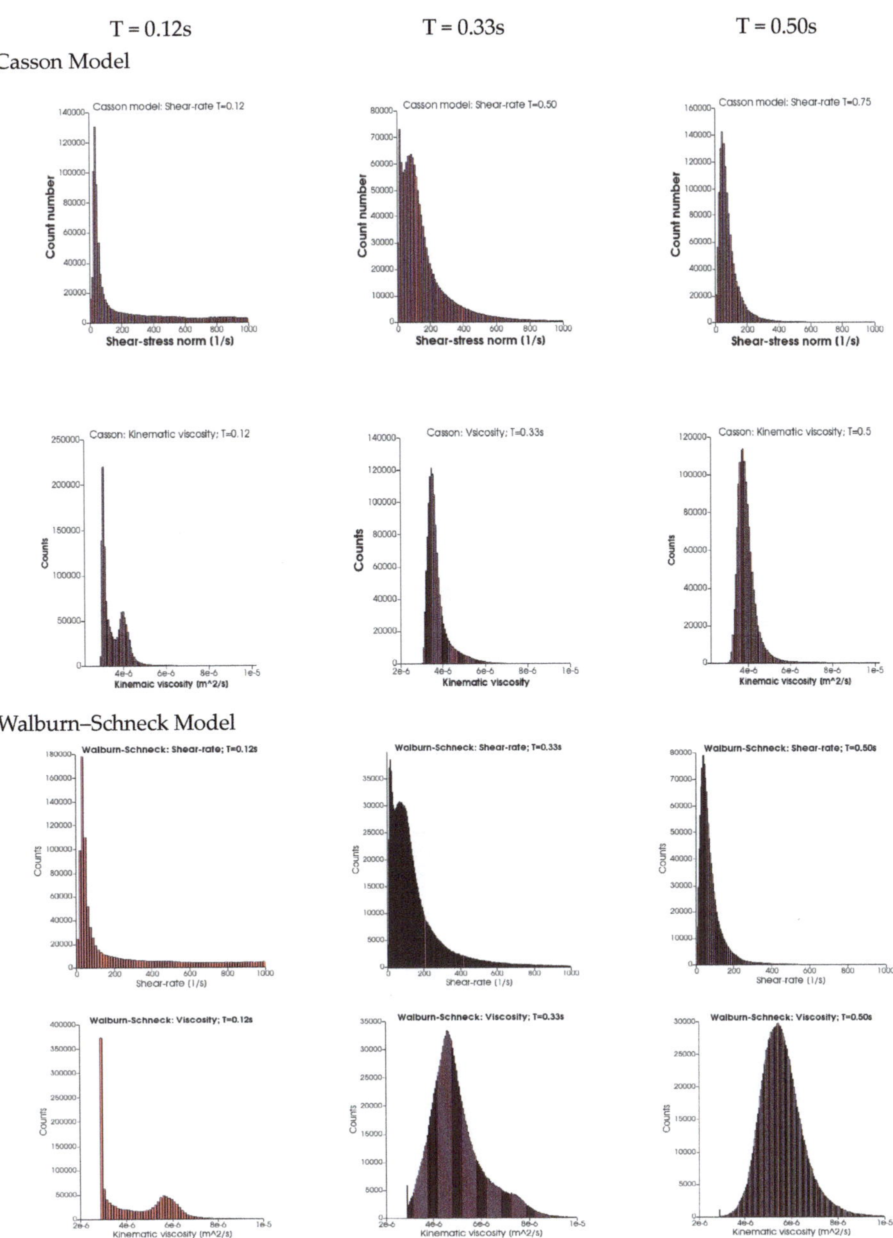

Figure 5. The distribution of shear rate (SR) and kinematic viscosity for the different rheological models at different instants in the cardiac cycle: T = 0.12 s (peak systole), end-systole (T = 0.33 s) and mid-diastole (T = 0.50 s).

3.5. Impact of Viscosity Model on WSS and Its Temporal Derivative

The impact of blood rheology on the WSS and its temporal derivative is determinantal for the development of vascular disease. Thus, the surface average of the WSS and its temporal derivative as a function of the cardiac cycle was studied to shed light on the dynamics of these quantities. Figure 6 depicts the instantaneous and surface mean values of the WSS and its temporal derivatives during the cardiac cycle for the four different rheological models. Note that the geometry in Figure 6 excludes the branching arteries from the aortic arch in contrast to the averages in Tables 2 and 3. The instantaneous WSS with the four rheological models is shown in Figure 6a (near peak systole). Although the figures are qualitatively similar, a clear local variation in terms of peak values is shown. These variations reduce significantly when taking the average over the whole aorta wall, an effect clearly demonstrated in Tables 2 and 3. The temporal development of the surface averaged WSS during the cardiac cycle is depicted in frames Figure 6b. The shapes of the curves are similar though the amplitudes differ: the Quemada model has a peak larger than 1.6 Pa compared to about 1.4 for the Casson and Walburn–Schneck models. The instantaneous and surface averages of the temporal derivatives of WSS are depicted in Figure 6c,d respectively. The rheological model has a clear qualitative impact in the ascending aorta and near the branching of the left subclavian artery. The surface-averaged behavior (Figure 6d) shows non-monotone behavior in systole. There is a dip during the initial acceleration in the systole, followed by a quick increase and oscillations as the flow rate is reduced after peak-systole. The initial dip reflects the reduction in viscosity following the initial increase in the shear rate. As the flow increases further, the near-wall shear layer becomes sharper, leading to an increase in the WSS. A more detailed study of the shape of the shear layer shows that after the peak in the systole, multiple inflection points in the axial velocity profiles may be observed (not shown here). Some of these inflection points may lead to instability with exponential growth which is believed to be related to the observed oscillation in the temporal derivative of the WSS. The dynamics of the flow allows a short time for any growth in the oscillations, which are dissipated during diastole. The oscillations in Figure 6d contain different frequencies: higher frequencies for the Newtonian and WS cases and lower frequencies (smoother behavior) for the Quemada case. The observed oscillations in the averages weakly reflect the considerable variation in the spatial distribution of the WSS and its temporal derivatives.

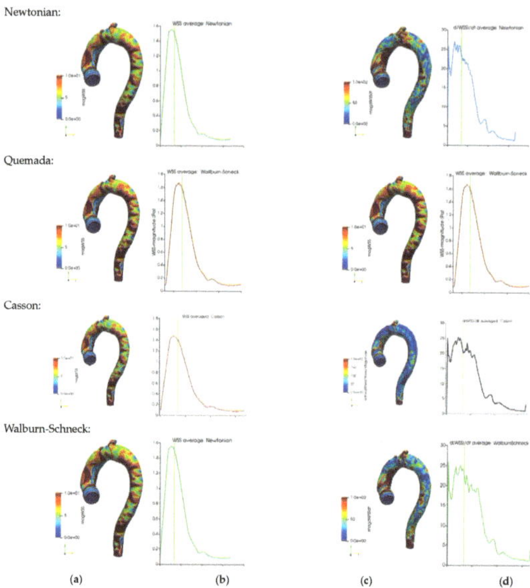

Figure 6. Wall shear stress (Pa) and its temporal derivative (Pa s^{-1}) for the four rheological models: instantaneous WSS distribution at near peak systole (frames (**a,c**)). Surface average of the time-derivative of WSS vs. the cardiac cycle (frames (**b,d**)). The vertical green lines in frames b and d mark the time instant of the WSS in frames b.

3.6. Impact of Viscosity Model on Particle Transport

To assess the accumulated effects of the local viscosity, the impact of blood rheology on particle transport in the thoracic aorta was considered. The residence time of macromolecules, cells and thrombus/emboli plays a central role in certain pathological processes (such as thrombus, formation, growth and transport). In each time step, two sets of spherical particles (2 mm and 2 mm, both with a density of 1200 kg/m^3) were randomly injected at the inlet (aortic valve) plane. Solid, spherical particles were tracked by integrating Newton's second law in time, assuming models for the forces acting instantaneously on each particle. Hence, the particle paths reflect the accumulated effects of forces acting on them. It is assumed that the particles are subject to drag, (Saffman's) lift, added mass and pressure forces. The local fluid viscosity enters in the former three force terms, thereby directly affecting the particle histories. The injected particles were tracked over 10 cardiac cycles. The total number of particles was between 400,000 and 3,000,000. The particle motion and history were characterized by (i) the particle residence time ("age") in the flow field and (ii) the shortest particle exit times.

Table 5 shows the "age" of particles leaving the four exits from the thoracic aorta after ten cardiac cycles. Particle ages correspond to the time the particle spent in the thoracic aorta since the instance of particle injection. The mean and standard deviation (std) values for both particle sizes are close to each other for each exit and the different rheological models. Similarly, the averages of particle ages (mean values in the table) are similar for each particle group. The differences between the mean values for the two particle groups are also relatively small. The differences in mean values are smaller the longer the particle paths are, measured from the inlet to the exit planes (Tables 6–8). However, the rather large std values indicate a significant spread in the data. The shortest particle residence time (denoted by "min" in the tables) is about one cardiac cycle before particles reach the exit plane of the thoracic aorta (EXT). The effect of the rheological models in the shortest residence time is small for each of the exit planes except for the BC exit. The shortest residence time for the 2 mm particles with the Newtonian model is twice as long as for the Quemada and Casson models. This difference is not observed for the larger particles. It may also be noted that for 2 mm particles the mean values are similar and have a relatively large variance. For the LC exit and the larger particle size, the Walburn–Schneck yields a significantly (about 1.5 times) longer minimal residence time as compared to the other rheological models. The same model also deviates (about 60%) from the other models for particles existing in the BC.

Table 5. The "age" of particles leaving at the distal exit of the thoracic aorta (entrance to the abdominal aorta, denoted as EXT). The age is given in seconds. As the heart rate was 90 BPM, the corresponding values in terms of cardiac cycle may be obtained by multiplying the figures in the table by a factor of 1.5.

Model	Particle Diameter = 2 mm (s)				Particle Diameter = 2 mm (s)			
	Mean	Std	Max	Min	Mean	STD	Max	Min
Newtonian	1.196	0.269	3.022	0.654	1.150	0.339	3.755	0.680
Quemada	1.211	0.277	2.637	0.660	1.157	0.348	3.929	0.667
Casson	1.209	0.277	2.845	0.663	1.143	0.334	3.933	0.673
Walburn–Schneck	1.197	0.270	3.115	0.654	1.138	0.327	3.873	0.679

Table 6. The "age" of particles leaving at the distal exit of the brachiocephalic artery (BC).

Model	Particle Diameter = 2 mm (s)				Particle Diameter = 2 mm (s)			
	Mean	Std	Max	Min	Mean	Std	Max	Min
Newtonian	0.916	0.268	2.724	0.205	0.895	0.308	3.696	0.325
Quemada	0.871	0.350	5.265	0.287	0.843	0.277	2.856	0.160
Casson	0.879	0.310	3.993	0.278	0.866	0.265	2.768	0.167
Walburn–Schneck	0.895	0.267	2.909	0.175	0.850	0.307	3.742	0.256

Table 7. The "age" of particles leaving at the distal exit of the left common carotid artery (LC).

Model	Particle Diameter = 2 mm (s)				Particle Diameter = 2 mm (s)			
	Mean	Std	Max	Min	Mean	Std	Max	Min
Newtonian	0.762	0.290	2.129	0.141	0.634	0.307	3.657	0.194
Quemada	0.706	0.276	2.134	0.147	0.641	0.324	5.292	0.182
Casson	0.751	0.290	2.397	0.137	0.639	0.287	3.562	0.189
Walburn–Schneck	0.739	0.271	2.055	0.147	0.628	0.294	3.373	0.188

Table 8. The "age" of particles leaving at the distal exit from the left subclavian arteries (LS).

Model	Particle Diameter = 2 mm (s)				Particle Diameter = 2 mm (s)			
	Mean	Std	Max	Min	Mean	Std	Max	Min
Newtonian	1.164	0.368	3.819	0.429	1.072	0.356	3.749	0.644
Quemada	1.149	0.344	3.884	0.240	1.103	0.409	5.970	0.635
Casson	1.132	0.308	3.604	0.472	1.071	0.331	3.771	0.637
Walburn–Schneck	1.134	0.312	3.175	0.612	1.069	0.329	3.905	0.640

Long residence times are of clinical significance as they may affect tendencies for thrombus formation or its transport. The effect of the rheological model on the maximal residence time is larger for the larger particles with the Quemada model and the BC exit. The same model yields the longest residence time for the LC and LS exits. Moving into the thoracic aorta, the particles are subject to varying viscosity, with the particle ages reflecting an averaging effect. This aspect is demonstrated by the results for the exit plane EXT: the particles stay in the domain for almost five and six cardiac cycles for the larger and smaller particles, respectively. The corresponding differences between the models are therefore less than 5% and 20%, respectively.

4. Discussion

Studies related to the importance and significance of different rheological models for blood flow simulations have been discussed extensively in the literature. The conclusions do not fully agree in terms of the importance of using non-Newtonian models versus using a constant viscosity value. Several authors have concluded that the flow was significantly affected by certain rheological models and raised a caution (Jahangiri et al. [19]). Other studies reported that the computed flow field general characteristics exhibit no major differences when using Newtonian and non-Newtonian models (cf. Akherat et al. [2]), while other publications find the Newtonian approximation reasonable for high shear flows (Skiadopoulos et al. [23]). The significance and importance of the models have been determined by the visual judgment of graphical data (e.g., surface values of the OSI, TAWSS and RRT). In this study, several different parameters were used to assess the impact of four rheological models to provide clear indications of in which *situations* the rheological approach is mostly influential. The forces acting on the endothelium through the WSS (and most importantly its space and time derivatives) are known to affect the development of arterial wall pathologies. Similarly, mass transport (such as platelets, VWF and lipoproteins), which represents the accumulated effect of the local rheology, plays a central role in the development of wall pathologies and in the risk of the formation and transport of thrombi/emboli.

The direct comparison of the relative viscosity and the RMS deviation from a reference (Newtonian case) value shows that the Walburn–Schneck model yields the largest viscosity values throughout the cardiac cycle, particularly during flow acceleration with the lowest values during flow deceleration. The latter effect is directly related to the formation of retrograde flow in a substantial part of the thoracic aorta. The importance factor (I_g) reaches the lowest values at local peaks in flow rate (i.e., peak systole and a local peak at early diastole). The probability distributions of the time-averaged I_L and I_g show clear differences between the non-Newtonian viscosity models. Moreover, the Casson model has the lowest increase in viscosity (near the peak with a smaller range) followed by the Quemada model. The largest increase in viscosity as compared to the reference value is noted for the Walburn–Schneck model. The impact of the models on the temporal development of the space-averaged kinetic energy and the corresponding viscous dissipation rate shows small differences among the models (not shown in the text). This effect may be observed for larger blood vessels but not for vessels where the relative wall region is large. The near wall region is characterized by slow flow and for small vessels the shear rates may be below 100 s^{-1}, requiring more refined rheological modeling. The vorticity in the aorta, WSS and its temporal derivative are affected by viscosity and/or alter the blood composition. However, WSS-related quantities (such as the OSI, TAWSS and RRT) are computed as time and averages, which "filters out" any significant differences among the rheological models. Averaged WSS-related parameters may be misleading for assessing the impact of viscosity and in particular time-dependent viscosity on the endothelium and arterial wall re-modeling.

The impact of the rheological models must be evaluated for relevant parameters at the blood flow of relevance, namely the geometrical conditions, boundary conditions, including flow rate variations and the specific aims with the numerical simulations. In the thoracic aorta, the flow is characterized by regions with relatively high shear rates, mostly greater than the order of 10 s^{-1}. For large shear rates (depicted in Figure 2a), the differences among the models are considerably smaller than for shear rates below O (100 s^{-1}). Hence, for predominantly high shear rate flows, assuming that the blood mixture is a Newtonian fluid is a natural choice. The sensitivity of the rheological models (i.e., the derivative of viscosity with respect to shear rate) is less than 0.01 Pa s/Pa for the Quemada model and shear rates in the range of 100 s^{-1} to 1000 s^{-1}. In contrast, the sensitivity of the Walburn–Schneck model is an order of magnitude larger with shear rates of about 100 s^{-1} (not shown here). The increase in viscosity due to the different non-Newtonian models may be as much as an order of magnitude. However, vorticity generation within the aorta is dominated by vortex stretching and with only a marginal, if at all, contribution of viscosity. The WSS, on the other hand, is directly dependent on the local value of the viscosity. Hence, the viscosity model can be expected to be mostly important for medium and smaller size arteries where the flow rate and total wall area are relatively high, hence viscosity plays a more important role compared to large arteries. This effect is clearly shown in arteries often subject to atherosclerosis, such as the coronary, carotid and renal arteries, as shown by van Wyk et al. [39] for a generic aortic bifurcation. For smaller arterioles, the shear rate is low and flow losses are viscosity dominated, hence the impact of the rheological models could be more important than for large arteries. Flow stagnation, independent of the size of the artery, may also require using non-Newtonian rheological models.

Another situation where viscosity plays a central role is in computing the forces acting on particles immersed in the flowing blood. Viscosity enters the formulation of some of the forces acting on small (spherical) particles. These include (viscous) drag, shear-dependent lift (Saffman's lift force) and added mass. As discussed above, the rheological models may affect the residence times of small particles. Long residence times are relevant for assessing the risk for thrombus formation. Particle transport is also of interest in assessing the risks of embolus transport from the thoracic aorta into vessels leading to the brain. This paper shows the impact of blood rheology on transport and indirectly also on the mixing processes in the thoracic aorta. Prolonged residence time has been believed to have an impact on the metabolism in the endothelium. Similarly, the instantaneous WSS and its temporal variation are believed to affect the endothelium. Unfortunately, the amount of experimental and clinical quantitative data on these processes is very limited and hence it is difficult to assess particular rheological models in these respects.

The results above are limited to a single geometry, single hematocrit and a single heart rate and flow rate. The extent of retrograde flow and particle residence times depends strongly on the geometry of the thoracic aorta and the geometrical details of the arteries branching from the aortic arch and the cardiac output. Therefore, the results and conclusions related to the impact of rheological models as reported here are of qualitative value. However, the conclusions are generally valid whenever viscosity has an accumulative effect (arterial wall remodeling, and transport of blood components). On the other hand, it is well established that time- and space-averaged quantities are not appropriate markers/descriptors for clinical applications due to low specificity. A major limitation of the rheological models used herein is due to neglecting the viscoelastic response of cells and macromolecules. Linderkamp et al. [40] noted that RBC deformability leads to lowering blood viscosity. They also found that RBC (extensional) response time is of the order of 0.1 s, implying a strong interaction between RBC cell response and the temporal variations in the aortic flow, especially during diastole (low shear rate).

Author Contributions: Conceptualization, L.P.W.; methodology, N.B., A.F., L.F. and L.P.W.; validation, A.F. and N.B.; formal analysis, N.B., A.F., L.F. and L.P.W.; resources, A.F. and L.P.W.; data curation, A.F.; writing—original draft preparation, A.F.; writing—review and editing, A.F., L.F., N.B. and L.P.W.; software, N.B.; visualization, A.F. and N.B.; supervision and project administration, L.P.W.; funding acquisition L.P.W. and A.F. (partially). All authors have read and agreed to the published version of the manuscript.

Funding: This research received no external funding.

Institutional Review Board Statement: All patients had been diagnosed and treated at the University Hospital in Linköping, Sweden. The research was performed in accordance with the Declaration of Helsinki and was approved by the Swedish regional ethical vetting board in Linköping (project identification code: DNR 2017/258-31, Prof. Anders Persson). The CT images as well as all patient-related data were anonymized.

Informed Consent Statement: Informed consent was obtained from all subjects involved in the study.

Data Availability Statement: Data are available upon request to the corresponding author.

Acknowledgments: We thank Chunliang Wang and Örjan Smedby for allowing us to use the segmentation software (Mialab). The computations were carried out on computer resources at NSC at Linköping University and HPC2N at Umeå University. A.F. acknowledges the support from the Department of Radiology at Linköping University Hospital.

Conflicts of Interest: The authors declare no conflict of interest.

References

1. Lippi, G.; Bovo, C.; Buonocore, R.; Mitaritonno, M.; Cervellin, G. Red blood cell distribution width in patients with limb, chest and head trauma. *Arch. Med. Sci.* **2017**, *3*, 606–611. [CrossRef]
2. Danese, E.; Lippi, G.; Montagnana, M. Red blood cell distribution width and cardiovascular diseases. *J. Thorac. Dis.* **2015**, *7*, E402–E411. [CrossRef] [PubMed]
3. Ananthaseshan, S.; Bojakowski, K.; Sacharczuk, M.; Poznanski, P.; Skiba, D.S.; Wittberg, L.P.; McKenzie, J.; Szkulmowska, A.; Berg, N.; Andziak, P.; et al. Red blood cell distribution width is associated with increased interactions of blood cells with vascular wall. *Sci. Rep.* **2022**, *12*, 13676. [CrossRef]

4. Cowan, A.Q.; Cho, D.J.; Rosenson, R.S. Importance of Blood Rheology in the Pathophysiology of Atherothrombosis. *Cardiovasc. Drugs Ther.* **2012**, *26*, 339–348. [CrossRef] [PubMed]

5. Agarwal, R.; Sarkar, A.; Paul, S.; Chakraborty, S. A portable rotating disc as blood rheometer. *Biomicrofluidics* **2019**, *13*, 064120. [CrossRef] [PubMed]

6. Yamamoto, H.; Kawamura, K.; Omura, K.; Tokudome, S. Development of a Compact-Sized Falling Needle Rheometer for Measurement of Flow Properties of Fresh Human Blood. *Int. J. Thermophys.* **2010**, *31*, 2361–2379. [CrossRef]

7. Wang, S.; Boss, A.H.; Kensey, K.R.; Rosenson, R.S. Variations of whole blood viscosity using Rheolog scanning capillary viscometer. *Clin. Chim. Acta* **2003**, *332*, 79–82. [CrossRef] [PubMed]

8. Mirza, A.; Ramaswamy, S. Importance of Non-Newtonian Computational Fluid Modeling on Severely Calcified Aortic Valve Geometries—Insights From Quasi-Steady State Simulations. *J. Biomech. Eng.* **2022**, *144*, 114501. [CrossRef]

9. Hund, S.J.; Kameneva, M.V.; Antaki, J.F. A Quasi-Mechanistic Mathematical Representation for Blood Viscosity. *Fluids* **2017**, *2*, 10. [CrossRef]

10. Marcinkowska-Gapińska, A.; Gapinski, J.; Elikowski, W.; Jaroszyk, F.; Kubisz, L. Comparison of three rheological models of shear flow behavior studied on blood samples from post-infarction patients. *Med Biol. Eng. Comput.* **2007**, *45*, 837–844. [CrossRef]

11. Gallagher, M.; Wain, R.; Dari, S.; Whitty, J.; Smith, D. Non-identifiability of parameters for a class of shear-thinning rheological models, with implications for haematological fluid dynamics. *J. Biomech.* **2019**, *85*, 230–238. [CrossRef] [PubMed]

12. Carreau, P.J. Rheological Equations from Molecular Network Theories. *Trans. Soc. Rheol.* **1972**, *16*, 99–127. [CrossRef]

13. Boyd, J.; Buick, J.M.; Green, S. Analysis of the Casson and Carreau-Yasuda non-Newtonian blood models in steady and oscillatory flows using the lattice Boltzmann method. *Phys. Fluids* **2007**, *19*, 093103. [CrossRef]

14. Jedrzejczak, K.; Makowski, L.; Orciuch, W. Model of blood rheology including hemolysis based on population balance. *Commun. Nonlinear Sci. Numer. Simul.* **2023**, *116*, 106802.

15. Arzani, A. Accounting for residence-time in blood rheology models: Do we really need non-Newtonian blood flow modelling in large arteries? *J. R. Soc. Interface* **2018**, *15*, 20180486. [CrossRef]

16. Akherat, S.M.J.M.; Cassel, K.; Boghosian, M.; Dhar, P.; Hammes, M. Are Non-Newtonian Effects Important in Hemodynamic Simulations of Patients With Autogenous Fistula? *J. Biomech. Eng.* **2017**, *139*, 0445041–0445049. [CrossRef]

17. Karimi, S.; Dabagh, M.; Vasava, P.; Dadvar, M.; Dabir, B.; Jalali, P. Effect of rheological models on the hemodynamics within human aorta: CFD study on CT image-based geometry. *J. Non-Newton. Fluid Mech.* **2014**, *207*, 42–52. [CrossRef]

18. Johnston, B.M.; Johnston, P.R.; Corney, S.; Kilpatrick, D. Non-Newtonian blood flow in human right coronary arteries: Steady state simulations. *J. Biomech.* **2004**, *37*, 709–720. [CrossRef]

19. Jahangiri, M.; Haghani, A.; Ghaderi, R.; Harat, S.M.S. Effect of Non-Newtonian Models on Blood Flow in Artery with Different Consecutive Stenosis. *Int. J. Adv. Des. Manuf. Technol. (ADMT)* **2018**, *11*, 89–96.

20. Apostolidis, A.J.; Moyer, A.P.; Beris, A.N. Non-Newtonian effects in simulations of coronary arterial blood flow. *J. Non-Newton. Fluid Mech.* **2016**, *233*, 155–165. [CrossRef]

21. Mendieta, J.B.; Fontanarosa, D.; Wang, J.; Paritala, P.K.; McGahan, T.; Lloyd, T.; Li, Z. The importance of blood rheology in patient-specific computational fluid dynamics simulation of stenotic carotid arteries. *Biomech. Model. Mechanobiol.* **2020**, *19*, 1477–1490. [CrossRef]

22. Liepsch, D.; Sindeev, S.; Frolov, S. An impact of non-Newtonian blood viscosity on hemodynamics in a patient-specific model of a cerebral aneurysm. *J. Phys. Conf. Ser.* **2018**, *1084*, 012001. [CrossRef]

23. Skiadopoulos, A.; Neofytou, P.; Housiadas, C. Comparison of blood rheological models in patient specific cardiovascular system simulations. *J. Hydrodyn.* **2017**, *29*, 293–304. [CrossRef]

24. Molla, M.; Paul, M. LES of non-Newtonian physiological blood flow in a model of arterial stenosis. *Med. Eng. Phys.* **2012**, *34*, 1079–1087. [CrossRef]

25. Walburn, F.J.; Schneck, D.J. A constitutive equation for whole human blood. *Biorheology* **1976**, *13*, 201–210. [CrossRef] [PubMed]

26. Casson, N. *Rheology of Disperse Systems*; Pergamon Press: London, UK, 1959; Volume 84.

27. Quemada, D. Rheology of concentrated disperse systems and minimum energy dissipation principle. *Rheol. Acta* **1977**, *16*, 82–94. [CrossRef]

28. Quemada, D. Rheology of concentrated disperse systems ii. a model for non-newtonian shear viscosity in steady ows. *Rheol. Acta* **1978**, *17*, 632–642. [CrossRef]

29. Cokelet, G.; Merrill, E.; Gilliland, E.; Shin, H.; Britten, A.; Wells, R., Jr. The rheology of human blood{measurement near and at zero shear rate. *J. Rheol.* **1963**, *7*, 303–307. [CrossRef]

30. Perktold, K.; Karner, G.; Leuprecht, A.; Hofer, M. Influence of Non-newtonian Flow Behavior on Local Hemodynamics. *Zamm* **1999**, *79*, 187–190. [CrossRef]

31. Fuchs, A.; Berg, N.; Wittberg, L.P. Stenosis Indicators Applied to Patient-Specific Renal Arteries without and with Stenosis. *Fluids* **2019**, *4*, 26. [CrossRef]

32. Benim, A.C.; Nahavandi, A.; Assmann, A.; Schubert, D.; Feindt, P.; Suh, S.H. Simulation of blood flow in human aorta with emphasis on outlet boundary conditions. *Appl. Math. Model.* **2011**, *35*, 3175–3188. [CrossRef]

33. Suo, J.; Oshinski, J.N.; Giddens, D.P. Blood flow patterns in the proximal human coronary arteries: Relationship to atherosclerotic plaque occurrence. *Mol. Cell. Biomech.* **2008**, *5*, 9.

34. Williams, B.; Kabbage, M.; Kim, H.J.; Britt, R.; Dickman, M.B. Tipping the balance: Sclerotinia sclerotiorum secreted oxalic acid suppresses host defenses by manipulating the host redox environment. *PLOS Pathog.* **2011**, *7*, e1002107. [CrossRef] [PubMed]

35. He, X.; Ku, D.N. Pulsatile Flow in the Human Left Coronary Artery Bifurcation: Average Conditions. *J. Biomech. Eng.* **1996**, *118*, 74–82. [CrossRef] [PubMed]

36. Rikhtegar, F.; Knight, J.A.; Olgac, U.; Saur, S.C.; Poulikakos, D.; Marshall, W.; Cattin, P.C.; Alkadhi, H.; Kurtcuoglu, V. Choosing the optimal wall shear parameter for the prediction of plaque location—A patient-specific computational study in human left coronary arteries. *Atherosclerosis* **2012**, *221*, 432–437. [CrossRef] [PubMed]

37. Gallo, D.; Steinman, D.A.; Morbiducci, U. Insights into the co-localization of magnitude-based versus direction-based indicators of disturbed shear at the carotid bifurcation. *J. Biomech.* **2016**, *49*, 2413–2419. [CrossRef]

38. Fuchs, A.; Berg, N.; Parker, L.P.; Wittberg, L.P. The Impact of Heart Rate and Cardiac Output on Retrograde Flow in the Human Thoracic Aorta. *J. Eng. Sci. Med Diagn. Ther.* **2022**, *5*, 031107. [CrossRef]

39. van Wyk, S.; Wittberg, L.P.; Fuchs, L. Wall shear stress variations and unsteadiness of pulsatile blood-like flows in 90-degree bifurcations. *Comput. Biol. Med.* **2013**, *43*, 1025–1036. [CrossRef]

40. Linderkamp, O.; Nash, G.B.; Wu, P.Y.K.; Meiselman, H.J. Deformability and Intrinsic Material Properties of Neonatal Red Blood Cells. *Blood* **1986**, *67*, 1244–1250. [CrossRef] [PubMed]

bioengineering

Article

Modeling Anisotropic Electrical Conductivity of Blood: Translating Microscale Effects of Red Blood Cell Motion into a Macroscale Property of Blood

Alireza Jafarinia [1,*,†], Vahid Badeli [2,†], Thomas Krispel [1], Gian Marco Melito [3], Günter Brenn [4], Alice Reinbacher-Köstinger [2], Manfred Kaltenbacher [2] and Thomas Hochrainer [1]

[1] Instiue of Strength of Materials, Graz University of Technology, Kopernikusgasse 24/I, 8010 Graz, Austria; thomas.krispel@tugraz.at (T.K.); hochrainer@tugraz.at (T.H.)

[2] Institute of Fundamentals and Theory in Electrical Engineering, Graz University of Technology, Inffeldgasse 18, 8010 Graz, Austria; vahid.badeli@tugraz.at (V.B.); alice.koestinger@tugraz.at (A.R.-K.); manfred.kaltenbacher@tugraz.at (M.K.)

[3] Institute of Mechanics, Graz University of Technology, Kopernikusgasse 24/IV, 8010 Graz, Austria; gmelito@tugraz.at (G.M.M.)

[4] Institute of Fluid Mechanics and Heat Transfer, Graz University of Technology, 8010 Graz, Austria; guenter.brenn@tugraz.at (G.B.)

* Correspondence: alireza.jafarinia@tugraz.at (A.J.)

† These authors contributed equally to this work.

Citation: Jafarinia, A.; Badeli, V.; Krispel, T.; Melito, G.M.; Brenn, G; Reinbacher-Köstinger, A.; Kaltenbacher, M.; Hochrainer, T. Modeling Anisotropic Electrical Conductivity of Blood: Translating Microscale Effects of Red Blood Cells Motion into a Macroscale Property of Blood . *Bioengineering* **2024**, *11*, 147. https://doi.org/10.3390/bioengineering11020147

Academic Editor: hung-Hao Lee

Received: 24 December 2023
Revised: 28 January 2024
Accepted: 30 January 2024
Published: 1 February 2024

Abstract: Cardiovascular diseases are a leading global cause of mortality. The current standard diagnostic methods, such as imaging and invasive procedures, are relatively expensive and partly connected with risks to the patient. Bioimpedance measurements hold the promise to offer rapid, safe, and low-cost alternative diagnostic methods. In the realm of cardiovascular diseases, bioimpedance methods rely on the changing electrical conductivity of blood, which depends on the local hemodynamics. However, the exact dependence of blood conductivity on the hemodynamic parameters is not yet fully understood, and the existing models for this dependence are limited to rather academic flow fields in straight pipes or channels. In this work, we suggest two closely connected anisotropic electrical conductivity models for blood in general three-dimensional flows, which consider the orientation and alignment of red blood cells (RBCs) in shear flows. In shear flows, RBCs adopt preferred orientations through a rotation of their membrane known as tank-treading motion. The two models are built on two different assumptions as to which hemodynamic characteristic determines the preferred orientation. The models are evaluated in two example simulations of blood flow. In a straight rigid vessel, the models coincide and are in accordance with experimental observations. In a simplified aorta geometry, the models yield different results. These differences are analyzed quantitatively, but a validation of the models with experiments is yet outstanding.

Keywords: anisotropic electrical conductivity; electrical conductivity of blood; computational fluid dynamics; bioimpedance signals; impedance cardiography; impedance plethysmography

1. Introduction

The early detection of cardiovascular diseases (CVDs) is an important healthcare objective, given that they are the leading global cause of mortality. While conventional diagnostic methods, such as CT or MRI scans and invasive procedures, have been effective, they are costly and potentially risky for patients [1,2]. There is a need for a rapid, non-invasive, and cost-effective alternative to enable the early detection and personalized clinical decision-making for CVDs. Bioimpedance signals promise a compelling solution with these advantages. This study introduces an innovative modeling approach for the electrical conductivity of blood to enable the investigation of bioimpedance measurements.

Bioimpedance signals, including impedance plethysmography (IPG) and impedance cardiography (ICG) signals, offer insights into the physiological and pathophysiological processes within blood vessels. Unlocking the diagnostic potential of bioimpedance signals depends on the ability to classify these signals [3]. The quantitative interpretation of bioimpedance signals presents a significant challenge, and this field remains largely unexplored, explaining its absence in current clinical practices. This challenge becomes more apparent when addressing CVDs characterized by altered local hemodynamics and blood flow disruptions, such as stenosis, aneurysms, aortic dissection, atherosclerosis, and thrombosis.

In clinical practice, bioimpedance signals are obtained by injecting a low-frequency electric current into the body. Since blood is a highly conductive material when compared with the surrounding tissues and organs (lungs, bones, muscles, etc.), the electric current travels preferentially through the blood vessels. Therefore, bioimpedance signals are highly sensitive to changes in the electrical conductivity of blood [4–6]. The electrical current primarily flows through the blood plasma, as red blood cells (RBCs) are electrically non-conductive at the frequencies relevant to impedance measurements, i.e., up to electrical frequencies of 1 MHz [7,8]. Because of their non-conductive nature, RBC concentration and complex motions significantly influence the electrical conductivity of blood. A higher concentration of RBCs results in an increased presence of non-conductive material within the blood, reducing the overall conductivity. When red blood cells form channel-like pathways within blood vessels, it facilitates an easier flow of electrical current through the conductive plasma, increasing the overall conductivity.

Local hemodynamics affect the RBCs motion, particularly their orientation, alignment, and deformation, which depend on the state of shear rate and shear stress. Given that RBCs are non-conductive and non-spherical, their orientation, alignment, and deformation affect the anisotropic electrical conductivity of blood [7,9,10]. Consequently, RBC motion plays a pivotal role in the analysis of bioimpedance signals linked to CVDs.

The study of the electrical conductivity of blood was started by Maxwell [11], Rayleigh [12], and Fricke [13]. The first two initially developed a basic theory for determining the size and shape of spherical particles in a suspended fluid. Successively, Fricke [13] extended their work to consider the electrical conductivity of randomly distributed ellipsoidal particles in a fluid. In the literature, this theory is known as the Maxwell–Fricke theory, and it allows for the computation of the electrical conductivity of stationary blood based on the blood temperature, volume fraction, and the shape of RBCs. Many years later, Edgerton [9] extended the Maxwell–Fricke theory by including the probability distribution of RBCs orientation in the blood flow and showing that their distribution is a function of shear rate. Ref. Visser [14] confirmed the result of Edgerton and also indicated that the orientation of RBCs is the main cause of blood conductivity changes.

Recently, Hoetink et al. [7] developed a model based on the Maxwell–Fricke theory for computing the blood conductivity of a diluted suspension of ellipsoidal particles, which simulates blood in a steady flow in a rigid vessel. Considering RBCs as ellipsoidal particles, their results showed that, due to high values of shear stress, the RBCs deform and orient such that one of their long axes is parallel to the streamlines of the blood flow. This new configuration causes a substantial change in the electric current path through the blood [7,8]. In particular, the electrical conductivity of blood in the direction of flow increases due to channel-like paths available between the aligned RBCs. Channel-like paths are shown in Figure 1.

Gaw et al. [10] extended the investigation and reported the effects of pulsatile blood flow on electrical conductivity in a rigid vessel. Theoretically and experimentally, it was shown that, when the velocity increases during systole, there is a robust linear relationship between the average velocity and the conductivity of the blood. Similarly, a decrease in impedance is observed when blood velocity decreases, i.e., during diastole.

The recent modeling approaches developed by Hoetink et al. [7] and Gaw et al. [10] inspired several application studies for the simulation of bioimpedance signals [15,16], in particular in the field of aortic dissection [3,4]. However, these approaches focused on the analytical and numerical solutions for one-dimensional (1D) computations of the electrical conductivity of blood, and none offered the possibility to model and compute the 3D anisotropic electrical conductivity of blood as a field variable. For example, a previous study by Badeli et al. [4] showed the important application of the ICG method in detecting aortic dissection. However, the isotropic assumption imposed by the 1D formulation led to ignored changes in the blood conductivity values due to varying flow direction, local flow hemodynamics, and disturbances due to pathology.

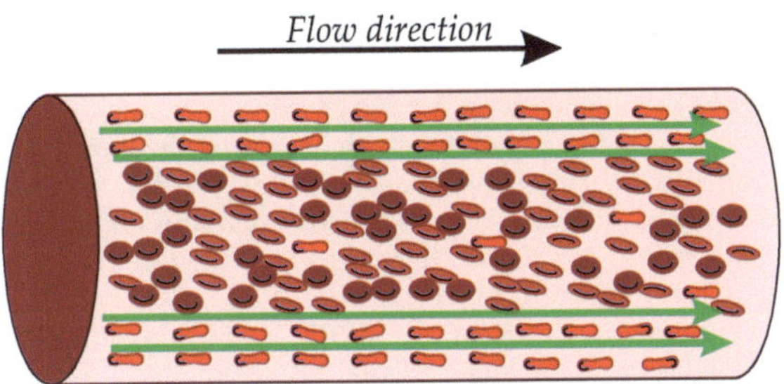

Figure 1. An idealized schematic of how RBCs align in the flow direction, creating channel-like paths near the vessel wall. The green arrows indicate the passage of the electrical current.

An initial attempt toward a more accurate description of the conductivity field was performed by Jafarinia et al. [17] and Badeli [18]. A two-dimensional (2D) model for computing anisotropic conductivity as a time-dependent field variable was developed. The studies initiated the use of computational fluid dynamics (CFD) simulations for the computation of blood flow conductivity. Through a multiphysics approach, combining the electromagnetic and CFD simulations, the authors showed that, by tracking ICG signals, it is feasible to specify the changes in false lumen thrombosis in the case of aortic dissection. Recently, Badeli et al. [19] extended the 2D conductivity model to be used in three-dimensional (3D) multiphysics CFD simulations with some simplifications, including the assumption that the RBCs are prolate spheroid. As a result, the simulated bioimpedance signals confirmed that physiological changes, such as thrombosis, are trackable by monitoring the impedance changes. However, it was noted that the simplifications had caused some inaccuracies in the simulated bioimpedance signals.

The presented study aims to develop a 3D anisotropic electrical conductivity model capable of translating the microscopic effect of RBCs orientation and alignment into the macroscopic property of blood in a general 3D blood flow. The motion of RBCs is influenced by hemodynamic conditions, and the new model shall be based on experimental observations related to the orientation of RBCs in shear flows, which will be discussed in the next section. However, there are two alternative interpretations for what hemodynamic characteristic determines the orientation of the RBCs, which happen to coincide in the experimentally studied flow conditions. We therefore developed two 3D models based on either characteristic. These models share their root in the previous conductivity model proposed by Gaw et al. [10] and further studied by Melito et al. [20], which will likewise be introduced in Section 2. In Section 3, we validate the models in a pipe flow and compare their results to a flow in a simplified aorta geometry. The differences between the models are discussed in detail in Section 4, and we offer conclusions in Section 5.

2. Materials and Methods

Since the orientation and deformation of RBCs are the cause of anisotropic blood conductivity, it is essential to understand their complex motion in the blood flow. Several experimental studies have investigated the behavior of RBCs in Couette and Poiseuille flow fields, including Fischer et al. [21], Goldsmith et al. [22], Keller and Skalak [23],

Bitbol [24], Schmid-Schönbein and Wells [25]. These studies suggest that RBCs exhibit two types of motion. First, unsteady motions like flipping, tumbling, and rolling, in which the biconcave shape of RBCs remains unchanged. Second, a steady motion where the RBCs undergo deformation into ellipsoidal particles due to high shear stress. In the steady motion, the RBCs maintain a steady orientation with their membrane circulating around the interior viscous fluid (cytoplasm) [26]. This motion is called tank-treading, where the rotating motion of the membrane transfers the tangential stresses of the flow to the cytoplasm. In fact, the cytoplasm recirculates and dissipates the energy transferred from the external flow, which allows the RBCs to keep a steady orientation and shape [26].

The two motions can occur simultaneously in a blood vessel. The probability of finding differently oriented RBCs varies depending on the shear rate and shear stress. The unsteady rigid-body-like motions are seen in low shear rates. There exists a gradual transition to tank-treading steady motion with increasing shear rate, where eventually the RBCs orient approximately in the direction of the flow [21–23]. According to Gaw et al. [10], the RBCs in unsteady motions are categorized as randomly oriented, while in the tank-treading motion, they are aligned in a preferred orientation determined by characteristics of the local blood flow.

Knowing the orientation of RBCs allows for computing the anisotropic blood conductivity in a fully aligned state. Therefore, in the following, the tank-treading motion is investigated in detail in order to extract the required information to determine the orientation of RBCs. Later, the computation of conductivity will be combined with the conductivity of blood with randomly oriented RBCs in order to treat different degrees of RBCs alignment.

The experiments of Fischer et al. [21] and Minetti et al. [27] showed that the RBCs in the tank-treading motion are triaxial ellipsoidal particles with a short (minor), an intermediate, and a long (major) axis, see Figure 2. According to their experiments, two primary conclusions are derived:

- The RBCs are oriented such that the intermediate and major axes are in a plane of maximum shear stress, which we shall call a 'shear plane' in the following. There exist two perpendicular shear planes because the viscous stress tensor τ is symmetric. The intermediate and major axes of the RBCs are found to lie in the shear plane, which mostly contains the flow direction. The major axis of ellipsoidal tank-treading RBCs is found to be parallel to the flow direction;
- The intermediate axis is parallel to the vorticity vector of the flow. Also, in this case, the major axis of the RBCs is parallel to the flow direction. Figure 3 shows an idealized schematic of a tank-treading RBC in shear flows.

As will be discussed in more detail in Section 4, in the case of the experimentally studied Couette and Poiseuille flows, it so happens that the shear plane, which contains the velocity vector, coincides with the plane spanned by the vorticity vector and the velocity vector. These planes do not coincide in general 3D flow fields and from the literature it is not clear whether shear stresses or the vorticity vector chiefly determine the orientation of tank treading RBCs.

Before introducing the models, we note that Edgerton [9], Bitbol [24], and Gaw et al. [10] advocated that for the computation of blood electrical conductivity in the cardiovascular system, it is reasonable to consider the RBCs as ellipsoidal particles with two equal long axes and one short axis, i.e., a symmetry axis, meaning that the RBCs are assumed to be oblate spheroids, see Figure 2b. We adopt this simplification in the current study. With this assumption, the orientation of RBCs is fully specified by the symmetry axis.

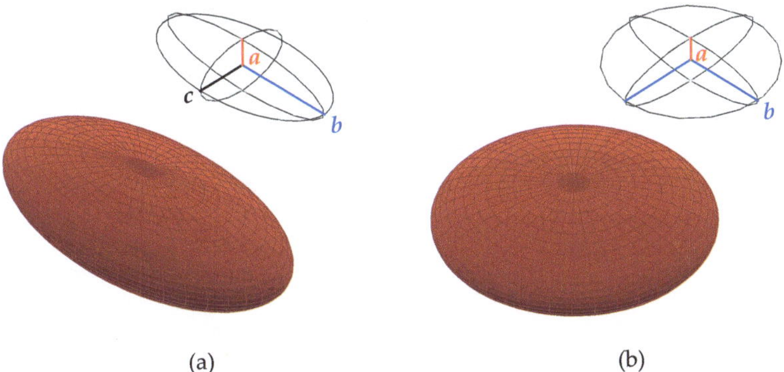

(a)   (b)

Figure 2. A triaxial ellipsoidal particle (**a**) characterized by a short ($2a$), an intermediate ($2c$), and a long ($2b$) axis and an oblate spheroid (**b**) with a short ($2a$) and two equal long axes ($2b$). See Table 1 for the values of a and b.

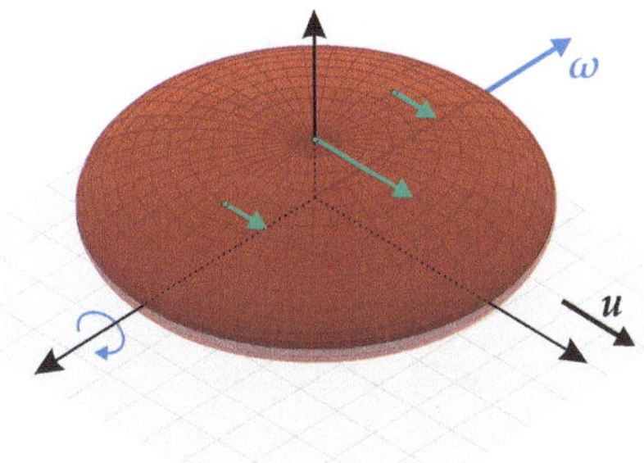

Figure 3. An idealized schematic of a tank-treading ellipsoidal RBC near a vessel wall with a high shear rate. The RBC is assumed to be an ellipsoidal particle with two equal long axes with the length $2b$ and one short axis of length $2a$. The shear plane, shown in gray, is the plane of maximum shear stress containing the velocity vector u. The vorticity vector ω, is shown in a blue vector. The curved blue arrow indicates the cavity flow of the cytoplasm. The green arrows on the RBCs membrane indicate the local membrane speed due to the tank-treading motion.

2.1. Modeling RBCs Motion

Considering the conclusions derived from the mentioned experiments and assuming RBCs to be oblate spheroids, the tank-treading RBCs are either considered to be oriented such that their flat surface (the surface spanned by the long axes) is parallel to the shear plane; such that the symmetry axis of RBCs is normal to the shear plane. Or such that the long axes of the RBCs are in the direction of flow and vorticity, suggesting that the symmetry axis of RBCs is normal to the plane spanned by the velocity and vorticity vectors. Figure 3 depicts these considerations at an idealized schematic of a tank-treading RBC with two long equal axes $2b$ and the symmetry axis $2a$. The tank-treading motion of the RBC is indicated by the green arrows on the RBCs membrane. In Figure 3, the gray plane is the shear plane containing the velocity vector u, while the RBCs symmetry axis $2a$ is parallel to the normal of this shear plane. The long axes $2b$ are parallel to the shear plane, with one of them parallel to the velocity vector u and the other one to the vorticity vector ω.

Based on these observations, we propose the following two models to find the direction of the symmetry axis of RBCs and, consequently, their orientation in the tank-treading motion:

- ***The eigenvector model***: The direction of the symmetry axis is determined by the normal vector of the shear plane, which contains or mostly contains the velocity vector. The normal vector can be computed using the eigenvectors of the viscous stress tensor;
- ***The velocity–vorticity model***: the direction of the symmetry axis is determined by computing the cross product of vorticity and velocity, i.e., the so-called Lamb vector [28].

Up to now, our focus has been on tank-treading steady motion of RBCs occurring at high shear rates. However, in low shear rates, the orientation of RBCs is random [21]. There exists a gradual transition between complete randomness and full alignment in a steady orientation of RBCs, which is a function of shear rate. In principle, in the case of the blood flow in a vessel, the RBCs near the wall, where the shear rate is high, are assumed to be aligned to the flow direction [10]. They are randomly oriented in the middle of the vessel where the shear rate is zero [7,10]. For intermediate shear rates, Gaw et al. [10] assumed a gradual transition from random orientation in low shear rates and full alignment at high shear rates. In the current study we adopt this assumption together with the shear rate dependent transition function introduced by Gaw et al. [10], compare Equation (15) in Section 2.3, where we also specify what we consider as high, intermediate, and low shear rates.

Next, the computation of the direction of RBCs symmetry axis using eigenvector and velocity–vorticity models is explained. Knowing the orientation of RBCs, eventually, the electrical conductivity tensor of blood in a general three-dimensional (3D) flow is defined.

2.1.1. Eigenvector Model

The global Cartesian coordinates system, which will be used in the numerical simulations of blood flow, is defined by x, y, and z axes, with the unit basis vectors e_x, e_y, and e_z. We assume that the viscous stress tensor τ has three distinct eigenvalues and define the orthonormal basis vectors e_1, e_2, and e_3 as the eigenvectors corresponding to maximum, intermediate, and minimum eigenvalue, respectively. These basis vectors form the viscous stress tensor's principal coordinate system and are locally defined because τ is a field variable. The maximal shear stress in the system is given by the difference between the maximum and minimum eigenvalues of the stress tensor. The maximal shear stress is obtained as the magnitude of the in-plane component of the stress vector in surfaces with normal vectors along 'the diagonal' between the first and the third eigenvector. Note that since the directions of the eigenvectors are only defined up to a sign, there are two maximum shear planes with orthogonal normal vectors. These normal vectors may be obtained by rotating e_1 (or e_3) around the e_2-axis in positive or negative $45°$. In the eigenvector model, we assume that the minor axis of the RBC e_α is aligned with the normal of the shear plane, which ideally contains the velocity vector (at high shear rates) or which contains the larger component of the velocity vector upon projection into the shear plane.

Equivalently, e_α (i.e., the sought shear plane normal) shall have a larger angle to the velocity vector u than the normal of the alternative shear plane. The two (up to a sign) $45°$ rotations of the eigenvector e_1 around e_2, are the unit vectors $\frac{1}{\sqrt{2}}(e_1 + e_3)$ and $\frac{1}{\sqrt{2}}(e_1 - e_3)$. Accordingly, we define the direction of the short axis of the RBCs as follows:

$$e_\alpha^{EV} := \begin{cases} \frac{1}{\sqrt{2}}(e_1 - e_3) & \text{if } |\langle u, \frac{1}{\sqrt{2}}(e_1 + e_3)\rangle| > |\langle u, \frac{1}{\sqrt{2}}(e_1 - e_3)\rangle| \\ \frac{1}{\sqrt{2}}(e_1 + e_3) & \text{otherwise,} \end{cases} \tag{1}$$

where the superscript EV refers to the eigenvector model.

Note that also these unit vectors need only to be defined up to a sign, which does not matter in defining the symmetric conductivity tensor below Equation (6). Figure 4 shows a 3D representation of the unit basis vectors. In Figure 4, because the angle θ between u

and $\frac{1}{\sqrt{2}}(e_1 + e_3)$ is an acute angle, the first condition in Equations (1) is satisfied, hence the following:

$$e_\alpha^{EV} := \frac{1}{\sqrt{2}}(e_1 - e_3).\qquad(2)$$

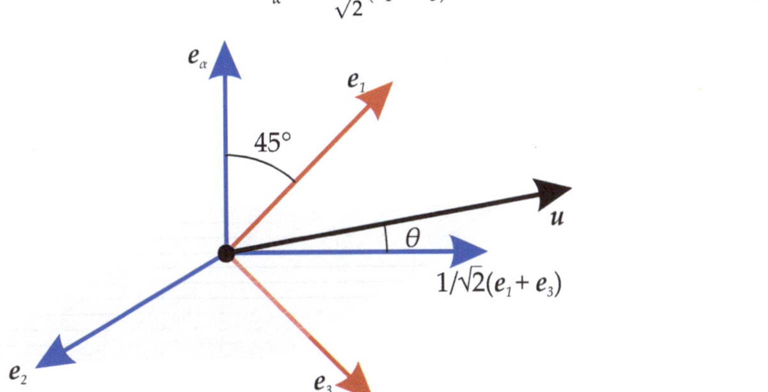

Figure 4. 3D representation of the unit basis vectors. The unit basis vectors e_1, e_2, e_3 are the eigenvectors of the viscous stress tensor τ and correspond to maximum, intermediate, and minimum eigenvalues, respectively. The plane in gray is the shear plane. The unit basis vector e_α is normal to the shear plane. The angle θ is the angle between velocity vector u and $\frac{1}{\sqrt{2}}(e_1 + e_3)$.

2.1.2. Velocity–Vorticity Model

In the velocity–vorticity model, the unit basis vector e_α, which defines the direction of the symmetry axis of the RBC, is computed by the normalized cross product of velocity u and vorticity ω vectors as follows:

$$e_\alpha^{VV} = \frac{u \times \omega}{|u \times \omega|},\qquad(3)$$

where the superscript VV refers to the velocity–vorticity model.

The vorticity vector ω is defined as the curl of the velocity vector u as follows:

$$\omega = \nabla \times u.\qquad(4)$$

2.2. Definition of the Conductivity Tensor

We begin this section by briefly recalling that in the case of materials with anisotropic electrical conductivity, this needs to be described by a conductivity tensor σ of second order. Ohm's law, connecting the electrical current density J and the electrical field E then attains the form [29] as follows:

$$J = \sigma \cdot E,\qquad(5)$$

where the central dot indicates matrix–vector multiplication. In the isotropic case, the conductivity tensor is a multiple of the unit matrix I, $\sigma = \sigma I$, with the scalar conductivity σ. In general, the conductivity tensor is symmetric and therefore represented by a symmetric matrix in every coordinate system. The diagonal elements of this matrix connect the strength of the electrical current in the direction of the coordinate axes with the electrical field in the same direction. The non-diagonal elements are called in-plane conductivities and they account for electric currents induced perpendicular to the electrical field in one of the coordinate directions. Because of the symmetry of σ, there exists a local orthonormal coordinate system of eigendirections, with regard to which no in-plane conductivities occur, such that in this coordinate system the matrix has diagonal form. The diagonal entries with regard to this coordinate system are the eigenvalues of the matrix and are called the principal conductivities.

Both models introduced above define one distinguished eigendirection, e_α, of the conductivity tensor and assume that the conductivity is isotropic in the plane perpendicular to this direction. The conductivity tensor may therefore be defined by two quantities, the principal conductivity in the direction of the short axis of the RBCs, σ_α, and the transverse conductivity σ_β. Having specified e_α, either by the eigenvector model or the velocity–vorticity model, the conductivity tensor is therefore defined as follows:

$$\sigma = \sigma_\alpha (e_\alpha \otimes e_\alpha) + \sigma_\beta (I - e_\alpha \otimes e_\alpha). \tag{6}$$

Note that e_α determines the dominating short axis direction, but depending on the shear stress we expect various degrees of alignment of the normals. At high shear stresses, the RBCs are expected to be strongly aligned, while at low shear stresses the orientations are mostly random. How these conductivities are derived from the local shear stress and shear rates is described in the next section.

2.3. Calculation of the Average Conductivities

The Maxwell–Fricke theory [13], with the formulation introduced by Hoetink et al. [7] and Gaw et al. [10], is adopted in this study for the calculation of components of the conductivity tensor.

The Maxwell–Fricke theory provides the anisotropic conductivity of a fluid suspension with completely aligned spheroidal particles based on the volume fraction and the aspect ratio $\lambda = a/b$ of the particles. That is, the principal conductivities in the direction of the short axis σ_a and in either of the long axes σ_b are calculated as follows:

$$\frac{\sigma_{a,b}(\lambda)}{\sigma_{\text{pl}}} = \frac{1 - H}{1 + (C_{a,b}(\lambda) - 1)H}, \tag{7}$$

where σ_{pl} is the conductivity of the blood plasma and H is the volume fraction of RBCs in the blood, i.e., the hematocrit value. See Table 1 for the values of model parameters. The orientation factors C_a and C_b depend on the aspect ratio λ via a function $M(\lambda)$ through the following [10]:

$$C_a(\lambda) = 1/M(\lambda), \tag{8}$$

and

$$C_b(\lambda) = 2/(2 - M(\lambda)). \tag{9}$$

The function $M(\lambda)$ in turn is computed for $a < b$ as follows [10]:

$$M(\lambda) = \frac{\phi(\lambda) - \frac{1}{2}\sin(2\phi(\lambda))}{\sin^3 \phi(\lambda)} \cos\phi(\lambda), \tag{10}$$

with

$$\phi(\lambda) = \arccos(\lambda) = \arccos\left(\frac{a}{b}\right). \tag{11}$$

Table 1. Constant parameters of the conductivity and blood rheology models.

Description	Symbol	Value	Units	References
Particle aspect ratio	λ	0.38	[-]	[7,10,20,30]
Conductivity of the blood plasma	σ_{pl}	1.3	S m^{-1}	[31]
Volume fraction of RBCs in the blood	H	45	%	[31]
Short particle semiaxis	a	1.52×10^{-6}	m	[31]
Long particle semiaxis	b	4×10^{-6}	m	[31]
Membrane shear modulus	μ	10^{-5}	kg/s^2	[7,32]
Orientation/Disorientation constant	k	1	$\text{s}^{-1/2}$	[31,33]
Dynamic viscosity of the blood plasma	η_{pl}	4.8×10^{-2}	$\text{kg m}^{-1}\text{s}^{-1}$	[7,10,34]
Blood density	ρ_{bl}	1060	kg/m^3	[35,36]

In the flowing blood, RBCs deform due to shear stresses [7], most notably in the tank treading motion based on the maximum shear stress τ_{max}. Given an undeformed aspect ratio λ in stationary blood, Hoetink et al. [7] derived the shear stress-dependent aspect ratio λ_d of the deformed RBCs as follows:

$$\lambda_d(\tau_{max}) = \lambda \left[1 + \frac{\tau_{max}b}{4\mu}\right]^{-3},$$ (12)

where μ denotes the membrane shear modulus of the RBCs, see Table 1.

Inserting the last relation in (7) yields the shear stress-dependent principal conductivities of blood with perfectly aligned RBCs as follows:

$$\frac{\sigma_{a,b}(\tau_{max})}{\sigma_{pl}} = \frac{1-H}{1+(C_{a,b}(\lambda_d(\tau_{max}))-1)H}.$$ (13)

However, full alignment of the RBCs is only expected at very high shear stresses, while in the absence of shear stresses, the orientation of the RBCs minor axis is supposed to be random. The conductivity model will thus also need an interpolation between these two extreme cases based on the fraction of aligned RBCs. A model for this interpolation based on the maximum shear rate $\dot\gamma_{max}$ is available from Gaw et al. [10]. The authors assume the conductivity of blood with randomly oriented RBCs is obtained from Equation (13) by substituting $C_{a,b}$ by the average orientation factor as follows:

$$C_r = 1/3(C_a + 2C_b).$$ (14)

The interpolation of the conductivities between the fully aligned and the randomly oriented case is performed by interpolating the orientation factors based on the fraction of aligned RBCs given as a function of the maximum shear rate by the following:

$$f(\dot\gamma_{max}) = \frac{\dot\gamma_{max}}{\dot\gamma_{max} + k\sqrt{\dot\gamma_{max}}}.$$ (15)

where k is a constant whose value is indicated in Table 1. Function $f(\dot\gamma_{max})$ is plotted in Figure 5. Since this function provides a gradual transition from no alignment at vanishing shear rate to full alignment at an infinite shear rate, it does not provide a clear cut distinction as to what are high, intermediate, and low shear rates with regard to alignment. We suggest considering a fraction of aligned RBCs below 20% as low, beyond 80% as high, and as intermediate in-between. This defines shear rates to be considered low below a value of 0.0625 s^{-1}, as high beyond 16 s^{-1}, and as intermediate in-between.

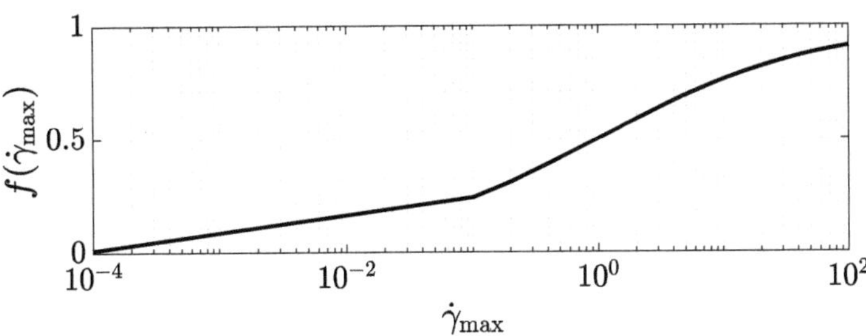

Figure 5. The fraction of aligned RBCs, computed by $f(\dot\gamma_{max})$, versus the maximum shear rate $\dot\gamma_{max}$.

The interpolated orientation factors C_α and C_β are then defined as follows:

$$C_{\alpha,\beta}(\dot\gamma_{max}, \tau_{max}) = f(\dot\gamma_{max})C_{a,b}(\tau_{max}) + (1 - f(\dot\gamma_{max}))C_r(\tau_{max}).$$ (16)

Eventually, the conductivities in the direction of the dominant alignment e_α, σ_α, and orthogonal to that σ_β are computed as follows:

$$\frac{\sigma_{\alpha,\beta}(\dot{\gamma}_{max}, \tau_{max})}{\sigma_{pl}} = \frac{1 - H}{1 + (C_{\alpha,\beta}(\dot{\gamma}_{max}, \tau_{max}) - 1)H}. \tag{17}$$

Note that the shear rate is a unique function of shear stress and vice versa, such that the conductivity may as well be regarded as being either solely a function of the shear rate or of the shear stress. For low shear rates $\dot{\gamma}_{max} \ll 1$, the aligned fraction becomes small, such that the two conductivities are approximately equal (and equal to the conductivity with randomly oriented RBCs) and the tensor is nearly isotropic. At high shear rates, when $\sigma_\alpha \approx \sigma_a$ and $\sigma_\beta \approx \sigma_b$ the anisotropy becomes maximal.

2.4. Computational Fluid Dynamics and Rheological Modeling

The Navier–Stokes equations model the blood flow as follow:

$$\rho \left[\frac{\partial u}{\partial t} + (u \cdot \nabla)u \right] = -\nabla p + \nabla \cdot \tau, \tag{18}$$

with pressure p and the viscous stress tensor τ. The blood is modeled as an incompressible fluid and a constant density ρ; therefore, the mass balance equation reduces to $\nabla \cdot u = 0$. Though blood is known to show shear thinning and a kind of flow stress [37], for large blood vessels it is admissible to model blood as a Newtonian fluid, resulting in a linear relation between the extra stress tensor τ and the rate-of-deformation tensor D.

$$\tau = 2\eta_{bl}D. \tag{19}$$

The rate-of-deformation tensor D is defined as the symmetric part of the velocity gradient ∇u and is therefore computed as follows:

$$D = \frac{1}{2}\left(\nabla u + \nabla u^T\right). \tag{20}$$

The dynamic viscosity of blood η_{bl} is defined as a function of the hematocrit H introduced in Merrill [34], which reads as follows:

$$\eta_{bl} = \eta_{pl}(1 + 2.5H + 7.32H^2). \tag{21}$$

Here, η_{pl} is the dynamic viscosity of the blood plasma. The kinematic viscosity of blood can then be computed as follows:

$$\nu_{bl} = \frac{\eta_{bl}}{\rho_{bl}}, \tag{22}$$

where ρ_{bl} is the density of blood. The values of η_{pl} and ρ_{bl} are constant, and their values are listed in Table 1.

The Newtonian modeling of blood as a suspension of cells and of the plasma as the carrier liquid of the cells means that well-known non-Newtonian rheological properties of blood are not accounted for. Non-Newtonian fluid behavior includes shear thinning, thixotropy, and viscoelasticity, which are seen in rheological material data of blood [38]. However, since electrical impedance changes due to blood flow variations can only be observed and measured for large vessels in which the effects of non-Newtonian behavior are small, the limitations imposed by the simplified model are acceptable.

2.5. Numerics

For CFD simulations of blood flow, the open-source CFD software OpenFOAM [39] is used. A new CFD solver is developed in OpenFOAM [39] to incorporate the conductivity models presented in Sections 2.1.1, 2.1.2 and 2.3. The implementation of all the equations is performed for a general 3D flow.

OpenFOAM provides a spectral analysis of the viscous stress tensor τ, yielding the principal stresses (eigenvalues) $\tau_1 > \tau_2 > \tau_3$ and the corresponding eigendirections e_1, e_2, and e_3, respectively. The maximum shear stress is given by $\tau_{\max} = \tau_1 - \tau_3$ and the maximum shear rate is obtained from the (isotropic) constitutive law as $\dot{\gamma}_{\max} = 1/(2\eta_{bl})\tau_{\max}$.

3. Results

In this section, two CFD simulations are performed from which the electrical conductivity of blood with a hematocrit value of $H = 45\%$ are calculated with the two suggested models. The first case is a fully developed steady-state laminar flow in a straight rigid vessel, while the second case regards the flow in a simplified aorta model, containing a rigid curved pipe imitating the aortic arch.

The first case is motivated by the extensive research and comprehensive understanding of blood conductivity within a straight rigid vessel, as reported in previous studies by Hoetink et al. [7], Gaw et al. [10]. This specific axisymmetric flow has distinctive characteristics. The shear planes are tangent to cylinders of constant radius, and, consequently, the normals of the shear planes point in radial direction. Moreover, besides the velocity vector pointing in axial direction, the vorticity vector points in circumferential direction in the cylindrical flow. Likewise, the cross-product of velocity and vorticity vector thus points in a radial direction, such that both models coincide in assuming the dominant short axis direction e_α in radial direction. For this simulation, the average inlet velocity $\bar{u}$ is $0.008\,\mathrm{m\,s^{-1}}$ and the vessel diameter D is $0.04\,\mathrm{m}$. With a hematocrit H of 45%, the kinematic viscosity of blood ν_{bl}, according to Equations (21) and (22), has a value of $4.59 \times 10^{-5}\,\mathrm{m^2/s}$. The Reynolds number Re is computed as follows:

$$Re = \frac{\bar{u}D}{\nu_{bl}} = 6.96.\tag{23}$$

Velocity and shear rate color contours in the cylindrical flow are illustrated in Figure 6. The velocity in the center is maximal, while at the wall it is zero due to the no-slip boundary condition. The shear rate is minimal at the vessel's centerline and maximal near the wall. The shear stress follows the same pattern due to the linear isotropic constitutive law. In order to analyze the predicted conductivity tensor, we display the conductivities σ_α and σ_β in Figure 7, which are, in the current case, the radial conductivity and, the conductivity in a tangential plane, i.e., equally, say, in the axial and circumferential direction, respectively. The conductivity σ_β has its maximum near the walls, and σ_α is minimum there.

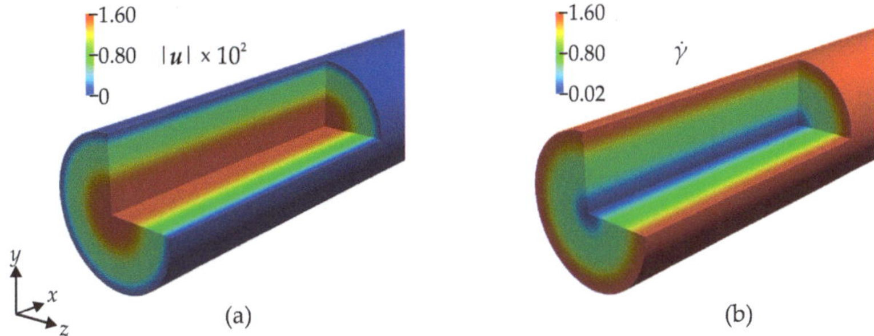

Figure 6. Color contour of (**a**) velocity magnitude $|u| \times 10^2$ in $\mathrm{m\,s^{-1}}$, and (**b**) shear rate $\dot{\gamma}$ in $\mathrm{s^{-1}}$.

In order to quantify the anisotropy of the conductivity, we introduce an indicator η, which relates the maximal in-plane conductivity to the average conductivity, i.e., the difference between the largest and smallest principal conductivity over the sum of the principal conductivities, and is as follows:

$$\eta = \frac{\sigma_\beta - \sigma_\alpha}{2\sigma_\beta + \sigma_\alpha}. \tag{24}$$

A contour plot of the anisotropy indicator is provided in the right-most picture in Figure 7. The conductivity is found to be nearly isotropic (low η) in the center of the cylinder but quickly reaches a high level away from the centerline. This is consistent with the expectation that the RBCs tend to have a more random orientation toward the center of the cylinder and are highly aligned in the higher shear stresses closer to the wall.

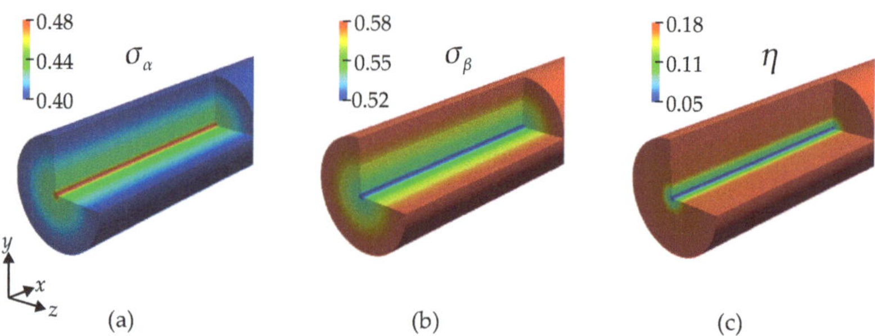

Figure 7. Color contours of (**a**) σ_α in $\mathrm{S\,m^{-1}}$, (**b**) σ_β in $\mathrm{S\,m^{-1}}$, and (**c**) anisotropic indicator η.

Model Comparison

In the cylindrical flow treated model so far, the two suggested models coincide. In the current section, we regard a case where the models yield different conductivity tensors and quantify these differences. The current CFD simulation is a laminar steady-state blood flow in an idealized 3D aorta. This case is chosen because the models are not intrinsically identical in this case. The inlet average velocity $\bar{u}$ of the aorta is $0.022\,\mathrm{m\,s^{-1}}$, and the diameter D is $0.024\,\mathrm{m}$. With a hematocrit H of 45%, the kinematic viscosity of blood ν_{bl}, according to Equations (21) and (22) is again $4.59 \times 10^{-5}\,\mathrm{m^2/s}$. The Reynolds number Re is computed as follows:

$$Re = \frac{\bar{u}D}{\nu_{\mathrm{bl}}} = 11.49. \tag{25}$$

Figure 8 shows the color contours of the velocity magnitude $|u|$ (a) in $\mathrm{m\,s^{-1}}$ and shear rate $\dot{\gamma}$ (b) in $\mathrm{s^{-1}}$. The velocity magnitude is zero at the wall due to the no-slip boundary condition at the wall, and it increases towards the center line of the aorta. The values of the shear rate $\dot{\gamma}$ close to the wall are higher than they are farther away from the wall. Also, in the vicinity of the inlet and the outlet of the aorta, the streamlines are analogous to a Poiseuille flow. However, in the arch itself, we observe that streamlines can move from the center towards the wall and vice versa due to the curvature of the geometry.

The difference between the calculated conductivity tensors originates from the determined orientation of the RBCs with the eigenvector (EV) and the velocity–vorticity (VV) models. Therefore, we compare how the models predict the orientation of RBCs, which is determined by e_α, and then how different orientation predictions impact the computation of conductivity tensor.

To quantify the difference between the models we regard the angle ψ_α between e_α^{EV} and e_α^{VV} on the one hand, and the 'angle' ψ_σ between the tensors on the other hand.

The angle between the differently defined symmetry axes is obtained as follows:

$$\psi_\alpha = \arccos\left(\left|\langle e_\alpha^{\mathrm{EV}}, e_\alpha^{\mathrm{VV}}\rangle\right|\right), \tag{26}$$

where $\langle \cdot, \cdot \rangle$ denotes the standard scalar product of two vectors.

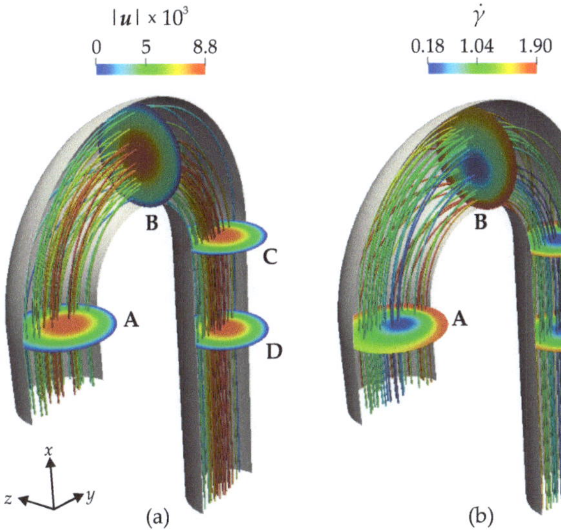

Figure 8. Color contours of (**a**) velocity magnitude $|u| \times 10^3$ in m s^{-1} and (**b**) shear rate $\dot{\gamma}$ in s^{-1}. The color contours are illustrated in the cross-sections (A–D) and on the streamlines.

To quantify the difference between the calculated conductivity tensors, the angle ψ_σ between the tensors σ^{EV} and σ^{VV} is defined as follows:

$$\psi_\sigma = \arccos\left(\frac{|\langle \sigma^{\mathrm{EV}}, \sigma^{\mathrm{VV}} \rangle|}{\|\sigma^{\mathrm{EV}}\| \times \|\sigma^{\mathrm{VV}}\|} \right), \tag{27}$$

where $\langle \cdot, \cdot \rangle$ now denotes the induced standard scalar product of two tensors and $\| \cdot \|$ the induced tensor norm, i.e., the Frobenius norm of the matrix representing σ. Note that since σ_α and σ_β are the same for both models, the norm of the tensors is the same as follows:

$$\|\sigma^{\mathrm{EV}}\| = \|\sigma^{\mathrm{VV}}\| = \sqrt{\sigma_\alpha^2 + 2\sigma_\beta^2}, \tag{28}$$

such that no difference in 'magnitude' occurs (which would not be quantified by the angle ψ_σ). The two angles are actually related to each other by the following relation:

$$\cos\psi_\sigma = \frac{1}{\sigma_\alpha^2 + 2\sigma_\beta^2}\left(\sigma_\alpha^2 \cos^2 \psi_\alpha + 2\sigma_\alpha \sigma_\beta \left(1 - \cos^2 \psi_\alpha \right) + \sigma_\beta^2 \left(1 + \cos^2 \psi_\alpha \right) \right), \tag{29}$$

which may be derived by inserting the e^{EV} and e^{VV} in Equation (6) and forming the scalar product. With the ratio between the two principal conductivities, $\lambda_\sigma = \sigma_\alpha / \sigma_\beta$, this relation reads as follows:

$$\cos\psi_\sigma = \frac{1}{\lambda_\sigma^2 + 2}\left(\lambda_\sigma^2 \cos^2 \psi_\alpha + 2\lambda_\sigma \left(1 - \cos^2 \psi_\alpha \right) + \left(1 + \cos^2 \psi_\alpha \right) \right). \tag{30}$$

From Equation (17), we see that the ratio λ_σ is a function of the shear rate and a monotonously decreasing one at that. The minimal value attained in the current simulation with a maximal shear rate of about 1.9 (compare Figure 8) is therefore $\lambda_{\sigma,\min} = \lambda_\sigma$ $(\dot{\gamma}_{\max} = 1.9) \approx 0.77$, and the maximal angle ψ_σ is obtained for the minimal ratio and the maximal angle between the symmetry axis of the two models, $\psi_\alpha = 90^\circ$, which yields $\psi_{\sigma,\max} \approx 11.6^\circ$.

Figure 9a displays the color contour of ψ_α in degrees. By definition, when $\psi_\alpha = 0$, the two anisotropy models predict the same orientation of the RBCs and when $\psi_\alpha > 0$, the models do not agree. In the proximity of the wall, where the viscous forces are dominant, $\psi_\alpha = 0$, and therefore, the models are identical. In the arch, moving away from the immediate vicinity of the wall, it is evident that $\psi_\alpha > 0$, reaching a peak at $\psi_\alpha \approx 85°$. This is highlighted in the cross-sections A, B, and C. Cross-section A is at the beginning of the arch, B is in the middle, and C is at the end of the arch. The cross-sections (A–C) and the rainbow-like volume color contour show that the areas where the $\psi_\alpha > 0$ expand from A to B and again shrink from B to C. Therefore, cross-section B highlights the maximum difference between the two models in predicting the orientation of the RBCs. At cross-section D, $\psi_\alpha = 0$, a bit downstream of the arch, the flow is similar to a Poiseuille flow and the models agree.

Figure 9b, displays the color contours of ψ_σ in degrees. The observed patterns closely resemble those of ψ_α. This similarity is expected because the discrepancy in conductivity tensor computed in Equation (6) arises from the different e_α directions of the models, compare Equation (30). As already deduced from Equation (30) the values of ψ_σ are much lower than those of ψ_α and only reach values of about $12°$.

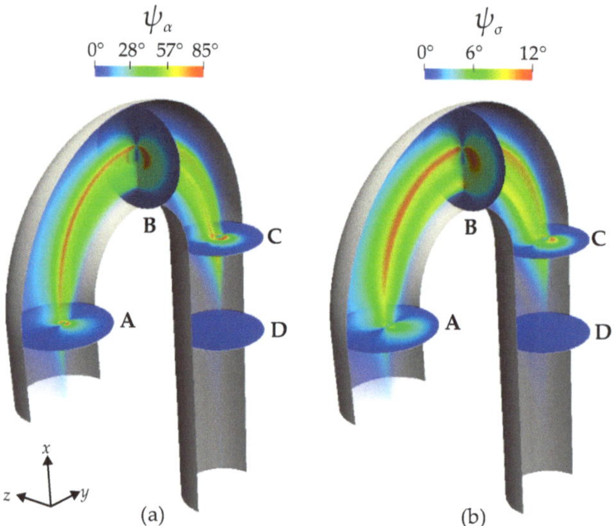

Figure 9. Color contours of (**a**) ψ_α in degrees and (**b**) ψ_σ in degrees. The color contours are illustrated in the cross-sections (A–D) and in the volume of the aorta. ψ_α is the angle between e_α^{EV} and e_α^{VV} vectors resulting from the two models. ψ_σ is the angle between the conductivity tensors σ^{EV} and σ^{VV} resulting from the two models.

4. Discussion

The morphology of bioimpedance signals exhibits a high degree of dependence on the variations in the electrical conductivity of the blood flow. These conductivity variations are affected by the dynamic motion of RBCs, which are linked to the local hemodynamic conditions and disturbances of the flow. This suggests that an electrical conductivity model, capable of describing and incorporating local hemodynamic conditions, will eventually contribute to a better understanding of signal morphology and classification of the bioimpedance signals.

The key to understanding and classifying the influence of local pathophysiological flow disturbances on bioimpedance signals is to translate the microscopic effect of RBC motions into the macroscopic property of blood, i.e., anisotropic electrical conductivity. This study did just that by considering the macroscopic hemodynamic quantities, i.e., velocity, vorticity, shear rate, and shear stress. Two novel models, i.e., eigenvector and velocity–vorticity models, were developed to predict a spatially inhomogeneous and unsteady orientation and a deformation of RBCs and the anisotropic blood electrical conductivity. The two models are based on different assumptions for the dominating direction of the short axis of the RBCs. The principal conductivities of the anisotropic conductivity tensor are in both models, computed by adopting the blood-specific modifications of the Maxwell–Fricke theory.

The new models in this study overcome the limitations of previous 1D analytical formulations of Hoetink et al. [7] and Gaw et al. [10], and are a significant step towards modeling and understanding the electrical conductivity of blood. The models allow computing anisotropic blood conductivity as a field variable and, therefore, enable us to compute the conductivity through CFD simulations. Although the simulations performed in this study covered the particular case of laminar steady-state blood flow, the models are developed regardless of these assumptions.

The presented results are in good qualitative agreement with the analytical and experimental findings of Hoetink et al. [7], Gaw et al. [10], and Fischer et al. [21]. In the following, we shed light on the findings that are in agreement with the literature. In the straight rigid vessel flow, the changes in the velocity field only occur in the radial direction, and, therefore, shear rate and shear stress are only functions of the radial distance. In such a flow, the shear planes are tangent to cylinders of constant radius. Considering the analogies used in previous studies of Gaw et al. [10], Melito et al. [20], in the straight vessel flow, the larger principal value, σ_β, is the conductivity in the flow direction (and the circumferential direction), and σ_α is the conductivity in the radial direction. The simulation results of conductivity are in alignment with the statements of Hoetink et al. [7] and Gaw et al. [10] in the sense that in high shear rate zones, the conductivity in the flow direction is maximal, and the conductivity perpendicular to the flow direction is minimal.

The eigenvector and velocity–vorticity models are both identical for the simulation of blood flow in the straight rigid vessel; however, the simulation of blood flow in the idealized aorta showed that the models are not equivalent. The discrepancy in the symmetry axis predicted by the two models is mended by the limited anisotropy of the conductivity tensor. The anisotropy increases with shear stress, and since the highest angles between the symmetry axes occurred in the low shear stress (and thus low shear rate) areas, the difference between the models does not seem very strong. Further research is necessary to find out how the two models would differ in the predicted ICG signal one may obtain by inserting the conductivity tensor field from the CFD simulations in a 3D electric simulations, as was performed in [19]. ICG measurements at suitable simplified geometries could be used to validate the models and to determine which of the models yields better predictions. The validation of the models might also be performed via spatially resolved measurements, such as electrical impedance tomography [40].

On the modeling part, a key assumption was that the RBCs are oblate spheroids. Even though in the previous studies by Gaw et al. [10] such an assumption was also made and proved useful, we still suggest an investigation on the possibility of considering RBCs as triaxial ellipsoidal particles. The oblate spheroid assumption implies that in the principal coordinate system of RBCs, the principal conductivities in the direction of the long axes are equal. However, the experiments of Fischer et al. [21] and Minetti et al. [27] showed that the RBCs in tank-treading motion are triaxial ellipsoidal particles with a short, an intermediate, and a long axis. The latter suggests distinct conductivity values in the principal directions. In reality, red blood cells exhibit a biconcave morphology and, under various pathological conditions, the shape of RBCs may undergo further alterations. Further studies are required to investigate whether considering more details of the shapes improves the modeling of the electrical conductivity of blood. As a further simplification, this study assumed a constant value for the hematocrit. However, in the circuitry systems, the hematocrit may change, for instance, due to collision between RBCs and blood elements, the accumulation and adhesion of RBCs, and the bifurcation of blood vessels [41]. Considering that the presented models allow for the computation of the electrical conductivity of blood as a field variable, we suggest that future studies explore defining hematocrit as a field variable to account for multicellular collisions and compression.

It is important to acknowledge that the models presented in this study only account for the pathologies associated with geometric changes in the blood vessels. The pathologies that affect the shape of erythrocytes, such as sickle cell anemia or the electrochemical properties of blood, are not considered. These electrochemical properties, specifically the zeta potential of erythrocytes, play a key role in the repulsion of cells from one another [42]. A decrease in the zeta potential increases red blood cell aggregation at low shear rates, leading to a higher viscosity. A preliminary attempt to numerically model these effects has been presented in [43], and further investigations will examine their impact on viscosity and, subsequently, electrical conductivity.

Besides desisting from the shape and electrochemical details of the RBCs, we also simplified blood as a Newtonian fluid. However, blood is well-known for displaying non-Newtonian characteristics, like shear thinning, thixotropy, and viscoelasticity. Significant progress has been recently achieved in hemorheology, compare, for example, Giannokostas and Dimakopoulos [44], Giannokostas et al. [45], and Beris et al. [38]. The models developed in the present study can be seamlessly integrated with any blood rheology model since they only require input from hemodynamics without providing any feedback to the rheology. Although a simplified rheological model sufficed for evaluating the models, incorporating a suitable blood rheology model is recommended for model validation.

Furthermore, in our simulations, a rigid vessel wall was considered. However, when aiming at experimental validations, it might be necessary to consider the compliance of the vessel wall in a fluid–structure interaction model.

We finally note a slight discrepancy in our modeling, since the alignment of the RBCs, which effectuates an anisotropic electrical conductivity will likely also cause other material properties of blood to be anisotropic. Most notably, this would apply to the viscosity of blood, which would more consistently be modeled by a transversely isotropic second-order tensor. By contrast, we employ an isotropic constitutive law in Equation (19). Anisotropic hemodynamic models emerge, for example, from the conformation tensor used in modeling thixotropy [45]. The relation of the conformation tensor to the preferred orientation of the RBCs modeled in the current work will have to be explored in future work.

5. Conclusions

This study presents a novel approach to computing the electrical conductivity of blood. The new approach stems from determining the 3D inhomogeneous and unsteady motion of RBCs contributing to the anisotropic nature of the conductivity. In other words, this study established that variations in the anisotropic electrical conductivity of blood can be characterized by spatially inhomogeneous and unsteady changes in the orientation and deformation of RBCs. Two models were developed to compute conductivity as a tensor field variable. The CFD simulation of blood flow in the straight, rigid pipe showed that the computed conductivity tensor is consistent with experimental observations of Gaw et al. [10], Wtorek and Polinski [46], and Fischer et al. [21]. The CFD simulations of blood flow in the aorta showed that despite the differences between the models in the computations of the RBCs orientation, either of the models may be taken for the calculation of the conductivity tensor due to small discrepancies. While all the results are in qualitative agreement with the literature, validation of the models with experiments is yet outstanding.

The presented models for computing anisotropic conductivity allow computing the blood flow-related conductivity distribution in different locations in arteries; thus, the outcome might be very beneficial in investigating the possibility of using bioimpedance measurements as an alternative method for detecting CVDs and various pathologies, such as aortic stenosis, aneurysm, and aortic dissection. In addition, the findings of this study will potentially increase the accuracy of the simulation of bioimpedance signals by considering the effects of detailed blood flow.

Author Contributions: Conceptualization, A.J., V.B., T.K. and T.H.; methodology, A.J., V.B., T.K. and T.H.; software, A.J. and T.K.; validation, A.J., V.B., T.K. and T.H.; formal analysis, A.J. and T.H.; writing—original draft preparation, A.J., T.H. and G.M.M.; writing—review and editing, A.J., V.B., T.K., T.H., G.B., A.R.-K., M.K. and G.M.M.; visualization, A.J. and G.M.M.; supervision, T.H., G.B. and M.K. All authors have read and agreed to the published version of the manuscript.

Funding: This work was funded by Graz University of Technology, Austria, through the LEAD Project on "Mechanics, Modeling, and Simulation of Aortic Dissection". This work is supported by Graz University of Technology Open Access Publishing Fund.

Institutional Review Board Statement: Not applicable.

Informed Consent Statement: Not applicable.

Data Availability Statement: No new data were created or analyzed in this study. Data sharing is not applicable to this article.

Conflicts of Interest: The authors declare no conflicts of interest.

Abbreviations

The following abbreviations are used in this manuscript:

1D	One-dimensional
2D	Two-dimensional
3D	Three-dimensional
CFD	Computational fluid dynamics
CVD	Cardiovascular disease
CT	Computed tomography
ICG	Impedance cardiography
IPG	Impedance plethysmography
MRI	Magnetic resonance imaging
RBC	Red blood cell

References

1. Semelka, R.C.; Armao, D.M.; Elias Junior, J.; Huda, W. Imaging strategies to reduce the risk of radiation in CT studies, including selective substitution with MRI. *J. Magn. Reson. Imaging* **2007**, *25*, 900–909. [CrossRef] [PubMed]
2. Kanal, E.; Barkovich, A.J.; Bell, C.; Borgstede, J.P.; Bradley, W.G.; Froelich, J.W.; Gilk, T.; Gimbel, J.R.; Gosbee, J.; Kuhni-Kaminski, E.; et al. ACR Guidance Document for Safe MR Practices: 2007. *Am. J. Roentgenol.* **2007**, *188*, 1447–1474. [CrossRef]
3. Badeli, V.; Ranftl, S.; Melito, G.M.; Reinbacher-Köstinger, A.; von der Linden, W.; Ellermann, K.; Biro, O. Bayesian inference of multi-sensors impedance cardiography for detection of aortic dissection. *COMPEL—Int. J. Comput. Math. Electr. Electron. Eng.* **2021**, *41*, 824–839. [CrossRef]
4. Badeli, V.; Melito, G.M.; Reinbacher-Köstinger, A.; Bíró, O.; Ellermann, K. Electrode positioning to investigate the changes of the thoracic bioimpedance caused by aortic dissection—A simulation study. *J. Electr. Bioimpedance* **2020**, *11*, 38–48. [CrossRef] [PubMed]
5. Wiegerinck, A.I.P.; Thomsen, A.; Hisdal, J.; Kalvøy, H.; Tronstad, C. Electrical impedance plethysmography versus tonometry to measure the pulse wave velocity in peripheral arteries in young healthy volunteers: A pilot study. *J. Electr. Bioimpedance* **2021**, *12*, 169–177. [CrossRef]
6. Reinbacher-Köstinger, A.; Badeli, V.; Bíró, O.; Magele, C. Numerical Simulation of Conductivity Changes in the Human Thorax Caused by Aortic Dissection. *IEEE Trans. Magn.* **2019**, *55*, 1–4. [CrossRef]
7. Hoetink, A.; Faes, T.; Visser, K.; Heethaar, R. On the Flow Dependency of the Electrical Conductivity of Blood. *IEEE Trans. Biomed. Eng.* **2004**, *51*, 1251–1261. [CrossRef]
8. Jaspard, F.; Nadi, M.; Rouane, A. Dielectric properties of blood: An investigation of haematocrit dependence. *Physiol. Meas.* **2003**, *24*, 137. [CrossRef]
9. Edgerton, R.H. Conductivity of Sheared Suspensions of Ellipsoidal Particles with Application to Blood Flow. *IEEE Trans. Biomed. Eng.* **1974**, *BME-21*, 33–43. [CrossRef]
10. Gaw, R.L.; Cornish, B.H.; Thomas, B.J. The Electrical Impedance of Pulsatile Blood Flowing Through Rigid Tubes: A Theoretical Investigation. *IEEE Trans. Biomed. Eng.* **2008**, *55*, 721–727. [CrossRef]
11. Maxwell, J.C. *A Treatise on Electricity and Magnetism*; Clarendon Press: Oxford, UK, 1873.

12. Rayleigh, L. LVI. On the influence of obstacles arranged in rectangular order upon the properties of a medium. *Lond. Edinb. Dublin Philos. Mag. J. Sci.* **1892**, *34*, 481–502. [CrossRef]
13. Fricke, H. A Mathematical Treatment of the Electric Conductivity and Capacity of Disperse Systems I. The Electric Conductivity of a Suspension of Homogeneous Spheroids. *Phys. Rev.* **1924**, *24*, 575–587. [CrossRef]
14. Visser, K.R. Electric properties of flowing blood and impedance cardiography. *Ann. Biomed. Eng.* **1989**, *17*, 463–473. [CrossRef]
15. Ulbrich, M.; Mühlsteff, J.; Leonhardt, S.; Walter, M. Influence of physiological sources on the impedance cardiogram analyzed using 4D FEM simulations. *Physiol. Meas.* **2014**, *35*, 1451. [CrossRef]
16. Voss, F.; Korna, L.; Leonhardtb, S.; Walterb, M. Modeling of flow-dependent blood conductivity for cardiac bioimpedance. *Int. J. Bioelectromagn.* **2021**, *23*, 21.
17. Jafarinia, A.; Badeli, V.; Melito, G.M.; Müller, T.S.; Reinbacher-Köstinger, A.; Hochrainer, T.; Biro, O.; Ellermann, K.; Brenn, G. False lumen thrombosis in aortic dissection and its impact on blood conductivity variations—An application for impedance cardiography. In Proceedings of the Book of abstract, Young Investigators Conference, Valencia, Spain, 7–9 July 2021; p. 297.
18. Badeli, V. Modelling and Simulation of Aortic Dissection by Impedance Cardiography. Ph.D. Thesis, Graz University of Technology, Graz, Austria, 2021.
19. Badeli, V.; Jafarinia, A.; Melito, G.M.; Müller, T.S.; Reinbacher-Köstinger, A.; Hochrainer, T.; Brenn, G.; Ellermann, K.; Biro, O.; Kaltenbacher, M. Monitoring of false lumen thrombosis in type B aortic dissection by impedance cardiography – A multiphysics simulation study. *Int. J. Numer. Methods Biomed. Eng.* **2023**, *39*, e3669. [CrossRef]
20. Melito, G.M.; Müller, T.S.; Badeli, V.; Ellermann, K.; Brenn, G.; Reinbacher-Köstinger, A. Sensitivity analysis study on the effect of the fluid mechanics assumptions for the computation of electrical conductivity of flowing human blood. *Reliab. Eng. Syst. Saf.* **2021**, *213*, 107663. [CrossRef]
21. Fischer, T.M.; Stöhr-Liesen, M.; Schmid-Schönbein, H. The Red Cell as a Fluid Droplet: Tank Tread-Like Motion of the Human Erythrocyte Membrane in Shear Flow. *Science* **1978**, *202*, 894–896. [CrossRef] [PubMed]
22. Goldsmith, H.L.; Marlow, J.; MacIntosh, F.C. Flow behaviour of erythrocytes—I. Rotation and deformation in dilute suspensions. *Proc. R. Soc. Lond. Ser. B. Biol. Sci.* **1972**, *182*, 351–384. [CrossRef]
23. Keller, S.R.; Skalak, R. Motion of a tank-treading ellipsoidal particle in a shear flow. *J. Fluid Mech.* **1982**, *120*, 27–47. [CrossRef]
24. Bitbol, M. Red blood cell orientation in orbit C = 0. *Biophys. J.* **1986**, *49*, 1055–1068. [CrossRef]
25. Schmid-Schönbein, H.; Wells, R. Fluid Drop-Like Transition of Erythrocytes under Shear. *Science* **1969**, *165*, 288–291. [CrossRef]
26. Abkarian, M.; Viallat, A. Vesicles and red blood cells in shear flow. *Soft Matter* **2008**, *4*, 653–657. [CrossRef]
27. Minetti, C.; Audemar, V.; Podgorski, T.; Coupier, G. Dynamics of a large population of red blood cells under shear flow. *J. Fluid Mech.* **2019**, *864*, 408–448. [CrossRef]
28. Lamb, H. *Hydrodynamics*; Cambridge University Press: Cambridge, UK, 1932.
29. Millikan, R.A.; Bishop, E.S. *Elements of Electricity: A Practical Discussion of the Fundamental Laws and Phenomena of Electricity and Their Practical Applications in the Business and Industrial World*; American Technical Society: Sacramento, CA, USA, 1917.
30. Goldsmith, H.L. Flow-induced interactions in the circulation. In *Advances in the Flow and Rheology of Non-Newtonian Fluids*; Rheology Series; Elsevier: Amsterdam, The Netherlands, 1999; Volume 8, pp. 1–62. [CrossRef]
31. Gaw, R.L. The Effect of Red Blood Cell Orientation on the Electrical Impedance of Pulsatile Blood with Implications for Impedance Cardiography. Ph.D. Thesis, Queensland University of Technology, Brisbane City, Australia, 2010.
32. Evans, E. New Membrane Concept Applied to the Analysis of Fluid Shear- and Micropipette-Deformed Red Blood Cells. *Biophys. J.* **1973**, *13*, 941–954. [CrossRef]
33. Bitbol, M.; Quemada, D. Measurement of erythrocyte orientation in flow by spin labeling. *Biorheology* **1985**, *22*, 31–42. [CrossRef] [PubMed]
34. Merrill, E.W. Rheology of blood. *Physiol. Rev.* **1969**, *49*, 863–888. [CrossRef] [PubMed]
35. Hinghofer-Szalkay, H. Method of high-precision microsample blood and plasma mass densitometry. *J. Appl. Physiol.* **1986**, *60*, 1082–1088. [CrossRef] [PubMed]
36. Jafarinia, A.; Müller, T.S.; Windberger, U.; Brenn, G.; Hochrainer, T. A Study on Thrombus Formation in Case of Type B Aortic Dissection and Its Hematocrit Dependence. In Proceedings of the 6th World Congress on Electrical Engineering and Computer Systems and Sciences (EECSS'20), Prague, Czech Republic, 13–15 August 2020. [CrossRef]
37. Carreau, P.J. Rheological Equations from Molecular Network Theories. *Trans. Soc. Rheol.* **1972**, *16*, 99–127. [CrossRef]
38. Beris, A.N.; Horner, J.S.; Jariwala, S.; Armstrong, M.J.; Wagner, N.J. Recent advances in blood rheology: A review. *Soft Matter* **2021**, *17*, 10591–10613. [CrossRef]
39. Weller, H. G.; Tabor, G.; Jasak, H.; Fureby, C. A Tensorial Approach to Computational Continuum Mechanics Using Object-Oriented Techniques. *Comput. Phys.* **1998**, *12*, 620–631. [CrossRef]
40. Brown, B. Electrical impedance tomography (EIT): A review. **2003**, *27*, 97–108. [CrossRef]
41. Lee, C.A.; Paeng, D.G. Effect of particle collisions and aggregation on red blood cell passage through a bifurcation. *Microvasc. Res.* **2009**, *78*, 301–313. [CrossRef]
42. Fontes, A.; Fernandes, H.P.; Barjas-Castro, M.L.; Thomaz, A.A.d.; Pozzo, L.d.Y.; Barbosa, L.C.; Cesar, C.L. Red blood cell membrane viscoelasticity, agglutination, and zeta potential measurements with double optical tweezers. In Proceedings of the Imaging, Manipulation, and Analysis of Biomolecules, Cells, and Tissues IV, San Jose, CA, USA, 21–26 January 2006; Volume 6088, pp. 296–305. [CrossRef]

43. Lee, C.A.; Paeng, D.G. Numerical simulation of spatiotemporal red blood cell aggregation under sinusoidal pulsatile flow. *Sci. Rep.* **2021**. *11*, 9977. [CrossRef]

44. Giannokostas, K.; Dimakopoulos, Y. TEVP model predictions of the pulsatile blood flow in 3D aneurysmal geometries. *J. Non-Newton. Fluid Mech.* **2023**, *311*, 104969. [CrossRef]

45. Giannokostas, K.; Moschopoulos, P.; Varchanis, S.; Dimakopoulos, Y.; Tsamopoulos, J. Advanced constitutive modeling of the thixotropic elasto-visco-plastic behavior of blood: Description of the model and rheological predictions. *Materials* **2020**, *13*, 4184. [CrossRef] [PubMed]

46. Wtorek, J.; Polinski, A. The contribution of blood-flow-induced conductivity changes to measured impedance. *IEEE Trans. Biomed. Eng.* **2005**, *52*, 41–49. [CrossRef] [PubMed]

MDPI AG
Grosspeteranlage 5
4052 Basel
Switzerland
Tel.: +41 61 683 77 34

Bioengineering Editorial Office
E-mail: bioengineering@mdpi.com
www.mdpi.com/journal/bioengineering

www.ingramcontent.com/pod-product-compliance
Lightning Source LLC
LaVergne TN
LVHW071930160726
843515LV00011B/2573